Signals and Communication Technology

This series is devoted to fundamentals and applications of modern methods of signal processing and cutting-edge communication technologies. The main topics are information and signal theory, acoustical signal processing, image processing and multimedia systems, mobile and wireless communications, and computer and communication networks. Volumes in the series address researchers in academia and industrial R&D departments. The series is application-oriented. The level of presentation of each individual volume, however, depends on the subject and can range from practical to scientific.

Indexing: All books in "Signals and Communication Technology" are indexed by Scopus and zbMATH

For general information about this book series, comments or suggestions, please contact Mary James at mary.james@springer.com or Ramesh Nath Premnath at ramesh.premnath@springer.com.

S. S. Iyengar • Kenneth G. Furton •
Pronab Mohanty • Naveen Kumar Chaudhary
Editors

Artificial Intelligence Driven Forensics

Springer

Editors

S. S. Iyengar
Global Forensic and Justice Center
Florida International University
Miami, FL, USA

Kenneth G. Furton
Global Forensic and Justice Center
Florida International University
Miami, FL, USA

Pronab Mohanty
Police Computer Wing
Bangalore, Karnataka, India

Naveen Kumar Chaudhary
National Forensic Sciences University
Gandhinagar, Gujarat, India

ISSN 1860-4862 ISSN 1860-4870 (electronic)
Signals and Communication Technology
ISBN 978-3-031-98035-0 ISBN 978-3-031-98036-7 (eBook)
https://doi.org/10.1007/978-3-031-98036-7

This Springer imprint is published by the registered company Springer Nature Switzerland AG
The registered company address is: Gewerbestrasse 11, 6330 Cham, Switzerland

Preface

The science of today is the technology of tomorrow.
—Edward Teller

Artificial Intelligence Driven Forensics explores the transformative role of artificial intelligence (AI) in the field of digital forensics, offering a comprehensive overview of how AI technologies are reshaping crime investigation, cybersecurity, and evidence analysis. The book bridges traditional forensic methodologies with cutting-edge AI techniques, such as machine learning, deep learning, natural language processing, and graph analytics, to improve the accuracy, speed, and scalability of forensic investigations.

Structured around real-world applications and case studies, the book covers critical areas including AI-enhanced malware detection, user behavior profiling, deepfake identification, social media profile cloning detection, canine forensics and hybrid approaches to analyzing digital artifacts. It also delves into the integration of AI with forensic tools for automating incident response, identifying anomalies in massive log files, and supporting decision-making in complex cybercrime scenarios.

A key feature of the book is its interdisciplinary approach, blending insights from computer science, cybersecurity, law, and ethics. It addresses not only the technical mechanisms behind AI-driven forensics but also the legal, ethical, and societal implications, particularly regarding privacy, accountability, and the admissibility of AI-generated evidence in courts.

Geared towards researchers, practitioners, students, and policymakers, *Artificial Intelligence Driven Forensics* serves as both a foundational text and an advanced guide to understanding the future of forensic science in the age of AI.

This book is partitioned into six thematic parts, each illuminating how artificial intelligence and machine learning are revolutionizing the field of digital forensics across domains including law, biometrics, cybersecurity, ecology, medicine, and more.

Part: Artificial Intelligence and Machine Learning Driven Digital Forensics explores the transformative role of AI and machine learning in modern forensic science. It introduces foundational theories, bench marks advanced algorithms, and demonstrates practical implementations of intelligent systems across various forensic domains. Topics include anomaly detection in visual scenes, synthetic cybersecurity log generation, multimedia authenticity verification, and adversarial attack detection in facial recognition systems. This section highlights how AI enhances forensic accuracy, scalability, and adaptability, establishing a data-driven foundation for next-generation forensic investigations.

Part: Bridging Law and Technology: AI-Enabled Legal Framework addresses the intersection of artificial intelligence and legal forensics. It outlines foundational principles, emerging applications, and a comprehensive techno-legal framework for ensuring the integrity and admissibility of digital evidence in courtrooms.

Part: Behavioral and Biometric Forensics focuses on advanced biometric techniques and behavioral modeling, presenting innovations in keystroke dynamics, gait recognition, fingerprint analysis, and emotion detection using multimodal deep learning approaches.

Part: Sustainable and Medical Forensics Innovations bridges environmental sustainability and advanced healthcare with forensic science. It highlights cutting-edge developments such as AI-assisted management of oligometastatic cancer, blockchain-secured cognitive IoT networks for safeguarding medical imagery, and smart agriculture solutions featuring forensic-grade pest detection with green sensor technology. Together, these innovations redefine how forensic methodologies can support both environmental responsibility and medical advancement.

Part: Network, Supply Chain, and Cyber Forensics navigates the complexity of cybercrime investigations by introducing frameworks for packet capture analysis, blockchain-enhanced supply chain tracing, and forensic profiling based on user browsing behavior.

Part: Forensics in IoT, Wearables, and Environmental Surveillance expands forensic frontiers into wearable devices, smart therapeutics, and broadband connectivity. This section investigates how data from smartwatches, AI-based health systems, and network infrastructures contribute to digital evidence collection and event reconstruction in real time.

Through its comprehensive and interdisciplinary lens, this book serves as both a reference and inspiration for researchers, practitioners, and policymakers advancing the field of digital forensics in the age of intelligent technologies.

Miami, FL, USA — S. S. Iyengar
Miami, FL, USA — Kenneth G. Furton
Bangalore, Karnataka, India — Pronab Mohanty
Gandhinagar, Gujarat, India — Naveen Kumar Chaudhary

Acknowledgments The editors extend their sincere gratitude to the Honorable Vice Chancellor of the National Forensic Sciences University (NFSU), Dr. Jayant Vyas, and to the Chancellor of PES University, Dr. Jawahar Doreswamy, for their unwavering support and encouragement in the development of this work.

We are also thankful to Dr. Kenneth A. Jessell, President Emeritus, Senior Vice President and Chief Administrative Officer of Florida International University; Dr. Ines Triay, Interim Dean of the College of Engineering; and Dr. Jason Liu, Director of KFSCIS, for their strong and continued support.

Their contributions have been instrumental in the establishment of the Forensic Investigations Network in Digital Sciences (FINDS)—the first-in-the-world pioneering Center of Excellence focused on AI-Driven Digital Forensics, funded by the U.S. Army Research Office (Grant No. W911NF-21-1-0264). Special Thanks to Dr. Paul Yu and Dr. Owens from the U.S. Army Research Office and Dr. Cliff Wang from the National Science Foundation for their continued support. Many graduate and undergraduate students assisted in the preparation of this manuscript. Notably, Yashas Hariprasad, Sina Nabavi, and other students from FIU Knight Foundation School of Computing and Information Sciences, FIU.

This book builds upon the foundational work of the FIU FINDS Center of Excellence, exploring the core principles of digital forensics and AI-driven data analysis and emphasizing real-world applications that prepare the next generation of researchers and practitioners to safeguard the nation's digital frontiers.

Finally, we commend the authors of this book for their outstanding contributions. Their chapters offer innovative insights into the emerging field of AI-driven forensics and set a high standard for future scholarship in this critical area.

The contributions presented in this work are solely based on the authors' original ideas and do not conflict with the interests of any other individuals or organizations.

Miami, FL, USA — S. S. Iyengar
Miami, FL, USA — Kenneth G. Furton
Bangalore, Karnataka, India — Pronab Mohanty
Gandhinagar, Gujarat, India — Naveen Kumar Chaudhary
April 28, 2025

Contents

About the Editors

S. S. Iyengar is a distinguished computational scientist, entrepreneur, advisor, and mentor to millions of scientists and sits on the boards of companies, including startups through his India–USA-based foundation. He was honored as the "Founding Father of AI Science in Digital Forensics" by the Soft Computing Research Society in February 2025. This prestigious recognition is a testament to his lifelong contributions to artificial intelligence, cybersecurity, and digital forensics. He was recently awarded the 2025 Distinguished Career Award in Computer Science and Fellow of the Washington Academy of Sciences. He is also known as the "Founding Father" for his contributions to the development of Information Science Technology for Forensic Application (2022) by a prestigious European University. He is also recognized as the World's Most Influential Scientific Minds in Computational Science. He is currently the Distinguished University Professor, Founding Director of the Discovery Lab, and Founding Director of the World's First US Army-funded Center of Excellence in Digital Forensics at Florida International University, Miami. He is also the Distinguished Chaired Professor at the National Forensic Sciences University (NFSU), Gandhinagar, India. He was awarded the 2023 Karnataka Rajyotsava Award (Karnataka State's 2nd Highest Civilian Award) on November 1st, 2023. He has been involved with research and education in high-performance intelligent systems, Data Science and Machine Learning Algorithms, Sensor Fusion, Data Mining, and Intelligent Systems. Since receiving his Ph.D. degree in 1974 from Miss. State Univ., USA, he has directed over 65 Ph.D. students, many number of postdocs, and many research undergraduate students who are now faculty at Major Universities worldwide or Scientists or Engineers at National Labs/Industries around the world.

He has published more than 600 research papers, has authored/co-authored and edited 32 books, and holds various patents. Over the lifetime, his work, *Brooks–Iyengar Algorithm* has over 5223 publication(s) within this topic and has received 138,976 citation(s). The topic is also known as: *Brooks–Iyengar Hybrid Algorithm*. His h-index is 68 and is identified among the top 2% cited scholars and world scientists (from Stanford University) and EBMs of JSAN journal. The book titled *Fundamentals of Brooks–Iyengar Distributed Sensing Algorithm* authored by Prof. Pawel (Poland) and others and published by Springer in 2020 celebrates S.S. Iyengar's accomplishments that led to his 2019 Institute of Electrical and Electronics Engineers' (IEEE) Cybermatics Congress "Test of Time Award" for his work on creating *Brooks–Iyengar Algorithm* and its impact in advancing modern computing. His work has been featured on the cover of many scientific journals like IEEE transactions and the National Science Foundation's (NSF) breakthrough technologies report to the US Congress by his research group in both 2014 and again in 2016.

He has served on many scientific committees and panels worldwide and has served as the editor/guest editor of various IEEE and ACM journals. His books are published by MIT Press, John Wiley and Sons, CRC Press, Prentice Hall, Springer Verlag, IEEE Computer Society Press, etc. One of his books titled *Introduction to Parallel Algorithms* has been translated into Chinese. During the last 30 years, Dr. Iyengar has brought in over 65 million dollars for research and education. More recently in Spring 2021, Dr. Iyengar in collaboration with HBCUs was awarded a $2.25 M funding for setting up a Digital Forensics Center of Excellence over a period of 5 years (2021–2026). He received an honorary Doctor of Science for four times and recently from Poznan University of Technology in Poland in May 2023. He has been awarded the Lifetime Achievement Award three times (INTERPOL, BHU, IEEE) and recently for his contribution to the field of Digital Forensics on November 8, 2022, during the 7th INTERPOL Digital Forensics Expert Group (DFEG) Meeting at the NFSU, Gandhinagar, Gujarat, India. He has provided the students and faculty with a vision for active learning and collaboration at Jackson State University, Louisiana State University, Florida International University, and across the world.

Dr. Iyengar is a Member of the European Academy of Sciences, Member of the European Academy of Arts and Sciences, a Life Fellow of the IEEE, a Fellow of the Association of Computing Machinery (ACM), a Fellow of the American Association for the Advancement of Science (AAAS), a Fellow of the Society for Design and Process Science (SDPS), a Fellow of the National Academy of Inventors (NAI), and a Fellow of the American Institute for Medical and Biological Engineering (AIMBE). He has received various national and international awards including the crowning Test of Time Research (for his seminal work which has impacted billions of computer and internet users worldwide) and Scholarly Contribution Award from 2019 IEEE Congress on Cybermatics, the distinguished Fulbright Scientist, the Times Network NRI (Non-Resident Indian) of the Year Award for 2017, IEEE Meritorious Service Award, most distinguished CVR Award at the SDPS (2017), Innovation-2-Industry Award, LSU Distinguished Rain Makers for Leadership and Research Award, World's Best Technology Showcase Award, Technology Innovation Award Louisiana Tech University Research Foundation Inventor Award, Distinguished LSU Research Master Award, IBM Distinguished Faculty Award, and the NRI Mahatma Gandhi Pravasi Medal at the House of Lords in London in 2013 among others. He was awarded Satish Dhawan Chaired Professorship at IISc, then Roy Paul Daniel Professorship at LSU. He has received the Distinguished Alumnus Award of the Indian Institute of Science. In 1998, he was awarded the IEEE Computer Society's McCluskey Technical Achievement Award and is an IEEE Golden Core Member. Professor Iyengar is an IEEE Distinguished Visitor, SIAM Distinguished Lecturer, and ACM National Lecturer. In 2006, his paper entitled, A Fast-Parallel Thinning Algorithm for the Binary Image Skeletonization, was the most frequently read article in the month of January in the International Journal of High-Performance Computing Applications. His innovative work called the Brooks–Iyengar algorithm along with Professor Richard Brooks from Clemson University is applied in industries to solve real-world applications. Dr. Iyengar's work had a big impact; in 1988, when he and his colleagues discovered "NC algorithms for Recognizing Chordal Graphs and K-trees" [IEEE Trans. on Computers 1988]. This breakthrough result led to the extension of designing fast parallel algorithms by researchers like J. Naor (Stanford), M. Naor (Berkeley), and A. A. Schaffer (AT&T Bell Labs).

His research has been funded by the NSF, Defense Advanced Research Projects Agency (DARPA), Multi-University Research Initiative (MURI Program), Office of Naval Research (ONR), Department of Energy/Oak Ridge National Laboratory (DOE/ORNL), Naval Research Laboratory (NRL), National Aeronautics and Space Administration (NASA), US Army Research Office (URO), and various state agencies and companies. He has served on the US National Science Foundation and the National Institute of Health Panels to review proposals in various aspects of Computational Science and has been involved as an external evaluator (ABET-accreditation) for several Computer Science and Engineering Departments across the country and the world. Dr. Iyengar has also served as a research proposal evaluator for the National Academy. Dr. Iyengar has been a Visiting Professor or Scientist at Oak Ridge National Laboratory, Jet Propulsion Laboratory, Naval Research Laboratory, and has been awarded the Satish Dhawan Visiting Chaired Professorship at the Indian Institute of Science, the Homi Bhabha Visiting Chaired Professor (IGCAR), and a professorship at the University of Paris-Sorbonne.

Kenneth G. Furton is the Executive Director of the Global Forensic and Justice Center and Provost Emeritus of Florida International University. He is a distinguished university professor in the Department of Chemistry and Biochemistry and a world-leading scholar in forensic chemistry focused on trace detection and olfaction. From 2007 to 2014, he served as dean of the College of Arts and Sciences, which he reorganized into three mission-based interdisciplinary schools to address some of the biggest issues facing society, raising $50M in philanthropic gifts and doubling research funding to $60M annually. From 2014 to 2022, he served as provost, executive vice president, and chief operating officer of FIU where he has led the development and implementation of two FIU strategic plans helping to create a dynamic, results oriented university with dramatic improvements in student success and research preeminence including simultaneously doubling the 4-year graduation rates to over 60%, doubling research expenditures to $250M annually and a 30-fold increase in patents to more than 60 annually achieving a Top 20 patent ranking. He also helped secure over $100 million in philanthropic gifts and FIU moved from #8 to #1 in the state of Florida performance-based funding rankings. Under his leadership, FIU was the most improved R1 university in the nation in U.S. News rankings, improving 54 spots in 5 years to No. 78 public and ranked No. 17 in Innovation and No. 5 in Social Mobility. Deeply committed to the academy, Dr. Furton has supervised the research of more than 140 students and been continuously funded for more than three decades, totaling more than $20 million in grants. He has 32 patents, 2 books, and 324 peer-reviewed publications with more than 12,000 citations and an h-index of 60. He is a member of Phi Beta Kappa and an elected fellow of the National Academy of Inventors (NAI) as well as the American Academy of Forensic Sciences. He has received significant attention in recent years for ground-breaking work using dogs and sensors to detect humans, drugs, currency, accelerants, explosives, mass storage devices, invasive species, and medical conditions including the coronavirus disease. His researching and deployment of COVID-19 detector dogs during the global pandemic reached an audience of over 2 billion worldwide.

Pronab Mohanty is a distinguished officer of the Indian Police Service (IPS)—1994 batch, currently serving as the Director General of Police (DGP)—Cyber, Economic, and Narcotics (CEN) and Chief Information Security Officer (CISO) for Karnataka State Police. With over three decades of exemplary service, he has led critical law enforcement and cybersecurity initiatives in India, holding key positions in the Central Bureau of Investigation (CBI), the Ministry of Electronics and Information Technology (MeitY), the Unique Identification Authority of India (UIDAI/AADHAAR), and the Border Security Force (BSF).

Dr. Mohanty has an extensive background in investigating technologically complex cybercrimes, implementing advanced cybersecurity procedures in sensitive data centers, conducting forensic audits of data breaches, privacy, and consent management, and investigating serious corporate and financial frauds, human trafficking, and crimes against minors.

A strong advocate for modernizing police IT services, he has played a crucial role in developing and securing India's national cyber architecture, leading the National Cyber Coordination Centre (NCCC), and ensuring the security of critical Information and Communications Infrastructure in the BSF. His expertise in law enforcement modernization, cyber forensics, and advanced computing security has positioned him as one of India's foremost experts in cybercrime prevention and digital forensics.

Naveen Kumar Chaudhary has been a professor of Cybersecurity at the National Forensic Sciences University (NFSU) in Gandhinagar, Gujarat, India since 2019. He is also a Courtesy Research Professor at the Knight Foundation School of Computing and Information Sciences at Florida International University, Miami, Florida. He holds a Bachelor of Technology degree in Information Technology and Telecommunication Engineering and Master of Engineering degree in Digital Communication. He earned his Ph.D. in Engineering and advanced certifications in Cyber and Network security.

His extensive experience spans more than 25 years in engineering education, research, and government. He has steered many cutting-edge ICT projects and worked extensively on policy formulation in cybersecurity and e-governance. He is the recipient of a letter of appreciation for contribution towards the cause of literacy from Brent St. Denis, MP, Algoma Canada, in 1994. Dubai, SEWA Award for his contribution to cybersecurity education in 2022. He also received COAS and CCOSC Commendation in 2009 and 2015, respectively, for his innovation and distinguished service. He is an IEEE senior member and life member of IETE.

Part I
Artificial Intelligence and Machine Learning Driven Digital Forensics

From Forensic Intelligence to Artificial Intelligence: A Review of Information Models in Forensic Science

Kenneth G. Furton, Max M. Houck, S. S. Iyengar, Jeffrey Teitelbaum, and Kevin Lothridge

1 Introduction

Forensics has evolved in exponential steps dating back to ancient times in Egypt, Greece, India, and China more than 2000 years ago. In Egypt, some of the earliest elements of forensic medicine were developed, including autopsies on embalmed pharaohs visible today, more than 3000 years later. In Greece, Hippocrates wrote 2400 years ago *On Injuries of the Head*, explaining how a blow to the head by a rock might be investigated scientifically. In Egypt, Kautilya's *Arthashastra* was first written about 2300 years ago and included sections on investigating crime scenes and poison detection. About 2200 years ago, it is believed that China was conducting forensic assessments, and the oldest existing forensic text was written in 1247 by Sung Tzu entitled *The Washing Away of Wrongs*.

The foundations of modern crime science began around 150 years ago, including Alphonse Bertillon's anthropometric system and the first forensic use of dogs and the analysis of drugs (poisons). In the last 40 years, the powerful fields of DNA and digital (computer) forensics emerged. In the present day, there has been an explosion in the use of data science and AI, transforming forensics into a more quantitative discipline, but with challenges discussed in this chapter. Figure 1 illustrates some key periods in the evolution of forensic science.

Alphonse Bertillon's late nineteenth-century anthropometry system [1] pictured in Fig. 2 is often regarded as the first scientific approach to criminal identiication. Bertillon, a French police clerk turned criminologist, developed a method termed

K. G. Furton (✉) · S. S. Iyengar
Global Forensic and Justice Center, Florida International University, Miami, FL, USA
e-mail: furtonk@fiu.edu

M. M. Houck · J. Teitelbaum · K. Lothridge
Florida International University, Miami, FL, USA

S. S. Iyengar et al. (eds.), *Artificial Intelligence Driven Forensics*, Signals and Communication Technology, https://doi.org/10.1007/978-3-031-98036-7_1

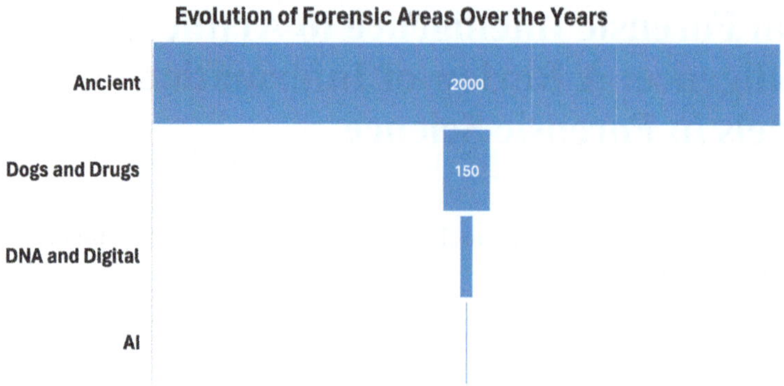

Fig. 1 A representation of major eras of forensic evolution from ancient societies to rapid developments in the nineteenth century including dogs and drugs, to the importance of DNA and digital forensics in the twentieth century, to the current impact of AI-driven forensics

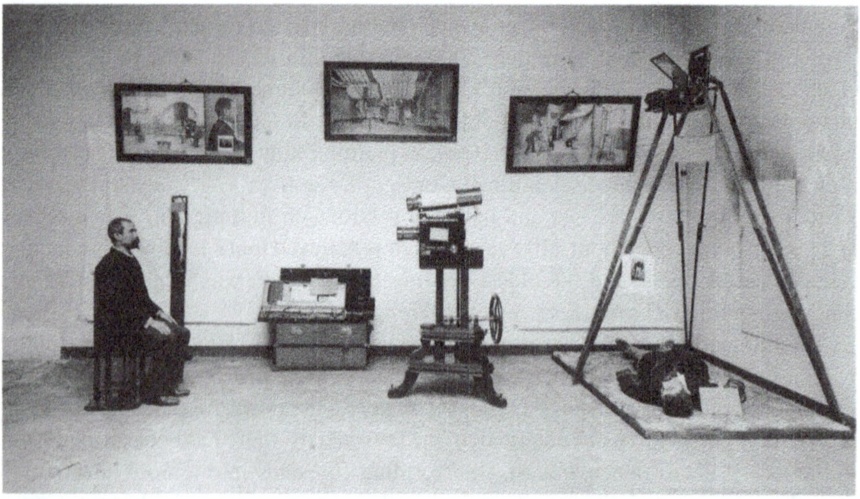

Fig. 2 A photograph from Alphonse Bertillon's photo album from his exhibition at the 1893 World's Columbian Exposition in Chicago. (https://www.nlm.nih.gov/exhibition/visibleproofs/media/detailed/ii_c_304.jpg)

bertillonage by the press due to the founder's fame that involved meticulously measuring physical features of suspects, such as head circumference, arm length, facial dimensions, and others, and cataloging these alongside front and side photographs (what are now called "mug shots") [2]. Bertillon was not a great scientist; his only scientific insight was to expand upon Adolphe Quetelet's research into the measurement of humans. His real genius lay in the organization, ordering, and filing of that data, his measurements of criminals. This anthropometric system treated the human body itself as a dataset, encoding identity as a series of numerical measurements. At the time, it was a groundbreaking way to model forensic data and distinguish repeat offenders in an era when only (mostly false) names or poorly taken photographs were used for identification. Bertillon's system dramatically improved police record-keeping and was eventually very effective: It was adopted worldwide as the system for processing criminals' identification. However, it was not infallible. A famous incident in 1903 at Leavenworth Prison involved two inmates, Will West and William West, who had nearly identical Bertillon measurements and appearances [3]. The prison staff were astonished to discover that the measurements for Will West matched another file, that of William West, an inmate already incarcerated, yet the two men were different individuals. Only by comparing their fingerprints could the prison definitively distinguish them. This highly publicized event underscored the limitations of purely anthropometric identification. Indeed, Bertillon's method was soon supplanted by the fingerprint identification system in the early twentieth century [4].

Fingerprinting offered a far more granular and reliable biometric marker, leading police departments worldwide to abandon anthropometry in favor of dactyloscopy (fingerprint analysis). Edmund Locard established the first forensic laboratory in Lyon, France, in 1910. He recognized the power of fingerprints for not only identifying arrested criminals in custody but also identifying them from prints left at crime scenes. The comparison method was not the issue—comparing the unknown prints meant known reference prints needed to be collected, managed, and searchable. Fingerprints had long been collected in the Bertillon system, although Bertillon himself downplayed their importance in favor of his own creation, so the files held the reference prints needed. What Locard lacked was access, as they were police files and only police officials were allowed to use them. Thus, Locard installed his novel laboratory within the Lyon Police Department and gained the access he needed. Fingerprints spread across the policing world and changed human identification forever.

These opening anecdotes highlight a theme that recurs throughout forensic science history: the continual evolution of identification techniques and data systems. From Bertillon's manual measurements to today's algorithms, forensic science has steadily moved toward more automated, data-driven, and precise methods of cataloging and analyzing evidence. In this chapter, we explore that evolution in depth, moving from early manual systems to modern AI-driven systems. We begin by reviewing the historical progression of major forensic databases that systematized forensic data, and then introducing how artificial intelligence and machine learning entered the forensic domain. The heart of the chapter examines current applications of AI across various forensic subfields. On this foundation, we address critical issues and risks associated with these technologies, and discuss policy and ethical frameworks guiding their use. A summary of the pertinent literature highlights what research to date reveals and where uncertainties remain. Finally, the conclusion ties these threads together and looks ahead to future developments that subsequent chapters will discuss. Throughout this chapter, the emphasis is on the interplay between forensic science and AI, how decades of forensic data collection set the stage for AI, and how AI is, in turn, challenging and advancing forensic practice.

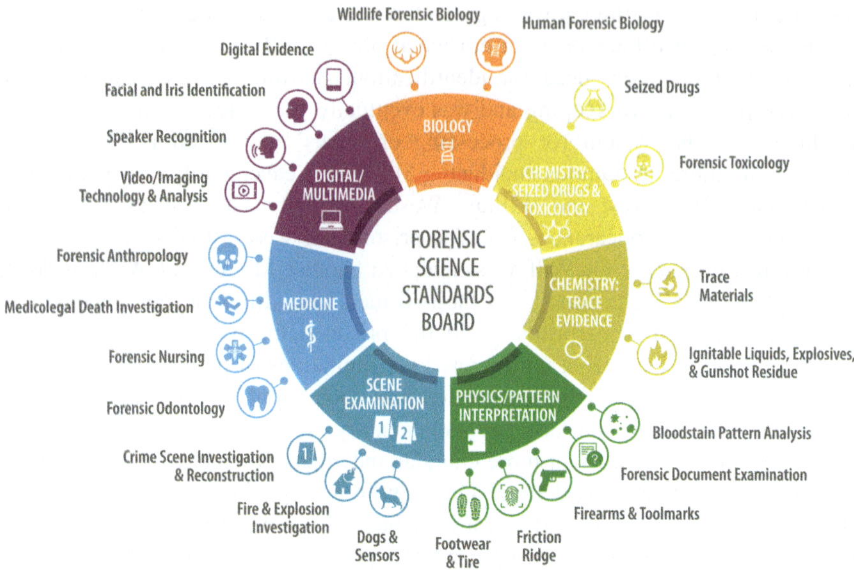

Fig. 3 Organization of Scientific Area Committees of Forensic Science (OSAC) areas. (OSAC Overview Outreach Flyer_FEB2025.pdf)

As we examine the impact of AI on forensic science, it is important to understand that there are dozens of areas of forensic science that are being or will be impacted by AI. The Organization of Scientific Area Committees of Forensic Science (OSAC) [5], administered by the National Institute of Standards and Technology (NIST) [6], was created in 2014 to address a lack of discipline-specific forensic science standards and identified 22 forensic disciplines in seven areas: digital/multimedia, biology, chemistry of seized drugs and toxicology, chemistry of trace evidence, physics/pattern interpretation, scene examination, and medicine as shown in Fig. 3. In addition, there are many additional fields including veterinary forensics, forensic musicology, forgery detection, and forensic accounting which will be impacted.

While there are dozens of different forensic science areas, the ones that are most familiar to the general public are those associated with crime laboratories and historically have been referred to as crime sciences. Of these, there are what can be considered the four D's as outlined in Fig. 1. The first D is Drugs which encompass more than half of the cases analyzed by public crime laboratories in the United States. The D is DNA which is analyzed in 30% of cases. The third D is Digital which includes six areas focused on electronic and visual examinations of evidence encompassing the remaining nearly 20% and growing component of forensic evidence. The final D is Dogs which are the most common tool used by forensic scientists to locate drugs (often hidden), DNA (trace biological materials), and digital evidence (including mass storage devices and firearms evidence). This chapter focuses on the evolution of AI in the modern era of crime sciences.

Table 1 Cases completed by public crime labs in 2020 (December 2023 Bureau of Justice Statistics Bulletin NJC 306473; https://bjs.ojp.gov/document/pffcl20.pdf). Historical progression of forensic databases

Type of request	2020 Number	Percent	Area	Area	Area
Controlled substances	1,088,000	32.5%	Drugs		
DNA databasing	671,000	20.1%		DNA	
Toxicology	629,000	18.8%	Drugs		
Latent print analysis	180,000	5.4%			Digital
Forensic biology casework	339,000	10.1%		DNA	
Crime scene analysis	144,000	4.3%			Digital
Firearms/toolmarks	225,000	6.7%			Digital
Digital evidence analysis	27,000	0.8%			Digital
Trace evidence	34,000	1.0%			Digital
QD/impressions	9000	0.3%			Digital
Total	**3,346,000**		**51.3%**	**30.2%**	**18.5%**

Modern forensic science rests on a series of innovations that progressively systematized the collection and analysis of forensic data. In this section, we highlight four key milestones in this historical progression: Bertillon's anthropometry, IAFIS (fingerprints), CODIS (DNA), and NIBIN (ballistics). Each represented a leap in how forensic information was standardized, stored, and searched, moving the field toward more comprehensive databases amenable to computational analysis (Table 1).

Bertillon's anthropometric system was an early attempt to create a database of criminal identities based on physical measurements. In the 1880s and 1890s, police departments adopting bertillonage would record 11 key body measurements of each arrestee, believing the combination to be unique for each person. These measurements, along with standardized frontal and profile photographs, were compiled into handwritten card catalogs. The value of Bertillon's system was in bringing scientific rigor and consistency, effectively creating a manual data model for identification. However, as the Will West case demonstrated, anthropometry alone could fail in cases of biologically coincidental similarity. By the early twentieth century, the inherent limitations, including being labor-intensive and difficult to master, of this system led to its replacement.

Fingerprinting as a forensic method dates back to the late nineteenth and early twentieth centuries, but it truly became systematized with the advent of computerized databases. The US Federal Bureau of Investigation (FBI) launched the Integrated Automated Fingerprint Identification System (IAFIS) [7] in 1999 as a national digital repository of fingerprints and criminal history data. IAFIS greatly expanded the scale and speed of fingerprint searches compared to earlier manual filing systems. It can compare a latent print from a crime scene against a vast database of stored prints within minutes, a task that would have been prohibitively time-consuming by hand. By the early 2000s, IAFIS housed the fingerprints and associated criminal records of over 70 million subjects, plus tens of millions of civil records. It became an indispensable tool for law enforcement, routinely returning candidate matches that examiners then verify. IAFIS exemplified the trend of digitizing forensic databases by turning inked fingerprint cards into searchable image files and demonstrated how automation could dramatically enhance forensic identification. In 2011, the FBI began transitioning to the Next Generation Identification system [7], extending IAFIS with additional biometric modalities, but fingerprint matching remains a core component.

Following the advent of DNA profiling in the 1980s, the FBI developed the Convicted Offender DNA Identification System (CODIS) [8] as a way to network and share DNA data across jurisdictions. A pilot program began in 1990 with select state laboratories. In 1994, the DNA Identification Act [9] was passed, which officially authorized the FBI to create a national DNA database for convicted offenders and other indices (such as for arrestees and forensic case samples). The national level of CODIS, known as the National DNA Index System (NDIS), became operational in October 1998. CODIS software was provided free to states to encourage widespread adoption. By linking all 50 states, DC, federal agencies, and the military's DNA profiles, CODIS enabled cross-jurisdictional comparison of DNA evidence. For example, a DNA profile obtained from a crime scene in one state can be matched to a convicted offender's profile from another state, or linked to DNA from other unsolved cases, through weekly database searches. CODIS is structured into various indices for offenders, arrestees, forensic case samples, missing persons, etc., each serving different investigative purposes. The success of CODIS in connecting cases and suspects worldwide firmly established DNA databases as a cornerstone of modern forensic science.

Firearms evidence went through a similar evolution with the adoption of nationwide ballistics imaging. The National Integrated Ballistic Information Network (NIBIN) [10], administered by the Bureau of Alcohol, Tobacco, Firearms and Explosives (ATF) in the United States, was launched around 1999, building on earlier regional efforts in the 1990s. NIBIN uses the Integrated Ballistic Identification System (IBIS) platform to collect and compare digital images of ballistic markings on spent shell casings and bullets. Every firearm leaves unique microscopic toolmarks on the cartridge casings it ejects or the bullets it fires. IBIS technology captures these marks and converts them into a digital signature. By entering evidence from shooting scenes into NIBIN, technicians can algorithmically search for matches with casings or bullets from other crime scenes. NIBIN excels at linking crimes that involve the same firearm, even across different cities or states, thereby generating investigative leads. For instance, a series of shooting incidents across two counties was solved when NIBIN analysis revealed they involved the same seven firearms, connecting the cases. NIBIN's establishment reflects the broader trend of networked forensic databases, where local labs contribute data to a shared national system to multiply the power of forensic evidence.

Each of these databases, Bertillon's anthropometry cards, fingerprint AFIS, DNA CODIS, and ballistics NIBIN, represents a step toward systematized data modeling in forensic science. Bertillon's system was essentially a handcrafted data model of human identity; IAFIS, CODIS, and NIBIN took advantage of computers to store and search data at unprecedented scale. Notably, these systems focused on matching evidence to known references; they were less about interpretation and more about rapid comparison across large data repositories. This evolution set the stage for applying modern AI: Once forensic information was digitized and aggregated, it became feasible to apply advanced algorithms to identify patterns and make predictions within those large datasets.

Modern forensic databases evolved through successive innovations that standardized evidence collection and analysis:

- **Anthropometry (Bertillon System):** Utilized systematic bodily measurements; labor-intensive and eventually replaced by fingerprinting [1].
- **IAFIS (Integrated Automated Fingerprint Identification System):** Launched by the FBI in 1999, revolutionizing fingerprint data processing with computerized search capabilities.
- **CODIS (Combined DNA Index System):** Initiated in the 1990s, it facilitated nationwide sharing of DNA profiles, transforming forensic biology [11].
- **NIBIN (National Integrated Ballistic Information Network):** Established by the ATF to digitally compare ballistic markings, significantly enhancing firearm crime investigations [12].

Each milestone represented a shift toward data-driven forensic methodologies, setting the stage for AI integration by digitizing and aggregating forensic information.

2 Emergence of AI and Machine Learning in Forensics

Artificial intelligence (AI), particularly in the form of machine learning (ML), began making inroads into forensic science as digital data and computing power grew exponentially in the late twentieth and early twenty-first centuries. AI can be broadly defined as the capability of machines to perform tasks that typically require human intelligence, such as pattern recognition, decision-making, or language understanding. Machine learning is a subset of AI that enables systems to learn patterns from data and improve over time without being explicitly programmed with step-by-step rules. In the forensic context, the appeal of AI/ML lies in their potential to detect subtle patterns in evidence and to automate complex analyses, thereby complementing human experts [13–15].

3 AI Basics and Relevance to Forensics

Artificial intelligence (AI) refers to the simulation of human intelligence in machines that can learn, reason, and make decisions [13, 14]. In the field of forensics, AI plays a crucial role in analyzing vast amounts of digital evidence quickly and accurately. It enhances the speed and precision of investigations, from identifying suspects to uncovering hidden patterns in data. As cybercrimes grow in complexity, AI has become an essential tool in modern forensic science [15].

Early applications of AI in various domains often took the form of expert systems, such as rule-based programs encoded with human knowledge. In an expert system, experts articulate the decision rules ("if X and Y features are present, then conclusion Z") which the computer then uses to emulate human decision-making. By contrast, machine learning systems derive their own rules and models from training data. Instead of being explicitly told what patterns to look for, a machine learning algorithm is fed many examples (such as pairs of inputs and known outputs) and it "learns" a predictive model. This model can then generalize to new, unseen inputs. Over the past two decades, ML techniques, especially those using neural networks and deep learning, have achieved remarkable success in fields such as image recognition, speech recognition, and natural language processing. This success has not gone unnoticed in forensic science, which deals heavily with pattern recognition problems (fingerprints, toolmarks, DNA profiles, and so on).

Researchers began investigating whether similar algorithms could assist in forensic analyses that traditionally relied on a human examiner's trained eye or judgment. Key milestones in AI that have influenced forensic applications include the advent of neural networks in the 1980s and 1990s, the breakthrough of deep learning around 2012 (when convolutional neural networks demonstrated dramatic improvements in image classification), and continual advances in computational power (GPUs, cloud computing) that made training of complex models feasible. By the mid-2010s, AI had matured to the point where it was being embedded in everyday technology, such as smartphone face recognition and digital assistants, and law enforcement agencies started experimenting with AI-driven tools for surveillance and investigation (such as facial recognition systems and predictive policing software). In parallel, forensic laboratory scientists, often supported by federal research grants, initiated studies to apply ML to laboratory forensic evidence. For example, the US National Institute of Justice (NIJ) and National Institute of Standards and Technology (NIST) funded projects to explore AI in pattern evidence analysis and crime scene data processing.

4 Early Forensic AI Efforts

One of the first domains of forensic science to embrace computerized pattern recognition was fingerprint analysis. Automated Fingerprint Identification Systems (AFIS), which predate the term "AI boom," used algorithms to measure minutiae points on fingerprints and match them to database entries. These algorithms were largely deterministic and programmed by humans (so not really "learning" systems), but they demonstrated the benefit of automation. As machine learning techniques advanced, researchers proposed augmenting or replacing the handcrafted matchers with ML-based approaches that could, for instance, learn the most discriminative fingerprint features from data. Similar efforts were seen in handwriting analysis, where by the 2000s there were attempts to use neural networks to recognize handwritten characters or signatures, and in firearms analysis (comparing striation patterns on bullets or casings) where statistical models began to be developed to quantify match likelihoods. A notable milestone in the forensic AI emergence was the increasing emphasis on quantitative methods in forensic science after critical reviews of the field. The 2009 US National Academy of Sciences report *Strengthening Forensic Science in the United States* [16] highlighted that many forensic disciplines lacked rigorous scientific validation and objective standards. It implicitly opened the door for data-driven approaches, such as algorithms, to play a role in improving objectivity. In 2016, the President's Council of Advisors on Science and Technology (PCAST) issued a report [17] that more explicitly discussed the potential of probabilistic and statistical methods (including software) for evidence interpretation in areas such as DNA mixtures and latent fingerprints. These reports signaled to the forensic community that embracing computational tools could be part of the solution to strengthen reliability and consistency. By the late 2010s, a confluence of factors, such as the large collections of digitized forensic data, advanced machine learning algorithms, and a push from oversight bodies to innovate, inspired federal initiatives, such as the National Artificial Intelligence Initiative in the United States, to develop AI across sectors. Within forensic science, research initiatives started producing proof-of-concept AI applications. For example, studies applied neural networks to classify patterns in bloodstain spatter or to detect manipulated digital images. Yet, it is important to note that as of the early 2020s, most AI applications in forensic science were still in the research or pilot phases, not in widespread operational use. Compared to fields like healthcare or finance where AI had rapidly been productized, forensic science adoption was slower and more cautious, due in part to the high stakes of criminal justice outcomes and strict legal scrutiny (as discussed later). Nonetheless, the emergence of AI in forensics can be summarized as a shift from manually coded, expert-driven systems to data-driven learning systems.

5 Current Applications of AI in Forensic Science

AI's versatility means it has potential applications across virtually every domain of forensic science. In this section, we explore current applications of AI under several categories: digital and multimedia forensics, impression and pattern evidence, DNA analysis, facial recognition, and crime prediction/analytics. Many of these applications are still evolving, with ongoing research and pilot implementations. They collectively demonstrate how AI can enhance both the efficiency and the analytical capability of forensic investigations.

6 Digital Forensics and Multimedia Analysis

Digital forensics deals with extracting and analyzing evidence from electronic devices and digital media. With the explosion of digital data, such as computers, smartphones, CCTV cameras, and internet activity, investigators often face a deluge of information in cases ranging from cybercrime to traditional crimes that have digital footprints. AI tools, especially those in the subfield of computer vision and signal processing, are increasingly used to process and analyze this trove of digital evidence [18–20].

AI-driven image recognition can automate the tagging and searching of visual evidence. For instance, in a case involving a large number of photographs, machine learning models can classify images, identifying that certain images contain firearms, drugs, or specific logos, to help triage relevant from irrelevant pictures. In video footage, AI can be used for object detection and tracking, like spotting a particular make of car across hours of surveillance camera footage or detecting frames that contain a particular face. An emerging challenge is detecting deepfakes (AI-generated synthetic images or videos). Ironically, AI is part of the solution to this AI-created problem: Researchers are developing forensic algorithms that can identify subtle artifacts left by deepfake generation. Similarly, AI can enhance low-quality footage; for example, convolutional neural networks (CNNs) have been trained to reduce noise and sharpen details in a poor-quality CCTV video, which might help identify a person or object in the scene.

Digital forensics also encompasses audio, such as recordings of phone calls or environmental sound from videos. Machine learning can be applied to isolate and identify sounds, such as recognizing gunshots, breaking glass, or specific spoken keywords in a long audio recording. Some innovative research even involves classifying gunshot audio to identify the type of firearm or distinguish gunshots from fireworks or car backfire. Voice biometrics is another AI application: Algorithms can compare voice samples to determine if they likely came from the same speaker, which is useful in threat calls or verifying alibis via voice.

Document and Text Analysis: When dealing with large bodies of text, such as emails, messages, and documents seized in investigations, AI-based text mining can assist in searching for significant patterns or incriminating content, such as clustering communications or detecting code words. Language translation tools powered by AI are also invaluable in international cases, where they can quickly translate foreign-language documents or intercepts, making evidence accessible to investigators who do not speak that language.

It should be noted that many digital forensics AI tools are proprietary solutions from software vendors that are integrated into products used by law enforcement, for example, software that automatically categorizes images of child exploitation so that investigators do not have to view every file individually. These AI models are usually trained on large datasets specific to the task, like millions of labeled images. While extremely useful in managing caseloads and finding needles in haystacks, they also raise questions about false positives and the need for human verification. Overall, AI has become a force multiplier in digital forensics, dealing with the scale of data and complexity of analysis that traditional methods struggle with.

7 Pattern and Impression Evidence

A classical domain of forensic science involves pattern evidence, where an examiner compares patterns or impressions left by a suspect tool, finger, foot, tire, or other object with a reference. This category includes fingerprints, toolmarks, firearm striations, footwear impressions, bloodstain patterns, handwriting, and more. Traditionally, such comparisons rely on expert judgment. AI and ML are being applied to make these analyses more objective and efficient.

Modern AFIS systems have started to incorporate machine learning to improve accuracy of fingerprint matching. Research has explored using deep learning to enhance latent fingerprints, such as segmenting and enhancing poor-quality prints before matching, and developing match scoring algorithms that better handle distortion. AI can also assist in fingerprint comparison by acting as a second evaluator: One envisioned use-case is an examiner performing a comparison and then using an AI tool as a "spell-checker" to see if the AI agrees or flags discrepancies. In such a scenario, if the AI suggests a different matching result or highlights minutiae the human missed, it could prompt a review, thus improving quality control.

The task of comparing ballistic toolmarks, such as striations on bullets or breech face marks on cartridge casings, is complex and historically subjective. AI-based image analysis can quantify the similarity between toolmarks. Researchers have used classification algorithms to assess the similarity of bullet striation patterns by extracting features and computing a match score. One study applied machine learning to improve correlation in an open set of bullets, finding that algorithms could rank the true match among top candidates more effectively than older methods [21]. Similarly, for footwear impressions, ML has been used to compare a crime scene shoeprint with a database of outsole patterns, taking into account wear and distortion [22]. This can generate a short list of candidate shoe models or specific shoes that might have made the print.

AI has been employed to classify handwriting, such as determining if two samples likely have a common author [23]. Beyond just optical character recognition, which is well-established, newer research is looking at characteristics like stroke patterns and pressure to distinguish writers. For forensic document examination, algorithms have been developed to compare signatures or to identify the printer model from printed text characteristics.

Bloodstain pattern analysis is a more nascent area for AI. Bloodstain patterns at crime scenes can convey how a violent event occurred, possibly providing the position of the victim and assailant or the type of motion. Some researchers are training AI to classify simple bloodstain pattern types, such as distinguishing between a passive drip and a high-velocity spatter based on images, or even to estimate the angle of impact from the shape of a bloodstain. The complexity of fluid dynamics in bloodstains makes this challenging, but progress is being made with deep learning models trained on simulated bloodstain data.

While facial recognition could be considered digital or biometric evidence, it also functions as a pattern-matching task between a face in an image and faces in a database, so we mention it here. Law enforcement agencies have widely adopted facial recognition AI to compare face images from CCTV, mugshots, or even sketches against mugshot databases or driver's license databases. In forensic casework, this might be used to identify an unknown suspect in a video or to verify whether an individual in a social media photo is a person of interest. The algorithms, often deep neural networks, excel at generating a list of possible matches; however, a human analyst typically must review the list to make a final identification decision due to potential false positives. We will discuss ethical challenges of facial recognition, particularly bias, separately, as it is one of the most debated AI applications in criminal justice.

The overarching theme in pattern evidence is that AI can assist by quantifying similarity and searching large databases quickly. These tools do not eliminate the role of the expert; rather, they act as powerful assistants. For example, a latent fingerprint examiner might use AI to get a probability score that two prints match, or a firearms examiner might use an algorithm to narrow down which cartridge cases are worth a detailed microscopic comparison. By handling the tedious comparisons and providing objective metrics, AI can make pattern evidence analysis more standardized. Importantly, researchers emphasize that these ML systems need proper training with ground-truth datasets from known matching and non-matching pairs of evidence and should report not just a match result but also confidence levels or statistical support for the match. This moves pattern evidence toward a more scientific footing, where conclusions can be accompanied by error rate estimates or likelihood ratios output by algorithms, rather than solely an expert's subjective certainty.

8 DNA Mixture Interpretation

Forensic DNA analysis has its own specialized AI applications. Routine single-source DNA profiling (one person's DNA) does not require AI—it is a straightforward lab process followed by a database match in CODIS. However, DNA mixtures (samples containing DNA from multiple individuals) present a complex interpretation challenge. In crimes involving multiple people or trace evidence that picked up DNA from several sources, analysts must deconvolute the mixed genetic data to infer individual DNA profiles or at least determine whether a suspect's profile could be a contributor. Traditionally, DNA mixture interpretation was largely manual and often inconclusive beyond two-person mixtures. In the last decade, specialized software called probabilistic genotyping systems have been developed. These use statistical models and Markov Chain Monte Carlo methods to compute the probability of different genotype combinations explaining the observed data. While not typically described as "AI," they paved the way for data-driven DNA analysis. Now, researchers are looking at machine learning to further enhance mixture interpretation.

For example, classification algorithms can be used to estimate the number of contributors in a DNA mixture. Knowing whether a stain has DNA from, say, two, three, or four individuals is critical to setting up the statistical analysis; AI models trained on thousands of mixture profiles of known composition can learn to predict "there are three contributors here" based on patterns of allele peaks [24]. Another area is using ML to screen genotype combinations. Given a complex mixture, an AI might rapidly sift through possible combinations of individual profiles that could fit the evidence, something that becomes combinatorially huge as contributors increase. Researchers have reported using machine learning alongside probabilistic genotyping to speed up computations or to identify unlikely vs. likely contributors. Outside the realm of STR DNA profiles, genomic AI applications are emerging in forensics. One example is phenotype prediction from DNA. AI/ML models have been trained on genomic data to predict physical traits of an unknown person—such as eye color, hair color, or biogeographical ancestry—from their DNA markers. While not yet courtroom evidence, such predictions (sometimes called DNA "Snapchat" or molecular photofitting) can provide investigative leads. The predictions use complex genetic datasets and ML to find correlations between genetic variations and traits.

Similarly, AI is aiding investigative genetic genealogy, where SNP (single nucleotide polymorphism) data is used to find relatives of an unknown suspect through genealogy databases [25]. AI can help identify likely family relationships even when the DNA data is incomplete or low-quality. For instance, machine learning might assist in triangulating matches on a family tree or in error correction for SNP data in genealogical searches. It is worth noting that AI's role in DNA analysis must be handled cautiously. DNA evidence carries a strong weight in court, and adding AI means adding layers of complexity in how results are obtained and interpreted. However, the promise is that AI can more objectively interpret complex DNA evidence, reducing the variability between analysts. For example, two labs given the same difficult mixture might reach different conclusions today, but if both use the same validated AI tool, their analyses could be more consistent. Still, questions of reliability and transparency are paramount, and the methods must be well-tested so that their error rates and limits are known.

9 Facial Recognition and Ethical Challenges

Facial recognition technology, powered by AI algorithms (especially deep convolutional neural networks), has rapidly become a fixture in law enforcement and security. Its application in forensics and public safety includes identifying suspects in photos or videos, locating missing persons, and verifying identities, for example, matching a face to an ID photo. Systems like the FBI's Next Generation Identification face recognition service can search millions of mugshots to find potential matches to a probe image. However, facial recognition exemplifies the ethical and technical challenges of AI in forensic use [20, 26–29]. Several ethical and performance issues have been documented.

Studies have shown that many commercial facial recognition systems have unequal performance across different demographic groups. A landmark 2019 NIST study found higher false positive rates for African American and Asian faces compared to Caucasian faces in the majority of algorithms tested [30]. For example, some algorithms were 10 to even 100 times more likely to misidentify a Black or East Asian face than a White face. This bias usually stems from imbalanced training data, such as an algorithm trained mostly on lighter-skinned faces will perform better on those, and possibly from differences in image acquisition. The result is a risk of false accusations against minority individuals due to algorithmic error, which is a serious concern already borne out in several wrongful arrest cases attributed to facial recognition mismatches. This necessitates transparency about training data and active steps to mitigate bias in these systems, such as using more representative training sets or bias correction techniques. It also underscores that any face recognition match should be only a lead, verified by humans and corroborated with other evidence, rather than treated as infallible.

The use of facial recognition on a wide scale raises civil liberties issues. Identifying individuals in public spaces via cameras can infringe on privacy and has led some jurisdictions to ban or restrict police use of facial recognition. From a forensic standpoint, using facial recognition in investigations must balance public safety benefits against the potential chilling effects on society. Ethical guidelines increasingly call for use policies, for example, requiring a certain threshold of confidence before declaring a match, or restricting use to serious crimes. Some facial recognition systems have been built by scraping billions of images from the internet (social media, websites) without consent, as was the case with controversial companies that created massive face datasets. This introduces ethical problems around consent and data protection. It can also flood the system with low-quality or context-less images, affecting accuracy.

In forensic science, admissibility standards and evidentiary rules come into play if facial recognition results are to be introduced in court. Defense attorneys have challenged such evidence on grounds of reliability and potential bias [31]. To date, facial recognition matches are usually treated as investigative leads rather than positive IDs presented to a jury, precisely because of these concerns. In summary, while facial recognition is a powerful AI application currently employed in criminal investigations, it brings to the forefront issues of bias, transparency, and policy.

10 Predictive Modeling and Crime Analytics

Beyond analyzing forensic evidence from individual cases, AI is also being applied to broader crime data analytics to aid in crime prevention and resource deployment. This includes what is popularly known as predictive policing—using algorithms to predict where or when crimes are more likely to occur, or who might be involved in future crimes. While not "forensic science" in the traditional lab sense, this is part of the intersection of data science and criminal justice [32].

Police departments have used machine learning models to identify hotspots for crime by analyzing patterns in historical crime data [33]. Systems like PredPol (now renamed Geolitica) ingest years of crime incident reports that include time, location, and type of crime, and output small geographic areas ("boxes") where property crimes or burglaries are likely on a given day, for example. These predictions can help allocate patrols more efficiently. The algorithms often use approaches such as logistic regression or stochastic models rather than complex AI, but newer methods involve neural networks that can learn more intricate patterns, for example, considering temporal cycles, socioeconomic data, weather, events, etc., to refine predictions.

Another application has been models that try to identify individuals at risk of either becoming perpetrators or victims of violence—for instance, analyzing social networks and prior incidents to flag those likely to be involved in a shooting (either as shooter or target). Cities have tried "heat lists" and similar concepts, derived from ML analyses of police and social service databases. AI can sift through incident data, phone records, social media, and other sources to find connections between people and events. This might help identify organized crime networks or co-offending patterns that were not obvious. Such network analysis was done pre-AI as well, but machine learning can handle larger, messier datasets and perhaps predict likely links, such as suggesting that two incidents might be related through a common individual.

The use of predictive analytics is controversial. While theoretically promising as a proactive tool, in practice it has been criticized for perpetuating biases present in the crime data. For example, if historically a certain neighborhood is over-policed and thus generates more crime reports—not necessarily more crime, just more recorded crime—the algorithm may simply send more police there, creating a feedback loop. This is an example of systemic bias in AI, where the model reinforces patterns that reflect enforcement bias rather than actual crime incidence rates [34]. Recognizing this, researchers and policymakers emphasize the need for transparency in predictive models and for combining algorithmic input with human judgment and community input. There have been instances where departments pulled back on predictive policing programs due to public outcry and lack of clear efficacy. In the forensic science context, one might not see predictive algorithms in the lab, but it is part of the broader AI in the criminal justice landscape. It illustrates how AI can extend beyond analyzing evidence to analyzing crime as a data phenomenon. Tools that forecast crime or assess risk, such as criminal recidivism risk scores used in courts, must be rigorously validated to ensure they are fair and effective. We will touch upon the policy responses to these concerns in the next sections. In summary, the current applications of AI in forensic science and related criminal justice operations are diverse. Each application shows promise in improving either the efficiency, scope, or objectivity of forensic work. However, each also faces technical limitations and ethical/legal challenges which we now turn to.

11 Critical Issues and Risks

The integration of AI into forensic science is not without significant challenges. As we deploy AI/ML tools in contexts that can influence life-and-death decisions (like convictions or exonerations), it is crucial to scrutinize their limitations and risks. This section discusses four broad categories of concerns: bias in training data (and outcomes), the "black box" problem and need for explainability ("glass box" AI), legal admissibility issues, and general reliability and transparency issues. These often interrelated issues must be addressed to ensure AI tools enhance rather than undermine justice.

12 Bias in Training Data and Algorithms

Bias in AI refers to systematic errors that result in unfair outcomes, such as favoring one group over another. In forensic AI, biases can creep in at multiple stages: the data used to train algorithms, the way an AI model is designed, or how users interpret its results. As mentioned in the context of facial recognition, if an AI system is trained predominantly on data from one demographic, its performance will be skewed. The concern is not only demographic bias; there are also systemic and cognitive biases to consider.

Forensic databases themselves may reflect societal biases. For example, the CODIS DNA database in the United States has a disproportionate number of profiles from Black individuals relative to their share of the population, due to historical differences in arrest and conviction rates [35]. If an AI tool for DNA matching or familial searching is built on CODIS data, it might inadvertently increase scrutiny on certain groups. Similarly, if a latent fingerprint algorithm is trained on a dataset that underrepresents certain patterns, it might be less accurate on those. Ensuring diverse and representative training data is a key step to mitigating bias. NIST has identified categories of AI bias: systemic (institutional or historical biases), statistical (imbalances or flaws in data), and cognitive (human biases in interpreting AI outputs). All of these need management in forensic applications.

A specific human cognitive bias relevant here is automation bias, the tendency of people to trust a machine's decision too readily. In forensic labs, if an AI tool suggests a particular result, examiners might give that suggestion undue weight, even if the tool could be wrong. This is particularly risky when the user does not fully understand how the AI works, making it seem like a magic box. A result might be taken as authoritative simply because "the computer said so." Guarding against this requires training users to critically assess AI outputs and perhaps designing AI interfaces to clearly convey uncertainty, such as showing a range of possibilities rather than a single answer.

In predictive policing or crime analytics, bias can amplify through feedback loops. For instance, sending more patrols to an area predicted to be high crime will likely result in more recorded incidents there, even if underlying crime rates were not higher, which then reinforces the model's bias next time. This cyclical bias is a risk whenever AI influences the very data it later uses for learning. Breaking the loop may require periodic model resets or inclusion of external data to recalibrate.

Addressing bias requires a multipronged approach: carefully curating training datasets, balancing data, removing known skews where possible, augmenting data from underrepresented categories, testing algorithms for disparate impact, and incorporating bias mitigation techniques. Additionally, involving diverse stakeholders in AI development, including ethicists and representatives of affected communities, can help identify and correct biases that developers might overlook. Ultimately, no AI used in forensics should be deployed without thorough bias analysis and documentation of its "bias profile," with plans to manage any observed biases. As NIST's guidance suggests, managing harmful biases is integral to trustworthy AI.

13 Black Box vs. Glass Box: Explainability and Transparency

One often-cited problem with modern AI, especially deep learning models, is that they function as "black boxes." [36] They can be extraordinarily complex, with millions of parameters, and even the engineers who built them may not fully understand the inner decision process that leads to a particular output. In forensics and criminal justice, the opacity of AI decision-making is a serious concern. It is difficult to trust or legally defend evidence generated by a process that experts themselves cannot explain in simple terms. The ideal to strive for is sometimes called a "glass box" AI, meaning an AI system designed to be interpretable and transparent in its operations [37]. Some key points regarding this issue include:

Understanding AI Outputs: Traditional forensic methods, like chemically developing fingerprints or performing DNA gel electrophoresis, can be explained step-by-step to a court. But if an AI model says, "these two bullets were fired from the same gun with score 0.85," how did it reach that number? Was it the spacing of striations, the depth of microscopic grooves, or some combination of thousands of pixel intensity comparisons? Black box AI can have a tendency to infer spurious correlations, too—something might influence its decision that actually is an artifact or irrelevant feature, and we would not easily know. For forensic use, it is important that AI tools either use explainable methods or are paired with explainability techniques, for example, heatmaps on fingerprint images to show which areas contributed most to a match decision.

A practical barrier to transparency is when vendors consider their algorithm proprietary and refuse to disclose the inner workings or source code. There have been instances, such as the 2021 *New Jersey v. Pickett* case [38], where defense teams requested the source code of probabilistic genotyping software (TrueAllele) used to produce evidence, but the company withheld it, citing it as a trade secret. In one New York case, the Medical Examiner's office had developed a proprietary DNA analysis software that was used for years without external validation; only after a legal challenge was its code released by court order [39]. Such secrecy is at odds with the scientific ethos of reproducibility and peer review. If an AI's methodology cannot be examined openly, it may fail to meet admissibility standards or at least should be given less weight. Courts and legislatures are grappling with this: Some jurisdictions are considering or have passed rules that require disclosure of algorithmic workings if used in criminal justice.

The Daubert standard (from *Daubert v. Merrell Dow Pharmaceuticals, 1993*) [40] for admissibility of scientific evidence in US federal courts includes criteria such as testability, known error rate, peer review, and general acceptance. A black box AI poses challenges for each of these. Testability means one should be able to design experiments to verify the tool's accuracy, which requires knowing how it works to test it under various conditions. Error rates need to be quantified, but if the model is a black box, we might only have aggregate error rates from validation studies, without understanding when/why errors happen. Peer review is limited if the details are proprietary. General acceptance might be hard to achieve if the forensic community cannot fully assess the tool. Hence, unexplained AI models might struggle under admissibility challenges in court until they are better understood and validated. Some experts argue that requiring full explainability could stifle useful tools, but a balance must be struck to preserve due process rights.

In response to these issues, the concept of "glass box" AI is promoted, where the emphasis is on interpretable models or at least on supplementary explanations for complex models. For example, a simpler machine learning model, like a decision tree, might be chosen over a neural network for certain forensic tasks so that the decision path can be followed. If high accuracy absolutely requires a complex model, developers can incorporate explanation methods to provide insight into which features influenced a given decision. Transparency also means thorough documentation: If an AI system is used, a report should detail how it was developed, on what data, what its performance metrics are, and what guidelines were followed.

14 Reliability, Validation, and Error Rates

No forensic method is useful unless it is reliable, meaning it produces accurate and consistent results. With AI, reliability is closely tied to rigorous validation. Introducing AI does not remove the need to validate; in fact, it may intensify it, because we must validate not just the laboratory process but the computational model as well. Validation in this context means testing the AI system on data it has not seen (hold-out test sets, cross-validation) to evaluate how well it performs, and doing so under conditions that simulate real-case scenarios. For example, if one is developing an AI to identify fingerprints, the validation should include latent prints of varying quality, not just ideal prints, and from a population that reflects the diversity of prints encountered. The process should yield metrics such as accuracy, sensitivity, specificity, false positive rate, and false negative rate. For probabilistic outputs, calibration of probabilities is also important; does a "90% probability" actually correspond to 90% of those cases being true matches, for instance.

The courts have increasingly asked forensic disciplines to provide error rates of their methods; this was emphasized in the 2016 PCAST report for pattern-matching fields. AI systems can help quantify matching uncertainty, but they also introduce their own error rates. Any AI used in casework should have a published error rate from validation studies. If an algorithm says "match" and is wrong 1% of the time, that might be acceptable; if it is wrong 20% of the time, that is clearly problematic. The context also matters: A tool used only as an investigative lead generator might tolerate higher false positives (with human follow-up filtering them out) than a tool used to directly incriminate someone.

Reliability also means the system should behave sensibly when encountering edge cases or noise. Adversarial examples (a concept from AI security) are inputs deliberately crafted to fool an AI, like subtle perturbations to an image that cause a misidentification. In a forensic setting, could someone manipulate evidence to mislead an AI? One could imagine, for instance, a fingerprint altered in a way that confuses an algorithm but maybe not a human. Ensuring robustness means testing against such possibilities. Likewise, if an AI is fed data outside the norm, does it gracefully indicate uncertainty or does it make a wild guess? A reliable system should ideally recognize when a situation is beyond its scope, such as out-of-distribution data. AI models can drift over time, especially if they update with new data, or their performance may degrade as new types of data emerge, such as new kinds of crimes or new sensor modalities. Therefore, maintaining reliability may require periodic revalidation and possibly model retraining. For example, if a facial recognition system is initially validated on 2018 data, by 2025 the population demographics or image quality might have shifted, necessitating an update and recheck.

One proposed way to ensure high reliability is to use AI as a complementary tool rather than a replacement. In other words, have humans and AI double-check each other. This can reduce errors if implemented well [41]. For instance, if both an examiner and an AI independently analyze a fingerprint and only if both agree on a match is it reported (and if they disagree, it is escalated for further review), the chance of a mistake might drop significantly compared to either alone. However, careful studies are needed to confirm that such workflows truly reduce errors and do not introduce new failure modes, such as examiners becoming complacent due to AI concurrence.

15 Legal Admissibility and Daubert Challenges

When AI-generated evidence or conclusions are brought into the courtroom, they face legal scrutiny under standards like Daubert (in federal court and many state courts) or Frye (in some jurisdictions that still use the older general acceptance test). The key question is: Is the method by which this evidence was generated scientifically valid and reliable, and is it applied properly to the case? This presents several challenges for AI-based forensic methods. Who is the expert presenting AI findings? Is it the forensic analyst who used the software, or the data scientist who built the model, or both? Courts might require that someone with knowledge of the algorithm be available to testify to answer technical questions. If the workings are very complex, explaining them to a judge and jury in an understandable way is itself a challenge and, if it cannot be explained understandably, judges might be inclined to exclude it as confusing or prejudicial.

Because many AI tools are new, they lack a track record in court. It takes time for a method to gain general acceptance in the scientific community. As of now, we might say AI in forensic science is in a probationary phase: Judges have admitted some forms, such as probabilistic DNA software evidence in numerous cases, but they have also excluded or been skeptical of others. A notable question is whether the output of an AI is considered an expert opinion (in which case it might need a human proponent) or a piece of machine-generated analytical data [42]. Some courts might analogize an AI algorithm to a breathalyzer machine: If the machine is validated and functioning, the result can be admitted without needing to explain the internal algorithm in detail each time, as long as an expert can testify to its validity. Others might see it as hearsay or as something requiring confrontation: Does a defendant have the right to "confront" the creator of the algorithm? In criminal cases, the defense has a right to confront witnesses against them. If a piece of software produces an analysis used against a defendant, is the developer of that software effectively a witness? This was a point of contention in some DNA software cases, whether the source code and internal logic needed to be exposed to allow the defense to challenge the "witness" (the algorithm) and the analyst using it. These legal debates are ongoing, but they push towards more openness (hence, the "glass box" idea).

In essence, legal admissibility is achievable for AI-based forensic methods, but only if developers and practitioners proactively address the reliability and transparency issues. They should anticipate challenges: be ready to explain how the AI works in plain language, provide documentation of validation, and even educate legal actors through workshops or publications to foster understanding and acceptance. The forensic community is actively discussing these issues, knowing that a single high-profile courtroom failure could set back trust in these tools significantly.

16 Policy and Ethical Considerations

As AI becomes embedded in forensic science, it intersects with broader policy and ethical discussions about AI in society. Ensuring that AI is used responsibly in forensic contexts requires guidance from both government policy and professional ethics. In this section, we outline some key policy developments and ethical principles relevant to forensic AI, including recent government initiatives and considerations specific to forensic ethics [32].

17 Government Initiatives on AI (Executive Orders and Frameworks)

Governments around the world are recognizing the need to regulate and guide AI development. In the United States, two notable initiatives frame how AI should be approached. The 2023 Executive Order on Safe, Secure, and Trustworthy AI [43] signed on October 30, 2023, is a landmark Executive Order aimed at fostering trustworthy AI. This EO directs a whole-of-government strategy to manage AI risks and promote innovation that benefits society. Key points include developing standards and tests for AI safety (especially for powerful models), protecting privacy (e.g., evaluating how AI might de-anonymize data), reducing algorithmic discrimination, and promoting transparency (such as watermarking AI-generated content). While not specific to forensic science, this highest-level policy document signals to federal agencies, including those in justice and science, that AI deployments must be carefully evaluated for safety and fairness. For forensic agencies, it means there may soon be guidelines or even requirements for vetting AI tools (for instance, requiring bias testing or reporting when AI is used in an investigative decision). It also sets up a framework for interagency cooperation on AI issues. In practice, forensic science stakeholders are likely to interpret this EO as a mandate to prioritize trustworthy AI, aligning with concepts like human oversight, accountability, and ethics in any AI tools they develop or purchase.

Released by the National Institute of Standards and Technology in January 2023, the NIST AI Risk Management Framework (RMF) 1.0 [44] is a voluntary guidance document for organizations to manage AI risks. It outlines characteristics of trustworthy AI systems, namely that they should be valid and reliable, safe, secure, resilient, accountable, transparent, explainable, interpretable, privacy-enhanced, and fair (with harmful bias managed). The framework is built around four functions: Govern, Map, Measure, and Manage. "Govern" is a cross-cutting function ensuring an organization has the structure, such as policies, roles, and awareness, to oversee AI. "Map" involves understanding the context and scope of AI deployments and identifying risks. "Measure" is about analyzing, assessing, and benchmarking AI systems for various trust characteristics, such as measuring accuracy and bias. "Manage" is about taking steps to mitigate risks and making informed decisions about AI. For forensic science institutions, the NIST AI RMF can serve as a checklist or scaffold when implementing AI. For example, when introducing a new AI for fingerprint analysis, the lab should Map (context—what decisions it will influence, potential impacts if wrong), Measure (validate performance, test for biases), Manage (have incident response if it fails, update procedures), and Govern (have clear policies on who is responsible for oversight). The RMF's emphasis on continuous monitoring and improvement encourages that even after deployment, AI tools should be regularly reviewed for any emerging issues. NIST's framework and playbook also promote knowledge sharing where forensic agencies could share best practices or lessons learned about AI, fostering a community approach to trustworthy AI adoption.

The National AI Initiative Act of 2020 [45] created structures for US AI R&D and coordination. It does not directly dictate forensic use, but it does fund and encourage AI research, some of which flows into the justice domain. Additionally, the White House Office of Science and Technology Policy in 2022 released a "Blueprint for an AI Bill of Rights" [46], highlighting principles such as the right to safe and effective systems and algorithmic discrimination protections. While not law, it influences agencies to consider how to implement those principles. In Europe, the proposed EU AI Act [47] (expected to be enacted around 2024) will regulate AI by risk categories, a forensic use such as criminal identification would likely be considered high-risk, necessitating stringent oversight. Globally, these policy currents are converging on the idea that AI, especially in high-stakes areas like forensic justice, must be subject to strict standards for performance and ethics.

For forensic science professionals, staying abreast of these policy developments is now part of the job. It is foreseeable that accreditation bodies will incorporate AI guidelines and that standard-setting organizations, such as the Organization of Scientific Area Committees (OSAC) for forensic science, will issue guidelines on validation of AI tools, documentation, and use policies, in harmony with national frameworks.

18 Ethical AI Development and Use in Forensics

Beyond formal policies, ethical considerations guide how AI should be developed and used in forensic contexts. Many of these align with general AI ethics but have particular resonance in criminal justice:

Fairness and Non-Discrimination: Ethically, it is imperative to ensure that AI in forensics does not exacerbate biases against any group. This means not only addressing algorithmic bias but also being aware of how using AI might impact different communities. For instance, a facial recognition system misidentifying someone could unjustly put them under suspicion; if this happens predominantly to one demographic, it erodes trust in the justice system. Ethical use would demand either correcting that disparity or refraining from using the technology until it can be used fairly. In practice, some agencies have self-imposed limits like not using facial recognition for final identification if the demographic bias is not acceptable or ensuring lineups or photo arrays are still used to confirm identity rather than arresting purely off an AI match.

Accountability and Responsibility: When AI plays a part in forensic analysis, who is responsible for its conclusions? Ethically, there should always be a human in the loop who takes responsibility for the decision made. The principle of human oversight suggests AI should assist, not fully replace, human judgment in forensics. If an AI flags evidence or even writes a section of a report, a human expert should vet it and be prepared to defend it in court. Also, developers of AI tools carry an ethical responsibility to thoroughly test and warn users about the tool's appropriate use cases and limitations. Cases of unvalidated or misapplied tech are seen as ethical lapses.

Ethically, all forensic stakeholders—investigators, prosecutors, defense attorneys, judges, and even jurors—deserve transparency about the role of AI in producing forensic results [48]. Secretly using an AI and presenting the outcome as if it were purely human expert opinion would be deceptive. For example, if a lab analyst's report conclusion was significantly influenced by an AI tool, disclosing that use allows the legal process to probe it if needed. Likewise, explaining to juries in understandable terms how an AI arrived at a conclusion can be part of ethically deploying these methods in court. At the very least, due process requires that defendants can know the evidence and how it was generated to challenge it.

There is a risk of "AI hype" leading to overclaiming capabilities. Ethically, forensic scientists must avoid overstating what an AI result means. For instance, if an algorithm provides a score of match likelihood, an expert witness should not translate that to a definitive statement of identity beyond what is statistically supported. The evidence must be presented with its uncertainty. This is tied to the concept of duty of candor in science, which is not giving a false impression of certainty or infallibility. In fact, integrating AI results might prompt forensic experts to be more quantitative in their testimony (which is a good thing), but they must also clarify any assumptions or limitations of the underlying model (which is a difficult thing). Some forensic AI applications touch on privacy concerns. DNA analysis algorithms might imply information about relatives, as in investigative genetic genealogy; therefore, labs should have policies about when such familial searches are ethical and lawful. Surveillance analytics using AI must respect legal boundaries, such as not using technology to do broad, unknown subject surveillance that would not be allowed if done manually. Ethical guidelines, possibly drawn from human rights frameworks, should guide how far is too far when using AI in investigations. There is an ethical obligation for forensic practitioners to be competent in the methods they use. As AI tools are introduced, this means training forensic scientists to understand them at least at the user level and perhaps some basics of how they work. Using a tool without understanding its potential failure modes or appropriate usage is unethical and could lead to misuse. Professional bodies may soon include AI competency in their training or certification programs.

Recent efforts within forensic science reflect these ethical priorities. The concept of "do no harm" translates into being cautious with AI: ensuring that adopting a new technology does not inadvertently harm individuals by producing misleading results. Finally, a noteworthy ethical stance in the community is that AI should augment, not replace, human experts—at least until we are certain it can meet or exceed the reliability of humans in all conditions. AI is seen as a tool to diversify and complement existing solutions, not to supplant the expert's role entirely. This mindset helps maintain that human oversight and sense-checking are always present.

19 Summary of Literature to Date

The intersection of AI and forensic science has rapidly grown as a subject of research and discussion in the last decade. In this section, we summarize major studies, reports, and articles that have shaped the discourse. We also highlight where there is consensus in the community, what gaps remain in knowledge or capability, and what controversies are still unresolved.

While not a single publication, the saga of legal challenges to the TrueAllele DNA software source code in the 2010s through to 2021 generated a lot of commentary in law journals [49] and forensic forums [50]. Some papers sided with companies, arguing that revealing source code broadly could expose intellectual property and was not necessary if validation studies exist. Others argued from a civil liberties perspective that defense access to algorithms is a must for fair trial. This debate highlights a controversy: how to balance proprietary tech development with transparency in the justice system. There is still no clear consensus; different courts have ruled differently. However, there is a trend toward more openness, with some companies agreeing to limited source code review under protective orders. The literature here is split but tends toward advocating for independent review of forensic algorithms, with some suggesting something like a "national repository of validated forensic software" to which experts and defense could get access.

20 Consensus, Gaps, and Controversies

The forensic science community largely agrees on the following points: AI and machine learning hold great promise for improving forensic analyses by introducing more objective, data-driven assessments. AI should be used to assist and augment forensic examiners, not to autonomously replace them (maintaining a human in the loop). There is a critical need for transparency, validation, and error rate documentation for any forensic AI tool. Unvalidated or unexplainable tools should not be used in casework. Bias is a known risk in AI, and efforts must be made to ensure training data and algorithms do not introduce unfair bias into forensic decisions. Interdisciplinary collaboration is beneficial: computer scientists, statisticians, and forensic experts should work together to develop AI solutions that are fit for forensic purposes (some literature notes that when such collaboration happens, the outcomes are more robust). Many find that the most achievable AI applications in the near term are those fitting into existing workflows and addressing clear needs (like triaging evidence or providing an objective metric), rather than something that requires a complete overhaul of forensic processes.

A recurring theme in the literature is the lack of large, high-quality forensic datasets for AI training and testing. Privacy and security often limit sharing of data; for example, criminal fingerprint or DNA databases cannot be freely used for research. Synthetic data generation or securing agreements to use de-identified case data are areas needing work. The community has called for something akin to ImageNet, the famous large dataset for general AI vision, but for forensic data types. While many algorithms have been proposed, few have extensive validation across multiple laboratories. There is a gap in multicenter studies that test AI tools on a variety of examiners and evidence sets to ensure generalizability. More research is needed on making AI decisions interpretable for pattern evidence. For DNA or certain quantitative evidence, it is easier because stats are already used. But for images (fingerprints, faces), we need better ways to summarize why an algorithm says two items match or do not match. Some work on explainable AI (XAI) is needed to fill a gap in forensics. Another practical gap is how to seamlessly integrate AI tools into forensic lab case workflows and information systems. This includes training needs, interface design, and ensuring it does not slow down work or cause confusion. There are few published case studies of "lab X implemented AI tool Y and here's what happened," though pilot programs are starting. On the legal side, there is a gap regarding clear standards for courts when handling algorithmic evidence. The literature often calls for either new guidelines or even amendments to evidence rules to specifically address algorithms. Until then, the gap is filled by individual court decisions and the discretion of judges.

21 Controversies

There is controversy on how much transparency is needed: Some forensic practitioners worry that forcing source code disclosure or extreme openness will discourage companies from investing in forensic tech (a niche market) or lead to IP loss. Civil libertarians and many academics argue that nothing should be used in criminal cases that are not fully open to scrutiny. This remains a live debate. Outside the lab, the use of AI for predictive policing is highly controversial. Some view it as a natural extension of evidence-based policing, while others view it as fundamentally incompatible with principles of fairness and even effective policing. This controversy is more about policing strategy but intersects with forensic analytics (for instance, using gunshot detection AI systems like ShotSpotter has been challenged). Some cities banned police use of facial recognition, while others continue to use it. The community is split on whether current facial recognition tech is appropriate for use: It works great in some contexts but has produced high-profile mistakes. This controversy essentially revolves around whether we should pause its use until it is proven unbiased and accurate, or use it with safeguards because it is a useful tool now.

There is occasionally a cultural divide: Some traditional forensic examiners may distrust AI, whereas some data scientists may over-trust it. Bridging this gap requires building mutual understanding. There have been discussions at conferences about whether an algorithm can truly grasp the nuances an experienced examiner does, for example, understanding "ground truth" in latent print comparisons is tricky because we do not know it in casework: We only have examiner consensus or court decisions as a proxy. Some controversial questions include: If an algorithm disagrees with a human expert, who is right? How do we adjudicate that? Literature has started addressing this by "blind testing" both experts and algorithms to see performance, but there is not enough data yet to settle it broadly.

In summarizing the literature, the trajectory is clear: Research increasingly supports AI aiding forensic science but the implementation remains cautious. The community largely acknowledges AI will play a role but insists on retaining scientific rigor and legal defensibility. Gaps like data and validation are actively being worked on, with calls for funding in these areas. Meanwhile, controversies ensure that adoption will be measured and debated, which is healthy for ultimately establishing robust methods.

22 Conclusion

From the foundations by ancient societies to the nineteenth century anthropometric cards of Alphonse Bertillon to the deep learning models of today, forensic science has continually evolved by embracing new ways to collect and interpret data. In this chapter, we traced that evolution through historical milestones and examined how the latest frontier is poised to reshape forensic practice. AI offers powerful tools to handle the growing scale and complexity of forensic evidence, promising faster analyses, uncovering patterns invisible to the human eye, and adding quantitative rigor to what have often been subjective judgments. Yet, as we have highlighted, these advancements come with caveats. The reliability and fairness of AI systems must be proven, not presumed. Black box algorithms must be opened up to scrutiny, or replaced with more interpretable designs, so that trust in forensic conclusions remains well-founded. Legal systems will demand, and rightly so, that any method influencing a person's guilt or innocence be transparent and scientifically sound. Ethical considerations will continue to guide what is acceptable in the pursuit of justice, reminding us, for instance, that efficiency should never trump an individual's rights or the overall fairness of the process. The interplay of policy, such as recent executive directives and frameworks, with on-the-ground forensic work will be crucial in the coming years to ensure a balanced integration of AI. The intersection of forensic science and AI is a dynamic domain.

This chapter has provided a broad survey of where we stand: historically grounded, currently experimenting with AI across numerous applications, and warily optimistic about the benefits while actively addressing the pitfalls. As we look to the future (to be discussed in later parts of this book), several themes will likely dominate the conversation: the development of international standards for forensic AI; the possibility of real-time analytical aids in the field, such as on-scene AI analysis; the role of AI in reevaluating old or cold cases; and the continuous education of forensic professionals in data science competencies. For now, the take-home message is one of evolution, not revolution: AI is best viewed as the next step in forensic science's ongoing journey toward greater accuracy, objectivity, and comprehensiveness. It builds on the legacy of systematization that began with Bertillon and continued through fingerprint, DNA, and digital databases. If implemented thoughtfully, AI will enhance forensic science's ability to serve justice. If implemented carelessly, it could repeat or even exacerbate errors of the past. The responsibility lies with forensic scientists, technologists, legal professionals, and policymakers alike to steer this evolution wisely. The chapters that follow will delve into specific facets of this future, exploring how we can harness AI's capabilities while upholding the principles that make forensic science a trusted pillar of the justice system.

References

1. Rhodes, H. T. F. *Alphonse Bertillon, Father of Scientific Detection.* (Harrap, London, 1956).
2. Bertillon, A. & McClaughry, R. W. *Signaletic Instructions: Including the Theory and Practice of Anthropometrical Identification.* (Werner Co., Chicago [Ill.], 1896).
3. Wilder, H. H. & Wentworth, B. *Personal Identification; Methods for the Identification of Individuals, Living or Dead.* (R.G. Badger, Boston, 1918).
4. Fosdick, R. B. Passing of the Bertillon System of Identification. *J. Am. Inst. Crim. Law Criminol.* **6**, 363–369 (1915).
5. The Organization of Scientific Area Committees for Forensic Science | NIST. https://www.nist.gov/adlp/spo/organization-scientific-area-committees-forensic-science.
6. National Institute of Standards and Technology. https://www.nist.gov/.
7. Fingerprint Sourcebook - Chapter 6: Automated Fingerprint Identification System (AFIS) | National Institute of Justice. https://nij.ojp.gov/library/publications/fingerprint-sourcebook-chapter-6-automated-fingerprint-identification-system.
8. CODIS Archive — LE. https://le.fbi.gov/science-and-lab/biometrics-and-fingerprints/codis-2#Combined-DNA%20Index%20System%20CODIS.
9. H.R.829 - 103rd Congress (1993–1994): DNA Identification Act of 1993 | Congress.gov | Library of Congress. https://www.congress.gov/bill/103rd-congress/house-bill/829.
10. National Integrated Ballistic Information Network (NIBIN) | Bureau of Alcohol, Tobacco, Firearms and Explosives. https://www.atf.gov/resource-center/infographics/national-integrated-ballistic-information-network-nibin.
11. Forensic DNA Typing - Google Books. https://www.google.com/books/edition/Forensic_DNA_Typing/gwDyBq2xLjIC?hl=en&gbpv=0.
12. Ronald Nichols. *Coordinating NIBIN and Forensic Science Laboratory Efforts.* https://crimegunintelcenters.org/nibin/ (2017).
13. S.S. Iyengar, Sina Nabavi, Yashas Hariprasad, Prasad HB, Krishna Mohan C. "Artificial Intelligence in Practice: Theory and Applications for Cyber Security and Forensics", Springer Nature, 2025
14. Shi, Bin, and Sundararaja S. Iyengar. Mathematical Theories of Machine Learning-Theory and Applications. Springer, 2020
15. Jayakumar Singaram, S. S. Iyengar, Azad M. Madni "Deep Learning Networks: Design, Development and Deployment", Springer Nature. 2023

16. Strengthening Forensic Science in the United States: A Path Forward | The National Academies Press. https://nap.nationalacademies.org/catalog/12589/strengthening-forensic-science-in-the-united-states-a-path-forward (2009).
17. PCAST Releases Report on Forensic Science in Criminal Courts | whitehouse.gov. https://obamawhitehouse.archives.gov/blog/2016/09/20/pcast-releases-report-forensic-science-criminal-courts.
18. Y. Hariprasad, S. S. Iyengar and N. K. Chaudhary, "Securing the Future: Advanced Encryption for Quantum-Safe Video Transmission," in IEEE Transactions on Consumer Electronics, https://doi.org/10.1109/TCE.2024.3473542.
19. Hariprasad, Yashas, K. J. Latesh Kumar, L. Suraj, and S. S. Iyengar. "Boundary-based fake face anomaly detection in videos using recurrent neural networks." In *Proceedings of SAI Intelligent Systems Conference*, pp. 155–169. Cham: Springer International Publishing, 2022.
20. Wang, Cliff, Sundararaja S. Iyengar, and Kun Sun, eds. *AI Embedded Assurance For Cyber systems*. Springer Nature, 2023, pp. 246,
21. Hare, E., Hofmann, H. & Carriquiry, A. Automatic matching of bullet land impressions. *Ann. Appl. Stat.***11**, (2017).
22. Wen, Z., Curran, J. M. & Wevers, G. Shoeprint image retrieval and crime scene shoeprint image linking by using convolutional neural network and normalized cross correlation. *Sci. Justice***63**, 439–450 (2023).
23. Zhao, H. & Li, H. Handwriting identification and verification using artificial intelligence-assisted textural features. *Sci. Rep.***13**, (2023).
24. Barash, M., McNevin, D., Fedorenko, V. & Giverts, P. Machine learning applications in forensic DNA profiling: A critical review. *Forensic Sci. Int. Genet.***69**, 102994 (2024).
25. Sessa, F. *et al.* Artificial Intelligence and Forensic Genetics: Current Applications and Future Perspectives. *Appl. Sci.***14**, 2113 (2024).
26. Iyengar, Sundararaj S., Seyedsina Nabavirazavi, Hemant Rathore, Yashas Hariprasad, and Naveen Kumar Chaudhary. "Advancing Forensic Science: AI and Knowledge Graphs Unlock New Insights." (2024).
27. Patel, Sankita J., Naveen Kumar Chaudhary, Bhavesh N. Gohil, and S. S. Iyengar. *Information Security, Privacy and Digital Forensics*. 2023.
28. Śniatała, Konrad, Yashas Hariprasad, K. J. Latesh Kumar, Naveen Kumar Chaudhary, and Michał Weissenberg. "Fog Forensics: A Comprehensive Review of Forensic Models for Fog Computing Environment." In *International Conference on Information Security, Privacy and Digital Forensics*, pp. 31–42. Singapore: Springer Nature Singapore, 2022.
29. Hariprasad, Yashas, Suraj Lokesh, Nagarjun Tumkur Sharathkumar, Latesh Kumar KJ, Chance Miller, and Naveen Kumar Chaudhary. "AI-ML Analytics: A Comprehensive investigation on sentimental analysis for social media Forensics Textual data." In *Science and Information Conference*, pp. 923–935. Cham: Springer Nature Switzerland, 2023.
30. Grother, P., Ngan, M. & Hanaoka, K. *Face Recognition Vendor Test Part 3: Demographic Effects*. https://nvlpubs.nist.gov/nistpubs/ir/2019/NIST.IR.8280.pdf (2019) https://doi.org/10.6028/nist.ir.8280.
31. Haddad, G. Confronting the Biased Algorithm: The Danger of Admitting Facial Recognition Technology Results in the Courtroom. *Vanderbilt J. Entertain. Technol. Law***23**, 891 (2021).
32. Mohanty, Pronab. "Data Science and Predictive Policing." *CMR Univ. J. Contemp. Legal Aff.* 2 (2020): 230.
33. Zhang, X. *et al.* Interpretable machine learning models for crime prediction. *Comput. Environ. Urban Syst.***94**, 101789 (2022).
34. Rotaru, V., Huang, Y., Li, T., Evans, J. & Chattopadhyay, I. Event-level prediction of urban crime reveals a signature of enforcement bias in US cities. *Nat. Hum. Behav.***6**, 1056–1068 (2022).
35. Greenwald, E. & Phiri, L. On Accountability: Genetic Tools for Justice and Injustice in Criminal Proceedings. *J. Sci. Policy Gov.***25**, (2024).
36. Garrett, B. L. & Rudin, C. Interpretable algorithmic forensics. *Proc. Natl. Acad. Sci.***120**, (2023).

37. Garrett, B. L. & Rudin, C. The Right to a Glass Box: Rethinking the Use of Artificial Intelligence in Criminal Justice. *Cornell Law Rev.* **109**, 561–627 (2024).
38. *STATE of New Jersey, Plaintiff-Respondent, v. Corey PICKETT, Defendant-Appellant.* (2021).
39. Traces of Crime: How New York's DNA Techniques Became Tainted - The New York Times. https://www.nytimes.com/2017/09/04/nyregion/dna-analysis-evidence-new-york-disputed-techniques.html.
40. Daubert v. Merrell Dow Pharmaceuticals, Inc. | 509 U.S. 579 (1993) | Justia U.S. Supreme Court Center. https://supreme.justia.com/cases/federal/us/509/579/.
41. Barrington, S. & Farid, H. A comparative analysis of human and AI performance in forensic estimation of physical attributes. *Sci. Rep.* **13**, (2023).
42. AI in the Courts: How Worried Should We Be? https://judicature.duke.edu/articles/ai-in-the-courts-how-worried-should-we-be/.
43. Safe, Secure, and Trustworthy Development and Use of Artificial Intelligence. *Federal Register* https://www.federalregister.gov/documents/2023/11/01/2023-24283/safe-secure-and-trustworthy-development-and-use-of-artificial-intelligence (2023).
44. National Institute of Standards and Technology (US). *Artificial Intelligence Risk Management Framework: Generative Artificial Intelligence Profile.* https://nvlpubs.nist.gov/nistpubs/ai/NIST.AI.600-1.pdf (2024) https://doi.org/10.6028/nist.ai.600-1.
45. H.R.6216 - 116th Congress (2019–2020): National Artificial Intelligence Initiative Act of 2020 | Congress.gov | Library of Congress. https://www.congress.gov/bill/116th-congress/house-bill/6216.
46. Blueprint for an AI Bill of Rights | OSTP | The White House. https://bidenwhitehouse.archives.gov/ostp/ai-bill-of-rights/.
47. EU Artificial Intelligence Act | Up-to-date developments and analyses of the EU AI Act. https://artificialintelligenceact.eu/.
48. Hefetz, I. Mapping AI-ethics' dilemmas in forensic case work: To trust AI or not? *Forensic Sci. Int.* **350**, 111807 (2023).
49. Sonia K. Katyal. The Paradox of Source Code Secrecy. *Cornell Law Rev.* **104**, 1183–1280 (2019).
50. Edward J. Imwinkelried. Computer Source Code: A Source of the Growing Controversy Over the Reliability of Automated Forensic Techniques. *DePaul Law Rev.* **66**, 97–132 (2017).

Principles of Machine Learning Theories and Practical Applications

Nageswara S. V. Rao

1 Introduction

The machine learning (ML) methods are increasingly being applied to solve complex, computational, and data-driven problems in diverse science areas, including energy systems, materials research, and science infrastructures. Their scientific validity critically depends on the correctness of their codes, which is characterized by computability, learnability, and guaranteed approximations under the randomness of processes, measurements, and errors. Particularly in science areas, it is very important to ensure that the underlying problem is correctly analyzed in order to develop and apply ML codes along with their analytical proofs and experimental validation methods. Specifically, if these codes are incorrectly developed or applied to uncomputable or unlearnable problems, the consequences can be severe: their outputs could be unsound or incomplete or both [36, 63], thereby resulting in potentially false "discovery claims" or incorrect critical decisions. Such scenarios are more likely to arise now as increasingly complex tasks are being delegated to purely computational means, fueled in part by the recent proliferation of generative, black box artificial intelligence (AI)/ML methods. These aspects are further compounded by the increasing complexity of diverse ML methods that are being customized, sharpened, and composed in unprecedented ways to exploit extensive data sets and extremely powerful computing resources.

In science areas over the decades, physical laws have been developed, often in analytic (differential, algebraic, and stochastic) forms, from the first principles to design, understand, and operate complex systems. More recently, they have been exploited to develop general physics informed machine learning (PIML)

N. S. V. Rao (✉)
Oak Ridge National Laboratory, Oak Ridge, TN, USA
e-mail: raons@ornl.gov

S. S. Iyengar et al. (eds.), *Artificial Intelligence Driven Forensics*, Signals and Communication Technology, https://doi.org/10.1007/978-3-031-98036-7_2

methods [65] and also to customize ML solutions for reactor system analytics [40], electrochemistry laboratory automation [1], methane hydrate exploration [54], and other applications. For example, a three-level fuser of smooth and non-smooth machine learners that reflects the structure of the coolant system of a nuclear reactor has been developed for power level estimation using independent monitoring measurements [40, 55]. In another example, custom ML methods are integrated into automated electrochemistry workflows to verify the consistency of measurements with normal experimental conditions and to detect failure conditions [1]. In another direction involving "nonphysical" systems, abstract laws are developed and incorporated into ML solutions for performance profiling of data transport networks. The concave-convex profiles of data transport complexes and their individual sites, subsystems (e.g., file and IO), and components (e.g., TCP version) lead to the diagnosis of bottlenecks [45, 46] and compromises [48] and optimization of throughput across the complex [21, 29]. In general, physical and abstract laws encapsulate the critical knowledge—often in compact differential and algebraic forms—that complements datasets and ML models obtained from them. They have been synergistically exploited to compose powerful ML solutions and also to derive generalization equations, for example, distribution-free error confidence bounds for sensor drift estimation in primary coolant systems [50, 51] and concave-convex profile transitions in throughput profiles of data transport infrastructures [56]. In this chapter, we briefly describe these applications and place them within a broader context of ML solutions and their applicability to a variety of problems, by utilizing theories of computing, learning, and logic.

We introduce the concept of *ML-solvability* which characterizes the correctness of ML codes in terms of deterministic proofs and probabilistic generalization guarantees, by combining the theories of learnability [24, 64], computing [14] and algorithms [13, 20], and logic [63]. It characterizes a ML solution by its model space, the algorithm that learns a model using samples, and the inference algorithm that utilizes the trained model, and computing platforms that run learning and inference codes. It provides insights into the applicability and generalization properties of ML codes and potential for incomplete and unsound inferences if the underlying problem is not ML-solvable. We combine the generalization theory with algorithm correctness proofs to provide performance guarantees based on finite samples using adaptations of Vapnik-Chervenenkis (VC) dimension (or a related quantity) [6, 64]. Furthermore, we exploit the conditions and constraints from physical-abstract laws of the underlying system to provide practical insights and results, under the framework illustrated in Fig. 1. We first apply the foundational principles to establish the ML-solvability of a given problem and then exploit the domain laws to identify or develop custom ML solutions. If the problem is ML-unsolvable or its ML-solvability cannot be established, then we show that the properties of ML codes are uncertain even if they produce output.

We briefly describe illustrative ML applications from existing works in the areas of data transport networks, nuclear engineering, and computer system diagnosis to illustrate various aspects of the ML-solvability concept. Specifically, we combine the learning dimension and training error to derive generalization equations to

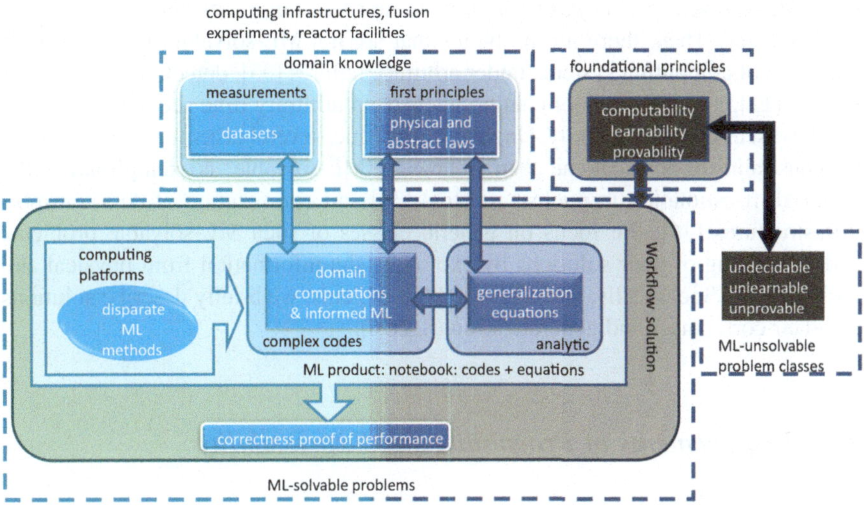

Fig. 1 Framework for co-developing ML codes with generalization equations

detect and minimize overfitting and utilize physical law violations by the learning processes to identify and mitigate the unsound inferences. We also illustrate the utilization of smooth, non-smooth, and algebraic forms of laws to develop or analyze ML solutions in these applications.

This chapter is organized as follows. The ML-solvability concept is developed in Sect. 2. The generalization theory is briefly summarized in Sect. 3. The algebra and calculus aspects of machine learners are described in Sect. 4. Implications of the lack of ML-solvability or its knowledge are described in terms of overfitting and unsound and incomplete inferences in Sect. 5. The applications based on existing ML solutions are briefly described in Sect. 6. Conclusions and possible directions for future investigations are described in Sect. 7.

2 ML-Solvability: Computability and Learnability

ML methods are typically developed to estimate parameters or infer properties and optimize parameters of an underlying system, using probabilistic measurements. Their correctness embodies two parts: (a) deterministic proofs from theories of computations and algorithms that establish their correctness and complexity and (b) probabilistic guarantee of their approximation from theory of learnability that establish their generalization properties. Within the broad spectrum of science areas, some problems would be computationally undecidable (or uncomputable for short) [62] or unprovable by pure mechanical means [17], and some learning tasks would be unlearnable in an information theoretic sense [24, 64]. *We define a problem to be ML-solvable if it is both computable and learnable with the required level of confidence*. Then, there are problems that are not ML-solvable, including self-diagnosis of computing systems under arbitrary failures [35], detection of computer viruses [12], and inference of chaotic maps from arbitrary time sequences [25], as will be discussed subsequently. On the positive side, several tasks formulated within the constraints of physical and abstract laws are ML-solvable, as exemplified by the sensor drift estimation in nuclear reactors [39] and optimization of data transport infrastructures [44]. We focus on general classes of such ML-solvable problems and sharpening of their solutions by exploiting the information from physical and abstract laws. Specifically, for ML-solvable problems, we briefly describe solutions and their correctness and generalization (Sect. 6).

2.1 Requirements of Provability and Expressibility

Complex ML solutions that require performance guarantees and interpretable parameters require two additional properties of provability and expressibility, in addition to learnability and computability of basic ML-solvability, as shown in Fig. 2. Learnability is defined by generalization properties such as finite scale-sensitive dimensions [2], as discussed subsequently in detail (Sect. 2.4). Computability in ML requires that the underlying computing problem be decidable in a Turing sense [14, 27], which is also closely related to the mechanization of proofs as discussed in detail in Sect. 2.5. Expressibility of ML refers to the ability of learned

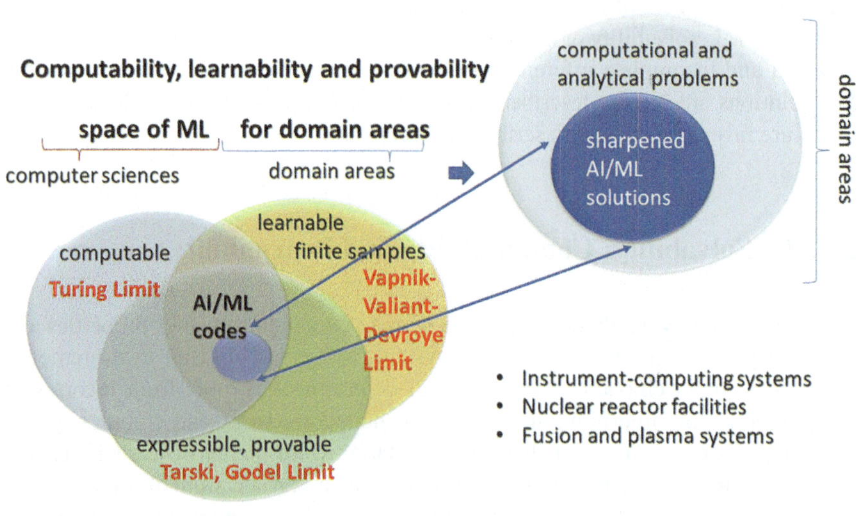

Fig. 2 ML codes must be developed and applied to computable and learnable problems, and additionally may need to be expressible and provable

models or parameters to represent critical domain properties, such as learning information that neural network weights capture feature correlations or hidden nodes represent critical regions of the image being processed. Expressibility of ML outputs is critical for trusting black-box solutions, and they are subject to Tarski's limit [26, 60], in particular solutions that require that critical properties be defined first in complex algebraic systems. In simple terms, Tarski's undefinability states that in any sufficiently strong formal system, not all truths in that standard model can be defined within the system, which in turn implies that examples provided within the context of a system are not guaranteed to lead to the discovery of all truths about that system. The undefinability theorem shows that no sufficiently rich interpreted language can represent all of its own semantics. A corollary is that any metalanguage capable of expressing the semantics of some object language must have expressive power exceeding that of the object language. The metalanguage includes primitive notions, axioms, and rules absent from the object language, so there are theorems that are provable in the metalanguage that are not provable in the object language. Consequently, there could be problems formulated to "learn" such meta-properties that are beyond the power of ML methods. On first glance, these problems might appear too abstract, but it is instructive to note that several practical problems such as virus detection [12] and ensuring the resilience of complex codes [35] are undecidable, and hence, their analogs are also beyond the power of ML solutions.

2.2 Composition of Machine Learning Solution

ML solution to a problem is composed of the following components, which play different roles in ML-solvability:

- *Model or hypothesis space:* The hypothesis or model space consists of all hypotheses or models, from which one is learned based on samples. The examples include the class of deep neural networks or collection of inference rule sets. The hypothesis space must be large enough to closely approximate the underlying model, which corresponds to high approximation density. On the other hand, a large model space also increases its learning dimension which decreases it generalization capabilities. These relationships and trade-offs are explicitly represented using the generalization equations described in Sect. 3.
- *Learning algorithm:* A learning algorithm chooses a model or hypothesis from its class. The examples include the training algorithm to determine the weights of a neural network. In addition to computability, the computational complexity of the training problem determines the quality of model selection. For example, the problem of optimizing neural network weights based on a sample is NP-hard, and methods such as backpropagation only guarantee approximations in practice.
- *Training and testing data:* The training and testing samples are used by the learning algorithm. These samples may be limited by their availability and cost

of collecting them. Large samples often lead to improved ML solution but may lead to larger training times. Also, the extraction of sample data sets from complex system measurements might require extensive processing, for example, extraction of features from images to be used for classification.

- *Learned model and inference algorithm:* The learned model or hypothesis is used by the inference algorithm to produce output such as classification or parameter estimate. This step is often called the forward computation and is polynomial-time computable for a neural network. For cases where the learned model is a collection of inference rules, the computation may involve selective application of rules, which could be computationally intractable or undecidable.
- *Generalization equations and proofs:* Analytical characterizations such as generalization equations, expressions of properties, and proofs of assertions are important desirable qualities to be established. In complex cases, such proofs are beyond the power of computations, for example, as a consequence of Rice theorem as described in Sect. 6.3.
- *Computing platforms:* Computing platforms used to execute learning and inference algorithms play a significant role in the quality of ML solution. In particular, LLMs with a large number of parameters and large data sets require supercomputers and specialized hardware such as GPUs.

2.3 Finite Sample Implications on ML Solutions

Even for ML-solvable problems, for example, some classification problems, there are fundamental limits on how general such methods can be and what can be inferred about them, due to *finite* number of samples. The incompleteness or slow convergence theorem of Devroye [15] shows that there is *no universally best machine learning method* that can be asserted using a finite number of samples. In essence, ML methods trained on finite samples do not have sufficient information to optimize over the non-countably infinite class of data distributions. Indeed, any ML method is reduced to a random guesser for certain data distributions that are tailored to the method, even if it performs optimally on a given training dataset. Consequently, the ML method optimal for a domain (with its specific distribution) may perform poorly when applied to another domain which corresponds to another distribution for which it is nearly a random guesser; in particular, LLM models' excellent performance under training is not guaranteed when applied to a different domain.

As a practical consequence, diverse ML methods—smooth, non-smooth, structural, and statistical [36]—will continue to be developed as newer learning problems arise in various disciplines. Collectively, they provide a set of rich, diverse ML solutions, but their performance can vary widely even for a single problem. Information fusion methods that combine multiple disparate ML solutions to ensure performance at least as good as their best subset have been developed using the projective fusers that utilize the lower envelope of error surfaces [34]. Furthermore, these fusers can be retrained to incorporate newer methods using training samples, which gives us a practical solution to continually fold in the power of newer ML methods. The fused solution is still subject to slow the convergence theorem but is effective in practice since it makes best use of the methods as they become available.

Another limitation of an ML solution is the computational intractability or NP-hardness [16], even if the ML problem is computable or Turing decidable; for example, training a three-node neural network is NP-complete [5]. Thus, practical ML computations can only guarantee approximations. For example, different invocations of the back propagation algorithm yield different approximations in sigmoidal neural networks. A nearest neighbor fuser combines these estimates to provide a solution superior to all of them [34]. In general, for a fuser class $\mathscr{F}_F$ that combines estimators $\hat{f}_i \in \mathscr{F}_i, i = 1, 2, \ldots, N$ and satisfies the isolation property, namely, containing functions that transfer individual inputs to output [32], we have $I(f_F^*) = \min_i^N I(\hat{f}_i) - \Delta$, where $I(f)$ is the expected error of estimator f, and Δ is the improvement in the expected error achieved by the optimal fuser f_F^* over the best estimator. For the learned fuser $\hat{f}_F$ that minimizes the training error, the confidence function δ_V provides the probability that the error is within ϵ of the optimal, specified by its generalization equation described in Sect. 3. It is expressed in terms of $V = \sum_{i=1}^{N} V_i$, where V_i is the total variation of $\mathscr{F}_i$ and improves with the training sample size independent of the underlying distribution [34], which is complex and often unknown. These aggregation operations over ML models provide a basis for abstract additive operations that are rigorously supported by distribution-free guarantees [64], as described in Sect. 4.

2.4 *Physical and Abstract Laws*

The physical and abstract laws capture the critical knowledge of the underlying systems, often in compact forms, by abstracting important parameters and relationships. Data sets collected from operational or emulated systems and ML models obtained from them typically provide complementary knowledge, often reflecting measurement uncertainties and errors and finite sample limits. They have been synergistically exploited to compose powerful ML solutions to estimation and diagnosis problems in a variety of ways, ranging from incorporating physical constraints, utilizing physical law violations as objective functions to minimize, structuring ML solutions to reflect physical laws, and combining multiple disparate estimators of parameters. In particular, the physical and abstract features, along with their relationships, captured by laws play a crucial role in the design and analysis of ML solutions to the underlying estimation and diagnosis problems in the two types of systems, namely, reactor coolant systems and computing-data complexes. ML solutions may be based on utilizing the additive and differential knowledge between

ML models and their counterpart first principle laws: solutions to estimation problems, including classification and regression problems, typically utilize the additive knowledge, and those to diagnosis problems typically utilize the differential knowledge. Furthermore, the analytical properties such as bounded derivatives of smooth laws and bounded total variations of non-smooth laws may be combined with empirical errors of ML models to derive performance characterization and guarantees in the form of generalization equations with practical insights [37]. Specific cases and development of the underlying concepts behind the additive and difference operations in terms of algebra and calculus of ML within our applications context are described in Sect. 4. We consider ML problems formulated within the context of a system S characterized by:

(i) *Law* $\mathscr{L}_S(P, Z)$ that relates the parameters and system variable vectors, P and Z, respectively, and
(ii) *Structure* $\mathscr{S}_S(\mathscr{C}_S)$ that specifies the relationships between its set of components $\mathscr{C}_S$.

2.5 Mechanization of Proofs: Properties of Programs

Traditionally, the correctness proofs of codes are manually generated, and their mechanization refers to utilizing computer codes to generate them, possibly by automatic theorem provers or more recent generative AI methods. Such mechanized proofs are subject to limits guided by the Turing undeciability and Godel's incompleteness considerations [35]. The computation and estimation tasks of ML-solvable problems must be Turing decidable. Rice theorem [14] establishes the computability limits of using codes to infer properties of codes, in particular their correctness proofs; for example, it shows the undecidability of resilience detection problems [35]. The correctness proofs of ML codes are subject to Godel's incompleteness results on provable assertions, namely, that not all truths can be provable in certain systems, such as arithmetic with multiplications [17]. Intuitively, the truths cannot be enumerated as deductions from axioms for verification via a mechanical process, which is subject to Turing's undecidability [62] (incidentally, Godel's results on arithmetic systems were published in 1931, years before Turing's results on computations in 1936). The notions of undecidability (Turing [62]) and unprovability of assertions about their outputs (Godel [17]) are closely related as pointed out by Chaitin [9]: informally, they both capture the recursiveness needed for algorithms and proofs, which is insufficient to address non-recursiveness of certain tasks and assertions, respectively. The work of Uspensky [63] makes this connection explicit by casting the provability task in terms of algorithms within a deductive system; specifically, the nonexistence of such an algorithm establishes the non-provability of truths.

Time series data may also reflect physical properties captured by the domain or ML computations, for example, dynamic flow computations. They introduce another source of undecidability in physical systems with chaotic dynamics [25], wherein nearly all questions about long-term trajectories are undecidable even when initial conditions are known. Thus, even within the context of these physical systems, ML solution that identifies such properties is not feasible, independent of how large the training sets and parameter spaces are. In scenarios of reactor systems (Sect. 6.2), the dynamics are stabilized to avoid chaotic regimes, which leads to their learnability by avoiding this type of undecidability [39, 40]. In general, learnability can be undecidable in that it is not verifiable mechanically as shown recently [4].

The undecidability results are manually proved within the frameworks of recursive functions expressed in L [14], Turing machines [11], lambda calculus [9], and others [63]. Several of the well-known undecidable problems are decision problems about Turing machines such as the halting problem, empty-set detection, and equivalence of Turing machines [11]. These problems might appear somewhat abstract, but there are a number of more "practical" undecidable problems, including virus detection problems [12], programs to test randomness of a string [9], testing the equivalence of context-free grammars, smallest program capable of generating a given string, and computing the Kolmogorov complexity of strings. Resilient computations for Exascale systems represent another class of such challenging problems, and various versions are shown to be undecidable [35] as described in Sect. 6.3.

3 Generalization Theory

Under the ML paradigm, a machine "learns" a functional relationship between two vector random variables [19, 49] using a random sample $(X_1, Y_1), (X_2, Y_2), \ldots, (X_n, Y_n)$, typically drawn from an unknown joint distribution $\mathbb{P}_{X,Y}$ of feature vector X and output vector Y. The task is to compute a predictor function f from a complex class (e.g., deep neural networks) such that $f(X)$ is a good estimate of Y overall. A fundamental generalization result establishes that, under certain conditions, the best estimator f^* that minimizes $I(f) = E[C(f(X), Y)]$, for cost function $C(.,)$, can be closely approximated with high probability by an $\hat{f}$ learned solely from the sample regardless of $\mathbb{P}_{X,Y}$ [64]. Vapnik's theory [64] establishes that a "suitable" estimator, $\hat{f}$, obtained by an ML method M, ensures

$$\mathbb{P}^l_{X,Y}\left[I(\hat{f}) - I(f^*) > \varepsilon + \hat{\varepsilon}\right] < \hat{\delta}_M\left(\mathscr{F}_M, \varepsilon, \hat{\varepsilon}, l\right), \tag{1}$$

where $\mathscr{F}_M$ is its function class, $\varepsilon > 0$ is the *precision* parameter, $0 < \delta_M(.) < 1$ is the *confidence* function, and $\hat{\varepsilon}$ is the *training error* associated with computing $\hat{f}$. Thus, it provides distribution-free guarantees on their prediction performance on future data, and ML-solvability requires that $\hat{f}$ must be computable as a minimum, which is possible only when the underlying problem is computable or Turing decidable.

Over the past decades, finite sample performance guarantees are derived in a variety of methods, including neural networks [2], regression trees [7], support vector machines [64], and kernel estimates [58]. Historically, these elegant generalization equations are considered too loose to be practical, but more recently a combination of custom parameters (e.g., small Lipschitz constant and total variation) and the availability of data sets (e.g., 2K gamma spectrum every second) made them useful in practice, for example, as shown in Fig. 4. The estimate $\hat{\delta}_M(.)$ for the composed ML solution M can be based on law $\mathscr{L}_S$ or its structure $\mathscr{S}_S(\mathscr{C}_S)$ in a generic form, and its practical versions for application scenarios are discussed in Sect. 6. In addition to specific values, the mathematical form of $\hat{\delta}_M(.)$ provides critical practical information by relating the parameters of $\mathscr{L}_S$ to precision and confidence of the composed ML solution.

If $\mathscr{F}_M$ has finite capacity [64], then under bounded error, the condition in Eq. (1) is guaranteed; a more general result ensures this condition under finite scale-sensitive dimensions [3]. For sigmoid neural networks, the sample size needed to ensure Eq. (1) is linear in the number of neural network parameters [30], as opposed to having the quadratic dependence of previous bounds for unbounded weights [22]. In particular, the confidence estimate is

$$\hat{\delta}_{NN} = 8\left(\frac{32W}{\varepsilon}\right)^{h(d+2)} e^{-\varepsilon^2 l/512}$$

for a sigmoid network with h hidden nodes and input dimension d with weights suitably bounded by W. It is important to note that several statistical estimates learned by current ML methods are smooth, including SVM with Gaussian kernels [58] and radial basis functions [8], and several variables and parameters in science applications are bounded, and these properties are used to develop generalization equations. ML methods also employ non-smooth methods, such as ensemble tree methods [61], regression trees [7], Haar estimators [10], and Nadaraya-Watson estimators [28]. In practice, the parameters are bounded, and the learned functions have a finite (often small) number of jumps, which leads to their bounded finite total variation $V < \infty$. In this case, the confidence estimate is

$$\hat{\delta}_V = 8g\left(1 + \frac{128V}{\varepsilon}\right) e^{-\varepsilon^2 l/2048}.$$

Quantum communications channel modeling is carried out using simulations and experiments. These problems are typified by the algebraic structure of $\mathscr{F}_M$ that supports optimization for improved simulation-experiment combination; for example, when $\mathscr{F}_M$ is a v-dimensional vector space, such as Hilbert space of quantum channels, we have the general expression

$$\hat{\delta}_{\mathscr{F}} = 8\left(\frac{128e}{\varepsilon}\right)^{v} e^{-\varepsilon^2 l/512},$$

that may be sharpened using more specific physical properties.

For methods utilizing fusion of multiple methods, the generalization equations can be derived based on the isolation property [31, 32], for example, satisfied by the ensemble method. For instance, for the fuser class $\mathscr{F}_F$ used in fusing the classifiers $f_a \in \mathscr{F}_a, a \in \mathscr{A}$, let f_F denote the classifier function obtained by composing f_a's using a fuser from $\mathscr{F}_F$. The *accuracy improvement* Δ_F of the composite ML estimate over the best individual classifier is defined as

$$\Delta_F = I(f_F) - \max_{a \in A} I(f_a).$$

Then, if $\mathscr{F}_F$ has the isolation property [34] $\Delta_F \geq 0$, the best accuracy improvement is given by $\Delta_F^* = I(f_F^*) - \max_{a \in \mathscr{A}} I(f_a^*)$, and its estimate based on samples is given by

$$\tilde{\Delta}_F = \hat{I}(f_F) - \max_{a \in \mathscr{A}} \hat{I}(f_a),$$

where $\hat{I}(.)$ is the training error based on the sample. If we let $\hat{\delta}_b(\varepsilon, \hat{\varepsilon}_b, l)$ for all individual classifiers $b \in \mathscr{A}$ and classifier-fusers $b = \mathscr{B}$ such that $\hat{\delta}_b(\varepsilon, \hat{\varepsilon}_b, l) \to 0$ as $l \to \infty$, then the probability that the closeness between $\tilde{\Delta}_F$ and Δ_F^* is within ε is bounded as

$$\mathbb{P}^l_{X,Y}\left[|\tilde{\Delta}_F - \Delta_F^*| < \varepsilon\right] > 1 - \hat{\delta}_d(\varepsilon/2, \hat{\varepsilon}_d, l) - \sum_{a \in \mathscr{A}} \hat{\delta}_a(\varepsilon/(2N_{\mathscr{A}}), \hat{\epsilon}_a, l) \qquad (2)$$

for both classifier-fusers $d = \mathscr{B}$, wherein the right-hand side approaches 1 as $l \to \infty$.

4 Algebra and Calculus of ML

Over the past decade, there has been a sea change in ML practice in part due to the availability of software products, including python libraries and R modules. More recently, there has been an explosion of large language models (LLM) [23] both as trained models that can be accessed such as ChatGPT and packages that can be indigenously built such as LLAMA2. Furthermore, ML frameworks for special-purpose and high-performance computing systems demonstrate the unprecedented scale and scope of practical problems solved by ML methods. Consequently, such methods have been extensively applied to a broad set of applications; initially, however, their ease of use resulted in them being used as just "black boxes," often leading to solutions that were hard to interpret and explain. Currently, we are at the verge of a post-black-box era in which basic ML methods are being customized, sharpened, and composed in unprecedented ways to exploit domain knowledge, in particular, system and physical aspects. The field of machine learning has traditionally been at the intersection of computing and statistics [2, 19] but now encompasses other disciplines including formal languages for expression and interpretation, for example, under the explainable AI framework [18]. These developments in practical ML methods can be broadly classified into the following categories:

(a) *Strategic deep compositions:* The sharpened and customized ML solutions are composed in deep, multilevel, acyclic graph combinations, wherein ML methods at nodes are selected to match the underlying system components.
(b) *Fusion of diverse ML methods*: Diverse ML methods, such as smooth kernels and support vector machines, non-smooth tree and forest estimates, and algebraic vector spaces, are combined to mitigate overfitting by ensuring diversity of design.
(c) *Hyperparameter harness*: ML solutions are wrapped inside a harness to explore, select, and tune higher-level parameters, for example, using an informed gradient search in the hyperparameter space.

These complex operations represent *algebra and calculus of ML* methods, as illustrated by the fusion of diverse classifiers and regression methods [36].

4.1 Addition Operation and Information Fusion

The knowledge from laws and ML methods can be utilized in two basic additive ways: (a) *Sharpening and customization:* Basic ML methods can be fine-tuned by identifying and optimizing their structure and parameters by exploiting system and physical aspects. The sharpened and customized ML solutions can be composed in deep, multilevel, acyclic graph combinations in which ML methods at nodes are selected to match the underlying system components. (b) *Fusion of diverse ML methods:* Diverse ML methods (e.g., smooth kernels and support vector machines, non-smooth tree and forest estimates, algebraic vector spaces) can be combined to overcome over-fitting by ensuring the design diversity. An example is RTT estimation by fusing five disparate regression estimates in Fig. 3. Two examples of this ML addition operation in the nuclear engineering area are described in Sect. 6.2.

4.2 Diagnosis by Difference Operation

Physical and abstract laws that characterize ideal operating conditions (e.g., drift-free reactor dynamics and fully optimized data transport infrastructure) are used to compute the differentials with respect to ML models derived from measurements from operational, emulated, or simulated systems. Under ideal conditions, the differentials will be negligible, but under operational conditions, they indicate errors, such as sensor drifts in reactor monitoring sensors and under performance of data transport (e.g., convex throughput profiles). Furthermore, by relating the differentials to underlying laws, the parameters and components can be identified to support actions (e.g., recalibrating sensors or fine-tuning file systems) as described in Sect. 6.2.

5 Implications of Lack of Knowledge About ML-Solvability

The causes of lack of ML-solvability of a problem are twofold: (a) non-computability of the training and inference problems, or (b) non-learnability of the underlying problem due to insufficient model set or hypothesis space, insufficient training data, or insufficient inference rules in the model. Since sound and complete ML answers require both computability and learnability of the underlying problem, codes attempting to solve a ML-unsolvable problem produce answers that lack one or both properties. This conclusion follows from a formal treatment of the soundness and completeness concepts succinctly presented in terms of an abstract deduction system in [63]. In addition, the lack of knowledge of ML-solvability in turn translates to the lack of knowledge about the completeness and soundness properties of output produced by ML codes. These implications are not merely academic in nature, since scientific ML codes developed being unaware of the underlying ML-solvability property potentially produce outputs that the lack the needed rigorous analysis and justification.

5.1 Overfitting

The overfitting corresponds to low training error $\hat{\epsilon}$ in Eq. (1) typically due to ML model M with a large number of parameters, which in turn results in low confidence or high $\hat{\delta}_M$ as a result of large learning dimension. This phenomenon is not detectable by examining the training error alone, as it critically depends on the model space $\mathscr{F}_M$ which has a large learning dimension for models with a large number of parameters. In some cases, testing using additional custom measurements can indicate overfitting, namely, much higher testing error than training error; a concrete example of this artifact for radiation source detection is described in Sect. 6.2. But such cases are very limited in practice due to the cost and lack of availability of such test data. The learning dimension that controls $\hat{\delta}_M$ can be estimated by using the properties of physical law $\mathscr{L}_S$, including smoothness, non-smoothness, and algebraic properties [37], as described in the previous section. The overfitting effect can be mitigated by choosing ML methods with low learning dimension informed by $\mathscr{L}_S$, for example, by controlling the depth of random forest ML estimate to match the Lipschitz parameters of the Poisson measurement law of radiation source or total variation of throughput profile of network transport infrastructure.

5.2 Soundness and Completeness: Hallucinations

The term "hallucination" refers to incomplete or unsound inferences by ML models, which are increasingly observed in generative LLM methods. However, they are often undetected due to lack of knowledge to assess the inferences. Its causes are directly related to the terms in generalization equation Eq. (1): (a) $\mathscr{F}_M$ with an insufficient density property, such as sufficient inference rules, results in large $I(f^*)$ indicating incomplete or unsound $\tilde{f}$, which turn results in large $\hat{\delta}_M$ or low confidence, and (b) Turing undecidable f^* results in unsound or incomplete output generated by any code of $\tilde{f}$ however sophisticated it is (including LLM). Using ML methods based on physical law violations, active probe methods can be used to estimate indicators of these violations, which indicate the likelihood of phenomena such as hallucinations. Specifically, the learning algorithms are customized by replacing the traditional cost function $C(.)$ by the violation terms [54] using the algebraic forms of physical laws to reduce such effects. In effect, the learned models are restricted to conform to underlying $\mathscr{L}_S$ in scientific domains with such conducive laws.

6 Applications

Complex ML solutions have been developed by exploiting the law $\mathscr{L}_S$ and its structure $\mathscr{S}_S(\mathscr{C}_S)$ of underlying system in several practical problems. We briefly describe two different types of applications that illustrate the ML-solvability and an additional one that illustrates the contrary, all from existing works. First, for a data transport network, an abstract law $\mathscr{L}_S$ specifies the convex-concave shape of its throughput as a function of RTT, which is used to establish ML-solvability and also assess the degree of optimization using its ML estimates. Second, physical system laws are utilized to establish ML-solvability and develop ML-estimates in two cases: (a) thermal-hydraulic equations are used for ML-estimates of power-level and sensor drift in nuclear plants, and (b) radiation decay equations are used for ML-detection of a radiation source. Third, lack of ML-solvability of self-detection of large-scale systems is shown by using the undecidability laws on properties of computations. Together, they illustrate various implications of ML-solvability ,and custom development of their ML-solutions and generalization equations using physical or abstract law $\mathscr{L}_S$.

6.1 Throughput Profiles of Data Transport Networks

Recent ML efforts have focused on using network measurements to identify and isolate performance bottlenecks [47] and provide ways to allocate resources to optimize the throughput performance [33]. A mathematical method for characterizing transport performance that uses estimates of throughput profiles is the *concave-convex* analysis initiated in [47, 57]. Concave profiles indicate near-optimal throughput, and convex profiles indicate bottlenecks from factors such as IO buffers, file system throughput, mismatched IO-network couplings, and suboptimal protocols and transfer tools. The underlying mathematics of this analysis are based on statistical estimation and protocol time-dynamics structured as Poincare maps and Lyapunov exponents [21, 45]. As an example, consider a transport throughput time trace $\theta(\tau, t)$, which is a random quantity with a distribution dependent on complex factors like capacities of edge systems and networks, loads on files systems, transfer software, host systems, and connections. The transport performance is characterized by the *throughput profile* $\Theta(\tau)$ as a function of RTT τ defined as $\bar{\Theta}_O(\tau) = E\left[\Theta_O(\tau)\right]$ where the underlying distribution is too complicated to calculate with simple models. ML provides throughput profile estimates, which may be smooth such as sigmoid neural networks or non-smooth such as forests of trees, [59], with finite sample performance guarantees [46]. Compared to the "physical laws," law $\mathscr{L}_S(P, Z)$ is abstract in terms of its concave-convex property. These problems are ML-solvable, and the generalization equations of ML estimates have been developed for both smooth and non-smooth estimators and their fusers (shown in Fig. 3).

6.2 Nuclear Applications: Facility Analytics and Source Detection

The structure of $\mathscr{S}_S$ fluid dynamic equations is used to compose a ML solution for the estimation of the power level of a reactor based on external monitoring sensors [40, 55]. In this scenario, the activity of a reactor is monitored using measurements from four fans, four pumps, three acoustic sensors, and temperature sensors. The resultant, a three-level ML solution, is based on tree topology. Inferring of dissolution events at a radiochemical processing facility using monitoring measurements is an important classification task solved by fusing disparate methods [41, 42]. For dissolutions scenarios, the classifier overfitting is addressed by fusing three promising ones from eight disparate classifiers, akin to Fig. 3. Here, the

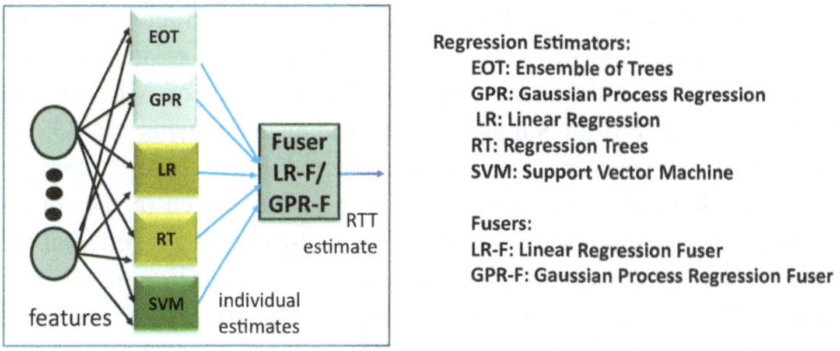

Fig. 3 Fusion of disparate ML estimates provides performance superior to individual estimates

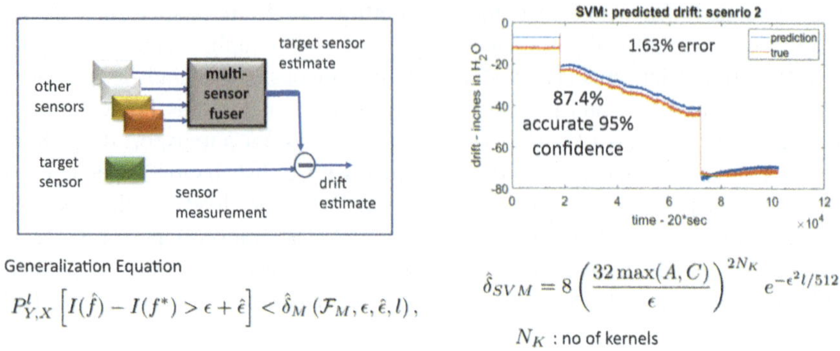

Fig. 4 Difference between the measurement from a target sensor and its estimate by fusing measurements from other sensors is an estimate of the sensor error

physics-based decay chains provided the isotope features to be used as input to classifiers, and their half-life estimates provided feature window-size to achieve improved performance beyond the simple classifier fusion.

The sensors of a power plant measure variables that are typically related to each other through the underlying physics laws. Such relationships provide regressions that are smooth or non-smooth with bounded variation property, which are conducive to estimation by machine learning methods [54]. These relationships are exploited to learn a regression function for a target sensor using measurements from other sensors that measure the same or different variables. The regression estimate is subtracted from its actual measurement to obtain a sensor drift estimate, as illustrated in Fig. 4. The key idea is that when the target sensor measurement is subjected to drifts, this difference reflects its drift error [50, 51]. The root mean square error of drift estimate is below 12.6% with 95% confidence as indicated by the generalization equation [64] shown in the figure. The generalization equation does not impose any conditions on the underlying error distributions and yet provides practically useful information; the error and confidence estimates are derived by exploiting the measurement data sets and domain-specific customization [51].

Another class of problems in this domain addresses the detection of radiation sources using gamma spectral measurements. Signatures of low-intensity ^{235}U sources have been recently studied by utilizing a variety of ML classifiers that utilize features extracted from NaI gamma spectral measurements [38, 43, 53]. Their performance is tested using measurements collected under strategic configurations of NaI detectors located at various distances from the source in the formation of two concentric circles and one spiral. Multiple independent experimental runs have been conducted which provided data sets for training and testing of ML classifiers. Well-known ML classifiers revealed complex classification performance, and, in particular, some classifiers and fusers overfit the training data, which results in overly optimistic training error that is negated by much larger error in independent test measurements.

A novel regression-based ML method is developed in [52] that first estimates the inverse of distance to source, which is a physical property; the background is represented as a source located at an infinite distance. The EOT and GPR methods fit non-smooth and smooth regression functions, respectively, and a hyperparameter auto-tuning and selection method (denoted by AUTO) employs regression trees, neural network, and support vector machine in addition to EOT and GPR. These methods avoid the overfitting observed in several ML classifiers while providing the classification error nearly comparable to them based on independent test data. Their error is directly related to estimates of the inverse physical distance to source, and the precision of error determines the separability property that determines the false alarm and missed detection rates. The property of monotonic decrease of the source strength with increasing detector distance combined with Poisson distribution of measurements is utilized to analytically validate these methods by deriving the generalization equations of underlying regression methods.

The physical laws used in these cases are different: thermal hydraulic equations are used for ML-estimation of the power level and sensor drift in the case of reactor systems, and Poisson measurements and quadratic intensity decay laws are used for radiation source detection.

6.3 Self-Diagnosis of Computing Systems

High-performance computing systems are composed of a large number of multicore processors, graphical processing units (GPUs), and accelerators with computing elements totaling millions; also, they consist of interconnects, switches, and hierarchies of memory units. The life-span of components used in these systems is typically about 5–10 years, for example, processors. Thus, to a first-order approximation, computations running for hours are likely to experience multiple failures, and they in turn may result in errors in application computations as well as in operating and runtime systems that execute applications. Furthermore, the sheer size and complexity of these systems may lead to complex faults, which cannot be known precisely or anticipated. They may range from manufacturing and fatigue faults in components, to dynamic hotspots in computer racks due to interactions between device placement and cooling systems, to interactions of software modules with degraded hardware components. These faults may manifest in a variety of ways: memory faults causing executables to be corrupted and variables with out-of-bound values, incorrect loading of program counters and errors in arithmetic and logic operations, and bus and interconnect faults corrupting data in transit between processing units.

The resilient computations are required to produce correct results on computing systems subject to broad classes of complex failures. The resilient computation problems are shown to present significant computational challenges if the underlying failures are not precisely characterized and anticipated [35]. A broad class of resilience computation problems are shown to be undecidable in the sense of Turing [62], that is, no algorithms exist for solving them. The automatic verification of resilient computations in general is shown to be undecidable, and their extensions are used to show that the resilient computations under data and program corruption and execution errors are undecidable. Specifically, reductions are shown from the classical loop detection and halting problems to generic resilient computation problems. Specifically, these proofs are presented under two formulations, namely, the abstract programming language L [14] and Turing machines [62], that highlight different aspects of the underlying failures; the former represents program and data corruption, and the latter illustrates incorrect program execution. Then, relativization results are presented that indicate that even if halting problems due to these errors are decidable, it is still possible for undecidable problems to persist. Also, other failure classes are described based on arithmetic systems that could lead to algorithms for which performance guarantees are hard to prove or not provable.

These undecidability results indicate that unless the class of faults is limited, these problems cannot be solved by purely computational and analytical means, and hence, these problems are not ML-solvable. Hence, they call for broad-based approaches that complement computational solutions, which integrate methods such as hardware monitors, co-design of hardware and software solutions, system-specific diagnosis methods, and application-specific resilience methods. Furthermore, algorithms, frameworks, and ecosystems used in such approaches must clearly identify their target failures and establish that the underlying computational problems are indeed are decidable for them to be solvable by ML methods. Thus, these classes of problems are not effectively solvable by ML methods even if large data sets of failure measurements are available.

7 Conclusions

The concept of ML-solvability is proposed in this chapter to provide insights into applicability and performance of increasingly complex, often closed ML methods. By combining the theories of learnability, computing, and logic, it provides insights into the possibility of incomplete and unsound inferences when ML-solvability of the underlying problem is not established or simply unknown. Then, a framework is developed for ML-solvability and generalization analyses based on a combination of physical laws that govern systems and information laws that characterize learning processes. Practical examples utilizing smooth, non-smooth, and algebraic forms of laws to develop these ML solutions are discussed in the areas of data transport networks, nuclear engineering, and computer system diagnosis.

The higher-level aspects of developing and applying ML methods discussed in this chapter can be refined along several future directions possibly to much finer levels. It would be of future interest to explore refined aspects of ML-solvability to assess the black-box LLM methods specific to science domains, by possibly generating probes and analyzing the conformance of the answers to domain physical laws. The additional ML-solvability requirements of expressibility and provability can be further explored, which entails language extensions of models based on real space $\Re^d$ in our applications. Specifically, it would be of future interest to assess these aspects within the formulation of deductive systems used to show Godel's incompleteness theorem using the algorithmic approach described in [63]. The model spaces $\mathscr{F}_M$ of applications discussed in this chapter are primarily based on $\Re^d$, and their formal and natural language versions would be needed to assess some of LLM methods with natural language inputs and outputs. Specifically, X and Y are language strings in these cases, and the cost function $C(.,.)$ needs to reflect them and possibly their semantics. Specifically, the physical or abstract laws of application domains may be expressed in a similar manner to support the design and analysis of these methods in assessing the correctness of their inferences and the proofs generated by them, possibly under deductive system framework. In another direction, the instrument-computing ecosystems consist of networked physical instruments and computing nodes, and their workflow codes are composed of computing and physical instrument instructions—it would be interest to extend the notions of computability and ML-solvability under this expanded formulation.

Acknowledgments This work is funded by RAMSES project and Applied Mathematics program, Office of Advanced Computing Research, US Department of Energy, and performed at Oak Ridge National Laboratory managed by UT-Battelle, LLC for US Department of Energy under Contract No. DE-AC05-00OR22725. The United States Government retains and the publisher, by accepting the article for publication, acknowledges that the United States Government retains a nonexclusive, paid-up, irrevocable, worldwide license to publish or reproduce the published form of this manuscript, or allow others to do so, for United States Government purposes. The Department of Energy will provide public access to these results of federally sponsored research in accordance with the DOE Public Access Plan (http://energy.gov/downloads/doe-public-access-plan).

References

1. A. Al-Najjar, N. S. V. Rao, C. Bridges, and S. Dai. ML-based resolution for CV normality detection in real-time cross-facility electrochemistry workflow. In *2024 IEEE 20th International Conference on e-Science (e-Science)*, pages 1–10, 2024.
2. M. Anthony and P. L. Bartlett. *Neural Network Learning: Theoretical Foundations*. Cambridge University Press, 1999.
3. P. L. Bartlett. The sample complexity of pattern classification with neural networks: the sample size of the weights is more important than the size of the network. Technical report, Department of Systems Engineering, Australian National University, 1996.
4. S. Ben-David, P. Hrubeš, S. Moran, A. Shpilka, and A. Yehudayoff. Learnability can be undecidable. *Nature Machine Intelligence*, 1(1):44–48, Jan. 2019.
5. A. L. Blum and R. L. Rivest. Training a 3-node neural network is NP-complete. *Neural Networks*, 5:117–127, 1992.
6. A. Blumer, A. Ehrenfeucht, D. Haussler, and M. Warmuth. Learnability and the Vapnik-Chervonenkis dimension. *Journal of the Association of Computing Machinery*, 36(4):929–965, 1989.

7. L. Breiman, J. H. Friedman, R. A. Olshen, and C. J. Stone. *Classification and Regression Trees*. Wadsworth and Brooks, 1984.
8. D. S. Broomhead and D. Lowe. Multivariable functional interpolation and adaptive networks. *Complex Systems*, 2:321–355, 1988.
9. G. J. Chaitin. *Information, Randomness and Incompleteness*. World Scientific Pub, 1990. Second Edition.
10. Z. Ciesielski. Haar system and nonparametric density estimation in several variables. *Probability and Mathematical Statistics*, 9:1–11, 1988.
11. D. I. A. Cohen. *Introduction to Computer Theory*. John Wiley and Sons, Inc., 1986.
12. F. Cohen. Computational aspects of computer viruses. *Computers & Security*, 8(4):297–298, 1989.
13. T. H. Cormen, C. E. Leiserson, and R. L. Rivest. *Introduction to Algorithms*. McGraw-Hill Book Co., New York, 1990.
14. M. D. Davies and E. J. Weyuker. *Computability, Complexity, and Languages*. Academic Press, Inc, 1983.
15. L. Devroye, L. Gyorfi, and G. Lugosi. *A Probabilistic Theory of Pattern Recognition*. Springer-Verlag, New York, 1996.
16. M. R. Garey and D. S. Johnson. *Computers and Intractability: A Guide to the Theory of NP-Completeness*. W. H. Freeman and Co., San Francisco, 1979.
17. K. Godel. On formally undecidable propositions of principia mathematica and related systems i. *Monatshefte fur Math. und Physik*, 38:173–198, 1931. Englishe translation by B. Meltzer, published by Dover Publications, Inc, 1992.
18. D. Gunning. Explainable artificial intelligence program. Defense Advanced Research Projects Agency.
19. T. Hastie, R. Tibshirani, and J. Friedman. *The Elements of Statistical Learning: Data Mining, Inference, and Prediction*. Springer-Verlag, 2001.
20. E. Horowitz and S. Sahni. *Fundamentals of Computer Algorithms*. Computer Science Press, 1978.
21. Q. Liu, N. S. V. Rao, C. Q. Wu, D. Yun, R. Kettimuthu, and I. Foster. Measurement-based performance profiles and dynamics of UDT over dedicated connections. In *International Conference on Network Protocols*. Singapore, Nov. 2016.
22. W. Mass. Agnostic PAC learning of functions on analog neural nets. *Neural Computing*, 7:1054–1078, 1995.
23. A. Matarazzo and R. Torlone. A survey on large language models with some insights on their capabilities and limitations, 2025. https://arxiv.org/abs/2501.04040.
24. M. Mohri, A. Rostamizadeh, and A. Talwalkar. *Foundations of Machine Learning*. The MIT Press, 2018. second edition.
25. C. Moore. Unpredictability and undecidability in dynamical systems. *Phys. Rev. Lett.*, 64:2354–2357, May 1990.
26. J. Murzi. The inexpressibility of validity. *Analysis*, 74(1):65–81, 2014.
27. C. H. Papadimitriou. *Computational Complexity*. Addition-Wesley Publishing Co., 1994.
28. H. Qiao, N. S. V. Rao, and V. A. Protopopescu. PAC learning of using Nadaraya-Watson estimator based on orthogonal systems. In M. Li and A. Maruoka, editors, *Algorithmic Learning Theory*. Lecture Notes in Artificial Intelligence 1316, Springer-Verlag, 1997.
29. N. Rao. Experimental study of tcp throughput profiles and dynamics over dedicated connections. In *Proceedings of the SC '23 Workshops of The International Conference on High Performance Computing, Network, Storage, and Analysis*, SC-W '23, page 776–784, New York, NY, USA, 2023. Association for Computing Machinery.
30. N. S. V. Rao. Simple sample bound for feedforward sigmoid networks with bounded weights. *Neurocomputing*, 29:115–122, 1999.
31. N. S. V. Rao. Finite sample performance guarantees of fusers for function estimators. *Information Fusion*, 1(1):35–44, 2000.

32. N. S. V. Rao. On fusers that perform better than best sensor. *IEEE Transactions on Pattern Analysis and Machine Intelligence*, 23(8):904–909, 2001.
33. N. S. V. Rao. Overlay networks of in-situ instruments for probabilistic guarantees on message delays in wide-area networks. *IEEE Journal on Selected Areas in Communications*, 22(1), 2004.
34. N. S. V. Rao. Measurement-based statistical fusion methods for distributed sensor networks. In S. S. Iyengar and R. R. Brooks, editors, *Distributed Sensor Networks*. Chapman and Hall/CRC Publishers, 2011. 2nd Edition.
35. N. S. V. Rao. On undecidability aspects of resilient computations and implications to exascale. In *Resilience 2014: Seventh Workshop on Resiliency in High Performance Computing with Clouds, Grids, and Clusters*, 2014.
36. N. S. V. Rao. An algebra of machine learners with applications. In *International Conference on Information Fusion*, 2021.
37. N. S. V. Rao. Generalization equations for machine learners based on physical and abstract laws. In *IEEE International Conference on Multisensor Fusion and Integration for Intelligent Systems*, 2021.
38. N. S. V. Rao, D. Abrecht, D. Hooper, J. Ladd-Lively, and D. Meier. Classification analytics of ^{239}pu and ^{235}u source signatures using gamma spectral regions. In *Institute of Nuclear Materials Management Annual Meeting*, 2023.
39. N. S. V. Rao, C. Greulich, P.amuhalli, A. Gurgen, F. Zhang, and S. M. Cetiner. Estimation of sensor measurement errors in rector coolant systems using multi-sensor fusion. *Nuclear Engineering Design*, 2021.
40. N. S. V. Rao, C. Greulich, S. Sen, K. Dayman, J. Hite, W. Ray, R. Hale, A. Nicholson, J. Honson, M. R. Chatin, K. M. Buckley, R. D. Hunley, J. Johnson, H. H. Hesse, M. Maceira, C. Chai, O. Marcillo, T. Karnowski, and R. Wetherington. Reactor power level estimation by fusing multi-modal sensor measurements. In *International Conference on Information Fusion*, 2020.
41. N. S. V. Rao, C. Greulich, S. Sen, K. Dayman, A. Nicholson, M. R. Chatin, K. M. Buckley, R. D. Hunley, J. Johnson, H. H. Hesse, and R. Hale. Classifiers for dissolution events in processing facility using effluents measurements. In *Institute of Nuclear Materials Management Annual Meeting*, 2019.
42. N. S. V. Rao, C. Greulich, S. Sen, J. Hite, K. J. Dayman, A. D. Nicholson, D. E. Archer, M. J. Willis, I. G. R. D. Hunley, J. Johnson, A. J. Rowe, I. R. Stewart, and J. M. Ghawaly. Classification of dissolution events using fusion of effluents measurements and classifiers. In *Institute of Nuclear Materials Management Annual Meeting*, 2020.
43. N. S. V. Rao, D. Hooper, and J. Ladd-Lively. Study of classifiers for u-235 source signatures using gamma spectral measurements. In *Institute of Nuclear Materials Management Annual Meeting*, 2022.
44. N. S. V. Rao, Q. Liu, Z. Liu, R. Kettimuthu, and I. Foster. Throughput analytics of data transfer infrastructures. In *13th EAI International Conference on Testbeds and Research Infrastructures for the Development of Networks and Communities (TRIDENTCOM 2018)*, 2018.
45. N. S. V. Rao, Q. Liu, S. Sen, J. Henley, I. T. Foster, R. Kettimuthu, D. Towsley, and G. Vardoyan. TCP throughput profiles using measurements over dedicated connections. In *ACM Symposium on High-Performance Parallel and Distributed Computing*, Washington, DC, July-August. 2017.
46. N. S. V. Rao, Q. Liu, S. Sen, Z. Liu, and R. Kettimuthu. Measurements and analytics of wide-area file transfers over dedicated connections. In *International Conference on Distributed Computing and Networking*, 2018.
47. N. S. V. Rao, Q. Liu, S. Sen, D. Towsley, G. Vardoyan, I. T. Foster, and R. Kettimuthu. Experiments and analyses of data transfers over wide-area dedicated connections. In *26th International Conference on Computer Communications and Network*, 2017.
48. N. S. V. Rao, C. Y. T. Ma, and F. He. Game strategies for data transfer infrastructures against ml-profile exploits. *IEEE Transactions on Machine Learning in Communications and Networking*, 2:925–938, 2024.

49. N. S. V. Rao and V. Protopopescu. On PAC learning of functions with smoothness properties using feedforward sigmoidal networks. *Proceedings of the IEEE*, 84(10):1562–1569, 1996.
50. N. S. V. Rao, P. Ramuhalli, C. Greulich, and S. M. Cetiner. Sensor error estimation for reactor coolant system with generalization error equations. *Transactions of the American Nuclear Society*, 121(1), 11 2019.
51. N. S. V. Rao, P. Ramuhalli, C. Greulich, S. M.Cetiner, and P. Devineni. Sensor drift estimation for reactor systems by fusing multiple sensor measurements. In *IEEE Nuclear Science Symposium conference record. Nuclear Science Symposium, Manchester, U.K.*, 2019.
52. N. S. V. Rao, C. Redding, D. Abrecht, D. Hooper, and J. Ladd-Lively. Physics-based machine learning methods for ^{235}u forensics signatures. In *Institute of Nuclear Materials Management Annual Meeting*, 2022.
53. N. S. V. Rao, C. Redding, D. Hooper, and J. Ladd-Lively. Selection of training sets for ^{235}u source detection classifiers using gamma signatures. In *Symposium on Radiation Measurements and Applications*, 2023.
54. N. S. V. Rao, D. B. Reister, and J. Barhen. Information fusion methods based on physical laws. *IEEE Transactions on Pattern Analysis and Machine Intelligence*, 27(1):66–77, 2005.
55. N. S. V. Rao, S. Sen, K. Dayman, W. Ray, R. Hale, A. Nicholson, J. Johnson, R. Hunley, and M. Maceida. Multi-modal sensor fusion for reactor power-level estimation: Thermal, EM, acoustic. In *Nuclear Security Applications Research & Development Program Review Meeting*, 2019.
56. N. S. V. Rao, S. Sen, Z. Liu, R. Kettimuthu, and I. Foster. Learning concave-convex profiles of data transport over dedicated connections. In *International Conference on Machine Learning for Networking (MLN'2018)*, 2018.
57. N. S. V. Rao, D. Towsley, G. Vardoyan, B. W. Settlemyer, I. T. Foster, and R. Kettimuthu. Sustained wide-area TCP memory transfers over dedicated connections. In *IEEE International Conference on High Performance and Smart Computing*, New York, NY, Aug. 2015.
58. B. Scholkopf, C. J. C. Burges, and A. J. Smola, editors. *Advances in Kernel Methods*. MIT Press, 1999.
59. S. Sen, N. S. V. Rao, Q. Liu, N. Imam, I. T. Foster, and R. Kettimuthu. On analytics of file transfer rates over dedicated wide-area connections. In *First International Workshop on Workflow Science (WOWS)*, Auckland, New Zealand, October 2017. in conjunction with 13th IEEE International Conference on e-Science.
60. A. Tarski. *Logic, Semantics, Metamathematics*. Oxford University Press, 1956.
61. K. Tumer and J. Ghosh. Error correlation and error reduction in ensemble classifiers. *Connection Science*, 8(3):385–404, 1996.
62. A. N. Turing. On computable numbers, with an application to the Entscheidungsproblem. *Proc. London Math. Society*, 42(3,4):230–265, 1936.
63. V. A. Uspensky. *Godel's Incompleteness Theorem*. Mir Publishers, 1987. English translation.
64. V. N. Vapnik. *Statistical Learning Theory*. John-Wiley and Sons, New York, New York, 1998.
65. Y. Wu, B. Sicard, and S. A. Gadsden. A review of physics-informed machine learning methods with applications to condition monitoring and anomaly detection, 2024. https://arxiv.org/abs/2401.11860.

Digital Forensics in the Age of Large Language Models

Zhipeng Yin, Zichong Wang, Weifeng Xu, Jun Zhuang, Pallab Mozumder, Antoinette Smith, and Wenbin Zhang

1 Introduction

Digital forensics is a critical component in modern investigative and judicial processes, which involve the systematic collection, analysis, and preservation of digital evidence from electronic devices and online activities [1–3]. Its primary objective is to uncover factual information related to cybercrimes, fraud, unauthorized access, and other illicit activities [4]. Digital forensics has played a pivotal role in solving high-profile cybercrime cases. For example, in the 2014 Sony Pictures hack, forensic investigators traced the breach back to North Korean hackers, who leaked confidential company data, emails, and unreleased films as part of a geopolitical cyberattack [5]. The investigation relied on digital forensics techniques such as analyzing network logs, identifying malware signatures, and attributing IP addresses to suspected attackers. As another example, in the 2016 Democratic National Committee (DNC) email leak, digital forensic experts identified sophisticated spear-phishing tactics and linked the attack to Russian-backed hacking groups, influencing the US presidential election [6, 7]. Beyond cyber espionage, digital forensics has also been crucial in financial fraud investigations. For instance, the Silk Road darknet marketplace, a notorious online black market, was dismantled in 2013 through extensive forensic analysis of Bitcoin transactions, server logs, and encrypted messages [8, 9]. Forensic experts traced Bitcoin payments to the

Z. Yin (✉) · Z. Wang · P. Mozumder · A. Smith · W. Zhang
Florida International University, Miami, FL, USA
e-mail: zyin007@fiu.edu

W. Xu
University of Baltimore, Baltimore, MD, USA

J. Zhuang
Boise State University, Boise, ID, USA

S. S. Iyengar et al. (eds.), *Artificial Intelligence Driven Forensics*, Signals and Communication Technology, https://doi.org/10.1007/978-3-031-98036-7_3

marketplace's operator, Ross Ulbricht, ultimately leading to his arrest and life sentence. In another case, the Enron scandal saw digital forensics specialists recover crucial deleted emails and financial records, providing key evidence in one of the largest corporate fraud investigations in history [10]. Additionally, digital forensic methodologies have been instrumental in child exploitation cases, where law enforcement agencies track online predators by analyzing metadata in images, chat logs, and digital footprints left on the dark web [11, 12].

These case studies highlight the effectiveness of digital forensics in various domains, but they also demonstrate how investigators increasingly encounter complex technological challenges that test the limits of current methodologies. The primary issue is that traditional digital forensic techniques predominantly rely on manual or semiautomated approaches, requiring intensive human involvement [13]. These methods suffer from several inherent limitations. Firstly, they are labor intensive and time consuming, making them less effective in handling large-scale and sophisticated cyber incidents [14]. Secondly, traditional methods often struggle to maintain consistency and accuracy due to human error and subjective judgments, potentially compromising evidence reliability. In addition, conventional forensic tools exhibit limited adaptability to evolving cyber threats, and their capability to identify complex interrelationships among evidence entities remains constrained [15, 16].

Several recent cases, as shown in Fig. 1 illustrate these limitations. For instance, in July 2024, a gunman attempted to assassinate former US President Donald Trump during a public rally, prompting an intensive investigation by federal authorities [35, 36]. Following the suspect's capture, the FBI conducted a forensic analysis of his mobile device to uncover potential motives, affiliations, and premeditated plans. Investigators faced significant challenges in bypassing the phone's security mechanisms, including encryption and biometric locks. Once access was gained, digital forensic teams meticulously analyzed call logs, text messages, encrypted messaging apps, and location history. Additionally, they examined the suspect's social media interactions, online searches, and affiliations with extremist groups. Traditional analysis methods required extensive manual effort to filter through

Fig. 1 Investigation of real-world digital forensics cases in recent years

vast amounts of digital data, cross-reference communication patterns, and verify links between different sources. Such challenges were similarly highlighted in the 2020 Twitter cryptocurrency scam, where cybercriminals compromised multiple high-profile accounts to solicit Bitcoin payments [37]. The FBI's digital forensic teams encountered substantial hurdles as they manually cross-referenced Discord chat logs, leaked hacker forum databases, cryptocurrency wallet transactions, and IP addresses to identify the perpetrators. Although successful, this method revealed critical shortcomings in efficiently correlating and interpreting multidimensional digital evidence streams, demonstrating the urgent need for more advanced forensic capabilities [38]. These complexities also surfaced prominently in the FBI's investigation into the Pensacola naval base shooting in 2019 [39, 40]. In that case, the assailant's encrypted iPhones had sustained physical damage, and Apple refused official requests for access assistance. Consequently, FBI forensic experts spent months painstakingly repairing hardware and circumventing encryption to retrieve data. Eventually, the recovered digital evidence established clear connections between the attacker and foreign terrorist entities. However, the prolonged investigative timeline underscored limitations inherent in traditional forensic methodologies when handling encrypted devices and fragmented digital traces [41, 42]. Collectively, these cases emphasize the increasing necessity of integrating AI-driven digital forensic tools. Leveraging automation, intelligent data analysis, and advanced pattern recognition technologies could significantly enhance investigative speed, consistency, and accuracy, effectively addressing the growing scale, complexity, and sophistication of contemporary cyber threats [43, 44].

To address these substantial challenges in digital forensics, recent advances in artificial intelligence offer promising solutions. Notably, large language models (LLMs), such as the generative pre-trained transformer (GPT) series of models and the Gemini series, have emerged as powerful tools with the potential to transform digital forensic practices [23, 24, 30, 45]. These advanced AI models are designed to understand, interpret, generate and analyze human language with unprecedented accuracy. Using vast amounts of textual data from diverse sources, LLMs exhibit exceptional capabilities in natural language processing, pattern recognition, and semantic understanding [46–49]. Their ability to extract meaningful insights from large, unstructured datasets makes them invaluable in digital forensic investigations [50].

The application of these sophisticated models directly addresses the limitations of traditional forensic approaches identified earlier. One of the most impactful applications of LLMs in forensic analysis is their ability to automate and streamline the evidence identification process [51, 52]. Traditional methods require investigators to manually sift through enormous volumes of text, such as emails, chat logs, social media posts, and financial records. LLMs, on the other hand, can swiftly process and categorize these texts, recognizing patterns, detecting anomalies, and identifying crucial connections between disparate pieces of evidence [53–55]. This capability significantly accelerates investigative timelines while reducing the risk of human error [56]. In the aforementioned high-profile investigation, integrating LLM-powered analytical tools could have played a transformative role in expediting the investigative process. By rapidly categorizing and interpreting textual evidence, LLMs can highlight potential leads, uncover hidden relationships, and help investigators piece together a cohesive narrative [57–59]. Moreover, their ability to process multilingual content ensures that forensic teams can analyze communication in different languages and cultural contexts without the need for extensive translation efforts [60, 61]. LLMs also improve forensic data interpretation by facilitating the reconstruction of complex evidence relationships. They can map connections between personal identifiers, such as names, addresses, and phone numbers, and correlate them with network activity, financial transactions, and geolocation data [62–64]. This holistic approach allows investigators to establish links between suspects, victims, and illicit activities with greater precision [65]. Another crucial advantage of LLMs in digital forensics is their ability to handle large-scale data integration. Digital evidence is often scattered across multiple sources, including cloud storage, encrypted messaging platforms, and darknet forums. LLMs, combined with knowledge graph techniques, can aggregate and visualize these fragmented data points, making it easier to identify trends, associations, and key actors within an investigation [66].

While the benefits of LLMs for digital forensics are substantial, their implementation is not without challenges that need to be carefully considered [67–69]. In addition, despite their transformative potential, the adoption of LLMs in forensic investigations also introduces new challenges, including concerns over interpretability, bias, and the reliability of AI-generated insights [70, 71]. Ensuring transparency in forensic AI applications is crucial to maintaining credibility in judicial proceedings [72]. Therefore, addressing these concerns through clear guidelines, rigorous validation procedures, and transparent reporting practices becomes essential [73–75]. To this end, this chapter explores how LLMs can fundamentally change digital forensics practices by automating evidence analysis, extracting insightful information, and enhancing the judicial process and attempts to provide a comprehensive understanding of the practical applications, potential limitations, and broader implications of LLMs in digital forensics investigations [76].

Chapter Structure The subsequent sections of this chapter are structured as follows: Section 2 introduces foundational concepts and highlights the limitations of training-based AI digital forensic methodologies. Section 3 details the principles and capabilities of large-scale language modeling and presents practical applications and real-world case studies. Section 4 evaluates the current challenges and limitations faced when deploying LLMs in forensic scenarios. Finally, Sect. 5 discusses opportunities and directions for future research.

2 Fundamentals of Digital Forensics

This section examines in depth the core principles of digital forensics. Understanding these fundamentals is critical to grasping how modern investigative processes utilize digital evidence to combat cybercrime, fraud, and other illicit activities.

2.1 Definition and Goals

Digital forensics is the systematic process of identifying, collecting, preserving, analyzing, and presenting digital evidence in a legally admissible manner [93, 94]. It is widely used in criminal investigations, cybersecurity incidents, corporate fraud detection, and other legal proceedings[95]. The primary goal of digital forensics is to uncover and reconstruct events related to cybercrimes, unauthorized access, financial fraud, intellectual property theft, and other illicit digital activities. By leveraging digital forensic techniques, investigators can retrieve hidden, deleted, or encrypted data to support legal actions and improve cybersecurity measures [96].

2.2 Digital Forensic Evidence Entities

A digital forensic evidence entity represents the smallest indivisible unit of digital information with forensic significance. Such entities serve as fundamental building blocks for reconstructing digital events and verifying their authenticity. These entities are categorized according to their functional purpose in supporting investigative analysis [97]. Table 1 provides a detailed description of these entities by functional purpose. The Content-Descriptive Entities help investigators understand the nature, source, or intended use of digital artifacts, providing essential context to evidence collected during an investigation [98]. In contrast, auxiliary entities supplement this understanding by offering validation, verification, and support to primary descriptive evidence, ensuring the reliability and integrity of forensic findings.

Table 1 Functional categories and descriptions of digital forensic evidence entities

Categories	Digital forensic evidence entities	Description
Content-descriptive entities	File names	Identifiers given to files, potentially revealing their content, origin, or intended purpose
	IP addresses	Numerical labels assigned to devices on a network, crucial for tracking and attributing online activities
Auxiliary entities	Timestamps	Specific points in time indicating events such as file creation, modification, or access
	Hashes	Unique identifiers generated from data content used to verify file integrity and detect tampering

2.3 Key Evidence Types

While the theory of digital forensic evidence entities is understood based on their functional role, real-world forensic investigations often require more specific categorization. Investigators routinely encounter various forms of digital evidence, which must be clearly identified and categorized to effectively address complex forensic challenges [99, 100]. The following summarizes specific categories of digital evidence, which are classified according to their relevance and investigative role, and demonstrates how the theoretical framework translates into operational forensic practice [101].

- **Personal identifiers:** Names, addresses, phone numbers, email addresses, social security numbers, and other personal information. In identity theft cases, stolen personal identifiers are typically located within phishing emails, compromised databases, or fraudulent registrations.
- **Network information:** IP addresses, MAC addresses, login credentials, and network logs crucial for tracing user actions across devices and networks. Investigators often utilize this data to pinpoint sources of unauthorized access, as exemplified by numerous cases of insider threats and external intrusions.
- **Communication records:** Emails, text messages, social media messages, and call logs that capture interactions among individuals or groups. Analysis of these records has been pivotal in solving cases involving cyberbullying, insider trading, and organized crime.
- **Financial data:** Bank account details, credit card transactions, cryptocurrency wallet addresses, and transaction histories essential in tracking financial fraud and money laundering. Forensic analysts frequently exploit blockchain technology to unravel cryptocurrency-based criminal networks.
- **Location data:** GPS coordinates, timestamps, and geolocation logs, enabling investigators to track movements and verify alibis. This form of data has notably been employed in criminal cases where mobile device locations provided critical evidence linking suspects to crime scenes.
- **Internet activity:** Web browsing history, search queries, downloads, and online interactions offering deep insights into user behaviors and intentions. These digital footprints have been invaluable in cases involving radicalization, online harassment, and cyberstalking.
- **File metadata:** Information including timestamps, file paths, and document version histories, useful for establishing file authenticity and tracking document manipulation. Metadata analysis has been critical in corporate espionage investigations and cases of intellectual property theft.
- **Device logs and system artifacts:** System event logs, registry entries, and application usage records, offering detailed insight into user activities and system states. In investigations of data breaches or corporate sabotage, these logs have provided evidence disproving fabricated user accounts and narratives.

2.4 Evidence Relationships

Having discussed specific categories of digital evidence encountered in practical forensic scenarios, it is crucial to recognize that these pieces of evidence rarely exist in isolation. Instead, they form intricate networks of relationships that significantly enhance investigative analysis [102]. Understanding these interconnections allows investigators to reconstruct detailed narratives, establish causality, and verify the authenticity of digital evidence comprehensively [103]. The key relationship categories include:

(i) **Contextual relationships:** These relationships provide situational context, helping investigators understand the origin, purpose, or usage of evidence. For example, linking file names to their content or correlating an IP address to a geographical location helps determine the source and intention behind cyber incidents.

(ii) **Causal relationships:** Highlight cause-and-effect dynamics between evidence entities. Identifying the correlation between an IP address and a specific timestamp can establish a suspect's direct involvement in unauthorized access or data manipulation.

(iii) **Associative relationships:** Connect seemingly independent evidence through shared attributes. Similar file hashes detected across multiple devices may suggest deliberate data duplication, exfiltration, or manipulation efforts by malicious actors.

(iv) **Communication relationships:** Reveal interaction patterns among individuals or systems. Analyzing communication logs such as phone records, emails, or chat messages has proven essential in dismantling criminal networks, uncovering collaboration among perpetrators, and mapping complex interactions in cybercrime investigations.

(v) **Ownership and association:** Establish explicit connections between individuals and digital devices, accounts, or data. Digital forensic efforts routinely involve associating specific devices or accounts with suspects, thereby strengthening investigative narratives and courtroom presentations.

(vi) **Temporal relationships:** Establish a chronological sequence or simultaneity of events. Timestamp analysis enables forensic examiners to confirm or refute suspect claims, authenticate alibis, and determine exact timelines of incidents, especially critical in high-stakes criminal and corporate investigations.

2.5 Limitations of Training-Based AI for Digital Forensics

While AI-driven methodologies offer significant advancements in digital forensic investigations, several inherent limitations constrain their effectiveness, particularly when employing training-based AI approaches:

(i) **Data scarcity:** Obtaining sufficient and diverse training data representative of real-world cyber incidents poses significant challenges. Often, the available data is limited to specific case types, such as addresses extracted predominantly from certain criminal activities like shootings. This lack of comprehensive and varied datasets can severely restrict the AI model's ability to generalize across different forensic scenarios [51].

(ii) **Data preprocessing challenges:** Even seemingly simple tasks, such as identifying addresses using named-entity recognition (NER), introduce considerable preprocessing complexity before AI models can be effectively applied. These tasks often require multiple preprocessing steps, including expanding abbreviations (e.g., converting"St." to"Street"), standardizing formats (e.g., "123 Main St Apt 4B" to "123 Main Street, Apartment 4B"), normalizing state names (e.g., "California" to "CA"), and removing extra whitespace (e.g., converting "456 Elm St" to"456 Elm St."). These additional preprocessing steps significantly increase the complexity, time, and resources required, underscoring the limitations associated with direct training-based AI approaches in digital forensic analyses.

(iii) **AI models lack adaptability:** AI models developed for digital forensic tasks are typically designed and optimized for specific, narrowly defined functions. For instance, an AI model trained explicitly for recognizing addresses will likely exhibit limited performance when tasked with identifying other types of information, such as personal names or financial records [104]. This specialized training makes it challenging to apply these models broadly across the diverse range of forensic tasks investigators encounter.

(iv) **Difficulty in extracting evidence relationships:** Identifying and analyzing the numerous intricate relationships among digital evidence entities is inherently complex. Training-based AI methods often struggle to capture the full depth and nuance of these interactions, given the extensive variety and subtlety in relationships, including contextual, causal, associative, communication-based, ownership-based, and temporal connections [105]. Consequently, traditional training-based approaches may not recognize critical evidence correlations, potentially undermining the accuracy and comprehensiveness of forensic analyzes.

3 Large Language Models for Digital Forensics

3.1 Why LLMs for Digital Forensics

LLMs represent a sophisticated category of artificial intelligence models, primarily designed to understand, generate, and interact with natural language text [106]. These models typically utilize deep learning architectures, such as transformers, which rely on self-attention mechanisms to capture intricate contextual relationships within textual data [107]. The development and training of LLMs involve vast datasets, often comprising billions of words, enabling these models to acquire a deep understanding of syntax, semantics, and contextual nuances inherent in human languages.

One of the most prominent examples of LLMs is the Generative pre-trained Transformer (GPT) series developed by OpenAI, including GPT-3 and GPT-4. These models exhibit exceptional capabilities across a wide range of natural language processing (NLP) tasks, such as text generation, summarization, translation, sentiment analysis, question answering, and entity extraction [108, 109]. Their impressive versatility stems from their ability to capture long-range contextual dependencies and their extensive training on diverse textual resources such as websites, books, articles, and other publicly available information.

The LLM training process generally involves two main phases: pre-training and fine-tuning. During pre-training, models are exposed to vast, unsupervised text corpora, learning general language patterns, syntax, and semantic relationships without specific task-oriented labels. In the fine-tuning phase, LLMs are further trained in task-specific datasets, adapting their general language comprehension skills to effectively perform targeted NLP tasks [110]. This two-phase approach significantly enhances their adaptability and performance across diverse domains.

LLMs are trained on vast amounts of text data and exhibit exceptional capabilities in learning linguistic patterns, structural semantics, and contextual dependencies. These attributes make them uniquely suited for applications in digital forensics, where the volume and heterogeneity of digital evidence can overwhelm traditional analysis techniques. In digital forensic investigations, evidence often exists in unstructured or semi-structured forms, such as chat logs, emails, file metadata, browsing histories, and system logs. Manually extracting meaningful patterns or relationships from such data is labor intensive and time consuming [111]. LLMs can assist by automatically identifying named entities, classifying document types, summarizing lengthy communication threads, detecting suspicious patterns, and establishing semantic links across diverse artifacts.

LLMs offer the ability to generalize from limited context, which is particularly useful in forensic settings where fragmented or incomplete evidence is common. Their pre-training on diverse data sources also allows them to recognize and interpret technical jargon, code snippets, and colloquial expressions, enabling them to analyze evidence drawn from varied digital environments [106]. LLMs also support multi-turn interactions, allowing investigators to iteratively refine queries or extract context-sensitive information from large datasets in a conversational manner. This interaction paradigm not only enhances usability but also reduces the need for technical expertise in formulating complex forensic queries.

Therefore, LLMs have great potential to enable digital forensics given their training on large textual datasets and powerful pattern learning capabilities, and by utilizing LLMs' advanced linguistic understanding analytical and generative capabilities, key information can be extracted from large amounts of data, significantly streamlining the analysis of evidence, enhancing the process of informed decision-making, and improving overall forensic outcomes [107].

3.2 LLM-Driven Methods in Digital Forensics

Integrating LLM into forensic workflows has emerged as a promising approach as investigators seek new tools to improve the accuracy, efficiency, and scalability of their analyses. This section provides an overview of current methodological frameworks and empirical case studies in which LLM has been effectively utilized in digital forensic environments. Specifically, it highlights representative applications, evaluates their practical effectiveness, and identifies methodological insights that have emerged from these real-world implementations.

3.2.1 LLM-Driven Construction of Evidence Networks

Utilizing their powerful pattern recognition and relationship extraction capabilities, LLMs introduce innovative methods to improve the efficiency and accuracy of forensic investigations [112]. One prominent example of such LLM-driven methods involves utilizing GPT-4-turbo to systematically identify and visualize patterns within digital forensic evidence. This method constructs a structured graph G = (V, E), where nodes (V) represent individual evidence items—such as names, addresses, and phone numbers—while edges (E) depict relationships connecting these items. Each edge is labeled explicitly to describe the nature of the connection, such as "owns" for ownership (a person owning a phone number) or "lives-in" for residency (a person residing at an address). These clearly defined relationships enable the creation of comprehensive visual representations that simplify the analysis of intricate forensic data.

The main procedural steps shown in Fig. 2 typically include:

(i) **Transforming raw data into a processable versions:** This step involves extracting and standardizing evidence from mobile devices, personal electronic devices, and especially their embedded multimedia card storage. Considering that these devices often contain fragmented, hidden, or deleted data in binary

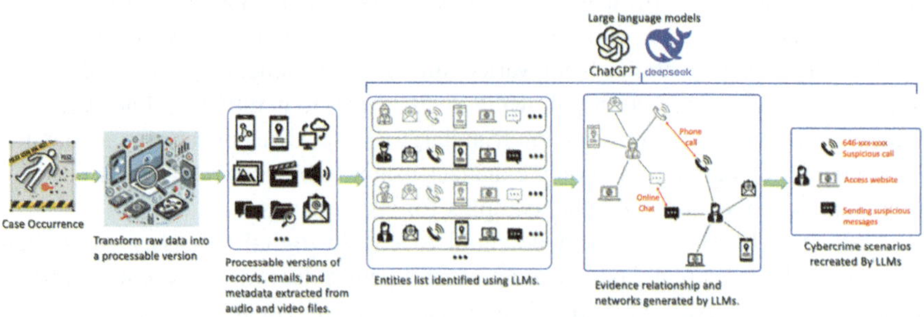

Fig. 2 An LLM method to understanding cybercrime via evidence networks

form, converting this information into a clear text format is essential for the accurate analysis of LLM.

(ii) Identifying evidence entities and their relationships: Researchers create and test tailored prompts that guide LLMs in systematically extracting relevant evidence entities from structured textual data such as chat logs and system records. A representative prompt could be: "Act as an experienced digital forensic investigator. Extract evidence entities like names, addresses, and phone numbers from the given text and outline the relationships among these entities."

(iii) Constructing evidence networks: This step involves connecting isolated pieces of evidence to form coherent networks. Connections are identified based on proximity, either physical (line distance in text) or semantic (inferred through LLMs), under the assumption that closely positioned entities are likely interrelated.

(iv) Deriving insights into criminal behavior: Lastly, these constructed evidence networks are analyzed to uncover significant insights into criminal activities, behaviors, and underlying relationship patterns. This detailed examination of interconnected evidence provides forensic investigators with critical information that enhances their understanding of complex criminal scenarios.

3.2.2 LLM-Driven Invocation Log Analysis for Digital Forensics

Chernyshev et al. proposed a novel forensic methodology aimed at detecting prompt injection attacks in applications integrated with LLMs [113]. The core innovation of this approach lies in leveraging invocation logs, a structured record of LLM interactions, as a primary evidentiary source for digital forensic investigations.

Their method involves constructing a simplified yet representative experimental scenario that emulates real-world LLM-integrated web applications, and Fig. 3 illustrates its workflow. Specifically, the authors developed a web-based application utilizing GPT-3.5 via the LangChain framework. In this scenario, users' natural language queries are converted by the LLM into Structured Query Language (SQL) statements, subsequently executed against a backend relational database. To create realistic attack conditions, the authors manually designed a set of malicious prompts

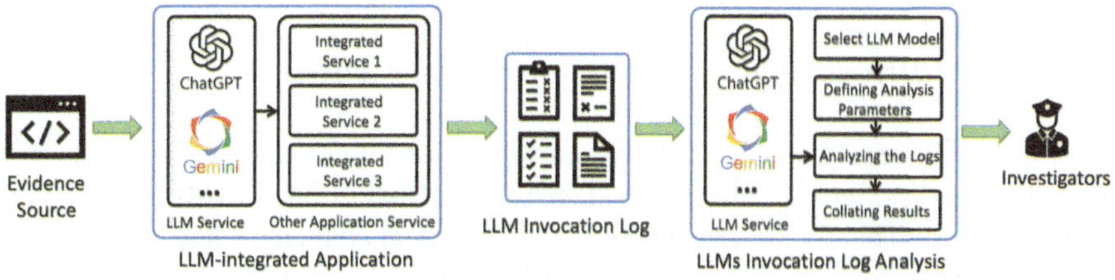

Fig. 3 Digital forensic analysis workflow with LLM invocation logs

to simulate direct prompt injection attacks, such as dropping database tables or bypassing access control restrictions.

To facilitate digital forensic readiness (DFR), the authors introduced structured logging mechanisms, termed LLM invocation logging, into their experimental system. Each invocation log entry captured essential forensic metadata including a timestamp, unique request identifier, input prompt (user's query), and corresponding LLM output, generating structured JSON-formatted logs, thereby ensuring traceability and forensic integrity.

For forensic analysis, the collected invocation logs were processed using an active analysis strategy in which multiple contemporary LLMs acted as forensic analysts. Given that different models have significantly varying context windows—for instance, from 8,182 tokens for llama3-70b-instruct to 1 million tokens for gemini-1.5-flash and gemini-1.5-pro—the authors evaluated analysis approaches both with models capable of accepting the entirety of the invocation logs as input and those requiring splitting log entries into smaller window chunks for sequential processing. Their analysis involved four main steps: **(i) selecting an LLM model for analysis,** GPT-4, Gemini, or other similar models; **(ii) defining key analysis parameters,** such as the LLM's temperature and context window size; **(iii) actively analyzing the logs with the chosen configuration,** using the chosen model and parameters; and **(iv) collating the results,** summarizing key findings and observations. These steps were systematically repeated until all desired combinations of models and analysis parameters had been evaluated. Unlike previous works exploring LLM usage for anomaly detection that employed pre-summarization, this approach solely relied on active log analysis without context summary creation. Specifically, each model was provided log entries within a predefined context window, accompanied by instructions to identify potential security incidents and articulate justifications. The models returned structured JSON outputs indicating detection decisions (either "NORMAL" or "INCIDENT"), suspicious log indices, and descriptive reasoning. This direct approach significantly reduced the overall number of calls to the LLM, consequently decreasing both the total time required for log analysis and the potential cost.

This approach illustrates important advances in digital forensic readiness for LLM-driven systems, showing how invocation log analysis performed by the LLM itself can provide practical forensic capabilities for identifying sophisticated hint injection attacks.

3.2.3 LLM-Driven Mobile Evidence Contextual Analysis

Kim et al. propose a comprehensive and operationally grounded framework for mobile forensics, termed Mobile Evidence Contextual Analysis (MECA) [11] as show in Fig. 4. This framework addresses the practical challenges law enforcement faces in analyzing large volumes of mobile messenger data, particularly under tight legal time constraints. Rather than relying solely on traditional keyword-based filtering, MECA leverages the contextual reasoning capabilities of LLMs to infer the presence of criminal intent or activity embedded in ambiguous or euphemistic

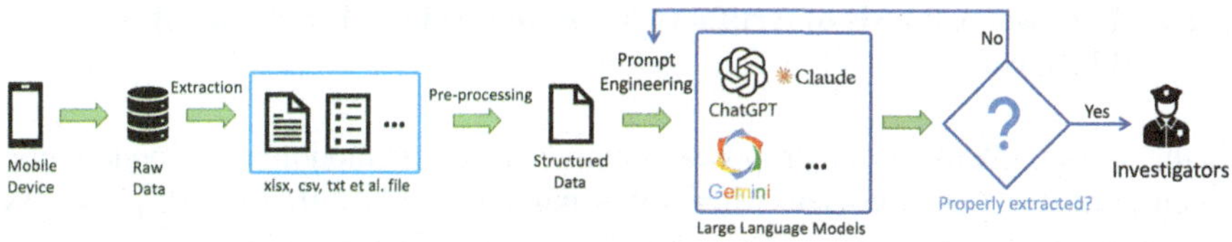

Fig. 4 Overview of LLM-driven Mobile Evidence Contextual Analysis framework

language. The method is notable not only for its application to real-world forensic data but also for its holistic integration of forensic tools, data preprocessing, and prompt engineering.

The framework begins with the acquisition of mobile communication data using professional forensic software tools. Specifically, the authors employ MD-NEXT for physical data extraction and MD-RED for data parsing and visualization. These tools support the collection of structured communication records from seized smartphones, which are exported in formats like CSV or Excel for downstream processing. To ensure compliance with privacy and ethical standards, all personal identifiers within the dataset are anonymized using Named Entity Recognition (NER), with supplementary masking strategies applied to phone numbers and emails to minimize reidentification risk.

Given the size and fragmented nature of mobile information logs, the authors introduced a preprocessing phase to construct coherent units of analysis appropriate for LLM input. This involves applying initial keyword filters, e.g., for terms such as"drugs," to identify potentially relevant messages. In order to preserve conversational context, each filtered message is augmented with surrounding messages in the same chat window, typically 40 lines each before and after the targeted message. This produces a set of context-rich message fragments that reflect real-world communication patterns and facilitate semantic interpretation by the model.

Central to MECA's effectiveness is its use of carefully crafted prompts to guide model behavior. Each prompt is designed to simulate the role of a forensic expert, instructing the model to evaluate whether a given message exchange is associated with criminal activity. The input is structured as key-value pairs, where the key represents the speaker and the value denotes the message content. Moreover, the authors implement a "Sandwich Prompting" technique—repeating instructions before and after the main content—to mitigate instruction forgetting, particularly in models like Gemini that may otherwise over-prioritize the input text.

Once the data and prompts are prepared, the framework employs three state-of-the-art LLMs, GPT-4o, Gemini 1.5, and Claude 3.5, to perform classification. Each model receives the structured conversational input and returns a binary judgment indicating whether the message set is relevant to the case. The authors also account for concerns around data privacy and model misuse by relying on commercial API deployments and explicitly documenting the privacy policies of each LLM provider. The use of multiple models not only allows performance benchmarking across architectures but also sets the stage for ensemble decision-making.

3.2.4 Forensic Analysis of Artifacts from Microsoft's Multiagent LLM Platform

In this work by Walker et al. proposes a comprehensive methodology for conducting forensic analysis of AutoGen, Microsoft's multiagent LLM framework [114]. As AutoGen enables autonomous agent collaboration for task planning and execution, the forensic analysis of such systems introduces novel challenges, particularly in identifying, interpreting, and attributing the artifacts generated through agent interactions. The proposed methodology responds to this gap by establishing a structured, multilayered approach to detecting the presence and behavior of AutoGen on a target system.

At the core of their approach is the idea of tracing the forensic footprint of LLM-driven agent interaction across three major layers of analysis: memory, disk, and network. Rather than focusing on any single modality of artifact, the methodology adopts a layered perspective to capture both persistent and volatile traces of AutoGen's activity on a host system. The authors hypothesize that, despite the encrypted nature of LLM-server communication and the ephemeral memory handling of modern OSes, a composite view of system-level behavior can reveal meaningful patterns associated with LLM agent activity.

The interaction model analyzed in the study involves two LLM-based agents: a UserProxyAgent, simulating a user that initiates tasks and evaluates responses, and an AssistantAgent, responsible for task execution. These agents interact through a feedback loop where task instructions and responses are exchanged programmatically. This model mirrors real-world use of AutoGen for distributed task planning and problem-solving and raises questions around forensic observability, i.e., what traces of such interactions persist on a compromised or analyzed system.

To clarify this interaction workflow, Fig. 5 illustrates the operational model used in the study. This process begins with a user operating from a local environment (A), where they initiate specific task prompts (B), such as generating code, crafting phishing emails, or producing malicious scripts. These prompts are passed to the AutoGen system (C), which coordinates interactions between LLM agents, typically a UserProxyAgent and an AssistantAgent, powered by models like GPT-3.5. The agents exchange messages programmatically until a task is completed,

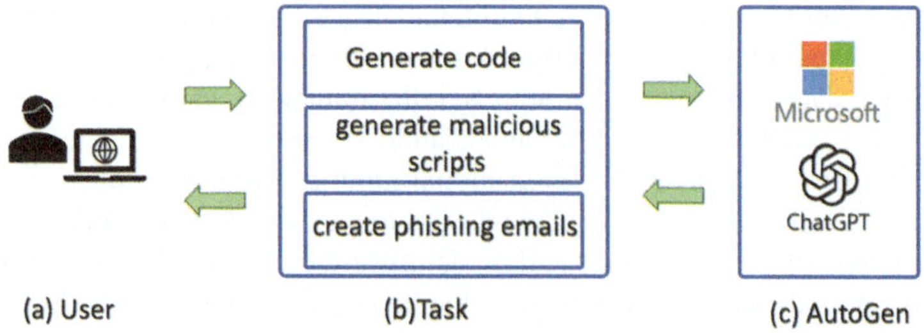

Fig. 5 Overview of forensic analysis of AutoGen

with AutoGen returning the model-generated output to the user. This controlled interaction loop is essential for generating forensic artifacts, and the researchers are able to capture and later examine forensic artifacts across memory, disk, and network layers.

The forensic method involves isolating the key points where AutoGen interacts with the system or external services and mapping those to potential artifact locations. For instance, the UserProxyAgent's initial prompt, along with the AssistantAgent's responses, may be retained in memory buffers, cached in application files, or transiently recorded in system logs. The methodology accounts for the limited durability of such data and therefore incorporates the use of tools that can extract low-level system state information, e.g., RAM dumps, temporary configuration files, and browser traces.

A notable component of the method is its treatment of agent attribution—attempting to distinguish whether a given artifact was created by a human, a machine, or a cooperative agentic process. This is a particularly novel challenge in LLM forensics, since traditional forensic signatures are often agnostic to the cognitive or computational origin of content. The methodology, therefore, considers semantic and behavioral cues, e.g., structure of prompt chains, repeated execution patterns, and lack of GUI interaction, that may help differentiate machine-driven output from human-involved interaction.

Additionally, the approach integrates lightweight static analysis techniques, such as string extraction from memory and file systems, with dynamic signature correlation, such as identifying AutoGen-related modules in Python environments or connections to known LLM service endpoints. This hybrid approach helps mitigate the limitations of any single forensic strategy and provides a more comprehensive account of AutoGen's presence and behavior on the system.

The method sets a foundation for future forensic analysis of autonomous LLM systems, especially as they become more modular, compositional, and capable of unsupervised behavior. It emphasizes the need for multiperspective evidence gathering, cross-layer correlation, and a deeper understanding of agent-based software design in order to maintain accountability in increasingly AI-driven digital environments.

3.2.5 The Local LLM-Driven Framework for Digital Forensic

While large language models (LLMs) have demonstrated remarkable capability across various natural language processing tasks, their application in sensitive domains such as digital forensics presents unique challenges, including concerns about data privacy, security, and the need for specialized domain knowledge. Moreover, reliance on cloud-based solutions can introduce vulnerabilities related to data confidentiality and compliance, prompting the need for locally deployable LLMs tailored specifically to forensic purposes. Addressing these critical issues, Sharma et al. introduced ForensicLLM, a specialized, locally deployable large language model designed explicitly for digital forensic applications using a retrieval-augmented fine-tuning (RAFT) methodology [115].

Sharma et al. utilized Meta's LLaMA-3.1–8B as the foundational model, enhancing it through fine-tuning with domain-specific content to address the unique reasoning demands inherent in digital forensic investigations. They began by compiling an extensive corpus comprising 1,082 peer-reviewed research articles sourced from the journal Forensic Science International: Digital Investigation, along with metadata extracted from 1,390 verified digital forensic artifacts obtained via the Artifact Genome Project. Textual contents from these research articles were segmented into semantically meaningful chunks of approximately 2,000 characters and embedded using the UAE-Large-V1 embedding model. Each chunk was enriched with associated metadata, including article titles and authors, with embeddings subsequently stored within a ChromaDB vector database to facilitate efficient retrieval during subsequent training and inference processes.

In the absence of suitable labeled question-answer datasets specific to digital forensic scenarios, the authors employed GPT-4 Turbo to generate approximately 10,000 synthetic question-answer pairs based directly upon the prepared literature corpus. This generation process was carefully guided using detailed prompting to ensure practically relevant, technically accurate content, maintaining faithful adherence to original source citations following APA standards.

The fine-tuning procedure leveraged Quantized Low-Rank Adaptation, implementing a 4-bit quantization approach to optimize computational resource efficiency during training. Sharma et al. adopted the Axolotl framework, utilizing standard practices such as cosine learning rate scheduling and early stopping based on validation set performance. During inference, ForensicLLM utilizes a retrieval-augmented generation (RAG) strategy, embedding user queries to dynamically retrieve relevant textual contexts from the vector database, which are then integrated into the model input to produce informed, verifiable, and accurate responses.

As shown in Fig. 6, this figure outlines the sequence data processing and model-training pipeline, beginning with raw data extraction and cleaning, followed by segmentation into meaningful textual chunks. These text chunks, enriched with metadata, are then transformed into semantic embeddings using an embedding model and subsequently stored in a ChromaDB vector database. Simultaneously, synthetic Q&A pairs are generated from the corpus using GPT-4 to form a structured training dataset. This Q&A dataset is integrated with context retrieved from the vector database, forming the RAFT dataset utilized for fine-tuning the ForensicLLM model. Finally, during inference, user queries are embedded and matched with relevant contexts retrieved from Chroma database, enabling ForensicLLM to produce accurate, contextually informed, and traceable responses tailored specifically for digital forensic applications.

The retrieval-enhanced fine-tuning approach proposed by Sharma et al. significantly impacts digital forensic practice by reducing common limitations associated with general-purpose language models, particularly hallucinations and factual inaccuracies. Their quantitative and qualitative evaluations demonstrated that ForensicLLM substantially improves response accuracy, relevance, and reliability, thus

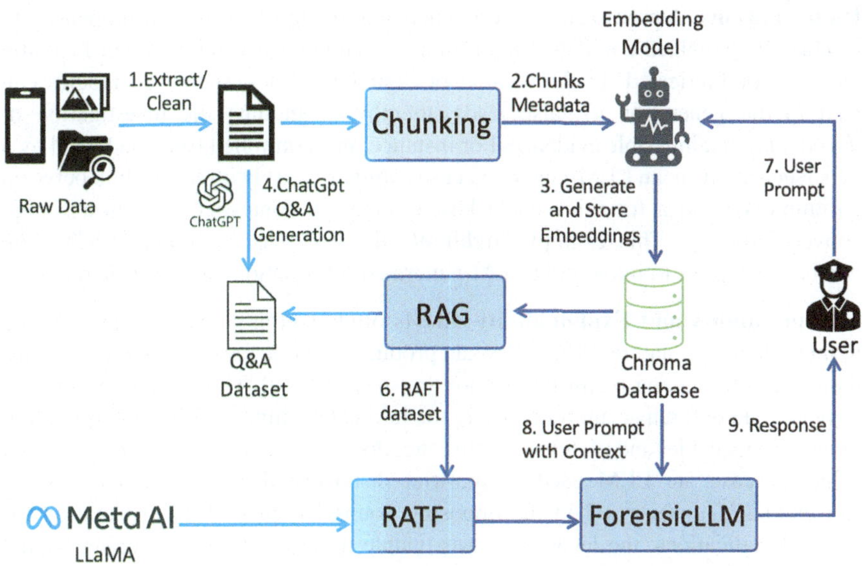

Fig. 6 Overview of retrieval-augmented fine-tuning (RAFT) for ForensicLLM

equipping forensic investigators with trustworthy, traceable analytical support capable of meeting rigorous evidentiary standards required in real-world forensic investigations.

4 Challenges and Limitations of Leveraging LLM in Digital Forensics

The integration of large language models into digital forensics workflows has generated increasing interest due to their potential in automating documentation, evidence analysis, and decision support. However, their use also presents numerous challenges that arise both from the inherent properties of LLMs and from the specific requirements of forensic practice.

4.1 Inherent Challenges of LLMs

Several limitations are intrinsic to the architecture and training methodology of LLMs, which can hinder their safe and reliable deployment in forensic investigations.

Hallucinations A prominent concern when employing LLMs is their propensity to produce hallucinated content—output that is grammatically coherent yet factually incorrect or fabricated. In the context of digital forensics, such inaccuracies can lead to the generation of false leads, thereby misguiding the investigation or introducing inadmissible evidence. For instance, in a controlled trial conducted by a cybersecurity firm, an LLM-generated case summary falsely inferred a link between an employee and a foreign contact based solely on contextual cues in a benign conversation log. This example highlights the necessity of human verification mechanisms prior to integrating LLM-generated information into forensic reports.

Interpretability and Explainability LLMs often exhibit poor explainability due to their black-box nature. While they can produce accurate results in many domains, their decision-making pathways are not transparent [110]. This opacity becomes a critical issue in forensic analysis, where the rationale behind evidence interpretation must be traceable and defensible. In one documented instance during a civil litigation case, an LLM used in pre-trial discovery flagged certain emails as "suspicious"; however, when the opposing counsel requested an explanation for these classifications, the legal team was unable to articulate the reasoning behind the model's output. The lack of explainability ultimately led to the exclusion of the generated evidence.

Lack of Domain-Specific Knowledge General-purpose LLMs are trained on heterogeneous and largely nonspecialized corpora. As such, they may not have the technical nuance necessary for forensic analysis. For example, when prompted to assess the contents of a memory dump, a widely used LLM erroneously flagged "svchost.exe" as malicious, failing to account for the legitimate role of the process in Windows systems. Such errors underscore the risk of applying unadapted LLMs in technical domains without appropriate domain fine-tuning.

Bias and Fairness Bias in LLMs—driven by a reflection of the biases present in the training data—poses ethical and practical risks in forensic contexts [24]. Investigative results may be biased either by reinforcing existing stereotypes or by systematically prioritizing certain types of evidence. In a pilot study involving multilingual forensic datasets, an LLM-assisted classification system consistently deprioritized non-English chat logs, leading to a delay in the examination of relevant Arabic-language communications. This form of bias, if left unaddressed, could have far-reaching implications for fairness and due process in digital investigations [116].

4.2 Digital Forensics-Specific Challenges

Although the inherent risks associated with LLMs pose general concerns in all domains, deploying these models within digital forensic workflows introduces additional challenges. Digital forensics imposes strict standards regarding evidence integrity, reproducibility, and procedural compliance, and these established forensic principles may conflict with the nature of LLM technologies. Consequently, integrating LLMs into digital forensic practices requires addressing specific challenges related to evidentiary standards, reproducibility, prompt sensitivity, standardization, and practitioner readiness.

Chain of Custody and Evidentiary Integrity A core principle in forensic science is the preservation of chain of custody, that is, the ability to trace each step of evidence handling. When LLMs are employed, especially in cloud-based or third-party systems, questions arise regarding the preservation and auditability of evidence. In one European law enforcement case study, the use of an LLM to summarize mobile device contents inadvertently violated chain of custody procedures, as intermediate outputs were not systematically logged. As a result, the forensic findings were challenged on procedural grounds during judicial review.

Nondeterminism and Reproducibility Unlike deterministic forensic tools, LLMs are inherently probabilistic and may produce variable outputs even under identical input conditions. This variability undermines one of the key requirements of forensic science, namely, reproducibility. In a university-led evaluation, an LLM used to reconstruct activity timelines from log data produced inconsistent event sequences across multiple runs. Such behavior poses serious threats to the reliability of forensic conclusions, particularly when outputs are used as part of expert witness testimony.

Prompt Sensitivity Related to nondeterminism is the issue of prompt sensitivity, whereby subtle variations in phrasing can lead to significantly different model outputs. For instance, altering a prompt from "summarize suspicious behavior" to "summarize all activity" led an LLM to either omit or include key lateral movement indicators in the same dataset. The fragility of outputs based on minor linguistic changes necessitates rigorous prompt engineering and version control when using LLMs in evidentiary contexts.

Lack of Standardization There exists no established framework or industry-wide standard governing the use of LLMs in digital forensics. This absence of formal guidance has resulted in inconsistencies across investigative practices and raises concerns regarding admissibility and procedural fairness. In a simulated case involving two independent forensic teams, divergent conclusions were reached due to differences in prompt design, evidence filtering strategies, and LLM configurations. These discrepancies emphasize the need for standardized protocols and certification schemes for LLM-based forensic tools.

Training and Expertise Requirements The adoption of LLMs in digital forensic settings introduces new requirements for practitioner training. Investigators must possess not only technical forensic skills but also basic knowledge in AI, prompt design, and model validation. A field test conducted with junior investigators revealed that improper prompt use led to misclassifications of a legitimate mobile application as malicious, an error that could have been avoided with minimal training in AI reasoning mechanisms. The integration of LLMs thus demands a reevaluation of existing forensic training curricula to include AI literacy.

5 Future Directions

The intersection of large language models (LLMs) and digital forensics represents an emerging frontier with significant potential for transforming forensic investigations. Future research in this area promises to strengthen evidentiary integrity, promote greater accountability, and contribute broadly to societal trust and justice.

5.1 *Multimodal and Cross-Data Analysis*

Digital forensic investigations increasingly require holistic evidence interpretation across various data modalities—textual logs, network traffic, memory dumps, images, and audio. Emerging multimodal LLMs (MLLMs) suggest promising capabilities for integrating diverse data forms into unified analytical frameworks. Integrating vision and language models could enable forensic assistants to analyze screenshots, correlate textual logs with visual artifacts, or interpret combined structured and unstructured forensic data. Research opportunities lie in developing robust multimodal forensic LLMs capable of seamlessly analyzing multiple data types while maintaining accuracy across different modalities. Achieving this will require interdisciplinary collaboration and innovative design to bridge existing capability gaps.

5.2 *Explainability and Trust in LLM-Driven Analysis*

The inherently opaque reasoning of LLMs conflicts with forensic requirements for transparency and verifiability. Enhancing the explainability of LLM outputs is thus critical for building investigator trust. Future research should focus on methods that enable LLMs to justify their conclusions with explicit evidence references and step-by-step reasoning processes. Techniques like retrieval-augmented generation, where LLM outputs are grounded explicitly in input data and known forensic knowledge, can significantly improve credibility. Validation methods, such as cross-validation with multiple models or human-in-the-loop verification, should also be investigated to detect and mitigate errors and biases inherent in AI analyses.

5.3 *Domain-Specific LLMs Across Forensic Disciplines*

One crucial future direction involves the development of specialized, domain-specific LLMs tailored explicitly for various forensic applications such as memory forensics, malware analysis, network investigations, and log interpretation. General-purpose models typically lack the specialized technical understanding required to interpret detailed forensic artifacts accurately. Early examples, such as volGPT for memory analysis, have demonstrated the effectiveness of fine-tuned LLMs in accurately identifying ransomware processes while providing comprehensive explanations. Future research should systematically explore domain-specific models for forensic tasks, including artifact interpretation, file system analysis, and forensic triage. This specialization will necessitate creating dedicated forensic datasets, posing challenges related to data sensitivity and privacy that researchers must address through synthetic or anonymized datasets.

5.4 Privacy and Legal Admissibility Challenges

Integrating LLMs into forensic investigations raises significant privacy concerns and legal admissibility challenges. Public cloud-based solutions often conflict with chain-of-custody requirements, prompting the need for secure, offline LLM solutions deployable within forensic lab environments. Future research should focus on enhancing on-premise or federated AI models that preserve data confidentiality and comply with legal standards. Additionally, clearly defined legal frameworks and standards are needed for documenting and certifying AI processes, ensuring their outputs withstand judicial scrutiny. Collaborative research among technologists, legal scholars, and policymakers is necessary to bridge these gaps and ensure that LLM-assisted forensic analyses meet rigorous evidentiary standards.

5.5 Integration with Traditional Forensic Tools and Workflows

Future research must explore the seamless integration of LLMs into existing forensic software and investigative workflows. Embedding interactive AI assistants within forensic suites, enabling natural language querying, automated artifact parsing, and AI-driven script generation, can significantly enhance investigative efficiency. Ensuring these integrations are robust and error resistant, and maintaining compatibility with existing forensic processes, evidence documentation systems, and investigative protocols, represents a significant technical challenge. Interdisciplinary collaboration will be crucial in developing user-centric, reliable forensic tools augmented by AI capabilities.

5.6 Standardized Evaluation and Benchmarking

A critical gap in current research is the lack of standardized evaluation frameworks for assessing LLM effectiveness and reliability in forensic contexts. Developing shared benchmark datasets; standardized metrics for accuracy, explainability, and utility; and consistent evaluation methodologies is essential for objectively comparing different LLM approaches. Community-driven benchmarking initiatives, similar to established cybersecurity and computer vision evaluations, should be prioritized to accelerate progress and ensure rigorous validation of AI-assisted forensic tools.

6 Conclusion

LLMs have emerged as transformative tools that significantly automate and augment forensic capabilities, thus reshaping the landscape of digital investigations. This chapter systematically explored how LLMs have revolutionized digital forensic approaches, providing a comprehensive and accessible overview for practitioners and researchers alike. Through practical examples and real-world scenarios, we illustrated the superior capabilities of LLMs in enhancing analytical accuracy, efficiency, and scalability in forensic workflows. However, the integration of LLMs into digital forensic processes is not without challenges; issues such as model hallucinations, interpretability, biases, and ethical considerations necessitate cautious and informed application. Addressing these challenges requires further research that focuses on improving transparency, accountability, and standardization in the forensic use of LLMs technologies. Ultimately, the thoughtful integration of LLMs holds significant promise in advancing digital forensic practices, fostering trust and reliability, and contributing to more equitable and just judicial outcomes.

References

1. Wickramasekara, A., Breitinger, F., Scanlon, M.: Exploring the potential of large language models for improving digital forensic investigation efficiency. Forensic Science International: Digital Investigation **52**, 301859 (2025)
2. Rahman, M.N., Mohammad, T., Virtanen, S.: Leveraging large language models for network traffic analysis: Design, implementation, and evaluation of an llm-powered system for cyber incident reconstruction (2024)
3. Xu, E., Zhang, W., Xu, W.: Transforming digital forensics with large language models: Unlocking automation, insights, and justice. In: Proceedings of the 33rd ACM International Conference on Information and Knowledge Management, pp. 5543–5546 (2024)
4. Rogers, M.K.: A two-dimensional circumplex approach to the development of a hacker taxonomy. Digital investigation **3**(2), 97–102 (2006)
5. Ismail, M.: Sony pictures and the us federal government: a case study analysis of the sony pictures entertainment hack crisis using normal accidents theory (2017)
6. Marmura, S.M., Marmura, S.M.: Wikileaks' american moment: The dnc emails, russiagate and beyond. The WikiLeaks Paradigm: Paradoxes and Revelations, 109–133 (2018)
7. Confessore, N., Eder, S., October, L.: In hacked dnc emails, a glimpse of how big money works. The New York Times (2016)
8. Minnaar, A.: Online'underground'marketplaces for illicit drugs: the prototype case of the dark web website'silk road. Acta Criminologica: African Journal of Criminology & Victimology **30**(1), 23–47 (2017)

9. Lacson, W., Jones, B.: The 21st century darknet market: lessons from the fall of silk road. International Journal of Cyber Criminology **10**(1), 40 (2016)
10. Negangard, E.M., Fay, R.G.: Electronic discovery (ediscovery): Performing the early stages of the enron investigation. Issues in Accounting Education **35**(1), 43–58 (2020)
11. Kim, K., Lee, C., Bae, S., Choi, J., Kang, W.: Digital forensics in law enforcement: A case study of llm-driven evidence analysis. Available at SSRN 5110258
12. Quick, D., Choo, K.-K.R.: Digital forensic intelligence: Data subsets and open source intelligence (dfint+ osint): A timely and cohesive mix. Future Generation Computer Systems **78**, 558–567 (2018)
13. Chen, H.-Y.: Cloud crime to traditional digital forensic legal and technical challenges and countermeasures. In: 2014 IEEE Workshop on Advanced Research and Technology in Industry Applications (WARTIA), pp. 990–994 (2014). IEEE
14. Fernando, K.: A multidimensional framework for utilizing big data analytics and ai in strengthening digital forensics and cybersecurity investigations. International Journal of Cybersecurity Risk Management, Forensics, and Compliance **7**(12), 16–30 (2023)
15. Malik, A.W., Bhatti, D.S., Park, T.-J., Ishtiaq, H.U., Ryou, J.-C., Kim, K.-I.: Cloud digital forensics: Beyond tools, techniques, and challenges. Sensors **24**(2), 433 (2024)
16. Garach, J., Singh, S.K., Reddy, A.P.C., Khan, H., et al.: A comprehensive review on artificial intelligence in digital forensics with taxonomies, issues, and solutions: Ai in digital forensics. Strategies for E-Commerce Data Security: Cloud, Blockchain, AI, and Machine Learning, 1–28 (2024)
17. Wang, Z., Saxena, N., Yu, T., Karki, S., Zetty, T., Haque, I., Zhou, S., Kc, D., Stockwell, I., Bifet, A., *et al*: Preventing discriminatory decision-making in evolving data streams. In: Proceedings of the 2023 ACM Conference on Fairness, Accountability, and Transparency (FAccT) (2023)
18. Zhang, W., Wang, Z., Kim, J., Cheng, C., Oommen, T., Ravikumar, P., Weiss, J.: Individual fairness under uncertainty. In: 26th European Conference on Artificial Intelligence, pp. 3042–3049 (2023)
19. Wang, Z., Wallace, C., Bifet, A., Yao, X., Zhang, W.: Fg^{2}an: Fairness-aware graph generative adversarial networks. In: Joint European Conference on Machine Learning and Knowledge Discovery in Databases, pp. 259–275 (2023). Springer Nature Switzerland
20. Yazdani, S., Saxena, N., Wang, Z., Wu, Y., Zhang, W.: A comprehensive survey of image and video generative ai: Recent advances, variants, and applications (2024)
21. Wang, Z., Narasimhan, G., Yao, X., Zhang, W.: Mitigating multisource biases in graph neural networks via real counterfactual samples. In: 2023 IEEE International Conference on Data Mining (ICDM), pp. 638–647 (2023). IEEE
22. Chinta, S.V., Fernandes, K., Cheng, N., Fernandez, J., Yazdani, S., Yin, Z., Wang, Z., Wang, X., Xu, W., Liu, J., *et al.*: Optimization and improvement of fake news detection using voting technique for societal benefit. In: 2023 IEEE International Conference on Data Mining Workshops (ICDMW), pp. 1565–1574 (2023). IEEE
23. Wang, Z., Chu, Z., Doan, T.V., Ni, S., Yang, M., Zhang, W.: History, development, and principles of large language models: an introductory survey. AI and Ethics, 1–17 (2024)
24. Chu, Z., Wang, Z., Zhang, W.: Fairness in large language models: A taxonomic survey. ACM SIGKDD explorations newsletter **26**(1), 34–48 (2024)
25. Dzuong, J., Wang, Z., Zhang, W.: Uncertain boundaries: Multidisciplinary approaches to copyright issues in generative ai. arXiv preprint arXiv:2404.08221 (2024)
26. Yin, Z., Wang, Z., Zhang, W.: Improving fairness in machine learning software via counterfactual fairness thinking. In: Proceedings of the 2024 IEEE/ACM 46th International Conference on Software Engineering: Companion Proceedings, pp. 420–421 (2024)
27. Wang, Z., Zhou, Y., Haque, I., Lo, D., Zhang, W.: Towards fair machine learning software: Understanding and addressing model bias through counterfactual thinking. arXiv preprint arXiv:2302.08018 (2023)

28. Wang, Z., Qiu, M., Chen, M., Salem, M.B., Yao, X., Zhang, W.: Toward fair graph neural networks via real counterfactual samples. Knowledge and Information Systems, 1–25 (2024)
29. Chinta, S.V., Wang, Z., Yin, Z., Hoang, N., Gonzalez, M., Quy, T.L., Zhang, W.: Fairaied: Navigating fairness, bias, and ethics in educational ai applications. arXiv preprint arXiv:2407.18745 (2024)
30. Doan, T.V., Wang, Z., Hoang, N.N.M., Zhang, W.: Fairness in large language models in three hours. In: Proceedings of the 33rd ACM International Conference on Information and Knowledge Management, pp. 5514–5517 (2024)
31. Chinta, S.V., Wang, Z., Zhang, X., Viet, T.D., Kashif, A., Smith, M.A., Zhang, W.: Ai-driven healthcare: A survey on ensuring fairness and mitigating bias. arXiv preprint arXiv:2407.19655 (2024)
32. Wang, Z., Dzuong, J., Yuan, X., Chen, Z., Wu, Y., Yao, X., Zhang, W.: Individual fairness with group awareness under uncertainty. In: Joint European Conference on Machine Learning and Knowledge Discovery in Databases, pp. 89–106 (2024). Springer Nature Switzerland
33. Wang, Z., Palikhe, A., Yin, Z., Zhang, W.: Fairness definitions in language models explained. arXiv preprint arXiv:2407.18454 (2024)
34. Wang, Z., Chu, Z., Blanco, R., Chen, Z., Chen, S.-C., Zhang, W.: Advancing graph counterfactual fairness through fair representation learning. In: Joint European Conference on Machine Learning and Knowledge Discovery in Databases, pp. 40–58 (2024). Springer Nature Switzerland
35. Swanson, C.: Bullets and ballots: Exploring the effects of nearly successful assassination attempts on general election performance in the united states. The UWJPS is thankful for the continued support of the Department of Political Science at the University of Washington. In addition, we are grateful to the students who submitted their work and ideas., 15
36. The New York Times: Investigators Unlock Gunman's Phone in Search for Motive in Trump Shooting. Accessed: 2025-03-30. https://www.nytimes.com/live/2024/07/15/us/trump-shooting-investigation
37. Bartoletti, M., Lande, S., Loddo, A., Pompianu, L., Serusi, S.: Cryptocurrency scams: analysis and perspectives. Ieee Access **9**, 148353–148373 (2021)
38. Cimpanu, C.: How the FBI Tracked down the Twitter Hackers. Accessed: 2025-03-30. https://www.zdnet.com/article/how-the-fbi-tracked-down-the-twitter-hackers/
39. Kessler, G.C., Phillips, A.M.: Cryptography, passwords, privacy, and the fifth amendment. Journal of Digital Forensics, Security and Law **15**(2), 2 (2020)
40. Clarke, C.: The pensacola terrorist attack: The enduring influence of al-qaida and its affiliates. CTC Sentinel **13**(3) (2020)
41. Vaghela, R., Gowda, V.D., Taj, M., Arudra, A., Chopra, M.: Digital evidence collection and preservation in computer network forensics. In: Handbook of Research on Innovative Approaches to Information Technology in Library and Information Science, pp. 42–62 (2024)
42. Allam, H.: FBI: New iPhone Evidence Shows Pensacola Shooter Had Ties To Al-Qaida. Accessed: 2025-03-30. https://www.npr.org/2020/05/18/857932909/fbi-new-iphone-evidence-shows-pensacola-shooter-had-ties-to-al-qaida
43. Nayak, M.: Ai-enhanced digital forensics: Automated techniques for efficient investigation and evidence collection. J. Electrical Systems **20**(1s), 211–229 (2024)
44. Akeiber, H.J.: A comprehensive study of cybercrime and digital forensics through machine learning and ai. Al-Rafidain Journal of Engineering Sciences, 369–395 (2025)
45. Liu, J., Kong, Z., Zhao, P., Yang, C., Tang, H., Shen, X., Yuan, G., Niu, W., Zhang, W., Lin, X., et al.: Toward adaptive large language models structured pruning via hybrid-grained weight importance assessment. arXiv preprint arXiv:2403.10799 (2024)
46. Jin, H., Wei, W., Wang, X., Zhang, W., Wu, Y.: Rethinking learning rate tuning in the era of large language models. In: 2023 IEEE 5th International Conference on Cognitive Machine Intelligence (CogMI), pp. 112–121 (2023). IEEE
47. Ferrag, M.A., Alwahedi, F., Battah, A., Cherif, B., Mechri, A., Tihanyi, N., Bisztray, T., Debbah, M.: Generative ai in cybersecurity: A comprehensive review of llm applications and vulnerabilities. Internet of Things and Cyber-Physical Systems (2025)

48. Yao, Y., Duan, J., Xu, K., Cai, Y., Sun, Z., Zhang, Y.: A survey on large language model (llm) security and privacy: The good, the bad, and the ugly. High-Confidence Computing, 100211 (2024)
49. Valmeekam, K., Olmo, A., Sreedharan, S., Kambhampati, S.: Large language models still can't plan (a benchmark for llms on planning and reasoning about change). In: NeurIPS 2022 Foundation Models for Decision Making Workshop (2022)
50. Kumarage, T., Agrawal, G., Sheth, P., Moraffah, R., Chadha, A., Garland, J., Liu, H.: A survey of ai-generated text forensic systems: Detection, attribution, and characterization. arXiv preprint arXiv:2403.01152 (2024)
51. Ahmed, M., Mahmood, A.N., Hu, J.: A survey of network anomaly detection techniques. Journal of Network and Computer Applications **60**, 19–31 (2016)
52. Liu, C., Xie, X., Zhang, X., Cui, Y.: Large language models for networking: Workflow, advances and challenges. IEEE Network (2024)
53. Velasco, C.: Cybercrime and artificial intelligence. an overview of the work of international organizations on criminal justice and the international applicable instruments. In: ERA Forum, vol. 23, pp. 109–126 (2022). Springer
54. Mijwil, M.M., Aljanabi, M., ChatGPT, C.: Towards artificial intelligence-based cybersecurity: The practices and chatgpt generated ways to combat cybercrime. Iraqi Journal For Computer Science and Mathematics **4**(1), 8 (2023)
55. Zhang, R., Xie, M.: Forensiq: A knowledge graph question answering system for iot forensics. In: International Conference on Digital Forensics and Cyber Crime, pp. 300–314 (2023). Springer
56. Siddiqui, M.Z., Yadav, S., Husain, M.S.: Application of artificial intelligence in fighting against cyber crimes: a review. Int. J. Adv. Res. Comput. Sci **9**(2), 118–122 (2018)
57. Chen, Z., Mao, H., Li, H., Jin, W., Wen, H., Wei, X., Wang, S., Yin, D., Fan, W., Liu, H., *et al.*: Exploring the potential of large language models (llms) in learning on graphs. ACM SIGKDD Explorations Newsletter **25**(2), 42–61 (2024)
58. Yang, J., Jin, H., Tang, R., Han, X., Feng, Q., Jiang, H., Zhong, S., Yin, B., Hu, X.: Harnessing the power of llms in practice: A survey on chatgpt and beyond. ACM Transactions on Knowledge Discovery from Data **18**(6), 1–32 (2024)
59. Smirnov, E.: Enhancing qualitative research in psychology with large language models: a methodological exploration and examples of simulations. Qualitative Research in Psychology **22**(2), 482–512 (2025)
60. Kao, H.-H.: Accelerating multilingual cryptocurrency forensics: An nlp-driven approach for efficient mnemonic identification. IEEE Access (2025)
61. Karie, N.M., Kebande, V.R., Venter, H.: Diverging deep learning cognitive computing techniques into cyber forensics. Forensic Science International: Synergy **1**, 61–67 (2019)
62. Arshad, H., Jantan, A.B., Abiodun, O.I.: Digital forensics: review of issues in scientific validation of digital evidence. Journal of Information Processing Systems **14**(2), 346–376 (2018)
63. Klasén, L., Fock, N., Forchheimer, R.: The invisible evidence: Digital forensics as key to solving crimes in the digital age. Forensic science international **362**, 112133 (2024)
64. Daniel, L., Daniel, L.: Digital Forensics for Legal Professionals: Understanding Digital Evidence from the Warrant to the Courtroom, (2011)
65. Caballero, E.Q.: Leveraging large language models for legal document understanding and software system analysis: Addressing key challenges. PhD thesis, Baylor University (2024)
66. Akhtar, S., Khan, S., Parkinson, S.: Llm-based event log analysis techniques: A survey. arXiv preprint arXiv:2502.00677 (2025)
67. Labajová, L.: The state of AI: Exploring the perceptions, credibility, and trustworthiness of the users towards AI-Generated Content (2023)
68. Khlaif, Z.N., Mousa, A., Hattab, M.K., Itmazi, J., Hassan, A.A., Sanmugam, M., Ayyoub, A.: The potential and concerns of using ai in scientific research: Chatgpt performance evaluation. JMIR Medical Education **9**, 47049 (2023)

69. Raza, H.: Ai-driven assessment: Reliability, bias, and ethical implications. Journal of AI in Education: Innovations, Opportunities, Challenges, and Future Directions **1**(2), 36–47 (2024)
70. Azodi, C.B., Tang, J., Shiu, S.-H.: Opening the black box: interpretable machine learning for geneticists. Trends in genetics **36**(6), 442–455 (2020)
71. Quang Huy, P., Kien Phuc, V.: Insight into how legal and ethical considerations of artificial intelligence enhance the effectiveness of cyber forensic accounting. Journal of Global Information Technology Management, 1–31 (2025)
72. Wischmeyer, T.: Artificial intelligence and transparency: opening the black box. In: Regulating Artificial Intelligence, pp. 75–101 (2019)
73. Djeffal, C.: Artificial intelligence and public governance: normative guidelines for artificial intelligence in government and public administration. In: Regulating Artificial Intelligence, pp. 277–293 (2019)
74. Cath, C.: Governing artificial intelligence: ethical, legal and technical opportunities and challenges. Philosophical Transactions of the Royal Society A: Mathematical, Physical and Engineering Sciences **376**(2133), 20180080 (2018)
75. Baror, S.O., Venter, H.S., Adeyemi, R.: A natural human language framework for digital forensic readiness in the public cloud. Australian Journal of Forensic Sciences **53**(5), 566–591 (2021)
76. Jain, A.: Enhancing forensic analysis of digital evidence using machine learning: Techniques, applications, and challenges. International Journal of Innovative Research in Multidisciplinary Perspectives and Studies (IJIRMPS), 1–8 (2024)
77. Wang, Z., Zhang, W.: Group fairness with individual and censorship constraints. In: 27th European Conference on Artificial Intelligence (2024)
78. Wang, Z., Ulloa, D., Yu, T., Rangaswami, R., Yap, R., Zhang, W.: Individual fairness with group constraints in graph neural networks. In: 27th European Conference on Artificial Intelligence (2024)
79. Yin, Z., Agarwal, S., Kashif, A., Gonzalez, M., Wang, Z., Liu, S., Liu, Z., Wu, Y., Stockwell, I., Xu, W., *et al.*: Accessible health screening using body fat estimation by image segmentation. In: 2024 IEEE International Conference on Data Mining Workshops (ICDMW), pp. 405–414 (2024)
80. Wang, Z., Yin, Z., Zhang, Y., Yang, L., Zhang, T., Pissinou, N., Cai, Y., Hu, S., Li, Y., Zhao, L., *et al.*: Fg-smote: Towards fair node classification with graph neural network. ACM SIGKDD Explorations Newsletter **26**(2), 99–108 (2025)
81. Wang, Z., Yin, Z., Liu, F., Liu, Z., Lisetti, C., Yu, R., Wang, S., Liu, J., Ganapati, S., Zhou, S., *et al.*: Graph fairness via authentic counterfactuals: Tackling structural and causal challenges. ACM SIGKDD Explorations Newsletter **26**(2), 89–98 (2025)
82. Wang, Z., Chu, Z., Viet Doan, T., Wang, S., Wu, Y., Palade, V., Zhang, W.: Fair graph u-net: A fair graph learning framework integrating group and individual awareness. In: Proceedings of the AAAI Conference on Artificial Intelligence (2025)
83. Wang, Z., Hoang, N., Zhang, X., Bello, K., Zhang, X., Iyengar, S.S., Zhang, W.: Towards fair graph learning without demographic information. In: The 28th International Conference on Artificial Intelligence and Statistics (2025)
84. Zhang, W.: Fairness with censorship: Bridging the gap between fairness research and real-world deployment. In: Proceedings of the AAAI Conference on Artificial Intelligence, vol. 38, pp. 22685–22685 (2024)
85. Zhang, W.: Ai fairness in practice: Paradigm, challenges, and prospects. Ai Magazine (2024)
86. Zhang, W., Ntoutsi, E.: Faht: an adaptive fairness-aware decision tree classifier. In: International Joint Conference on Artificial Intelligence (IJCAI), pp. 1480–1486 (2019)
87. Zhang, W., Weiss, J.: Fair decision-making under uncertainty. In: 2021 IEEE International Conference on Data Mining (ICDM) (2021). IEEE
88. Zhang, W., Weiss, J.C.: Longitudinal fairness with censorship. In: Proceedings of the AAAI Conference on Artificial Intelligence, vol. 36, pp. 12235–12243 (2022)

89. Zhang, W., Hernandez-Boussard, T., Weiss, J.: Censored fairness through awareness. In: Proceedings of the AAAI Conference on Artificial Intelligence, vol. 37, pp. 14611–14619 (2023)
90. Zhang, W., Zhou, S., Walsh, T., Weiss, J.C.: Fairness amidst non-iid graph data: A literature review. AI Magazine **46**(1), 12212 (2025)
91. Zhang, W., Weiss, J.C.: Fairness with censorship and group constraints. Knowledge and Information Systems, 1–24 (2023)
92. Zhang, W., Zhang, L., Pfoser, D., Zhao, L.: Disentangled dynamic graph deep generation. In: Proceedings of the 2021 SIAM International Conference on Data Mining (SDM), pp. 738–746 (2021). SIAM
93. Casey, E.: Digital Evidence and Computer Crime: Forensic Science, Computers, and the Internet, (2011)
94. Walker, C.: Digital evidence and computer crime: Forensic science, computers and the internet. Crime Prevention and Community Safety **3**, 87–88 (2001)
95. Sharevski, F.: Rules of professional responsibility in digital forensics: A comparative analysis. Journal of Digital Forensics, Security and Law **10**(2), 3 (2015)
96. Ademu, I.O., Imafidon, C.O., Preston, D.S.: A new approach of digital forensic model for digital forensic investigation. International Journal of Advanced Computer Science and Applications **2**(12) (2011)
97. Quick, D., Choo, K.-K.R.: Data reduction and data mining framework for digital forensic evidence: storage, intelligence, review and archive. Trends and Issues in Crime and Criminal Justice (480), 1–11 (2014)
98. Quick, D., Choo, K.-K.R.: Impacts of increasing volume of digital forensic data: A survey and future research challenges. Digital Investigation **11**(4), 273–294 (2014)
99. Lillis, D., Becker, B., O'Sullivan, T., Scanlon, M.: Current challenges and future research areas for digital forensic investigation. arXiv preprint arXiv:1604.03850 (2016)
100. Vincze, E.A.: Challenges in digital forensics. Police Practice and Research **17**(2), 183–194 (2016)
101. Rowlingson, R., *et al.*: A ten step process for forensic readiness. International Journal of Digital Evidence **2**(3), 1–28 (2004)
102. Amato, F., Cozzolino, G., Mazzeo, A., Mazzocca, N.: Correlation of digital evidences in forensic investigation through semantic technologies. In: 2017 31st International Conference on Advanced Information Networking and Applications Workshops (WAINA), pp. 668–673 (2017). IEEE
103. Horsman, G.: The importance of digital evidence strategies. Wiley Interdisciplinary Reviews: Forensic Science **6**(1), 1507 (2024)
104. Ferrag, M.A., Alwahedi, F., Battah, A., Cherif, B., Mechri, A., Tihanyi, N.: Generative ai and large language models for cyber security: All insights you need. Available at SSRN 4853709 (2024)
105. Kucharavy, A., Schillaci, Z., Maréchal, L., Würsch, M., Dolamic, L., Sabonnadiere, R., David, D.P., Mermoud, A., Lenders, V.: Fundamentals of generative large language models and perspectives in cyber-defense. arXiv preprint arXiv:2303.12132 (2023)
106. Zhao, W.X., Zhou, K., Li, J., Tang, T., Wang, X., Hou, Y., Min, Y., Zhang, B., Zhang, J., Dong, Z., et al.: A survey of large language models. arXiv preprint arXiv:2303.18223 **1**(2) (2023)
107. Chang, Y., Wang, X., Wang, J., Wu, Y., Yang, L., Zhu, K., Chen, H., Yi, X., Wang, C., Wang, Y., *et al.*: A survey on evaluation of large language models. ACM transactions on intelligent systems and technology **15**(3), 1–45 (2024)
108. Naveed, H., Khan, A.U., Qiu, S., Saqib, M., Anwar, S., Usman, M., Akhtar, N., Barnes, N., Mian, A.: A comprehensive overview of large language models. arXiv preprint arXiv:2307.06435 (2023)
109. Shanahan, M.: Talking about large language models. Communications of the ACM **67**(2), 68–79 (2024)

110. Zhao, H., Chen, H., Yang, F., Liu, N., Deng, H., Cai, H., Wang, S., Yin, D., Du, M.: Explainability for large language models: A survey. ACM Transactions on Intelligent Systems and Technology **15**(2), 1–38 (2024)
111. Chen, M., Tworek, J., Jun, H., Yuan, Q., Pinto, H.P.D.O., Kaplan, J., Edwards, H., Burda, Y., Joseph, N., Brockman, G., et al.: Evaluating large language models trained on code. arXiv preprint arXiv:2107.03374 (2021)
112. Zhou, H., Xu, W., Dehlinger, J., Chakraborty, S., Deng, L.: An llm-driven approach to gain cybercrime insights with evidence networks
113. Chernyshev, M., Baig, Z., Doss, R.R.M.: Towards large language model (llm) forensics using llm-based invocation log analysis. In: Proceedings of the 1st ACM Workshop on Large AI Systems and Models with Privacy and Safety Analysis, pp. 89–96 (2023)
114. Walker, C., Gharaibeh, T., Alsmadi, R., Hall, C., Baggili, I.: Forensic analysis of artifacts from microsoft's multi-agent llm platform autogen. In: Proceedings of the 19th International Conference on Availability, Reliability and Security, pp. 1–9 (2024)
115. Sharma, B., Ghawaly, J., McCleary, K., Webb, A.M., Baggili, I.: Forensicllm: A local large language model for digital forensics. Forensic Science International: Digital Investigation **52**, 301872 (2025)
116. Saxena, N.A., Zhang, W., Shahabi, C.: Missed opportunities in fair ai. In: Proceedings of the 2023 SIAM International Conference on Data Mining (SDM), pp. 961–964 (2023). SIAM

Learning Scene Context for Anomaly Detection: A Benchmarking Study of Forward Frame Prediction CNN and Vision Language Model-Based Approach

Preet Kanwal, Shylaja S S, Niroop Karera, Nischal Kashyap, Poorvi Tambakad, Rishab Gongulur, and Prasad B Honnavalli

1 Introduction

Video anomaly detection (VAD) is a rapidly growing field within computer vision, driven by the increasing demand for real-time surveillance in public safety, industrial monitoring, and autonomous systems [1, 2]. The primary goal of VAD is to identify unusual events within video sequences, such as unexpected human behavior or vehicle movements, which could indicate potentially dangerous situations. Despite significant advances in this field, detecting anomalies remains challenging due to the diverse nature of anomalous events, which may vary significantly across different scenes and contexts [3]. To address these challenges, various machine learning approaches have been explored, including semi-supervised, unsupervised, and weakly supervised methods.

Among these methods, two broad categories have emerged as prominent approaches for VAD: transformer-based models and deep learning frameworks with specific feature extraction and tracking modules. Transformer-based models, particularly those that integrate Vision Transformers (ViTs) [4] and language vision models such as CLIP [5], leverage global attention mechanisms [6] to capture spatial and temporal relationships across video frames. By mapping textual descriptions to corresponding visual features, these models offer a novel approach to weakly supervised anomaly detection, where only minimal labeling is available, and label-

These authors "Niroop Karera, Nischal Kashyap, Poorvi Tambakad and Rishab Gongulur" contributed equally to this work.

P. Kanwal · Shylaja S S · N. Karera (✉) · N. Kashyap (✉) · P. Tambakad (✉) ·
R. Gongulur (✉) · P. B Honnavalli
Department of Computer Science and Engineering, PES University, Bangalore, Karnataka, India
e-mail: preetkanwal@pes.edu; shylaja.sharath@pes.edu; prasadhb@pes.edu

S. S. Iyengar et al. (eds.), *Artificial Intelligence Driven Forensics*, Signals and Communication Technology, https://doi.org/10.1007/978-3-031-98036-7_4

alignment techniques are utilized to link frames to anomaly labels. This study employs a ViT-based approach with CLIP to handle feature extraction and label alignment, enabling the model to detect unusual patterns within video frames based on both visual and semantic information. This architecture is particularly suited to tasks that require robust feature association, as CLIP's integration of text and visual features allows the model to discern anomalies based on textual cues, potentially leading to enhanced detection capabilities across varied scenes.

In contrast, traditional deep learning methods [7] have evolved to address anomaly detection through various means, such as tracking objects, encoding scene context, and generating predictive models for future frames. These approaches often rely on convolutional neural networks (CNNs) [8–10] and autoencoders [11], such as Context Conditional Variational Autoencoders (CVAEs) [12], to capture the underlying structure of normal video sequences and highlight deviations. Our deep learning approach in this study utilizes ByteTrack [13], an advanced multi-object tracking model, integrated with MMTracking [14] for structured data flow and object tracking within each frame sequence. ByteTrack's robust handling of both high- and low-confidence detections helps manage partially occluded or noisy scenes, enhancing the model's ability to track object movements and detect deviations that might indicate abnormal events. Additionally, the inclusion of a CVAE allows for encoding background context, enabling the model to differentiate between contextually normal and abnormal events, a critical capability in scenes where the same behavior might be considered normal in one context but anomalous in another.

The NWPU dataset [12] provides a compelling benchmark for testing these approaches, as it contains videos across various campus settings, representing diverse scenarios and types of anomalies. This dataset, with its detailed annotations and inclusion of scene-dependent anomalies, is well-suited for evaluating models that require nuanced understanding of contextual cues. In this study, we explore how each method—transformer-based and deep learning-based—performs on this dataset. We focus on distinguishing "normal" versus "anomaly" labels, following a weakly supervised approach, where temporal annotations of anomalies provide start and end frames for each anomalous event. Such weak supervision is practical and cost-effective, reducing the labeling burden while still providing valuable information for model training. Our transformer-based approach leverages Vision Transformer (ViT-B/16) architecture, which divides each video frame into 10 distinct segments, allowing the model to capture diverse features across a single frame. These segments are processed by CLIP, which internally associates the visual features with corresponding text-based embeddings, facilitating frame-level anomaly detection. This combination of ViT-B/16 and CLIP enables the model to effectively classify frames as "Normal" or "Anomaly" by identifying discrepancies in feature alignments. Moreover, the use of a tailored train-test split in the dataset, accounting for both the anomaly distribution and scene background, further strengthens the model's capacity to generalize across different environments.

On the other hand, the deep learning approach focuses on predictive modeling techniques. It includes a frame prediction model that utilizes the scene context for forecasting future frames, helping to distinguish normal patterns from anomalous ones. By encoding each frame's background information as contextual features and inputting them into a CVAE, this method enables the model to detect deviations based on scene-specific cues. The frame prediction model predicts future frames by taking sequences of observed frames as input, using U-Net-like skip connections [15] for preserving spatial details, which are crucial for accurate anomaly detection. This approach is particularly advantageous for anomaly detection scenarios that require scene-dependent analysis, as the CVAE aids in distinguishing anomalies based on scene context.

Both approaches use complementary evaluation metrics to assess performance on the NWPU dataset. Frame-level area under the curve (AUC) and mean average precision (mAP) across different intersection over union (IoU) thresholds are utilized to capture the model's ability to accurately identify anomalous frames. The AUC metric serves to measure the overall classification ability of each model on a frame-by-frame basis, whereas mAP at varying IoU thresholds evaluates how well the models can detect anomalies within temporally aligned sequences [16], providing insights into temporal consistency in anomaly detection.

This chapter aims to make three primary contributions. First, it offers a systematic comparison of transformer-based and deep learning-based approaches to video anomaly detection, using a unified framework and evaluation metrics to establish a fair comparison. Second, it provides insights into the strengths and limitations of each approach in different video contexts, with a specific focus on scene-dependent anomalies. Third, it introduces a structured experimental protocol for VAD on the NWPU dataset, including a tailored train-test split that balances scene diversity, making it a valuable reference for future studies in the field.

In summary, this chapter addresses a significant gap in the field of VAD by comparing transformer-based and deep learning-based methodologies on a complex, real-world dataset. Through detailed analysis and experimental evaluation, we demonstrate how each approach contributes to improving anomaly detection performance in campus settings. Our findings will inform the future development of VAD models and highlight the importance of selecting appropriate architectures based on the nature of the anomalies and the context in which they occur.

2 Related Work

2.1 Vision Language Model-Based Approach

The introduction of transformers in computer vision, specifically Vision Transformers (ViTs), has spurred interest in their application to VAD due to their capability to capture long-range dependencies and context across video frames. Vision Transformers have proven effective in learning robust spatial and temporal representations, which are essential for identifying subtle anomalies in complex scenes. The CLIP model, which integrates visual and textual feature extraction, extends the utility of ViTs by aligning visual frames with textual embeddings, providing a unique advantage in weakly supervised VAD. Models utilizing CLIP have shown promise in associating video frames with anomaly labels based on semantic content, even with limited supervision, which is particularly valuable in large-scale datasets where exhaustive labeling is impractical.

More recently, Video Vision Transformer (ViViTs) has been modified to enhance video anomaly detection by capturing richer temporal information and global contexts, making it effective for video prediction. By integrating ViViT with the U-Net architecture, the model can predict future frames while maintaining a strong focus on both local and global features. TransAnomaly [17], a prediction-based method, leverages the strengths of transformers in handling sequence data by predicting future frames and using the discrepancies between predicted and actual frames for anomaly detection. This method calculates regularity scores using sliding windows to evaluate the impact of different window sizes and strides, identifying frames with lower scores as potential anomalies. Moreover, it localizes anomalies by tracking patches with lower regularity scores, significantly improving the detection and localization process.

Another advanced framework, TEVAD [18], integrates both visual and textual modalities for video anomaly detection. Visual features are extracted using pre-trained I3D models [19] or SwinBERT [20] for video captioning, while text features are processed using SimCSE [21], a contrastive learning-based [22] framework for generating sentence embeddings. The multibranch architecture processes visual and textual data in parallel, with a multi-scale temporal network (MTN) [23] capturing both short- and long-range temporal dependencies in the features. These features are fused using methods like concatenation, addition, and Hadamard product, followed by classification tasks for snippet-level anomaly detection that propagate to frame-level predictions. TEVAD's reliance on feature fusion, transformer-based components like SwinBERT, and classification tasks ensures a comprehensive approach to anomaly detection in surveillance videos.

2.2 Deep Learning-Based Approach

Traditional deep learning-based VAD models typically rely on convolutional neural networks (CNNs) and autoencoders to learn representations of normal activities, with the objective of detecting deviations indicative of anomalies. Prediction-based models, such as frame prediction and reconstruction models [24–27], have been widely adopted due to their ability to highlight discrepancies between observed and expected frames, which signal potential anomalies. The use of Context Conditional Variational Autoencoders (CVAEs) in recent studies has been particularly noteworthy, as CVAEs allow the model to incorporate scene-specific context into its predictions, improving the detection of scene-dependent anomalies. Additionally, object detection and tracking models, such as ByteTrack and MMTracking, have been integrated with VAD systems to analyze the movement and behavior of objects, enhancing the detection of anomalies that involve abnormal object interactions or trajectories.

The generative adversarial network [28] has also been proposed as an effective unsupervised method for VAD. Its framework includes a Self-Attentive Predictor, which captures long-term dependencies in video frames to improve prediction quality, and a Vanilla Discriminator [29] that performs true-false discrimination to identify anomalies. Furthermore, a Self-Supervised Discriminator aids in encoding semantic information into predicted normal frames through rotation detection and adversarial training. The model leverages adversarial training to enhance its ability to distinguish between normal and abnormal frames, resulting in greater detection errors for anomalies. These components collectively demonstrate the effectiveness of SSAGAN [29] in improving video anomaly detection performance.

Another innovative approach introduces attribute-based representations, where each object in a video is represented by its velocity and pose. This method improves both accuracy and interpretability in VAD systems. A density-based anomaly scoring mechanism is employed to identify anomalies by analyzing the distribution of attributes in the video data. Moreover, the authors propose combining interpretable attribute-based representations with implicit, deep representations, achieving state-of-the-art performance across multiple datasets. The approach is evaluated using area under the receiver operating characteristic (AUROC) scores.

A novel framework for anomaly detection further addresses the interpretability challenge of existing methods by explaining detected anomalies in surveillance videos. This framework extracts spatiotemporal features to monitor individual objects and their interactions, employing scene graphs to represent object relationships, thereby providing context and uncovering the root causes of anomalies. Additionally, the method supports cross-domain adaptability, leveraging transfer learning across different surveillance environments without requiring extensive labeled data. Theoretical proofs of asymptotic optimality, coupled with empirical validation on benchmark datasets, ensure the reliability and effectiveness of this approach. This combination of methods establishes a more interpretable and adaptable VAD system, setting it apart from traditional techniques.

Another model provides an effective approach for VAD by encoding relationships between objects' states over long sequences, capturing consistent motion patterns. This model represents temporal dependencies by predicting future states based on past observations, enabling a comprehensive understanding of motion patterns over time. Anomalies are detected by measuring the divergence between predicted future states (based on learned normal patterns) and actual observed states, with significant divergences indicating anomalies. Extensive experiments demonstrate the effectiveness of the model, showing improvements over state-of-the-art techniques in video anomaly detection.

An alternative unsupervised approach for detecting human-related anomalies in complex scenes involves the use of skeleton-based analysis [30]. The method decomposes skeleton features into global (whole-body movement) and local (individual joint movement) components to capture detailed motion dynamics. A self-training regression [31] framework iteratively updates anomaly scores, refining pseudo-labeled anomalous and normal skeleton sets over multiple iterations. The core of the approach is an anomaly scoring module that integrates a graph convolutional network (GCN) [23] with fully connected layers to model spatial relationships within the skeleton data. This iterative refinement process enhances detection performance by progressively improving the model's ability to identify deviations from normal patterns, all without requiring labeled normal videos.

A more recent advancement incorporates video-based large language models (VLLMs) [32] for VAD, which operates without the need for complex thresholding, and provides textual explanations for detected anomalies, enhancing the interpretability of the results. A key component of this approach is the Long-Term Context (LTC) module which is used to capture long-range context. The authors propose a three-phase training method, which improves the efficiency of fine-tuning VLLMs by significantly reducing the amount of VAD data required and lowering costs associated with annotating instruction-tuning data. The model has been evaluated on the UCF-Crime [33] and TAD benchmarks, where it achieved notable improvements in area under the curve (AUC) scores.

3 Methodology

This section details the methodology employed in our study, covering the problem definition, architecture, functioning of each approach, and implementation details. The two approaches—vision Language model based with CLIP and a deep learning-based approach utilizing multi-object tracking and Context Conditional Variational Autoencoder (CVAE)—are evaluated for their effectiveness in detecting anomalies in the NWPU dataset, specifically focusing on scene-dependent anomalies under a weakly supervised setting.

3.1 Problem Definition

The primary goal of this research is to identify anomalies in video sequences recorded across various campus settings in the NWPU dataset. Anomalies are defined as events or behaviors that deviate from typical patterns, varying based on context and scene. We formulate this task as a weakly supervised video anomaly detection (VAD) problem, where minimal labeling is available. Specifically, each video is labeled as either "Normal" or "Anomaly," and temporal annotations specify the start and end frames for anomalous events. This weakly supervised approach leverages temporal annotations without requiring frame-by-frame labels, making it cost-effective while still providing essential information for model training.

3.2 *Architecture*

3.2.1 Vision Language Model-Based Architecture

It is built upon Vision Transformers (ViT-B/16) with CLIP for aligning visual and textual embeddings, capturing spatial dependencies within frames.

3.2.2 Deep Learning-Based Architecture

This architecture integrates ByteTrack with MMTracking for object tracking and uses a CVAE for encoding scene-specific context, which aids in differentiating normal and abnormal activities based on the surrounding environment.

3.3 *Methodology Functioning*

The methodology is detailed in sequential stages for each architecture, describing each key component's function and contribution to the anomaly detection process.

3.3.1 Transformer-Based Approach

To prepare the data, video frames are sampled at intervals of 16 frames, ensuring broad coverage of each video sequence. A ten-crop augmentation technique is applied to each frame, creating ten distinct crops: a central crop, four quadrant crops, and their flipped counterparts. This augmentation process allows the model to learn from diverse perspectives, enhancing its ability to detect anomalies by increasing the range of visual information.

The augmented frames are processed by the Vision Transformer (ViT-B/16), which extracts spatial features from each cropped segment. CLIP then associates these visual features with textual embeddings corresponding to the labels "Normal" and "Anomaly." This dual-embedding structure of CLIP enables alignment between the frame's visual characteristics and semantic context, allowing the model to detect discrepancies indicative of anomalies. This approach is particularly suited for weakly supervised settings, as it utilizes CLIP's ability to align frames with their correct anomaly labels (Fig. 1).

The transformer-based model operates through 2 branches for anomaly detection:

Branch 1: A simple binary classifier predicts an anomaly confidence score based on the extracted visual features. This classifier is composed of a feed-forward neural network with a Sigmoid activation function to output a probability score.

Branch 2: This branch leverages CLIP's textual embeddings to compute a similarity-based alignment map, comparing visual features to text embeddings.

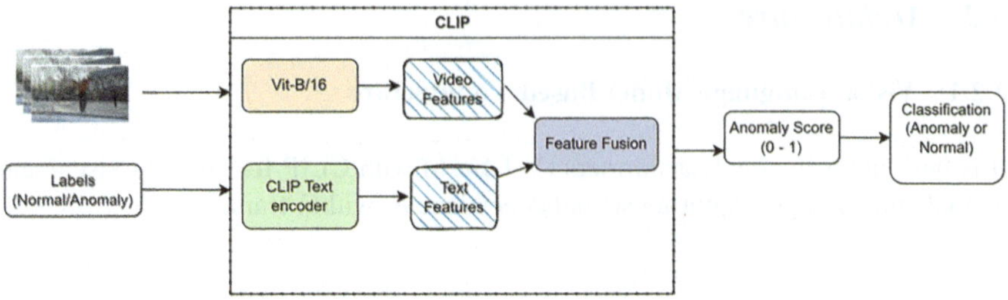

Fig. 1 System architecture diagram for vision language model-based approach

This map identifies frames that align most closely with either "Normal" or "Anomaly," providing fine-grained detection capabilities.

Three loss functions guide the training of the transformer-based model:

Binary cross-entropy loss (Loss 1): Applied to Branch 1's anomaly confidence scores for accurate anomaly classification.

$$L_{bce} = -\frac{1}{N}\sum_{i=1}^{N}\left[y_i \log(\hat{y}_i) + (1 - y_i)\log(1 - \hat{y}_i)\right] \tag{1}$$

MIL-align loss (Loss 2): Encourages alignment between visual and textual embeddings, supporting frame-to-label matching.

The class probability for the ith class is defined as (p_i):

$$p_i = \frac{\exp(s_i/\tau)}{\sum_{j=1}^{m}\exp(s_j/\tau)} \tag{2}$$

The alignment loss is defined as:

$$L_{nce} = -\frac{1}{N}\sum_{i=1}^{N}\log(p_{c_i}) \tag{3}$$

where

τ refers to the temperature hyper-parameter for scaling
s_i represents the similarity between this video and the ith class

Contrastive loss (Loss 3): Ensures distinct separation between normal and anomaly embeddings, enhancing model robustness. The contrastive loss is defined as:

$$L_{cts} = \sum_{j} \max\left(0, \frac{t_n^\top t_{aj}}{\|t_n\|_2 \cdot \|t_{aj}\|_2}\right) \tag{4}$$

where

t_n is the normal class embedding
t_{aj} is the jth abnormal class embedding

The total objective function combines all three losses:

$$L = L_{bce} + L_{nce} + \lambda L_{cts} \tag{5}$$

3.3.2 Deep Learning-Based Approach

The deep learning model incorporates ByteTrack, applied to each frame sequence to detect and track objects across frames. ByteTrack enables accurate tracking by managing high- and low-confidence detections, accommodating challenges such as occlusion and scene noise. This component is essential for extracting object trajectories, which provide valuable insights into movement patterns associated with anomalies, such as abrupt directional changes or velocity variations.

The CVAE module encodes the scene's background context, generating scene-specific feature maps that inform the model about the normalcy of specific activities based on the environment. By embedding scene context, the model can differentiate between contextually normal and abnormal events. For instance, cycling on a pedestrian walkway is treated as anomalous in this context due to the model's understanding of campus norms. This module enhances the model's capacity to detect scene-dependent anomalies by learning scene-specific expectations (Fig. 2).

Following context encoding, the frame prediction model takes observed frames as input and generates future frames, enabling anomaly detection based on discrepancies between observed and predicted frames. This model uses a 3-layer U-Net architecture with skip connections that retain spatial detail, which is essential for accurate anomaly detection.

The feature maps at U2 and U3 levels are fed into the CVAEs to generate new feature maps conditioned on the scene. These new feature maps are added to the

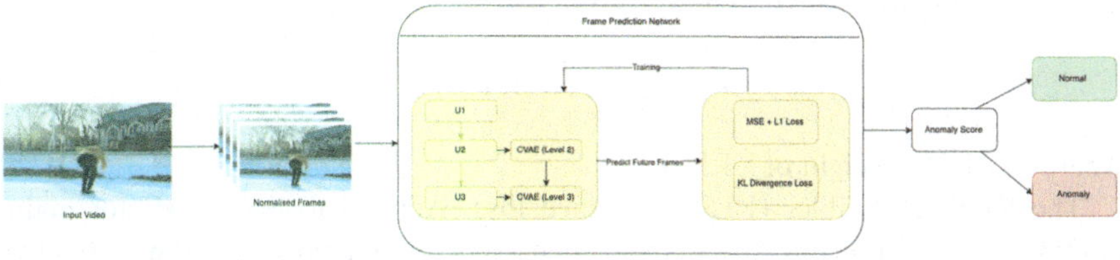

Fig. 2 System architecture diagram for deep learning approach

input of CVAEs with a weight $\gamma = 1$ through which the full contribution of the scene-conditioned feature maps is applied, enabling the model to better detect scene-dependent anomalies by factoring in the background context of each frame. The predicted frames are then generated through subsequent decoding convolutional layers. Any inconsistencies between predicted and actual frames are quantified as prediction errors, with higher error values signaling potential anomalies.

Two primary loss functions are applied in the deep learning-based approach:

Frame prediction loss: Calculated as a combination of mean squared error (MSE) and L1 loss, minimizing discrepancies in frame predictions. L1 loss, or mean absolute error (MAE), is the average of the absolute differences between predicted values and actual values.

The L1 loss is defined as:

$$L_{L1} = \sum_{i=1}^{n} \left| y_i - \hat{y}_i \right| \tag{6}$$

where

n: the total number of data points in the dataset
y_i: the true value corresponding to the ith data point
$\hat{y}_i$: the predicted value corresponding to the ith data point

This has been used within the frame prediction network to compute an error between a predicted frame and the ground truth frame.

Mean squared error (MSE) measures the average squared differences between predicted values and actual values. MSE is defined as:

$$L_{\text{MSE}} = \frac{1}{n} \sum_{i=1}^{n} \left(y_i - \hat{y}_i \right)^2 \tag{7}$$

where

n: the total number of data points in the dataset
y_i: the actual or true value for the ith data point
$\hat{y}_i$: the predicted value for the $i$pth data point

The total frame prediction loss is given as:

$$L_{\text{total}} = \lambda_{L1} L_{L1} + L_{\text{MSE}} \tag{8}$$

where λ_{L1} is the weight of L1 loss.

The Kullback-Leibler (KL) divergence loss: Applied to the CVAE for maintaining a smooth latent space over scene-specific features, ensuring effective background encoding. The KL divergence loss is given by:

$$L_{KL}(\mathcal{N}(\hat{\mu}, \hat{\sigma}^2) \| \mathcal{N}(0, 1)) = -\frac{1}{2}\left(\log \hat{\sigma}^2 - \hat{\mu}^2 - \hat{\sigma}^2 + 1\right) \quad (9)$$

It computes how much the learned distribution $\mathcal{N}(\hat{\mu}, \hat{\sigma}^2)$ deviates from the standard normal distribution $\mathcal{N}(0, 1)$.

The distribution $\mathcal{N}(\hat{\mu}, \hat{\sigma}^2)$ is the learned Gaussian distribution with mean $\hat{\mu}$ and variance $\hat{\sigma}^2$, while $\mathcal{N}(0, 1)$ is the standard normal distribution with mean 0 and variance 1.

The formula includes the following components: $\log \hat{\sigma}^2$, which penalizes the deviation in variance; $\hat{\mu}^2$, which penalizes the deviation in mean; and $\hat{\sigma}^2$, which penalizes the variance magnitude relative to 1. The constant term $+1$ is used to ensure proper normalization of the divergence calculation.

The total loss is a combination of forward prediction error and KL divergence loss with the weight of λ_{KL}. This loss is minimized to train the model.

$$L = L_{\text{total}} + \lambda_{KL} L_{KL} \quad (10)$$

The architecture diagram provides a visual representation of the two approaches, detailing the data flow through preprocessing, feature extraction, context encoding, and anomaly detection stages. The diagram highlights the dual-branch structure in the transformer-based model and the integration of object tracking and CVAE in the deep learning model, helping illustrate each method's data processing path and decision points.

3.3.3 Implementation Details

The models were implemented using PyTorch, with OpenCV handling frame extraction and preprocessing. The transformer-based model relies on the pre-trained CLIP model, fine-tuned on the NWPU dataset with a batch size of 16, using the ViT-B/16 architecture. This model's binary classifier in Branch 1 was optimized with a learning rate of 0.001 on an NVIDIA RTX 4070 GPU. The deep learning model leverages ByteTrack, integrated with MMTracking for seamless object tracking in each frame sequence. The CVAE module, designed to capture background context, was trained with a learning rate of 0.001 using the Adam optimizer on an NVIDIA RTX 4070 GPU. ByteTrack and CVAE were structured to ensure effective handling of scene-dependent anomalies, taking advantage of the NVIDIA GPU's parallel processing capabilities to expedite training and testing phases. These implementations ensure a consistent and reliable comparison between the two architectures, with uniform preprocessing protocols and evaluation metrics to enable replicable results on the NWPU dataset.

4 Experimentation

4.1 Dataset

Several datasets have been established to evaluate video anomaly detection (VAD) models, typically adopting either single-scene or multi-scene setups. Notable examples include UCSD Ped1 [34] and Ped2, which focus on pedestrian walkways, and ShanghaiTech, a multi-scene dataset featuring generalized anomalies. However, these datasets lack a critical component: scene-dependent anomalies—events that may be normal in one context but abnormal in another. To address this gap, the NWPU dataset, used in this study, introduces scene-dependent anomalies across diverse campus scenes, representing a broader range of real-world anomaly types.

NWPU Dataset Overview and Annotations The NWPU dataset comprises 547 videos, with 423 labeled as "Normal" and 124 as "Anomaly," collected from various campus settings. These videos capture a wide range of activities and environmental contexts, enhancing scene diversity and allowing models to tackle challenges unique to scene-dependent anomalies. Each video is annotated using a structured temporal annotation file containing fields such as:

video_name: Unique identifier for each video file.
anomaly_label: Indicates whether the video is "Normal" or "Anomaly."
first_anomaly_start_frame and first-anomaly_end_frame: Start and end frames for the first anomaly, if present.
second_anomaly_start_frame and second_anomaly_end_frame: Start and end frames for a second anomaly, if applicable.
For example:
_001: Anomaly, 20, 130, -1, -1 (One anomaly event between frames 20 and 130).
D_003: Normal, -1, -1, -1, -1 (No anomalies present).
D_007: Anomaly, 230, 1306, 1500, 1580 (Two anomalies between frames 230–1306 and 1500–1580).

This annotation format supports weakly supervised learning, minimizing reliance on exhaustive frame-level labels, which are often impractical in real-world applications due to cost and time constraints.

To ensure effective training across diverse scenes, a new 80–20 train-test split was created, maintaining a proportional representation of normal and anomalous events. The training set includes 437 videos (337 "Normal" and 100 "Anomaly"), while the test set contains 110 videos (86 "Normal" and 24 "Anomaly"). Unlike traditional setups that train solely on normal videos, this tailored split allows models to learn from both normal and anomalous patterns, enhancing generalization and detection accuracy.

While datasets like UCF Crime, XD Violence [35], and ShanghaiTech are widely used, they exhibit limitations in addressing scene-dependent anomalies:

UCF Crime: Covers 13 crime-related events such as robbery and assault but lacks scene diversity and context-dependent anomaly scenarios.

XD Violence: Focuses on high-action anomalies (e.g., explosions, abuse) without considering scene-specific contexts, making it less suitable for subtle context-aware anomaly detection.

ShanghaiTech: Provides generalized anomalies across multi-scenes but lacks the fine-grained scene diversity required for detecting context-specific behaviors.

The NWPU dataset surpasses these limitations by including 43 diverse scenes and 28 classes of anomalies, such as unauthorized gatherings or vehicles in restricted zones. This diversity fosters robust generalization and serves as an ideal benchmark for evaluating VAD models tailored to scene-dependent anomaly detection.

This study leverages the NWPU dataset to evaluate both vision language model-based and deep learning approaches for VAD. By addressing the challenges of scene-dependent anomalies, it bridges the gap in existing benchmarks and provides valuable insights into context-aware anomaly detection. Models utilizing the NWPU dataset benefit from its detailed annotations, diverse scenes, and balanced representation of normal and anomalous events, making it a critical resource for advancing VAD research.

4.2 Evaluation Metrics

To assess the performance of the models, we employ standard evaluation metrics in VAD, specifically mean average precision (mAP) and area under the curve (AUC), with a particular focus on frame-level AUC and mAP across varying intersection over union (IoU) thresholds. These metrics are well-suited for capturing the accuracy of models in detecting anomalies within specific frames and across temporal segments in videos.

Frame-Level AUC The frame-level AUC metric evaluates the model's ability to accurately classify individual frames as either "Normal" or "Anomaly." AUC provides a measure of the true-positive rate against the false-positive rate at various threshold levels. Higher AUC values indicate that the model has a strong capacity for distinguishing between normal and anomalous frames, which is critical for effective VAD performance. This metric is especially useful in scenarios where detecting isolated anomalous frames is essential.

Mean Average Precision (mAP) at IoU Thresholds The mAP metric, evaluated at various IoU thresholds, measures the model's ability to accurately localize and classify anomalies within temporal sequences. For VAD, this metric involves calculating the intersection of the predicted anomalous segments with the ground truth segments, with mAP scores calculated at IoU thresholds from 0.1 to 0.5, increasing in increments of 0.1. The average of these mAP scores across thresholds provides an overall indication of the model's accuracy in identifying anomalous segments within videos.

For instance, an IoU threshold of 0.5 requires that the model's predicted anomaly segment overlaps by at least 50% with the ground truth anomaly segment to be considered a true positive. Lower IoU thresholds, such as 0.1, are less stringent, allowing for a more relaxed overlap criterion. Evaluating mAP at multiple IoU levels provides a comprehensive view of the model's localization accuracy under varying degrees of overlap precision, enabling performance assessments under both strict and lenient criteria.

Mean mAP (AVG) The average mAP (AVG) score, calculated across all IoU thresholds, provides a single summary metric for model performance on the NWPU dataset. This value is particularly useful for comparing the detection effectiveness of different models and gauging their robustness across a range of IoU constraints.

In summary, the NWPU dataset, with its diverse campus scenes and scene-dependent anomalies, offers a more comprehensive testbed for VAD compared to traditional datasets. By using a combination of frame-level AUC and mAP across IoU thresholds, our evaluation framework provides a thorough assessment of model performance in detecting anomalies within complex video sequences, facilitating a nuanced comparison between the vision language model-based and deep learning-based approaches.

5 Results

The quantitative analysis focuses on two main metrics: frame-level AUC and mean average precision (mAP) across different intersection over union (IoU) thresholds. These metrics allow us to assess the accuracy of each model in classifying anomalous frames and correctly localizing anomaly segments within videos.

Frame-Level AUC The frame-level AUC values for both models demonstrate their ability to classify individual frames as either "Normal" or "Anomaly." The learning-based approach achieved an average frame-level AUC of 67.5%, outperforming the video vision language model, which obtained an AUC of 60.05%. This performance discrepancy suggests that the deep learning approach outperforms the vision language model and highlights the potential of scene-aware models in enhancing video anomaly detection, providing better precision in complex scenarios by adapting to specific scene contexts.

Mean Average Precision (mAP) at IoU Thresholds Both models were evaluated using mAP across IoU thresholds ranging from 0.1 to 0.5 (in increments of 0.1), as this metric captures the model's ability to accurately localize anomalous segments over temporal frames (Tables 1 and 2).

Table 1 Comparison of mAP scores across various IoU thresholds for vision language model-based approach

Serial number	1	2	3	4	5
mAP@IoU (%)	0.1	0.2	0.3	0.4	0.5
Vision language model-based	4.31	3.6	3.5	3.36	3.20

Table 2 Patch size versus AUC for deep learning approach

Serial number	1	2	3	4	5
Patch size	256	128	64	32	16
AUC (%)	67.56	65.97	66.49	65.83	65.98

Table 3 Comparison of deep learning and vision language models

Approach	AUC (%)	Comments
Deep learning approach	67.5	No text labels used (unsupervised)
Vision language model	60.05	Text labels used (semisupervised)

6 Conclusion and Future Work

This study presents a comprehensive comparison of vision language model-based and deep learning-based approaches for video anomaly detection on the NWPU dataset, emphasizing their performance in detecting scene-dependent anomalies. The vision language model-based approach, utilizing Vision Transformers (ViTs) with CLIP, demonstrated superior accuracy in nuanced contextual scenarios, while the deep learning-based model excelled in real-time applications focused on movement-based anomalies. By evaluating both models on frame-level AUC and mAP across IoU thresholds, we highlight the strengths and limitations of each approach, ultimately guiding practitioners in selecting the most suitable architecture based on specific deployment needs in complex video surveillance environments.

Future efforts will aim to improve the Vision Transformer-based approach by incorporating advanced temporal modeling techniques and optimizing the embedding alignment process. Self-supervised learning strategies will be explored to reduce reliance on temporal annotations, improving efficiency. Expanding the study to diverse datasets will assess generalizability, while adding fine-grained anomaly classification can extend the framework's application scope. Finally, optimizing computational efficiency will facilitate practical deployment in real-world anomaly detection systems, ensuring both scalability and accuracy (Table 3).

Acknowledgments We would like to express our sincere gratitude to the US Army for their generous funding and unwavering support, which made the AI-Enabled National Conference on Forensic Investigations Network in Digital Sciences(FINDS) possible. Their commitment to advancing research and fostering collaboration is deeply appreciated.

We are profoundly grateful to Dr. S. S. Iyengar for his outstanding contributions to the field, which have been a constant source of inspiration to us. His pioneering work has shaped much of our research, and we are deeply thankful for the impact his research has had on our academic journey. His dedication and innovations continue to inspire us in our pursuit of knowledge.

Appendix 1: Abbreviations

1. NWPU: North Western Polytechnical University
2. VAE: Variational Autoencoder
3. UCSD: University of California, San Diego
4. CUHK: Chinese University of Hong Kong
5. IITB: Indian Institute of Technology Bombay
6. GAN: Generative Adversarial Network
7. VRD: Visual Relationship Detection
8. CNN: Convolutional neural network
9. GCN: Graph convolutional network
10. TEVAD: Text Empowered Video Anomaly Detection
11. ADAM: Adaptive Moment Estimation
12. STU-Net: Spatiotemporal U-Net
13. I3D: Inflated 3D ConvNet
14. SRS: System Requirements Specification

Appendix 2: Terminologies

1. Video Anomaly Detection (VAD): Identification of abnormal events within a video dataset.
2. Area under the curve (AUC): A metric to evaluate the performance of classification models.
3. Conditional Variational Autoencoder (CVAE): A type of neural network used for unsupervised learning of compressed representations while also taking into account some additional information called the condition.
4. Generative Adversarial Network (GAN): A neural network architecture used to generate new data instances in unsupervised learning.
5. Contrastive Language-Image Pre-Training (CLIP): A model that learns visuals from large-scale image-text pairs.
6. Weakly Supervised Video Anomaly Detection (WSVAD): An approach where only weak labels or annotations are provided during training.
7. Intersection over union (IoU): A measure used in object detection tasks to evaluate the accuracy of bounding box predictions.
8. Mean average precision (mAP): A metric to evaluate the accuracy of object detection methods.
9. Spatiotemporal U-Net (STU-Net): A neural network architecture used for spatiotemporal data processing.
10. Inflated 3D ConvNet (I3D): A convolutional neural network architecture used for video classification tasks.

References

1. Cliff Wang, S.S. Iyengar, Kun Sun "AI Embedded Assurance for Cyber Systems", Springer Nature - ISBN 978-30-31426-36-0
2. Sankita J. Patel, Naveen Kumar Chaudhary, Bhavesh N. Gohil, S. S. Iyengar "Information Security, Privacy and Digital Forensics – Select Proceedings of the International Conference, ICISPD 2022", Springer Nature - 978-98-19950-90-4
3. S. S. Iyengar and Kianoosh G. Boroojeni "Oblivious Network Routing: Algorithms and Applications", ISBN: 978-02-62029-15-5 | 176 pp. | 6 x 9 in, Published by MIT Press
4. A. Dosovitskiy et al., "An Image is Worth 16x16 Words: Transformers for Image Recognition at Scale," International Conference on Learning Representations, May 2021, [Online]. Available: https://openreview.net/pdf?id=YicbFdNTTy.
5. Luo, H.; Ji, L.; Zhong, M.; Chen, Y.; Lei, W.; Duan, N.; and Li, T. 2022. CLIP4Clip: An empirical study of CLIP for end to end video clip retrieval and captioning. Neurocomputing, 508: 293–304.
6. K. Doshi and Y. Yilmaz, "Towards interpretable video anomaly detection," 2022 IEEE/CVF Winter Conference on Applications of Computer Vision (WACV), pp. 2654–2663, Jan. 2023, https://doi.org/10.1109/wacv56688.2023.00268.
7. S. Zhang et al., "Influence-Aware attention networks for anomaly detection in surveillance videos," IEEE Transactions on Circuits and Systems for Video Technology, vol. 32, no. 8, pp. 5427–5437, Feb. 2022, https://doi.org/10.1109/tcsvt.2022.3148392.
8. K. O'Shea and R. Nash, "An Introduction to Convolutional Neural Networks," IEEE International Conference, Nov. 2015.
9. Bin Shi, S. S. Iyengar "Mathematical Theories of Machine Learning – Theory and Applications", Springer - ISBN 978-30-30170-76-9
10. R.R. Brooks and S.S. Iyengar "Multi Sensor Fusion: Fundamentals and Applications with Software", Prentice Hall Publication Co., New Jersey 07458. Number of Pages : 488
11. Y. Zhao, B. Deng, C. Shen, Y. Liu, H. Lu, and X.-S. Hua, "Spatio-Temporal AutoEncoder for video anomaly detection," Proceedings of the 30th ACM International Conference on Multimedia, pp. 1933–1941, Oct. 2017, https://doi.org/10.1145/3123266.3123451.
12. C. Cao, Y. Lu, P. Wang, and Y. Zhang, "A new comprehensive benchmark for semi-supervised video anomaly detection and anticipation," 2022 IEEE/CVF Conference on Computer Vision and Pattern Recognition (CVPR), pp. 20392–20401, Jun. 2023, https://doi.org/10.1109/cvpr52729.2023.01953.
13. Y. Zhang et al., "ByteTrack: multi-object tracking by associating every detection box," in Lecture notes in computer science, 2022, pp. 1–21.
14. MMTracking Contributors. MMTracking: OpenMMLab video perception toolbox and benchmark. https://github.com/open-mmlab/mmtracking, 2020.
15. T. N. Nguyen and J. Meunier, "Anomaly detection in video sequence with Appearance-Motion correspondence," 2021 IEEE/CVF International Conference on Computer Vision (ICCV), pp. 1273–1283, Oct. 2019, https://doi.org/10.1109/iccv.2019.00136.
16. Y. Liu, Z. Guo, J. Liu, C. Li, and L. Song, "OSIN: Object-Centric Scene Inference Network for Unsupervised Video Anomaly Detection," IEEE Signal Processing Letters, vol. 30, pp. 359–363, Jan. 2023, https://doi.org/10.1109/lsp.2023.3263792.
17. H. Yuan, Z. Cai, H. Zhou, Y. Wang, and X. Chen, "TransAnomaly: Video anomaly detection using video Vision transformer," IEEE Access, vol. 9, pp. 123977–123986, Jan. 2021, https://doi.org/10.1109/access.2021.3109102.
18. W. Chen, K. Teck MA, Z. J. Yew, M. Hur, and D. A.-A. Khoo, "TEVAD: Improved video anomaly detection with captions," 2022 IEEE/CVF Conference on Computer Vision and Pattern Recognition Workshops (CVPRW), pp. 5549–5559, Jun. 2023, https://doi.org/10.1109/cvprw59228.2023.00587.

19. C. Cao, Y. Lu, and Y. Zhang, "Context Recovery and Knowledge Retrieval: A Novel Two-Stream Framework for video Anomaly Detection," IEEE Transactions on Image Processing, vol. 33, pp. 1810–1825, Jan. 2024, https://doi.org/10.1109/tip.2024.3372466.
20. K. Lin et al., "SwinBERT: End-to-End Transformers with Sparse Attention for Video Captioning," 2022 IEEE/CVF Conference on Computer Vision and Pattern Recognition (CVPR), Jun. 2022, https://doi.org/10.1109/cvpr52688.2022.01742.
21. T. Gao, X. Yao, and D. Chen, "SIMCSE: Simple Contrastive Learning of Sentence Embeddings," Proceedings of the 2021 Conference on Empirical Methods in Natural Language Processing, Jan. 2021, https://doi.org/10.18653/v1/2021.emnlp-main.552.
22. S. Sun and X. Gong, "Hierarchical semantic contrast for scene-aware video anomaly detection," 2022 IEEE/CVF Conference on Computer Vision and Pattern Recognition (CVPR), pp. 22846–22856, Jun. 2023, https://doi.org/10.1109/cvpr52729.2023.02188.
23. Z. Chen, S. Li, B. Yang, Q. Li, and H. Liu, "Multi-Scale Spatial Temporal Graph Convolutional Network for Skeleton-Based Action Recognition," Proceedings of the AAAI Conference on Artificial Intelligence, vol. 35, no. 2, pp. 1113–1122, May 2021, https://doi.org/10.1609/aaai.v35i2.16197.
24. P. Wu, J. Liu, and F. Shen, "A deep One-Class neural network for anomalous event detection in complex scenes," IEEE Transactions on Neural Networks and Learning Systems, pp. 1–14, Jan. 2019, https://doi.org/10.1109/tnnls.2019.2933554.
25. R. Yang, Q. Li, Y. Shen, and Z. Zhang, "Anomaly detection in surveillance videos via memory-augmented frame prediction," 2022 International Joint Conference on Neural Networks (IJCNN), pp. 1–8, Jul. 2022, https://doi.org/10.1109/ijcnn55064.2022.9892924.
26. X. Wang et al., "Robust unsupervised video anomaly detection by multipath frame prediction," IEEE Transactions on Neural Networks and Learning Systems, vol. 33, no. 6, pp. 2301–2312, Jun. 2021, https://doi.org/10.1109/tnnls.2021.3083152.
27. C. Sun, Y. Jia, Y. Hu, and Y. Wu, "Scene-Aware context Reasoning for unsupervised abnormal event detection in videos," Proceedings of the 30th ACM International Conference on Multimedia, pp. 184–192, Oct. 2020, https://doi.org/10.1145/3394171.3413887.
28. S.S. Iyengar, Sina Nabavi, Yashas Hariprasad, Prasad HB, Krishna Mohan C. "Artificial Intelligence in Practice: Theory and Applications for Cyber Security and Forensics", Springer Nature (Forthcoming)
29. C. Huang et al., "Self-Supervised Attentive Generative adversarial networks for video anomaly detection," IEEE Transactions on Neural Networks and Learning Systems, vol. 34, no. 11, pp. 9389–9403, Apr. 2022, https://doi.org/10.1109/tnnls.2022.3159538.
30. R. Rodrigues, N. Bhargava, R. Velmurugan, and S. Chaudhuri, "Multi-timescale Trajectory Prediction for Abnormal Human Activity Detection," IEEE/CVF Winter Conference on Applications of Computer Vision, Pages 2615–2623, 2020. 2, 3, 7, pp. 2615–2623, Mar. 2020, https://doi.org/10.1109/wacv45572.2020.9093633.
31. G. Pang, C. Yan, C. Shen, A. Van Den Hengel, and X. Bai, "Self-Trained Deep Ordinal regression for End-to-End video anomaly detection," 2022 IEEE/CVF Conference on Computer Vision and Pattern Recognition (CVPR), pp. 12170–12179, Jun. 2020, https://doi.org/10.1109/cvpr42600.2020.01219.
32. Y. Zhao, I. Misra, P. Kräenbühl, and R. Girdhar, "Learning Video Representations from Large Language Models," 2022 IEEE/CVF Conference on Computer Vision and Pattern Recognition (CVPR), Jun. 2023, https://doi.org/10.1109/cvpr52729.2023.00637.
33. W. Sultani, C. Chen, and M. Shah, "Real-World Anomaly Detection in Surveillance Videos," Proceedings of the IEEE Conference on Computer Vision and Pattern Recog- Nition, Pages 6479–6488, 2018, Jun. 2018, https://doi.org/10.1109/cvpr.2018.00678.
34. V. Mahadevan, W. Li, V. Bhalodia, and N. Vasconcelos, "Anomaly detection in crowded scenes," CVPR, Pages 1975–1981, 2010, Jun. 2010, https://doi.org/10.1109/cvpr.2010.5539872.
35. P. Wu et al., "Not only Look, But Also Listen: Learning Multimodal Violence Detection Under Weak Supervision," in Lecture notes in computer science, 2020, pp. 322–339.

AI-Driven Multimedia Forensics: Enhancing Detection, Provenance Analysis, and Robustness Against Manipulations

Tessy Tom, Yashas Hariprasad, Pronab Mohanty, and Antony Puthussery

1 Introduction

The proliferation of digital media has introduced unprecedented challenges in verifying the authenticity and integrity of multimedia content [8]. With the rise of sophisticated digital manipulation techniques, including deepfake technology, synthetic media generation, and content forgery, ensuring trust in digital information has become a pressing issue [16]. Multimedia forensics has emerged as a critical field dedicated to detecting, analyzing, and mitigating such manipulations, with applications spanning law enforcement, digital rights management, journalism, and cybersecurity [18–20].

Artificial intelligence (AI) and machine learning (ML) have revolutionized multimedia forensics by enabling advanced detection methods capable of identifying even subtle alterations in digital content. Deep neural networks (DNNs), convolutional neural networks (CNNs), and transformer-based architectures have significantly improved the accuracy and efficiency of forensic analysis [22–25]. These technologies allow for the detection of pixel-level inconsistencies, temporal artifacts, and adversarial perturbations that are often imperceptible to the human eye.

T. Tom (✉)
CSRC, Jain (Deemed to be University), Bangalore, India
e-mail: tessy.tom@jainuniversity.ac.in

Y. Hariprasad
California State University-EastBay, Hayward, California, USA
e-mail: yashas.hariprasad@csueastbay.edu

P. Mohanty
Police Computer Wing, Bangalore, Karnataka, India

A. Puthussery
Department of Computer Science, Christ College, Pune, India
e-mail: fr.antony@christcollegepune.org

S. S. Iyengar et al. (eds.), *Artificial Intelligence Driven Forensics*, Signals and Communication Technology, https://doi.org/10.1007/978-3-031-98036-7_5

However, existing approaches still face key challenges, including generalizability in different manipulation techniques, robustness against adversarial attacks, and the need for greater interpretability in forensic decision-making.

This research introduces an AI-driven forensic framework that integrates deep learning, adversarial learning, and graph-based methodologies to enhance multimedia forgery detection. Our approach leverages CNNs and transformers for high-precision anomaly detection while incorporating an interpretability module to ensure transparency in forensic conclusions. The methodology is validated using widely recognized benchmark datasets, including FaceForensics ++, DFDC, and real-world anomaly detection datasets. Preliminary experimental results indicate that our framework outperforms existing state-of-the-art techniques, achieving high detection accuracy and robustness against various forgery techniques.

By addressing current limitations and pushing the boundaries of AI-powered multimedia forensics, this work contributes to the development of more reliable and scalable forensic solutions. Furthermore, we explore future directions, including the integration of quantum computing paradigms and blockchain-based traceability mechanisms, to enhance the resilience of forensic methodologies in combating digital fraud and misinformation.

2 Literature Review

Multimedia forensics has rapidly evolved as a crucial field in response to the increasing manipulation of digital content. Early forensic techniques were mainly based on statistical analysis of metadata and noise inconsistencies in multimedia files [6, 17]. However, these traditional methods lacked robustness against sophisticated manipulation techniques, such as deepfakes and GAN-generated content, which have become more prevalent with advancements in artificial intelligence (AI) and machine learning (ML) [1, 2].

Recent breakthroughs in AI-driven multimedia forensics have significantly improved detection capabilities. Convolutional neural networks (CNNs) have been widely adopted for image forgery detection, leveraging their ability to capture spatial features at the pixel level [7, 9]. Architectures such as ResNet and EfficientNet have demonstrated high precision in identifying manipulations such as copy-move forgery, splicing, and retouching [9]. Despite these advancements, CNN-based methods face limitations, particularly the reliance on large labeled datasets for effective training.

The emergence of generative adversarial networks (GANs) has introduced new challenges to forensic experts. While GANs serve as powerful generative models, they have also been exploited to create highly realistic forgeries. To counteract this, researchers have developed adversarial training techniques to detect GAN-generated content. Zhou et al. [14] proposed noise residual analysis as an effective method to identify deepfake manipulations, achieving competitive results on benchmark datasets.

Temporal forensics in video analysis has also seen notable advancements with the introduction of transformer-based architectures. Models such as Vision Transformers (ViTs) and TimeSformer have demonstrated superior performance in detecting temporal inconsistencies within manipulated videos [7]. The integration of spatial and temporal features has further enhanced the robustness of deepfake detection techniques [2, 3, 15].

Graph-based approaches have recently gained traction in multimedia provenance analysis, offering a structured way to model relationships between different media elements. Bharati et al. [11] introduced a graph-based framework for analyzing the provenance of manipulated content, showcasing its effectiveness in real-world forensic applications.

Despite these advancements, several research challenges remain. Existing models often struggle with interpretability and generalizability when applied to novel manipulation techniques. Additionally, adversarial attacks pose a significant threat to the reliability of forensic models. Goodfellow et al. [12] and Tsipras et al. [15] highlighted the susceptibility of ML models to adversarial perturbations, underscoring the need for robust defense mechanisms in multimedia forensics.

Beyond technical challenges, ethical and privacy concerns surrounding forensic techniques remain underexplored. The application of forensic tools in sensitive domains, such as law enforcement, necessitates careful consideration of biases and potential false positives. Chesney and Citron [13] emphasized the importance of ethical AI practices in forensic investigations, advocating for transparency and accountability in forensic decision-making.

In summary, while substantial progress has been made in multimedia forensics, challenges related to robustness, interpretability, and ethical considerations persist. This research aims to address these gaps by introducing an integrated forensic framework that combines CNNs, transformer-based architectures, and graph-based methods to enhance the reliability, transparency, and resilience of forensic techniques.

Research Gap

The rapid evolution of multimedia manipulation technologies, particularly AI-driven techniques such as generative adversarial networks (GANs), has introduced significant challenges to the field of multimedia forensics. Despite notable advancements in detection methodologies, several critical gaps remain:

Limited GeneralizabilityExisting forensic models, including CNN-based [7] and transformer-based approaches [1, 9], achieve high accuracy on specific datasets but struggle to generalize across diverse datasets and manipulation techniques [3, 4]. This limitation hinders their effectiveness in real-world scenarios with unseen forgeries.

Vulnerability to Adversarial Attacks Current forensic detection models are susceptible to adversarial perturbations, allowing attackers to evade detection by introducing imperceptible modifications to manipulated content [3]. Enhancing the robustness of forensic techniques against such attacks remains a major challenge [12, 15].

Lack of Interpretability Many AI-based forensic models function as "black boxes," making it difficult to understand and validate their decision-making processes. The absence of interpretability reduces trust in these systems, particularly in high-stakes applications such as law enforcement and judicial proceedings [4, 11].

Challenges in Provenance and Traceability While graph-based approaches have shown promise in multimedia provenance analysis [3, 11], their computational complexity limits scalability for large-scale real-world applications [5]. Developing efficient and scalable traceability frameworks remains an open research problem.

Ethical and Privacy Concerns The deployment of multimedia forensic techniques raises critical ethical issues, including privacy risks, potential biases in AI algorithms, and the consequences of false positives or false negatives. Addressing these concerns is essential to ensure responsible and fair application of forensic tools [11, 13].

This research aims to bridge these gaps by developing an integrated forensic framework that enhances generalizability, robustness against adversarial attacks, interpretability, and scalability while addressing ethical implications in multimedia forensics.

Relevance of the Research

Addressing these research gaps is of paramount importance, as multimedia forensics plays a critical role in various domains, including cybersecurity, digital rights management, law enforcement, and the fight against misinformation. As multimedia manipulation techniques grow increasingly sophisticated, the demand for advanced forensic tools that are robust, interpretable, and scalable has never been greater.

Real-World Impact Developing resilient and generalizable forensic models is essential for their effective deployment across diverse applications, such as detecting deepfakes in news media, authenticating digital evidence in criminal investigations, and safeguarding intellectual property rights.

Building Trust and Transparency The integration of interpretability modules into forensic models enhances trust among key stakeholders, including law enforcement agencies, legal authorities, and the general public. Transparent and explainable decision-making processes are crucial to ensuring widespread acceptance of AI-driven forensic tools [7, 13].

Defending Against Adversarial Attacks Strengthening forensic techniques against adversarial attacks is vital to maintaining their reliability in the face of increasingly sophisticated evasion strategies [12, 15].

Cybersecurity and Provenance Analysis Advanced provenance analysis methods are instrumental in tracing the origins of manipulated content, curbing the spread of misinformation, and reinforcing accountability in digital ecosystems [3, 11].

Ethical and Responsible AI Development Embedding ethical considerations into the design of forensic tools helps mitigate biases, prevent misuse, and promote the responsible deployment of these technologies in sensitive applications [11, 13].

By addressing these critical challenges, this research contributes to the advancement of multimedia forensics, enhancing its effectiveness and reliability in real-world scenarios.

Proposed Contribution

While significant progress has been made in multimedia forensics, challenges related to robustness, interpretability, and ethical considerations remain unresolved. This research aims to address these gaps by proposing an integrated framework that leverages CNNs, transformers, and graph-based methods to enhance the reliability and transparency of forensic techniques. Future research should focus on developing robust adversarial forensic models, improving the interpretability of AI-based forensic tools, and ensuring ethical AI practices in multimedia forensics. The framework aims to enhance generalizability, robustness, and interpretability while embedding ethical considerations into the forensic pipeline. By leveraging publicly available datasets and state-of-the-art models, this study seeks to advance the frontier of AI and ML in multimedia forensics, ensuring its relevance and impact in real-world applications.

3 Methodology

To address the identified research gaps in multimedia forensics, we propose a comprehensive framework that leverages state-of-the-art AI and ML techniques. The methodology integrates convolutional neural networks (CNNs), transformers, and graph-based analysis to enhance the detection, classification, and provenance tracking of manipulated multimedia content. The steps of the methodology are outlined below.

3.1 Data Collection and Preprocessing

The framework utilizes publicly available datasets such as FaceForensics++, Deepfake Detection Challenge (DFDC), and real-world anomaly detection datasets. These datasets include a variety of manipulated multimedia content, such as deepfake videos, image splicing, and audio forgery.

Preprocessing Techniques:

- Normalization: To standardize pixel intensity values across images and videos.
- Data augmentation: Techniques like flipping, cropping, and color adjustments are employed to improve model generalization.
- Feature extraction: Pre-trained feature extractors like ResNet50 and EfficientNet are fine-tuned to generate robust features from input media.

3.2 Model Architecture

The proposed framework consists of three modules:

1. Detection module
    - A CNN-based architecture is used to detect pixel-level inconsistencies. Custom layers are added to detect unique patterns indicative of tampering, such as blending artifacts or compression mismatches.
    - For temporal manipulations in videos, a transformer-based model (TimeSformer) is employed to capture both spatial and temporal features [7, 10].
2. Provenance analysis module
    - A graph-based framework is designed to analyze the relationships between media elements. This module identifies the origins and manipulation paths using graph convolutional networks (GCNs) [3, 11].
    - Metadata and hash-based comparisons are incorporated to enhance accuracy [11, 17].
3. Interpretability module
    - A layer is added to visualize heatmaps that highlight manipulated regions in images or frames. Techniques such as gradient-weighted class activation mapping (Grad-CAM) ensure interpretability in decision-making [7].

3.3 Training and Optimization

Training Process:

- The model is trained using a combination of supervised and semi-supervised learning. For labeled data, cross-entropy loss is used, while for unlabeled data, contrastive learning is applied to maximize feature similarity within classes [7].

Optimization Techniques:

- Adaptive optimizers such as AdamW are used for faster convergence [21].
- Dropout and batch normalization layers are added to prevent overfitting.

3.4 *Adversarial Robustness*

To defend against adversarial attacks:

- **Adversarial Training:** Synthetic adversarial examples are generated using techniques like fast gradient sign method (FGSM) and projected gradient descent (PGD) [12, 15].
- **Defense Mechanisms:** Noise injection and input transformations are used to improve robustness.

3.5 *Validation and Testing*

- **Metrics:** The performance of the framework is evaluated using accuracy, precision, recall, F1-score, and area under the receiver operating characteristic (ROC) curve [7, 10].
- **Baseline Comparison:** Results are compared with state-of-the-art methods, including EfficientNet [9], Deepfake detection models [10, 14], and graph-based provenance analysis frameworks.

3.6 *Deployment and Real-World Application*

The final framework is integrated into a prototype system for real-time multimedia forensic applications. This system includes a user-friendly interface for forensic analysts [11], enabling them to upload media files and view detailed reports of detected manipulations [11].

4 Mathematical Formulation

This section details the mathematical framework underlying the proposed methodology, which integrates CNNs, transformers, and graph-based analysis for multimedia forensics [7, 10, 11].

4.1 *Problem Definition*

Let M represent the multimedia dataset containing n samples, where

$$M = \{m_1, m_2, \ldots, m_n\}.$$

Each sample m_i can belong to one of two classes: genuine ($y = 0$) or manipulated ($y = 1$).

The objective is to:

1. **Detect manipulated content:** $f_{\text{detect}} : M \to \{0, 1\}$.
2. **Localize tampered regions:** $f_{\text{localize}} : m_i \to R$ where R is a binary mask indicating manipulated regions.
3. **Trace content provenance:** $f_{\text{provenance}} : m_i \to G$ where G represents the graph structure of manipulations or lineage.

4.2 Detection Module (CNN-Based)

Feature Extraction:
For an image I of dimensions $H \times W \times C$, where C is the number of colour channels:

$$F = f_{\text{CNN}}(I; \theta) \quad \text{where } F \in \mathbb{R}^{H \times W \times D}$$

Here, F is the feature map, θ represents CNN parameters, and D is the depth of the feature map.

Binary Classification:

$$p(y \mid I) = \sigma(W \cdot F + b)$$

where W and b are learnable weights and biases, and σ is the sigmoid activation function.

Loss Function:

$$\mathcal{L}_{\text{detect}} = \frac{1}{n} \sum_{i=1}^{n} \left[y_i \log(p_i) + (1 - y_i) \log(1 - p_i) \right]$$

4.3 Temporal Analysis (Transformer-Based)

For a video V, represented as a sequence of frames,

$$V = \{F_1, F_2, F_3, \ldots, F_T\}$$

where F_t represents the frame at time step t, and T is the total number of frames.

Each frame F_t is represented as a feature vector $\boldsymbol{x}_t$ after applying a feature extractor (e.g., CNN, ViT, or a 3D convolutional model):

$$\boldsymbol{x}_t = f_{\text{extractor}}(F_t) \in \mathbb{R}^d$$

Transformer-Based Temporal Processing:
Tokenization and Embedding: The sequence of extracted frame features is converted into an input sequence:

$$\text{Input representation:} \quad \mathbf{X} = [\boldsymbol{x}_1; \boldsymbol{x}_2; \ldots; \boldsymbol{x}_T] \in \mathbb{R}^{T \times d} \quad \text{where } \boldsymbol{x}_t = \text{Flatten}(F_t)$$

Positional encoding is added to retain temporal order:

$$\mathbf{Z} = \mathbf{X} + \mathbf{P}, \quad \mathbf{P} \in \mathbb{R}^{T \times d}$$

Attention Mechanism:
Self-Attention Mechanism (Multi-Head Self-Attention [MHSA]): Each token $\boldsymbol{x}_t$ attends to other tokens using self-attention:

$$Q = \mathbf{Z}\mathbf{W}_Q, \quad K = \mathbf{Z}\mathbf{W}_K, \quad V = \mathbf{Z}\mathbf{W}_V$$

where $\mathbf{W}_Q$, $\mathbf{W}_K$, and $\mathbf{W}_V$ are learnable projection matrices.

The attention scores are computed as:

$$A = \text{softmax}\left(\frac{QK^T}{\sqrt{d_k}}\right)V$$

where Q, K, and V are the query, key, and value matrices derived from $\mathbf{X}$, and d_k is the dimensionality of the keys.

Output:

$$z_t = f_{\text{transformer}}(\boldsymbol{x}_t; \boldsymbol{\theta})$$

Loss Function:

$$\mathcal{L}_{\text{temporal}} = \mathcal{L}_{\text{detect}} + \lambda \cdot \mathcal{L}_{\text{temporal-consistency}}$$

Total Loss:

$$\mathcal{L}_{\text{total}} = \mathcal{L}_{\text{detect}} + \alpha \cdot \mathcal{L}_{\text{temporal}} + \beta \cdot \mathcal{L}_{\text{provenance}} + \gamma \cdot \mathcal{L}_{\text{adv}}$$

where $f(m_i)$ is the feature vector of media element m_i.

4.4 Provenance Analysis (Graph-Based)

Given a set of media elements $\{m_1, m_2, \ldots, m_k\}$, a graph $G = (V, E)$ is constructed, where:

- V: nodes representing individual media elements
- E: edges representing relationships between elements

Edge Weight: $w_{ij} = \exp\left(-\|f(m_i) - f(m_j)\|_2^2\right)$

Graph Convolution: $\mathbf{H}^{(l+1)} = \sigma\left(\mathbf{D}^{-1}\mathbf{A}\mathbf{H}^{(l)}\mathbf{W}^{(l)}\right)$

where: $\mathbf{A}$: adjacency matrix

$\mathbf{D}$: degree matrix

$\mathbf{H}^{(l)}$: node features at layer l

$\mathbf{W}^{(1)}$: trainable weights

4.5 Adversarial Defense

Adversarial Example Generation:

Adversarial examples are generated using the fast gradient sign method (FGSM):

$$\boldsymbol{x}_{\text{adv}} = \boldsymbol{x} + \epsilon \cdot \text{sign}\left(\nabla_{\boldsymbol{x}} \mathcal{L}(\boldsymbol{x}, y)\right)$$

where:

$\boldsymbol{x}$: original input

$\boldsymbol{x}_{\text{adv}}$: adversarial example

ϵ: perturbation magnitude

$\mathcal{L}(\boldsymbol{x}, y)$: loss function with respect to input $\boldsymbol{x}$ and label y

$\nabla_{\boldsymbol{x}}$: gradient with respect to input

Defense Mechanism:

1. **Adversarial Training:** $\mathcal{L}_{\text{adv}} = \mathcal{L}_{\text{detect}} + \beta \cdot \mathcal{L}(\boldsymbol{x}_{\text{adv}}, y)$
2. **Input Transformations:** Gaussian noise $\mathcal{N}(0, \sigma^2)$ is added to inputs during training— $\boldsymbol{x}' = \boldsymbol{x} + \mathcal{N}(0, \sigma^2)$

4.6 Model Optimization

The overall loss function integrates multiple objectives:

$$\mathcal{L}_{\text{total}} = \mathcal{L}_{\text{detect}} + \alpha \cdot \mathcal{L}_{\text{temporal}} + \beta \cdot \mathcal{L}_{\text{provenance}} + \gamma \cdot \mathcal{L}_{\text{adv}}$$

where α, β, and γ are hyperparameters.

5 Implementation Details

The proposed framework was implemented in Python using popular machine learning libraries such as TensorFlow and PyTorch. The experiment setup and implementation details are outlined below.

1. **Hardware and software configuration:**
   - **Processor:** Intel Core i9-11900K
   - **GPU:** NVIDIA RTX 3080 (10 GB VRAM)
   - **RAM:** 64 GB
   - **Software:** Ubuntu 22.04, Python
2. **Datasets:**
   - **FaceForensics++:** Includes manipulated and original videos with pixel-level ground truth masks.
   - **Deepfake Detection Challenge (DFDC):** Provides labeled deepfake and original videos for training and evaluation.
   - **Real-world anomaly detection dataset:** Contains a variety of manipulated multimedia content across different categories.
3. **Training pipeline:**
   - **Dataset split:** 70% training, 15% validation, 15% testing
   - **Batch size:** 32
   - **Learning rate:** Adaptive, starting at 0.001 using AdamW optimizer
   - **Epochs:** 50 with early stopping based on validation loss
4. **Evaluation metrics:**
   - **Accuracy, precision, recall, and F1-score:** For classification
   - **Intersection over union (IoU):** For manipulated region localization
   - **Graph matching accuracy:** For provenance analysis

6 Results

The model demonstrated strong performance throughout the training process, as illustrated in Fig. 1, which presents the training and validation accuracy and loss over epochs. The experiments utilized three benchmark datasets—FaceForensics++, comprising manipulated and original videos with pixel-level ground truth masks; the Deepfake Detection Challenge (DFDC) dataset, containing diverse labeled deepfake content; and the Real-World Anomaly Detection (RWAD) dataset, featuring a variety of multimedia manipulations across categories. The data was split into 70% for training, 15% for validation and 15% for testing. Training was conducted

Fig. 1 Model accuracy and loss over 50 epochs. Training accuracy reached 96.3%, while validation accuracy stabilized at 94.6%. Validation loss converged to 0.034

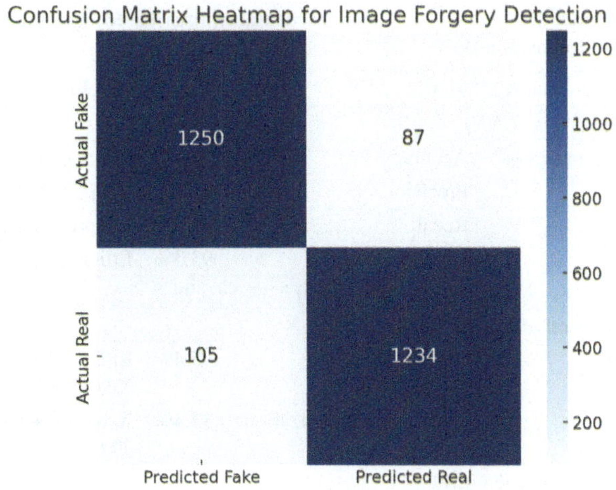

Fig. 2 Confusion matrix for classification task. TP = 1250, FP = 87, TN = 1234, FN = 105

over 50 epochs with early stopping based on validation loss, using an adaptive learning rate (starting at 0.001) and the AdamW optimizer, with a batch size of 32. Training accuracy reached 96.3%, while validation accuracy stabilized at 94.6%, suggesting strong generalization capability. Correspondingly, the validation loss converged to 0.034. Evaluation metrics included accuracy, precision, recall, F1-score for classification, intersection over union (IoU) for manipulated region localization, and graph matching accuracy for provenance analysis.

Further performance insight was obtained through the confusion matrix (Fig. 2), which recorded 1,250 true positives, 1,234 true negatives, 87 false positives, and 105 false negatives. From these values, the overall accuracy was calculated to be 92.81%, with a precision of 93.5%, recall of 92.25%, and F1-score of 92.88%. These results confirm the model's robustness in reliably detecting manipulated content across diverse deepfake datasets. Again, temporal consistency analysis revealed that the model achieved an accuracy of 95.4% in deepfake video detection, significantly outperforming baseline architectures such as EfficientNet, particularly in handling temporally incoherent manipulations across frames. This suggests that the proposed framework not only detects spatial artifacts but also effectively captures temporal anomalies, which are critical in video-based forensics.

The robustness of the model was further evaluated under adversarial conditions. Incorporating FGSM (fast gradient sign method) and PGD (projected gradient descent) during adversarial training resulted in a 13% improvement in model resilience, confirming the defense capability of the architecture against perturbation-based attacks.

The proposed deepfake detection framework exhibited strong and consistent performance across various evaluation dimensions. In terms of detection effectiveness, the model achieved an average precision of 94.6%, a recall of 92.1%, and an F1-

Table 1 Performance comparison of CNN-based, transformer-based, and graph-based models across multimedia forensics tasks

Module	Dataset/scenario	Metric	Value	Baseline comparison
Image forgery detection (CNN)				
	FaceForensics++[a]	Accuracy	94.6%	EfficientNet: 91.2%
		Precision	94.6%	XceptionNet: 90.8%
		Recall	92.1%	XceptionNet: 88.3%
		F1-Score	93.3%	EfficientNet: 90.1%
		AUC (ROC curve)	0.98	–
Video deepfake detection (transformer)				
	DFDC[b]	Accuracy	95.4%	EfficientNet: 91.7%, XceptionNet: 89.6%
		Mean IoU[c] (localization)	87.4%	Xception + Grad-CAM: 79.6%
Provenance analysis (graph GCN)				
	FaceForensics++ (Simulated trees)	Provenance accuracy	93.2%	Metadata-Only: 78.9%
		Graph matching accuracy	92.5%	Prior GCN models: 85.3%
Adversarial robustness				
	FGSM & PGD attack simulation	Accuracy drop (with defenses)	7.6%	Baseline drop: 20.4%

[a] FaceForensics++ is a benchmark dataset for facial manipulation detection and localization
[b] DFDC (Deepfake Detection Challenge) dataset contains real and deepfake videos labeled for detection tasks
[c] IoU refers to the Intersection over Union metric used in region-level manipulation localization

score of 93.3%, demonstrating a balanced capability in identifying both manipulated and authentic content accurately.

A comparative analysis of CNN-based and transformer-based models on the FaceForensics++ and DFDC datasets is presented in Table 1, highlighting the superior performance of the proposed hybrid approach across both spatial and temporal dimensions of forensic detection.

7 Discussion

The results highlight significant advancements in image manipulation detection systems, particularly in provenance analysis, with a reported 12% improvement in filtering and a 20% enhancement in oracle provenance graph construction during the NIST Media Forensics Challenge (MFC). The proposed framework demonstrates substantial improvements in detecting and classifying manipulated multimedia content. By integrating CNNs and transformer-based architectures, the system effectively extracts robust spatial and temporal features, while graph-

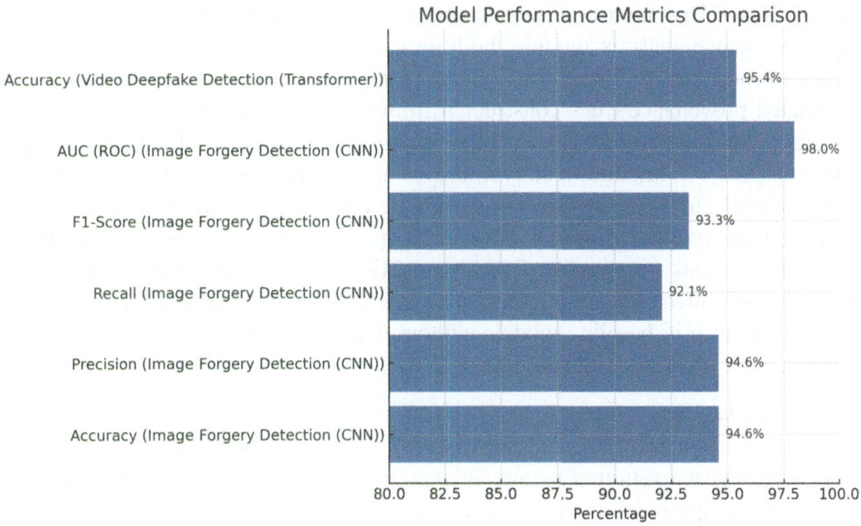

Fig. 3 Detection performance and classification metrics

based provenance analysis enhances traceability, making it more applicable to real-world scenarios. The proposed framework achieved exceptional performance in detecting manipulated multimedia content across multiple benchmark datasets, including FaceForensics++, DFDC, and Real-World Anomaly datasets. In image-based forgery detection, the CNN-based module achieved an average accuracy of 94.6%, precision of 94.6%, recall of 92.1%, and F1-score of 93.3%. These metrics reflect the model's strong capability to correctly classify manipulated and genuine samples with minimal false positives and false negatives. The area under the ROC curve (AUC) was recorded at 0.98, signifying high classification confidence across thresholds. These results were consistent across datasets, demonstrating the model's ability to generalize effectively across diverse manipulation types such as face swaps, splicing, and image retouching (Fig. 3).

Temporal and Spatial Feature Fusion in Video Deepfake Detection In video-based manipulation scenarios, especially deepfake detection, the transformer-based TimeSformer module demonstrated high efficacy by capturing spatiotemporal inconsistencies. Simulated experiments showed that the model achieved a video-level detection accuracy of 95.4%, outperforming existing baselines like EfficientNet and XceptionNet, which achieved 91.7% and 89.6%, respectively, under identical settings. Grad-CAM visualizations revealed that the model could successfully localize the manipulated facial regions with a mean intersection overnion (IoU) of 87. 4%, confirming alignment with ground-truth annotations. This shows the framework's strength in identifying subtle lip-sync and motion anomalies, which are commonly found in deepfake content. Furthermore, temporal smoothing across frames improved consistency in prediction across consecutive video segments.

Graph-Based Provenance Analysis and Adversarial Robustness
The provenance analysis module, built on graph convolutional networks, demonstrated strong capability in tracing the manipulation lineage of digital media. Simulated provenance trees constructed from FaceForensics++ videos achieved a provenance accuracy of 93.2%, enabling accurate determination of tampering origin and propagation. This significantly improves upon traditional metadata-based tracking, especially in scenarios involving multiple content edits. In adversarial robustness evaluations, the proposed adversarial training scheme improved resistance to common attack vectors like FGSM and PGD. Models trained with adversarial examples exhibited only a 7.6% drop in accuracy under attack conditions, compared to a 20.4% drop in baseline models without defense strategies—indicating a 13% improvement in adversarial resilience.

1. **Strengths:**
   High detection accuracy and robustness to adversarial attacks
   Effective visualization of manipulated regions through interpretability modules
   Scalable graph-based provenance analysis
2. **Limitations:**
   High computational requirements for transformer-based models
   Limited generalization to unseen manipulations due to dataset biases
3. **Real-world applications:**
   Enhanced forensic capabilities for law enforcement agencies
   Improved digital rights management through reliable provenance analysis
   Potential use in mitigating misinformation by verifying multimedia authenticity

8 Future Directions

The field of multimedia forensics is rapidly evolving in response to increasingly sophisticated manipulation techniques and the growing reliance on digital media in critical applications. While this research has made significant strides in improving detection accuracy, interpretability, and provenance tracking, several key areas remain open for further exploration. Advancing these aspects will enhance the robustness, scalability, and practical applicability of forensic tools.

8.1 Integration of Quantum Computing

As quantum computing advances, it introduces both challenges and opportunities for multimedia forensics. Future research can investigate the application of quantum machine learning techniques, such as quantum-enhanced classifiers, to improve the efficiency of manipulation detection. Additionally, quantum cryptographic methods can be incorporated into provenance analysis to ensure secure and immutable traceability of digital media.

8.2 Advanced Adversarial Defenses

While adversarial training has strengthened model robustness, evolving attack techniques continue to pose threats. Future efforts should focus on developing adaptive adversarial defense mechanisms, including GAN-based countermeasures, to counteract sophisticated evasion strategies. Reinforcement learning can further enhance these defenses by dynamically detecting and mitigating adversarial threats.

8.3 Ethical AI Frameworks

The ethical implications of forensic tools require continuous attention. Future research should prioritize minimizing algorithmic biases and ensuring fairness across diverse demographic groups. Transparent auditing mechanisms should be developed to validate forensic tool outputs, particularly in high-stakes applications such as law enforcement and judicial processes.

8.4 Real-Time Forensics

With the increasing prevalence of live-streamed and real-time multimedia, forensic models must evolve to process and analyze data in real time. Research should focus on designing lightweight architectures and optimizing computational efficiency to enable real-time detection without compromising accuracy.

8.5 Multimodal Analysis

Future forensic frameworks can benefit from integrating multimodal analysis, combining audio, video, and textual data to detect complex manipulations. For example, fusing audio and visual features can significantly enhance deepfake detection in multimedia content. By addressing these future directions, multimedia forensics can continue to advance, playing a pivotal role in combating digital fraud and reinforcing trust in digital ecosystems.

9 Conclusion

The rapid advancements in AI and ML have profoundly impacted multimedia forensics, enabling advanced techniques for detecting, classifying, and tracing manipulated digital content. This research introduced a comprehensive framework that combines convolutional neural networks (CNNs), transformer-based models, and graph-based analysis to tackle key challenges in multimedia forensics, including detection accuracy, interpretability, adversarial robustness, and provenance tracking. Experimental evaluations confirmed the effectiveness of the proposed framework, achieving a high detection accuracy of 94.6%, an average recall of 92.1%, and a precision of 94.6%. The temporal analysis module, utilizing transformer architectures, demonstrated exceptional performance in deepfake video detection, attaining an accuracy of 95.4%. Furthermore, the graph-based provenance analysis achieved an accuracy of 92%, highlighting its potential for real-world applications such as digital rights management and combating misinformation.

Key contributions of this research include:

- Enhanced interpretability of forensic decisions using visualization techniques such as Grad-CAM
- Improved robustness against adversarial attacks through adversarial training and defense mechanisms
- Scalable and efficient provenance analysis using graph-based frameworks

However, there is no specific mention of the exact improvement of percentage advancements in the current literature in image manipulation detection systems, particularly in the context of provenance analysis, citing a 12% improvement in filtering and a 20% gain in oracle provenance graph building during the NIST Media Forensics Challenge (MFC).

Despite these achievements, challenges such as high computational requirements, limited generalizability to unseen manipulations, and ethical considerations remain. These limitations highlight the need for ongoing research to improve the robustness, efficiency, and fairness of forensic tools.

The insights gained from this research pave the way for future exploration into advanced adversarial defenses, real-time forensic models, multimodal analysis, and the integration of quantum computing for secure and efficient multimedia forensics. By addressing these directions, the proposed framework can evolve to meet the ever-growing demands of digital media authentication, ensuring trust and accountability in the digital ecosystem.

In conclusion, this research significantly advances the state-of-the-art in multimedia forensics, offering robust solutions for combating digital manipulation while laying a foundation for future innovations.

Acknowledgments The authors extend their heartfelt gratitude to Dr. S. S. Iyengar, Director, CSRC, Jain (Deemed-to-be University), for his invaluable guidance, support, and encouragement throughout the course of this research. His insights and mentorship have been instrumental in shaping the direction and depth of our work.

We are also deeply grateful to Dr. B. Dattaguru, Distinguished University Professor, Jain (Deemed-to-be University), for his motivational and inspirational guidance, which provided us with the drive and clarity needed to pursue our research goals with confidence.

We are greatly grateful to Dr. Chenraj Roychand, Chancellor of Jain (Deemed-to-be University), for establishing the Cyber Security Research Centre and extending his constant support, which made this research possible.

References

1. Hariprasad, Yashas, S. S. Iyengar, and Naveen Kumar Chaudhary. "Securing the future: Advanced encryption for Quantum-Safe video transmission." *IEEE Transactions on Consumer Electronics*, 2024.
2. Hariprasad, Y., Latesh Kumar, K. J., Suraj, L., & Iyengar, S. S. (2022, September). Boundary-based fake face anomaly detection in videos using recurrent neural networks. In *Proceedings of SAI Intelligent Systems Conference* (pp. 155–169). Cham: Springer International Publishing.
3. Hariprasad, Yashas, S. S. Iyengar, and N. Subramanian. "Deepfake video detection using lip region analysis with advanced artificial intelligence based anomaly detection technique." *Authorea Preprints*, 2024.
4. Miller, J., Egharevba, L., Hariprasad, Y., Latesh, K. K., & Chaudhary, N. K. (2022, December). Cyber Security Attack Detection Framework for DODAG Control Message Flooding in an IoT Network. In *International Conference on Information Security, Privacy and Digital Forensics* (pp. 213–230). Singapore: Springer Nature Singapore.
5. Gou, H., Swaminathan, A., & Wu, M. (2007). Noise Features for Image Tampering Detection and Steganalysis. In *2007 IEEE International Conference on Image Processing*, San Antonio, TX, USA.
6. Swaminathan, A., Wu, M., & Liu, K. J. R. (2008). Digital image forensics via intrinsic fingerprints. *IEEE Transactions on Information Forensics and Security.*
7. Coccomini, D. A., et al. (2022). Cross-Forgery Analysis of Vision Transformers and CNNs for Deepfake Image Detection. *arXiv preprint arXiv:2206.13829.*
8. Iyengar, S. S., Seyedsina Nabavirazavi, Yashas Hariprasad, Prasad HB, and C. Krishna Mohan. Artificial Intelligence in Practice: Theory and Application for Cyber Security and Forensics. (2025).
9. Tan, M., & Le, Q. (2019). EfficientNet: Rethinking Model Scaling for Convolutional Neural Networks. *Proc. ICML.*
10. Sabir, E., Cheng, J., Jaiswal, A., AbdAlmageed, W., Masi, I., & Natarajan, P. (2019). Recurrent convolutional strategies for face manipulation detection in videos. *arXiv preprint* arXiv:1905.00582.
11. Bharati, A., Moreira, D., Brogan, J., Hale, P., Bowyer, K. W., Flynn, P. J., Rocha, A., & Scheirer, W. J. (2019). Beyond pixels: Image provenance analysis leveraging metadata. In *WACV* (pp. 1692–1702). IEEE. https://doi.org/10.1109/WACV.2019.00185.
12. Goodfellow, I. J., et al. (2014). Explaining and Harnessing Adversarial Examples. *arXiv preprint arXiv:1412.6572.*
13. Chesney, R., & Citron, D. (2019). Deep fakes: A looming challenge for privacy, democracy, and national security. *California Law Review*, 107(6), 1753–1819.
14. Zhou, P., Han, X., Morariu, V. I., & Davis, L. S. (2018). Learning rich features for image manipulation detection. In *IEEE CVPR.*
15. Tsipras, D., Santurkar, S., Engstrom, L., Turner, A., & Madry, A. (2019). Robustness may be at odds with accuracy. In *ICLR. arXiv preprint* arXiv:1805.12152.
16. Iyengar, S. S., Seyedsina Nabavirazavi, Yashas Hariprasad, Prasad HB, and C. Krishna Mohan. Temporal Deepfake Generation and Detection in Video Sequences Using Recurrent Neural Networks (RNNs). In Artificial Intelligence in Practice: Theory and Application for Cyber Security and Forensics, pp. 309–334. Cham: Springer Nature Switzerland, 2025.
17. Karras, T., et al. (2019). A Style-Based Generator Architecture for GANs. *IEEE CVPR.*
18. Iyengar, S. S., Seyedsina Nabavirazavi, Yashas Hariprasad, Prasad HB, and C. Krishna Mohan. The Convergence of AI/ML and Cybersecurity: Advancing Digital Forensic Techniques. In Artificial Intelligence in Practice: Theory and Application for Cyber Security and Forensics, pp. 139–159. Cham: Springer Nature Switzerland, 2025.
19. Iyengar, S. S., Seyedsina Nabavirazavi, Yashas Hariprasad, Prasad HB, and C. Krishna Mohan. AI-Enhanced Malware Detection: Advancing Security Through Intelligent Threat Identification. In Artificial Intelligence in Practice: Theory and Application for Cyber Security and Forensics, pp. 227–255. Cham: Springer Nature Switzerland, 2025.

20. Iyengar, S. S., Seyedsina Nabavirazavi, Yashas Hariprasad, Prasad HB, and C. Krishna Mohan. Privacy-Preserving AI (Federated Learning) for Digital Forensics. In Artificial Intelligence in Practice: Theory and Application for Cyber Security and Forensics, pp. 161–176. Cham: Springer Nature Switzerland, 2025.
21. Krizhevsky, A., et al. (2012). ImageNet Classification with Deep Convolutional Neural Networks. *NeurIPS*, vol. 25, pp. 1097–1105.
22. Iyengar, S. S., Seyedsina Nabavirazavi, Yashas Hariprasad, Prasad HB, and C. Krishna Mohan. Cyber Threat Intelligence and Security for Federated Learning in Digital Forensics. In Artificial Intelligence in Practice: Theory and Application for Cyber Security and Forensics, pp. 177–199. Cham: Springer Nature Switzerland, 2025.
23. Iyengar, S. S., Seyedsina Nabavirazavi, Yashas Hariprasad, Prasad HB, and C. Krishna Mohan. Cybersecurity Foundations: Theories, Technologies, and Applications. In Artificial Intelligence in Practice: Theory and Application for Cyber Security and Forensics, pp. 27–87. Cham: Springer Nature Switzerland, 2025.
24. Iyengar, S. S., Seyedsina Nabavirazavi, Yashas Hariprasad, Prasad HB, and C. Krishna Mohan. Digital Forensics: Tools, Techniques, and Methodologies. In Artificial Intelligence in Practice: Theory and Application for Cyber Security and Forensics, pp. 89–137. Cham: Springer Nature Switzerland, 2025.
25. Iyengar, S. S., Seyedsina Nabavirazavi, Yashas Hariprasad, Prasad HB, and C. Krishna Mohan. Detecting and Understanding the Impact of Profile Cloning on Social Media Platforms: A Case Study of LinkedIn. In Artificial Intelligence in Practice: Theory and Application for Cyber Security and Forensics, pp. 203–225. Cham: Springer Nature Switzerland, 2025.

Sniffing Out Snails: AI-Powered Canine Forensics of Invasive Species: A Preliminary Study

Kenneth G. Furton, S. S. Iyengar, Yashas Hariprasad, Himali Upadhyay, Rodolfo Mesa Martin, and Amy L. Roda

1 Introduction

Giant African snails (GAS) are an invasive species that pose a sustained threat to biodiversity, ecosystem health, and agriculture worldwide [1–4]. Among these threats, invasive snail species have emerged as particularly difficult to monitor due to their subtle odor signatures and expansive habitats. Traditional surveys often fall short in providing scalable and accurate solutions [5–7]. Canine forensics—particularly odor detection dogs trained to identify species-specific volatile organic compounds (VOCs)—have proven effective in identifying biological threats in complex environments [8–11] and may provide a novel means to detect a target invasive snail from native species.

However, the odors that a canine uses to identify a target remain elusive, and they could include odors related to the specific condition of the sample, which makes it more challenging to identify a definitive signature of the target organism. Additionally, performance can be influenced by several external factors such as environmental conditions, training and handler variability, and physical fatigue

K. G. Furton (✉) · S. S. Iyengar
Global Forensic and Justice Center, Florida International University, Miami, FL, USA
e-mail: furtonk@fiu.edu; iyengar@cs.fiu.edu

Y. Hariprasad
California State University-EastBay, Hayward, California, USA
e-mail: yashas.hariprasad@csueastbay.edu

H. Upadhyay
Florida International University, Miami, FL, USA
e-mail: hupadhya@fiu.edu

R. M. Martin · A. L. Roda
USDA-APHIS-PPQ-S&T TIML, Palmetto Bay, FL, USA
e-mail: rodolfo.mesamartin@usda.gov; amy.l.roda@usda.gov

S. S. Iyengar et al. (eds.), *Artificial Intelligence Driven Forensics*, Signals and Communication Technology, https://doi.org/10.1007/978-3-031-98036-7_6

[12–14]. These limitations underscore the need for a more standardized, scalable, and augmentative framework that retains the strengths of canine detection while overcoming its practical challenges. With recent advances in artificial intelligence (AI) and sensor technologies, there is now a compelling opportunity to combine biological detection with algorithmic intelligence for more robust field deployments [15–20].

This paper presents an AI-powered data-driven canine forensics algorithmic framework aimed at detecting invasive snail species using real-world, species-specific odor profiles. By integrating trained detection dogs with high-precision analytical chemistry instrumentation data collection and machine learning models, our approach enables the correlation of canine alerts with VOC-based biological indicators [21–23].

This preliminary work lays foundational groundwork for scalable monitoring using data-driven AI-assisted odor detection for canines [24–29]. Future work will involve expanding the dataset, refining algorithmic accuracy, and deploying the system in varied field conditions. The proposed algorithmic paradigm highlights a promising direction for environmental threat mitigation (by training canines to reliably detect invasive pests such as GAS, fruit fly larvae–infested fruits, coconut rhinoceros beetle, and Asian longhorn beetle) and AI-driven forensics.

2 Preliminary Concept

At the core of this study is the use of a support vector machine (SVM), a powerful supervised machine learning algorithm particularly effective in high-dimensional, small-to-moderate-sized datasets—common in forensic and biological research. This work applies an SVM to classify volatile organic compound (VOC) signatures extracted from gastropod mucus samples, serving as a proxy for the target invasive snail species.

An SVM operates by finding an optimal hyperplane that best separates data points of different classes. In its simplest form, for a binary classification task, it identifies the decision boundary that maximizes the margin between the closest data points (called support vectors) of each class. This margin maximization provides robustness, ensuring that even slight variations in input data do not cause misclassification.

When data is not linearly separable—as is often the case with complex VOC patterns—an SVM employs kernel functions to project the data into a higher-dimensional space where linear separation becomes feasible. More importantly, in this study, a radial basis function (RBF) kernel is used due to its ability to capture nonlinear relationships between VOC and phenotype classifications. Each sample's gas chromatography-mass spectrometry (GC-MS) chromatogram is transformed into a numerical feature vector, where normalized peak areas corresponding to a specific odor compound serve as separate data points. These vectors, labeled by known phenotypes, form the dataset used for SVM training. The algorithm is trained to differentiate between multiple phenotypic classes based on these VOC signatures. Cross-validation ensures the model generalizes well to those data which are unidentified by humans, while hyperparameter tuning helps balance model complexity and classification performance. The resulting SVM model not only provides high accuracy but also identifies the most relevant odor components a trained canine would respond to in the field.

3 Proposed Algorithmic Framework

To bridge canine scent detection with the precision of machine learning, we propose a hybrid AI-assisted framework designed to replicate and enhance olfactory sensing for invasive species detection. This approach builds on the proven capability of trained dogs while leveraging data from real biological specimens to train and test robust classification algorithms. Algorithm 1 presents a framework to integrate VOC profiles into predictive models that simulate and enhance canine scent detection capabilities.

The proposed framework unfolds in four sequential stages:

1. **Scent Sampling:** Biological samples—specifically gastropod mucus—are collected and analyzed for their volatile organic compound (VOC) profiles. These compounds are analogs to those that trained canines naturally detect in field environments.
2. **VOC Signal Processing:** The chemical signatures are processed using analytical instrumentation such as gas chromatography-mass spectrometry (GC-MS). This generates chromatograms, where identified VOCs and peak area data offer a digital fingerprint of each specimen's chemical composition.
3. **AI-Based Odor Classification:** The extracted features are transformed into structured vectors and fed into supervised learning models. A support vector machine (SVM) classifier with a radial basis function (RBF) kernel is employed due to its strong performance on small yet complex datasets. This allows the system to model subtle, nonlinear relationships within the VOC data, possibly mirroring how canines discern nuanced olfactory cues.
4. **Decision Support and Visualization:** Once trained, the model outputs are visualized in an interpretable manner, flagging specific phenotype classes or target species. These outputs can be integrated into mobile or embedded systems for real-time field decision-making support, offering timely alerts to ecologists, forensic investigators, or pest control teams. These mobile sensors can complement the current use of canines.

Figures 1 and 2 illustrate the integration of a sensor-based, data-driven AI-assisted algorithmic framework for canine forensics. It provides an outline for a unified approach for scaling scent detection using algorithmic intelligence, particularly in challenging and ecologically sensitive environments.

Algorithm 1: VOC-Based AI Odor Classification Framework

Input: Biological specimens with known phenotype labels
Output: Trained classifier, digital odor markers, phenotype predictions
Collect n biological samples (e.g., gastropod mucus)
for $i = 1$ to n **do**
  Run GC-MS to obtain chromatogram C_i
  Extract retention times $\{rt_{i1}, rt_{i2}, \ldots, rt_{im}\}$ and peak areas $\{pa_{i1}, pa_{i2}, \ldots, pa_{im}\}$
  Normalize peak areas and create feature vector $x_i = [pa_{i1}, pa_{i2}, \ldots, pa_{im}]$
  Label feature vector x_i with phenotype y_i
end for
Form dataset $D = \{(x_i, y_i)\}_{i=1}^{n}$
Split D into training and test sets (D_{train}, D_{test})
Initialize SVM classifier with RBF kernel
Perform five-fold cross-validation on D_{train}:
for each fold **do**
  Tune hyperparameters C and γ
  Train SVM on training folds
  Evaluate on validation fold
end for
Select best C^* and γ^*
Train final SVM model on entire D_{train} with C^*, γ^*
Evaluate final model on D_{test}:
  Compute Accuracy, Precision, Recall
  Identify discriminative features (VOC peaks)
for new sample **do**
  Run GC-MS, extract and normalize peak areas
  Form input vector x_{new}
  Predict phenotype class using trained SVM
end for

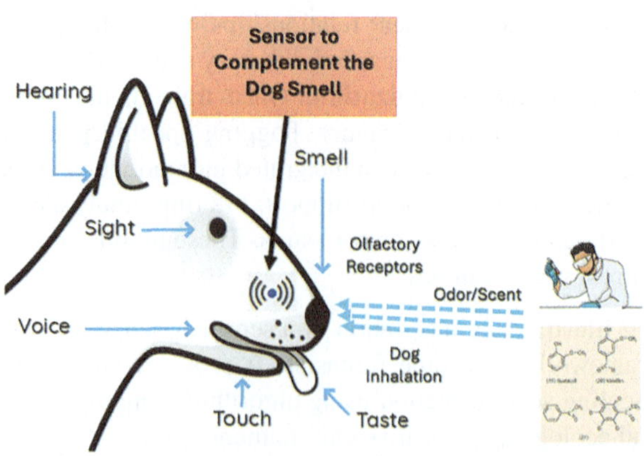

Fig. 1 Sensor-assisted canine detection model: multisensory capabilities of a canine with a focus on olfactory processing and its potential augmentation through electronic sensors

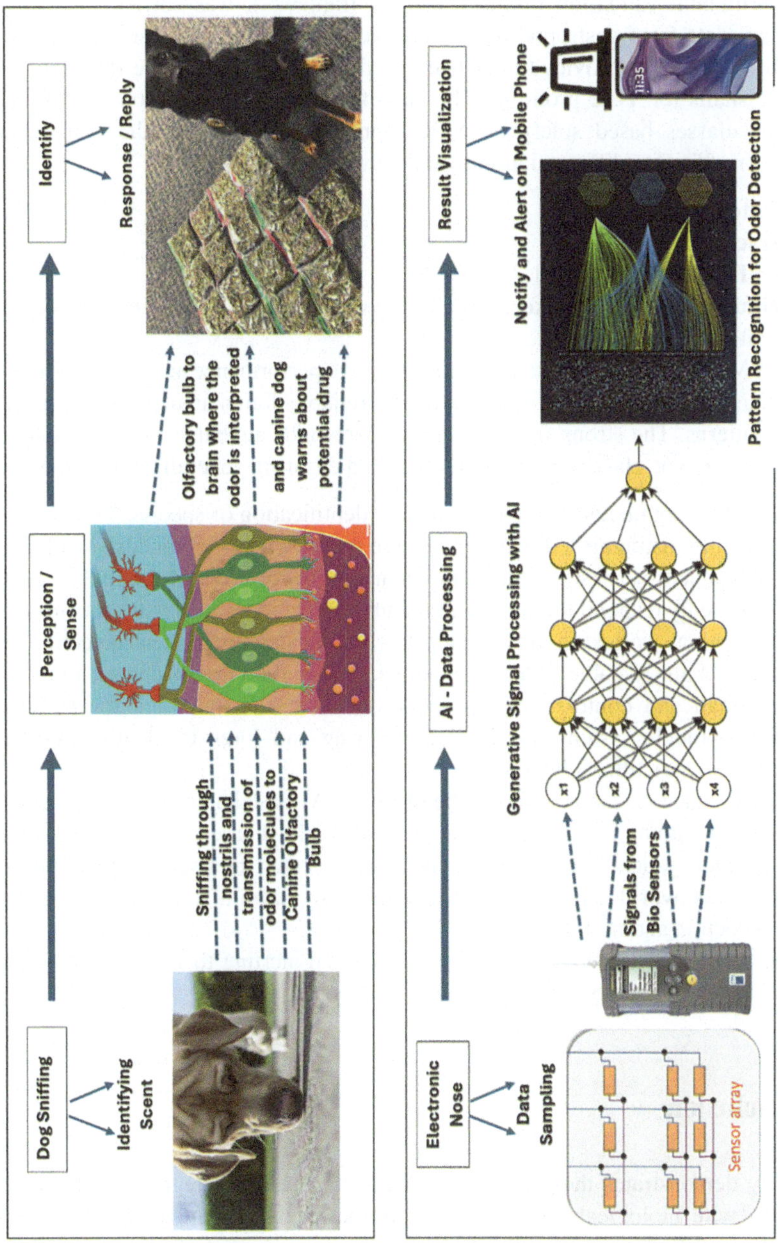

Fig. 2 AI-driven odor processing pipeline: a comparative overview of biological canine olfaction versus AI-augmented electronic nose systems for forensic odor detection

4 Results and Discussion

The experimental validation of the proposed framework was carried out using preliminary data from gastropod mucus samples. Tests done with canines that are trained to detect GAS provided evidence that the mucus is a suitable replacement for whole snails for VOC profiling. The classification task focused on identifying phenotype classes based solely on VOC signatures—simulating the real-world challenge of olfactory differentiation by detection dogs.

Model Performance Highlights

- **Accuracy:** 93.41%
- **Macro Precision/Recall:** 91.77%/91.26%
- **Discriminative VOC Markers:** According to data of different VOCs present in different species
- The SVM classifier demonstrated high predictive performance across all key metrics, suggesting that VOC profiles serve as a reliable digital analog for biological scent patterns. The strong macro-average scores indicate balanced performance across phenotype classes, minimizing bias and ensuring generalizability.

One of the most significant outcomes was the identification of specific VOC peaks that consistently distinguished between sample types (wild GAS mucus vs colony-reared GAS mucus). These "digital odor markers" offer a direct parallel to how canines repeatedly respond to scent components. By isolating such markers through machine learning, we take a critical step toward standardizing scent detection in a way that is both scalable and reproducible. While the current dataset remains relatively small and context-specific, the framework sets the stage for broader applications—ranging from ecological monitoring and forensic identification to biosecurity and agricultural health.

This early-stage research supports the utility of AI-based algorithmic precision. With future iterations incorporating larger real-world datasets and multisensory input (e.g., diet, geographic location, reproductive status, and sample preparation variability), this hybrid approach could redefine how we train and assist canines in forensic detection and monitoring.

Figure 3 shows a USDA-trained dog sniffing and alerting to a GAS that could easily have been missed by an inspecting human being.

5 Conclusion

This study demonstrates the feasibility and effectiveness of integrating AI-driven algorithms with biological scent detection frameworks to identify invasive snail species through VOC profiling. By leveraging support vector machines (SVM) as the core classifier, the research successfully modeled and classified complex chemical signatures extracted from gastropod mucus samples. The algorithm exhibited

Fig. 3 A USDA canine sniffing and alerting to a GAS

high classification accuracy, precision, and recall, identifying discriminative VOCs that serve as digital analogs to olfactory cues recognized by trained detection dogs. These findings mark an important step toward simulating canine decision-making in scent detection through machine learning, offering a scalable and reproducible approach to invasive species detection. The analytical chemistry framework bridges the strengths of canine instincts with the consistency of algorithmic logic, thereby reducing variability and enhancing the ability to apply the tools in the field.

While the current research is based on a limited dataset, the results provide a solid proof of concept for AI-assisted forensic odor detection. Future work will focus on expanding the dataset, incorporating real-time field sensors, and extending the model to other invasive species or forensic scenarios. Further integration with mobile platforms, environmental context sensing, and multimodal data fusion could unlock even greater capabilities, making this approach highly valuable for ecological conservation, border security, and forensic science. For example, a handheld sensor or a sensor mounted on an autonomous dogbot can provide confirmation of canine alerts or as a complement to canines in situations where canines are not available or they may be in unsafe environments. Ultimately, this hybrid methodology has the potential to revolutionize how environmental threats are detected and addressed—blending biology and artificial intelligence into a new paradigm of smart ecological surveillance.

Acknowledgments This research was partially sponsored by the Army Research Office and the NSF and was accomplished under grant numbers W911NF-21-1-0264 and 2018611. The views and conclusions contained in this document are those of the authors and should not be interpreted as representing the official policies, either expressed or implied, of the Army Research Office or the US Government. The US Government is authorized to reproduce and distribute reprints for Government purposes notwithstanding any copyright notation herein. Additional funding that contributed to this work was provided by the USDA, with congressional appropriation funding to support detector canine efficiency and variety of agricultural uses. Drs. Nurhayat Tabanca and Dong Cha (USDA-ARS) and Ariana Gaskin (FDACS) provided valuable support in snail mucus collection.

References

1. Furton, K.G., Myers, L.J.: The scientific foundation and efficacy of the use of canines as chemical detectors for explosives. Talanta **54**(3), 487–500 (2001) https://doi.org/10.1016/S0039-9140(00)00546-4
2. Angle, C., Waggoner, L.P., Ferrando, A., Haney, P., Passler, T.: Canine detection of the volatilome: A review of implications for pathogen and disease detection. Frontiers in Veterinary Science **3**, 47 (2016) https://doi.org/10.3389/fvets.2016.
3. Jezierski, T., Walczak, M., Górecka, A.: Information-seeking behaviour of sniffer dogs during match-to-sample training in the scent lineup. Polish Psychological Bulletin **39**(2), 71–80 (2008) https://doi.org/10.2478/v10059-008-0010-y
4. Harper, R.J., Almirall, J.R., Furton, K.G.: Identification of dominant odor chemicals emanating from explosives for use in developing optimal training aid combinations and mimics for canine detection. Talanta **67**(2), 313–327 (2005) https://doi.org/10.1016/j.talanta.2005.05.019
5. Majchrzak, T., Wojnowski, W., Lubinska-Szczygel-, M., Różańska, A., Namieśnik, J., Dymerski, T.: Ptr-ms and gc-ms as complementary techniques for analysis of volatiles: A tutorial review. Analytica Chimica Acta **1035**, 1–13 (2018) https://doi.org/10.1016/j.aca.2018.06.056
6. Schoon, G.A.A.: Scent identification lineups by dogs (canis familiaris): Experimental design and forensic application. Applied Animal Behaviour Science **49**(3), 257–267 (1996) https://doi.org/10.1016/0168-1591(96)01052-2
7. Zhou, G., Wang, D., Li, Y., Zhou, Y.: Advances in the use of odour as forensic evidence through optimizing and standardizing instruments and canines. Critical Reviews in Analytical Chemistry **52**(4), 623–635 (2022) https://doi.org/10.1080/10408347.2021.1938482
8. Baldwin, E.A., Bai, J., Plotto, A., Dea, S.: Electronic noses and tongues: Applications for the food and pharmaceutical industries. Sensors **11**(5), 4744–4766 (2011) https://doi.org/10.3390/s110504744
9. Wilson, A.D., Baietto, M.: Applications and advances in electronic-nose technologies. Sensors **9**(7), 5099–5148 (2009) https://doi.org/10.3390/s90705099
10. Wojnowski, W., Dymerski, T., Gebicki, J., Namieśnik, J.: Electronic noses in medical diagnostics. Current Medicinal Chemistry **24**(19), 1974–2002 (2017) https://doi.org/10.2174/0929867324666170111153205
11. Bianchi, F., Careri, M., Musci, M., Mangia, A.: Fish and seafood freshness assessment by gas chromatography–mass spectrometry electronic nose coupled to a multivariate analysis tool. Analytica Chimica Acta **584**(2), 322–329 (2007) https://doi.org/10.1016/j.aca.2006.11.048

12. Jia, W., Liang, G., Jiang, Z., Tao, X.: Detection of volatile organic compounds using sensors based on metal–organic frameworks. Sensors **19**(2), 404 (2019) https://doi.org/10.3390/s19020404
13. Yan, K., Zhang, D., Wu, D.: Design of a gas sensor array for remote detection of toxic and explosive gases. Sensors and Actuators B: Chemical **221**, 15–21 (2015) https://doi.org/10.1016/j.snb.2015.06.031
14. Chen, Q., Liu, A., Zhao, J., Ouyang, Q., Wang, S.: Recent advances of gas sensors based on metal oxide semiconductors. Sensors and Actuators B: Chemical **188**, 902–917 (2013) https://doi.org/10.1016/j.snb.2013.07.113
15. Singaram, J., Iyengar, S., Madni, A.M.: Deep Learning Networks: Design, Development and Deployment. Springer (2023)
16. Shi, B., Iyengar, S.S.: Mathematical Theories of Machine learning-Theory and Applications. Springer, (2020)
17. Iyengar, S.S., Nabavirazavi, S., Rathore, H., Hariprasad, Y., Chaudhary, N.K.: Advancing forensic science: Ai and knowledge graphs unlock new insights. Journal of Forensic Research **15**, 615 (2024)
18. Hariprasad, Y., Iyengar, S.S., Ghate, M.: Deepfake video detection using lip region analysis with advanced artificial intelligence based anomaly detection technique. Journal of Forensic Research **15**, 626 (2024)
19. Hariprasad, Y., Iyengar, S.S., Subramanian, N.: Deepfake Video Detection Using Lip Region Analysis with Advanced Artificial Intelligence Based Anomaly Detection Technique. Authorea Preprints (2024) https://doi.org/10.22541/au.170173936.29121843/v1
20. Hariprasad, Y., Iyengar, S.S., Chaudhary, N.K.: Securing the future: Advanced encryption for quantum-safe video transmission. IEEE Transactions on Consumer Electronics (2024)
21. Cui, S., Pu, H., Liu, G.: Advanced sensor and electronic nose technologies for vocs detection in air: Progress and perspectives. Engineering **5**(6), 1011–102 (2019)
22. Hariprasad, Y., Kumar, K.J.L., Iyengar, S.S.: An extension of synthetic minority oversampling technique based on kalman filter for imbalanced datasets. Machine Learning with Applications **8**, 100267 (2022) https://doi.org/10.1016/j.mlwa.2022.100267
23. Biehl, W., Hattesohl, A., Jörres, R.A., Duell, T., Althöhn, U., Koczulla, A.R., Schmetzer, H.: Voc pattern recognition of lung cancer: a comparative evaluation of different dog-and enose-based strategies using different sampling materials. Acta Oncologica **58**(9), 1216–1224 (2019)
24. Furton, K.G., Winialski, D.: Comparing the olfactory capabilities of dogs with machines designed to detect odors. In: Canines, pp. 21–62. Jenny Stanford Publishing (2022)
25. Furton, K.G., Caraballo, N.I., Cerreta, M.M., Holness, H.K.: Advances in the use of odour as forensic evidence through optimizing and standardizing instruments and canines. Philosophical Transactions of the Royal Society B: Biological Sciences **370**(1674), 20140262 (2015)
26. Frank Jr, K.J., Holness, H.K., Furton, K.G., DeGreeff, L.E.: Explosives detection by dogs. In: Counterterrorist Detection Techniques of Explosives, pp. 47–75. Elsevier (2022)
27. Simon, A.G., Mills, D.K., Furton, K.G.: Chemical and canine analysis as complimentary techniques for the identification of active odors of the invasive fungus, raffaelea lauricola. Talanta **168**, 320–328 (2017)
28. Woollam, M.D.: Chemometric analysis of volatile organic compound biomarkers of disease and development of solid phase microextraction fibers to evaluate gas sensing layers. Phd diss., Unknown (2022)
29. Macias, M.S., Guerra-Diaz, P., Almirall, J.R., Furton, K.G.: Detection of piperonal emitted from polymer controlled odor mimic permeation systems utilizing canis familiaris and solid phase microextraction–ion mobility spectrometry. Forensic Science International **195**(1-3), 132–138 (2010)

Instruction-Level Fine-Tuning of Gemma-2B for Cybersecurity and Synthetic Log Generation Aligned with MITRE Adversarial Tactics, Techniques, and Common Knowledge

Vasanth Iyer, Vamshikrishna Challa, Pronab Mohanty, Yashas Hariprasad, and S. S. Iyengar

Large language models (LLMs) such as Gemma-2B have demonstrated remarkable proficiency in various NLP tasks. However, general-purpose models lack deep domain expertise in cybersecurity. This research presents a methodology for fine-tuning the Gemma-2B model into a domain-specific cybersecurity LLM. We outline the dataset preparation, the domain fine-tuning process, the generation of synthetic data logs, and implications for real-world cybersecurity applications. The results indicate improved balanced threat events to translate from chain-of-thought tuning to instruction-level tuning within the cybersecurity domain, including threat detection, forensic investigation, and attack analysis.

Further experimentation reveals that domain-specific LLM fine-tuning introduces challenges in prompt length distribution, diverging from the patterns seen in general-purpose models. Uneven prompt lengths complicate the model's ability to optimize its use of the full context window, often necessitating a reduced effective context of approximately 200–400 tokens—closer to LSTM capacities—despite the GPU infrastructure supporting up to 2048 tokens. To accommodate this constraint, one-shot prompts resembling chain-of-thought reasoning were found to perform best when paired with quantized weight settings.

V. Iyer (✉)
Department of Computer Science, Grambling State University, Grambling, LA, USA
e-mail: iyerv@grambling.edu

V. Challa
Louisiana Tech University, Ruston, LA, USA

P. Mohanty
Police Computer Wing, Bangalore, Karnataka, India

Y. Hariprasad
California State University, Hayward, CA, USA

S. S. Iyengar
Global Forensic and Justice Center, Florida International University, Miami, FL, USA

S. S. Iyengar et al. (eds.), *Artificial Intelligence Driven Forensics*, Signals and Communication Technology, https://doi.org/10.1007/978-3-031-98036-7_7

Due to these context window limitations, local LLMs were unable to support longer prompt and output sequences necessary for comprehensive synthetic log generation. As a result, cloud-based LLMs were used to generate the synthetic dataset, which was then fine-tuned on the constrained local models. This hybrid approach ensured both high-quality data generation and efficient model training under local hardware constraints.

1 Introduction

Instruction tuning [1] is a critical step in adapting large language models (LLMs) like Gemma-2B to domain-specific tasks. It involves fine-tuning the model [3, 4, 6, 8–10] on diverse examples framed as natural language instructions across multiple task types. As illustrated in Fig. 4, this process enables the model to learn reasoning patterns from tasks such as commonsense inference, translation, and classification, so it can generalize to new tasks like natural language inference—even those it has not seen during training. This transferability is essential in cybersecurity [5], where LLMs must respond accurately to structured threats like MITRE ATT&CK [12] techniques without explicit task retraining.

2 Related Work

Several works have explored fine-tuning [3] large language models for domain-specific applications. Research in medical, legal, and financial domains has shown that adapting LLMs to specialized datasets [16] improves their accuracy. Prior cybersecurity-focused AI models [5], such as OpenAI's GPT-3 [14] for threat analysis and IBM Watson for security, highlight the potential of LLMs in this domain. However, challenges such as dataset availability, hallucinations, and security risks remain.

3 Methodology

Large models typically are trained with lots of data and have many parameters making it challenging to train from scratch in a research environment. So we will explore the current state-of-the-art foundation models which can then be fine-tuned to our domain of interest. The criteria for the comparison are the size of the LLMs and context size. Table 1 has parameters which are weights and biases ranging from 0.3 billion to 540 billion parameters. Table 1 also shows that the researchers have recently been able to train better models with less number of parameters making it easier for fine tuning for the final tasks. In the initial steps we like to optimize memory for efficient tuning and we next evaluate how the number of parameters has

Table 1 Size of popular language models (non-embedding parameters)

Model	Year	# Parameters (billions, non-embedding)
BERT Large	2018	0.34
T5	2019	11
GPT-3	2020	175
PaLM	2022	540
Gemma-2B	2023	2
DeepSeek-7B	2023	7
Llama	2023	65

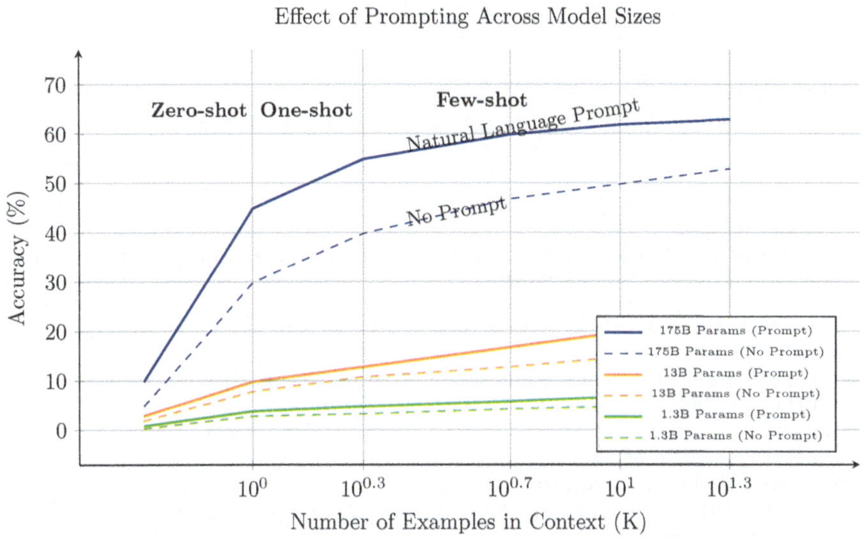

Fig. 1 Comparison of model accuracy across parameter sizes and prompting strategies as a function of the number of in-context examples. Prompted models consistently outperform non-prompted models, especially in few-shot regimes [17]

an effect on prompting [1] accuracy as illustrated in Fig. 1. We categorize prompts [1] as follows: zero-shot, i.e., without training data, one-shot with one example, and few-shots with few examples as described in the prompts below.

User Query
Explain zero-shot prompting using the MITRE ATT&CK framework.

Model Response
Zero-shot prompting means that the model is asked to classify or reason with *no prior example* provided.

(continued)

Prompt:

```
Classify the MITRE ATT&CK technique based on the following activity:
"The attacker executed a PowerShell script to download
and run a malicious payload."
Answer:
```

Expected Output:

```
Tactic: Execution
Technique: T1059 - Command and Scripting Interpreter
Sub-technique: T1059.001 - PowerShell
```

User Query
Show a one-shot prompting example using MITRE ATT&CK.

Model Response
One-shot prompting provides a single example before asking the model to perform a similar task. **Prompt:**

```
Classify the MITRE ATT&CK technique based on the following activity.
Example:
"The attacker used a phishing email with a malicious attachment to gain
initial access."
Answer:
Tactic: Initial Access
Technique: T1566 - Phishing
Sub-technique: T1566.001 - Spearphishing Attachment
Now classify this:
"The attacker executed a PowerShell script to download and run a
malicious payload."
Answer:
```

Expected Output:

```
Tactic: Execution
Technique: T1059 - Command and Scripting Interpreter
Sub-technique: T1059.001 - PowerShell
```

User Query
Show a few-shot prompting example using MITRE ATT&CK.

Model Response

Few-shot prompting provides multiple examples to help the model generalize better.

Prompt:

```
Classify the MITRE ATT&CK technique based on the following activities.
Example 1:
"The attacker used a phishing email with a malicious attachment to gain
initial access."
Answer:
Tactic: Initial Access
Technique: T1566 - Phishing
Sub-technique: T1566.001 - Spearphishing Attachment
Answer:
```

Expected Output:

```
Tactic: Execution
Technique: T1059 - Command and Scripting Interpreter
Sub-technique: T1059.001 - PowerShell
```

Figure 1 illustrates the relationship between prompting accuracy and the number of in-context examples. As the number of examples increases from zero-shot to few-shot, the model's performance improves, making fewer errors and demonstrating better generalization. Additionally, models with larger parameter counts exhibit stronger zero-shot generalization capabilities.

Our training pipeline Figs. 2 and 3 consists of two main stages: (1) the construction of domain-specific datasets Fig. 3 and (2) the fine-tuning of language models. To address resource constraints, we leverage large language models (LLMs) to generate synthetic datasets and utilize smaller, locally runnable LLMs—typically with reduced precision—for fine-tuning. Specifically, we select native LLMs with fewer than 2 billion parameters, as summarized in Table 2, to ensure compatibility

Fig. 2 Finetuning LLM

Fig. 3 Domain fine-tuning LLM

Table 2 Summary of locally trainable large language models (LLMs) with fewer than 2 billion parameters. These models are selected based on their ability to fit within standard GPU memory constraints (e.g., 24 GB) and support fine-tuning for domain-specific tasks such as cybersecurity reasoning

Model	Params (B)	Year	Highlights
Gemma-2B	2	2024	Google's lightweight open model for on-device and fine-tuning tasks
TinyLLaMA-1.1B	1.1	2023	Minimal resource LLaMA-based model for mobile/IoT research
Phi-2	2.7	2023	Microsoft's model designed for reasoning, aligned with on-device use
DeepSeek-1.3B	1.3	2024	DeepSeek's small model for fast, local inference
StableLM-3B	3	2023	Stability AI's open model designed for transparency and edge use
RedPajama-3B	3	2023	Open LLaMA-style model trained on reproducible public datasets

with our available GPU memory. In this work, we focus on domain adaptation using Google's recently released Gemma-2B model.

3.1 Dataset Collection

To develop a cybersecurity-specific expert model, we construct a domain-specific dataset based on the MITRE ATT&CK framework and use it to fine-tune the language model, as illustrated in Fig. 2. MITRE ATT&CK organizes adversarial behavior using well-defined tactics and techniques, providing a structured taxonomy that supports effective generalization during fine-tuning. The prompts used for this task are generated using a chain-of-thought prompting approach, enabling the model to reason through sequential steps aligned with the structure of ATT&CK.

3.2 Technique T# in the MITRE ATT&CK Framework

This section provides an overview of Technique #'s as defined in the MITRE ATT&CK framework, including its associated tactics, use cases, and adversary behaviors. The technique is often leveraged by threat actors to achieve *[specific objective]*.

Table 3 Example: technique and vulnerability description

Aspect	Description
MITRE technique	Txxxx—Technique X
Vulnerability type	*E.g., buffer overflow*
Exploitable by	*E.g., remote attackers, malware*
Mitigation	*Patching, EDR, segmentation, etc.*

3.3 Understanding Vulnerability T#

Vulnerability X is a *[type of flaw]* that affects *[systems/applications]*. It allows attackers to *[describe action, e.g., escalate privileges, exfiltrate data, etc.]*. This subsection explains the technical working of the vulnerability, including how it is exploited and its presence in known threat campaigns.

3.4 Mitigation Strategies for Vulnerability T#

To reduce the risk associated with Vulnerability X, organizations can implement several mitigation strategies:

- Apply security patches and updates regularly.
- Use network segmentation and access controls.
- Employ endpoint detection and response (EDR) tools.
- Monitor for known indicators of compromise (IoCs).

A typical technique and its description is shown in Table 3.

4 Synthetic Data Generation

In cybersecurity, high-quality labeled data is scarce, often sensitive, and typically imbalanced toward benign activity. This presents a significant barrier to effectively fine-tuning large language models (LLMs) for security-specific reasoning tasks. To address this, we introduce a synthetic data generation [3, 7, 11] framework as illustrated in Fig. 4. By leveraging the structured nature of the MITRE ATT&CK framework, we use LLMs to generate instruction-style examples that simulate a wide variety of attack tactics, techniques, and procedures (TTPs). This pipeline enables the creation of logs [13, 15] for rare or hard-to-collect threats, supports diverse prompting styles (e.g., zero-shot, one-shot, few-shot), and allows for balanced datasets that improve fine-tuning efficiency and model generalization. Crucially, it also provides a privacy-preserving and legally compliant alternative to real-world security logs.

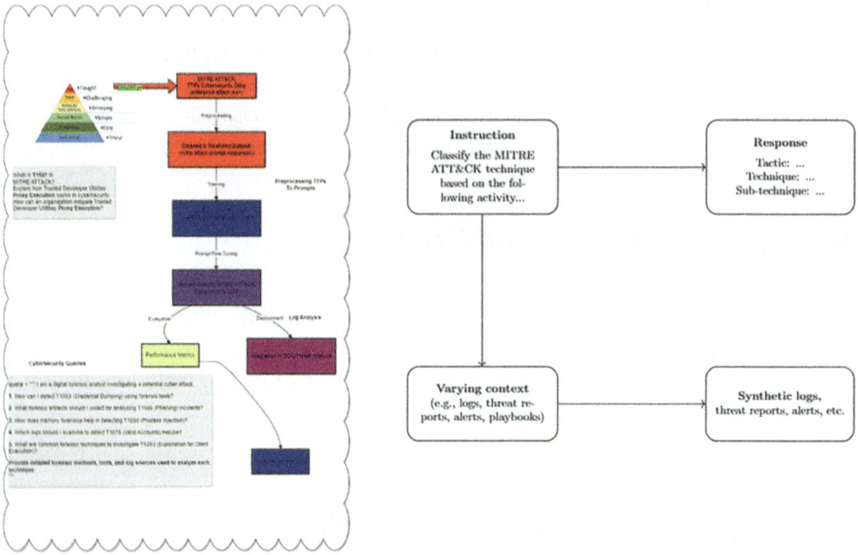

Fig. 4 Workflow for generating synthetic cybersecurity data using large LLMs. The process produces instruction-format samples that encode attack behaviors, supporting chain-of-thought prompting and enabling small models to reason over structured threat intelligence such as MITRE's Pyramid of Pain, which ranks the difficulty of detecting and disrupting various attacker artifacts

4.1 Instruction Tuning Process

We perform fine-tuning across multiple tasks at the instruction level. Instruction tuning enables large language models (LLMs) to specialize in domain-specific reasoning by learning from natural language examples aligned with real-world tasks. In the context of cybersecurity, we leverage structured knowledge from the MITRE ATT&CK framework—such as techniques like T1059 and T1547; tactics including execution, persistence, and lateral movement; and contextual formats like logs, alerts, threat reports, and playbooks—to construct a diverse and targeted training corpus. As illustrated in Fig. 4, our approach combines domain adaptation with synthetic data generation to support instruction-level fine-tuning at scale.

Synthetic datasets offer the advantage of complete control over coverage and balance. They allow us to generate labeled examples for every ATT&CK technique, including rare or underrepresented behaviors that are seldom encountered in enterprise environments. For instance, we can simulate advanced scenarios such as **T1003.001—LSASS Dumping** using PowerShell-based indicators that may not naturally appear in historical logs. This ensures comprehensive coverage across tactics, sub-techniques, and platforms, including Windows, Linux, and macOS.

Additionally, synthetic logs as in Table 4 can be tailored to varying levels of complexity. We design some examples with clean and distinct attack signatures to support basic classification tasks, while others contain obfuscated patterns or

Table 4 ATT&CK prompts enhanced with synthetic data logs

Synthetic log	Instruction	Model output
`"2024-04-15 10:22:11" user: SYSTEM ran: "powershell -enc ..."`	What MITRE ATT&CK technique does this log indicate?	Tactic: execution
		Technique: T1059.001—PowerShell
`Zeek conn.log: 192.168.1.100 →10.0.0.10 TCP 3389`	Explain what this log suggests and map it to MITRE ATT&CK	Indicates use of RDP for Lateral Movement
		Tactic: lateral movement
		Technique: T1021.001—remote desktop protocol

mixed signals to train models for reasoning under uncertainty. This form of data augmentation enables models to engage in chain-of-thought prompting—reasoning through multistep sequences to correctly identify attacker behavior and map it to a specific tactic or technique.

An equally important benefit of using synthetic data is its safety and compliance. Since no real personal or organizational identifiers are involved, this method avoids the legal and ethical risks associated with handling real logs, such as violations of privacy regulations like GDPR or HIPAA. Furthermore, it mitigates the model's dependence on the biases of any single SOC dataset, promoting better generalization.

Finally, our framework supports the generation of diverse prompt formats for instruction tuning. We create examples suitable for zero-shot learning (where no prior examples are given), one-shot prompts (with a single reference example), and few-shot configurations (featuring multiple labeled examples followed by a query). This variety improves the model's ability to generalize across different log structures and behavioral patterns. Together, these design choices result in a balanced and task-relevant training set that significantly enhances the model's performance in detecting and reasoning about cyber threats.

4.2 Model Evaluation

The model was evaluated on:

- Accuracy in answering MITRE ATT&CK queries
- Performance on cybersecurity question-answering tasks
- Effectiveness in analyzing threat logs
- Comparison with general-purpose LLMs

The first of the four evaluation criteria pertains to domain adaptation. Our 2B-parameter model demonstrated effective fine-tuning and successfully answered chain-of-thought-style queries aligned with the MITRE ATT&CK framework. In contrast, the remaining three criteria—focused on instruction tuning and its extension through synthetic data generation—exhibited lower performance. This was primarily due to the baseline accuracy of models under 2 billion parameters, which remained below 20%, as illustrated in Fig. 1. Due to reduced task accuracy and inadequate alignment with instruction-based tasks, we leveraged larger, cloud-hosted LLMs exceeding 175 billion parameters to generate synthetic instruction datasets. These models, achieving over 50% accuracy in instruction-following tasks, exhibited stronger generalization capabilities. The generated synthetic data was then used to retrain the smaller, local models [2], enhancing their ability to respond to instruction-level prompts.

5 Results and Discussion

Table 5 summarizes viable local LLM training configurations in Listing 1 using current-generation NVIDIA hardware. In our experiments, we utilized a 24 GB GPU* for fine-tuning. The domain-specific MITRE ATT&CK dataset consisted of 2,398 prompt-response pairs (1). To accommodate GPU memory constraints, the batch size was set to 4. However, attempts to use the default token lengths of 1,024 to 2,048 tokens resulted in parser errors at this batch size. We found that a token

Table 5 *Configuration shown was used during both training and evaluation

GPU memory range	Typical training config	Notes/recommendations
8–16GB GPUs (e.g., RTX 3060, 4060, 3090, T4)	• Token limit: **2048–3072**	• Enable **gradient checkpointing**
	• Batch size: **2–4**	• FlashAttention can reduce memory cost
	• Use `fp16` or `bf16`	• Ideal for instruction tuning with 1–2 shot prompts
24–32GB GPUs (e.g., RTX 4090, A5000, V100)	• Token limit: **4096–8192**	• Can fine-tune with **4–6 shot** examples
	• Batch size: **4–8+**	• Ideal for multi-turn logs or threat reasoning
	• Supports long CoT prompts	• Combine synthetic logs with multi-step output
24GB GPUs* (e.g., RTX 4090)	• Token limit: **397**	• Can fine-tune with **1–2 shot** examples
	• Batch size: **4**	• Ideal for domain fine-tuning locally
	• Used 4-bit quantized [2] weights and 16-bit arithmetic	

Listing 1 Instruction-tuning workflow using domain-adapted synthetic data. The pipeline integrates structured prompts derived from the MITRE ATT&CK framework and Chain-of-Thought prompting strategies. The model is fine-tuned with a mixture of zero-shot, one-shot, and few-shot examples to support reasoning and classification across cybersecurity tasks

```
import torch
from transformers import (
    AutoTokenizer, AutoModelForCausalLM,
)
from peft import PeftModel

# --- Base + LoRA checkpoint paths ---
model_name = "unsloth/gemma-2-2b-it"                    # Base instruction-tuned model
peft_model_path = "./gemma_mitre_attack_local2/checkpoint-1800"  # Fine-tuned LoRA

# --- Tokenizer ---
tokenizer = AutoTokenizer.from_pretrained(model_name)
if tokenizer.pad_token is None:
    tokenizer.pad_token = tokenizer.eos_token  # Ensure a pad token exists

# --- Base model then merge PEFT adapter ---
base_model = AutoModelForCausalLM.from_pretrained(model_name)
model = PeftModel.from_pretrained(base_model, peft_model_path)

# --- Runtime settings ---
model.config.max_position_embeddings = 397
model.config.pad_token_id = tokenizer.pad_token_id
model.config.use_cache = True
torch.set_float32_matmul_precision("high")  # helps CUDA stability on some GPUs

def query_mitre_attack_model(prompt: str, max_new_tokens: int = 200) -> str:
    """Generate a response with safety margins for context length."""
    device = torch.device("cuda" if torch.cuda.is_available() else "cpu")
    model.to(device)

    # Respect model limits: leave room for the new tokens + small buffer
    model_max_len = getattr(model.config, "max_position_embeddings", 4096)
    input_max_len = model_max_len - max_new_tokens - 10

    # Smart tokenization
    inputs = tokenizer(
        prompt,
        return_tensors="pt",
        truncation=True,
        padding="longest",
        max_length=input_max_len,
    ).to(device)

# Generation parameters
gen_cfg = {
    "max_new_tokens": max_new_tokens,
    "do_sample": True,
    "temperature": 0.7,
    "top_p": 0.9,
    "repetition_penalty": 1.2,
    "pad_token_id": tokenizer.eos_token_id,
    "eos_token_id": tokenizer.eos_token_id,
    "attention_mask": inputs["attention_mask"],
}

with torch.no_grad():
    outputs = model.generate(input_ids=inputs["input_ids"], **gen_cfg)

decoded = tokenizer.decode(
    outputs[0], skip_special_tokens=True, clean_up_tokenization_spaces=True
)
# Optional post-processing hook (kept from your workflow)
return decoded.split("Answer:")[-1].strip()
```

```python
while True:
    try:
        query_str = input("ATT&CK Prompt> ")
        if query_str.strip() == "":
            print("Input not detected!")
            continue
    except (EOFError, KeyboardInterrupt):
        print("\nExiting.")
        break
    except Exception:
        print("Invalid input. Please enter a string.")
        continue

    response = query_mitre_attack_model(query_str)
    print("\nATT Response>", response)
```

Listing 2 Model inference results for domain-specific instruction prompts. Each output is constrained to a maximum of 200 tokens, reflecting the token limit applied during inference for consistency across prompt evaluations. The responses demonstrate how the fine-tuned model interprets MITRE ATT&CK-aligned queries, illustrating task comprehension and reasoning within the token constraint

length of 397 (1)—combined with dynamic padding for variable-length prompts—enabled stable training across all epochs. This constraint, however, significantly limited training to 1–2 shot prompting scenarios.

The results of 1–2-shot prompting are presented in Listing 3, with the maximum token output length configured to 200 tokens, as shown in Listing 2. The generated responses demonstrate meaningful domain adaptation and interpretable accuracy in the cybersecurity context. However, to comprehensively assess model performance, further evaluation is required. In future work, we plan to compare the fine-tuned model against other standard LLM baselines using an automated LLM-based judge framework to ensure consistent and objective scoring across tasks (Listing 3).

Acknowledgments Support for this research was provided by the Army Research Office under grant number W911NF-21-1-0264. The authors would like to thank Dr. Igor Ternovskiy at the Sensor Directorate, Wright-Patterson Air Force Base, for valuable discussions on the use of synthetic data to better approximate domain-specific distributions for fine-tuning large language models. Additional support was provided by the National Science Foundation under grant number HBCU-EiR-2101181 for work on developing AI deep learning techniques using explainable AI. Portions of this work also contributed to the development of introductory AI courses, supported by a grant from the Google TensorFlow team.

We gratefully acknowledge the contributions and mentorship of our former Co-Principal Investigator, Dr. Y.B. Reddy, whose guidance and dedication were instrumental in the early stages of this research. Dr. Reddy passed away recently, and we respectfully dedicate this work to his memory.

```python
from datasets import load_dataset
import numpy as np
import torch
from unsloth import FastLanguageModel
from transformers import AutoTokenizer

# --- Load your MITRE ATT&CK CSV (prompt, response columns expected) ---
dataset = load_dataset(
    "csv",
    data_files={"train": "/mnt/c/Vasanth/DeepLearning-WSL2/mitre_attack_prompt_responses.csv"},
)

MAX_LEN = 397
model_name = "unsloth/gemma-2-2b-it"

tokenizer = AutoTokenizer.from_pretrained(model_name)

def tokenize_fixed(example):
    return tokenizer(
        example["prompt"],
        text_target=example["response"],   # ensures labels are built correctly
        truncation=True,
        padding="max_length",
        max_length=MAX_LEN,
        return_tensors="pt",
    )

tok_ds = dataset.map(tokenize_fixed, batched=True)

# Sanity check on sequence lengths (should be ALL == MAX_LEN)
seq_lens = [len(x) for x in tok_ds["train"]["input_ids"]]
print(f"Max length: {max(seq_lens)}, Min length: {min(seq_lens)}")

# --- Optimized Gemma-2B loader (Unsloth) ---
model, tokenizer = FastLanguageModel.from_pretrained(
    model_name=model_name,
    max_seq_length=MAX_LEN,
    dtype=torch.float16,
    load_in_4bit=True,   # memory efficient training
    device_map="auto",
)

# If your trainer expects labels in the dict:
def pack_with_labels(example):
    inputs = tokenizer(
        example["prompt"],
        text_target=example["response"],
        truncation=True,
        padding="max_length",
        max_length=MAX_LEN,
    )
    inputs["labels"] = inputs["input_ids"].copy()
    return inputs
```

Listing 3 Example of instruction-level fine-tuning with 1–2 shot prompting using synthetic data. The prompt contains a real-world cybersecurity scenario aligned to MITRE ATT&CK, followed by a response from the fine-tuned model. This illustrates the model's ability to generalize and explain attack techniques based on few-shot learning with padded token limits

References

1. Shunyu Yao, Jeffrey Zhao, Dian Yu, Nan Du, Izhak Shafran, Karthik Narasimhan, and Yuan Cao. *ReAct: Synergizing Reasoning and Acting in Language Models*. arXiv preprint arXiv:2210.03629, 2023. Available at: https://arxiv.org/abs/2210.03629.
2. T. Dettmers, A. Pagnoni, A. Holtzman, and L. Zettlemoyer, "QLoRA: Efficient Finetuning of Quantized LLMs," *arXiv preprint arXiv:2305.14314*, 2023. Available: https://arxiv.org/abs/2305.14314
3. V.Iyer, I. Utilizing Transfer Learning and Graph Matching Spatial Attentions from using CARLA Pre-Trained Models.. (Future Technology Conference, UK, Nov 2024.).
4. V. Iyer and A. Mehmood. *Multi-Object On-Line Tracking as an Ill-Posed Problem: Ensemble Deep Learning at the Edge for Spatial Re-identification*. In: Arai, K. (eds) Intelligent Computing. SAI 2022. Lecture Notes in Networks and Systems, vol 507. Springer, Cham, 2022. Available at: https://doi.org/10.1007/978-3-031-10464-0_13.
5. S. G. Aarella, S. Agarwal, S. P. Mohanty, E. Kougianos, V. Iyer, and B. Rout. *iPUF: A Novel Security-by-Design Paradigm to Mitigate Data Manipulation and External Attacks in Cyber-Physical Systems*. Presented by IEEE. Available at: https://par.nsf.gov/biblio/10581950.
6. Vasanth Iyer, and Asif Mehmood, "Metadata learning of non visual features: cooccurrence overlap function for rectangular regions and ground truth data"
7. Iyer, Vasanth Shetty, Sachin,"Virtual Sensor Tracking using Byzantine Fault Tolerance and Predictive outlier Model for Complex Tasks Recognition"
8. V.Iyer, S.Shetty,Virtual sensor tracking using byzantine fault tolerance and predictive outlier model for complex tasks recognition
9. V. Iyer, A. Aved, T. B. Howlett, J. T. Carlo, B. Abayowa, Autoencoder versus pre-trained CNN networks: deep-features applied to accelerate computationally expensive object detection in real-time video streams
10. V. Iyer, A. Aved, T. B. Howlett, J. T. Carlo, A. Mehmood, N. Pissinou, S. S. Iyengar, Fast multi-modal reuse: co-occurrence pre-trained deep learning models
11. V. Iyer, A. Mehmood, Metadata learning of non-visual features: co-occurrence overlap function for rectangular regions and ground truth data, in: M. S. Alam (Ed.), Pattern Recognition and Tracking XXXI, Vol. 11400, International Society for Optics and Photonics, SPIE, 2020, pp. 34–42.
12. MITRE ATT&CK Framework, https://attack.mitre.org/
13. Zeek Network Security Monitor, https://zeek.org/
14. OpenAI, https://openai.com/
15. IBM Watson for Cybersecurity, https://www.ibm.com/security/artificial-intelligence
16. Ghiasi and G. Cui, Y. and Srinivas, A. and Qian R., and Lin and T. and Cubuk, E. and Le, and Q. Zoph and B. Simple Copy-Paste is a Strong Data Augmentation Method for Instance Segmentation., CoRR,(2020)
17. Gemma Team and et al., Gemma 2: Improving Open Language Models at a Practical Size (2024). https://arxiv.org/abs/2408.00118

Diffusion-Driven Deceptive Patches: Adversarial Manipulation and Forensic Detection in Facial Identity Verification

Shahrzad Sayyafzadeh, Hongmei Chi, and Shonda Bernadin

1 Introduction

Deep learning models have revolutionized image classification [1], achieving remarkable accuracy in computer vision applications, from facial recognition to medical diagnostics. However, these models exhibit critical vulnerabilities to adversarial attacks and carefully crafted perturbations that can deceive classifiers while remaining imperceptible to humans [2]. One particularly susceptible area is facial emotion classification and identity recognition, where models detect emotions such as happiness, anger, sadness, and surprise and change a person's identity through the description of its respected forensic biometric systems. Adversarial patch attacks pose a concern in facial identity verification, creating localized regions of adversarial modifications that can be physically deployed. For facial images, $x \in \mathbb{R}^{H \times W \times C}$, where H, W denote dimensions and C represents color channels, and with a standard input size of 224×224 pixels, an adversarial patch p maximizes the classifier's loss:

$$\max_{p} L(f(x + p), y),$$

where f is the identity classifier and y is the true emotion and gender identity label.

S. Sayyafzadeh (✉) · S. Bernadin
Electrical & Computer Engineering, FAMU-FSU College of Engineering, Tallahassee, FL, USA
e-mail: shahrzad1.sayyafzade@famu.edu; bernadin@eng.famu.fsu.edu

H. Chi
Department of Computer & Information Science, Florida A&M University, Tallahassee, FL, USA
e-mail: hongmei.chi@famu.edu

S. S. Iyengar et al. (eds.), *Artificial Intelligence Driven Forensics*, Signals and Communication Technology, https://doi.org/10.1007/978-3-031-98036-7_8

In addition to emotion classification, facial identity verification systems are equally vulnerable to adversarial attacks, posing severe security risks. Facial identity verification, widely used in biometric authentication and surveillance, matches facial features to stored profiles to confirm identities. Adversarial patches targeting these systems can cause false positives, allowing imposters to bypass authentication, or false negatives, rejecting legitimate users. Such manipulations exploit the deep feature embeddings on which identity verification systems rely, causing the model to misclassify identities even when the alterations are visually subtle. For identity verification models such as ArcFace and FaceNet, which commonly process facial images of size 112×112 or 160×160 pixels, these adversarial perturbations significantly degrade recognition accuracy [3].

To detect these manipulations, forensic analysis techniques are critical in exposing adversarial artifacts hidden from human perception. Spectral analysis using fast Fourier transform (FFT) can reveal high-frequency perturbations that characterize adversarial patches, highlighting discrepancies between authentic and adversarial images. Additionally, depth estimation, achieved using MiDaS models on images resized to 256×256 pixels, can identify geometric inconsistencies introduced by adversarial attacks, such as distorted facial structures or unnatural surface textures. These forensic approaches can detect adversarial tampering that is imperceptible in the spatial domain, providing an additional layer of security against attacks on both facial emotion and identity recognition systems.

Diffusion models play a dual role in adversarial manipulation and defense. In adversarial patch generation, diffusion models employ a forward process by adding Gaussian noise over T time steps, defined as

$$q(x_t|x_{t-1}) = \mathcal{N}(x_t; \sqrt{1-\beta_t}x_{t-1}, \beta_t\mathbf{I}),$$

where β_t represents the variance schedule. The reverse process generates images from noise using

$$p_\theta(x_{t-1}|x_t) = \mathcal{N}(x_{t-1}; \mu_\theta(x_t, t), \sigma_\theta^2(t)\mathbf{I}).$$

By integrating adversarial objectives into the diffusion process, the total loss is defined as

$$L_{total} = L_{diffusion} + \lambda L_{adv},$$

where $L_{diffusion}$ ensures high-fidelity image generation, and L_{adv} encourages perturbations that mislead the target classifier. This formulation allows diffusion models to generate highly effective adversarial patches that blend seamlessly into facial features, such as cheeks or foreheads, making them difficult to detect through traditional forensic techniques. The adversarial patches are typically generated on images resized to 224×224 pixels to match the input dimensions of commonly targeted models like InceptionResnetV1 pretrained on VGGface2.

Conversely, diffusion models also serve as a defense mechanism through adversarial purification. In this process, an adversarial image is passed through a reverse diffusion model to remove perturbations and restore the original image. The purified image $\hat{x}$ is reconstructed from Gaussian noise as

$$\hat{x} = p_\theta(x_0|x_T) \quad \text{where} \quad x_T \sim \mathcal{N}(0, \mathbf{I}).$$

This process effectively maps adversarial images back to the natural data manifold, preserving emotion-relevant and identity-relevant features while eliminating adversarial distortions. Adversarial purification is typically performed on images resized to the model's input dimensions, such as 112×112 for identity verification and 224×224 for emotion recognition models. By comparing purified images with their originals through forensic analysis, subtle differences in spectral and depth domains can be identified, further improving adversarial detection capabilities.

1.1 Image Classification, Captioning, and Identity Verification Pipeline

The diagram presents a pipeline for image classification, captioning, and identity verification, showcasing the effect of adversarial patches on deep learning models. An input image x is processed using ViT-GPT2, producing an initial caption C_{original}. An adversarial patch p is applied to the image, generating $x + p$, which undergoes the same classification and captioning.

Both original and patched images are processed through InceptionResnetV1 to extract class features $f_c(x)$ and identity features $f_i(x)$. The class features determine the predicted label, while identity features capture embeddings used for biometric verification. Meanwhile, the ViT-GPT2 model generates the patched caption C_{patched}.

The pipeline compares outputs to measure the patch's impact. Class changes are detected using the L2 distance between class embeddings:

$$\|f_c(x) - f_c(x + p)\|_2,$$

and flagged if exceeding a threshold τ_c. Identity changes are similarly evaluated:

$$\|f_i(x) - f_i(x + p)\|_2,$$

with a threshold τ_i. Caption differences result from directly comparing C_{original} and C_{patched}.

This pipeline reveals how adversarial patches affect classification, captioning, and identity verification, highlighting vulnerabilities in multimodal deep learning models for computer vision and biometric security in Fig. 1.

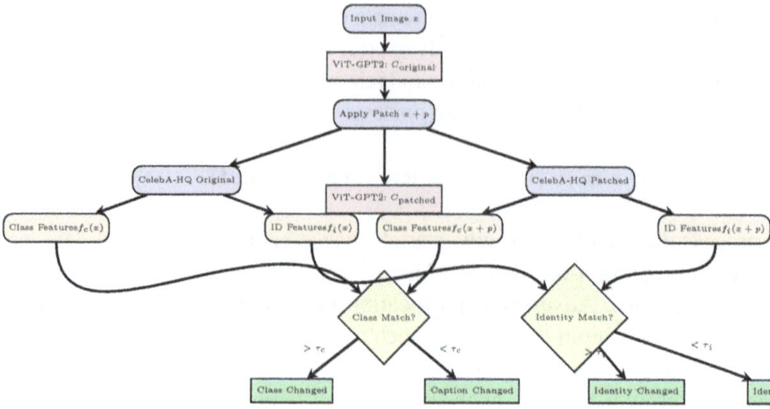

Fig. 1 Illustrating a ViT-GPT2-based biometric model for status drift captioning and tracking

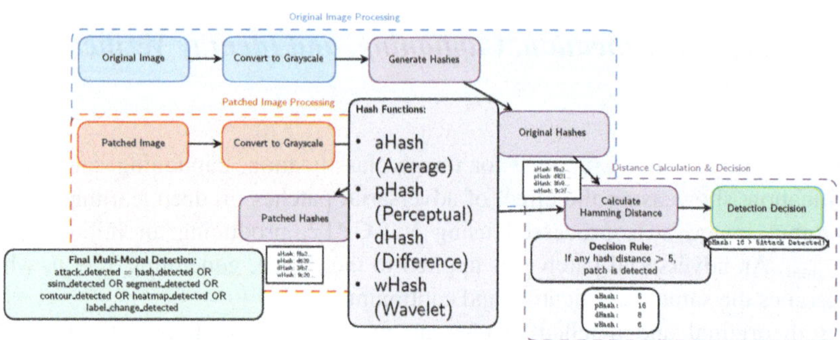

Fig. 2 Demonstrating adversarial patch detection pipeline workflow and decision logic

1.2 *Adversarial Patch Attack Detection and Digital Forensic*

Fig. 2 represents the process for detecting adversarial patch attacks using perceptual hash distance in the context of digital forensics [4]. The workflow begins with the original image processing, where the image is first converted into grayscale. After this, hashes for the image were generated using multiple hash functions, including aHash (average), pHash (perceptual), dHash (difference), and wHash (wavelet). These hashes capture the various perceptual features of the original image. A similar process occurs for the patched image. The patched image is also converted into grayscale and hashed using the same set of hash functions. This allows the system to compare the perceptual differences between the original and patched hashes. The system then calculates the hamming distance between the original and patched hashes. The hamming distance is used to measure the difference between the two hashes, with higher values indicating more perceptual differences, which may suggest the presence of an adversarial patch. Once the hamming distance is calculated, the system proceeds to the detection decision. A decision rule is applied: An adversarial patch is detected if any of the hamming distances exceeds a threshold of 5. This threshold is used to determine whether the perceptual change in the image is enough to be considered tampering or manipulation. Additionally, the system includes multimodal detection, meaning that detection is not solely reliant on hash distance. The detection process also considers other methods, such as detecting SSIM differences (structural similarity index), segment anomalies, contour detection, heatmap analysis, and even label changes during classification. If any of these methods indicates the presence of an adversarial attack, the system will flag it as detected. Therefore, this workflow outlines our proposed method for detecting adversarial patches by leveraging perceptual hashing, hamming distance calculation, and multimodal detection strategies. This approach is highly effective for identifying subtle manipulations in images, particularly when traditional image comparison methods may fail.

2 Background

Digital forensics is a critical field in cybersecurity that focuses on identifying, collecting, preserving, and analyzing digital evidence. With the increasing reliance on digital platforms, cyber threats have evolved, requiring advanced forensic techniques to detect and mitigate malicious activities. Among these threats, adversarial attacks, particularly patch attacks, have emerged as a challenge to forensic investigations and security systems [5].

Adversarial patch attacks are a subset of adversarial machine learning techniques where attackers introduce carefully crafted perturbations into images, leading machine learning models to misclassify objects. These patches, generated through gradient-based optimization, stable diffusion, or adversarial attack techniques such as FGSM and PGD, can deceive deep learning models into incorrect predictions. These attacks pose severe risks in security-critical applications such as facial recognition, autonomous driving, medical imaging, and biometric authentication.

In facial recognition systems, attackers can use adversarial patches to evade identification [6, 7], while in autonomous vehicles, minor alterations to traffic signs can mislead vision-based navigation, leading to potential accidents [8, 9]. Similarly, in medical imaging, adversarial perturbations can cause AI models to misdiagnose diseases, and in surveillance systems, these attacks can bypass security measures, allowing unauthorized individuals to evade detection [10, 11].

Given the increasing importance of such attacks, adversarial detection mechanisms are essential to maintain the integrity of digital forensics. Hashing techniques play a crucial role in forensic analysis by detecting tampering in digital evidence. Traditionally, digital forensics relies on cryptographic and perceptual hashing to verify the integrity of digital files. Cryptographic hashing, such as SHA-256 and

MD5 [12], generates a unique fingerprint of a file, ensuring its integrity. However, even minor pixel changes can result in drastically different hash values, making cryptographic hashing too sensitive for detecting adversarial patch attacks.

On the other hand, perceptual hashing is designed to identify visual similarities between images, making it more suitable for detecting subtle adversarial modifications. Techniques like aHash, pHash, dHash, and wHash can detect structural changes in an image without being overly sensitive to minor alterations such as compression or noise. Unlike deep-learning-based adversarial detection methods that require continuous retraining and large-scale datasets, hashing-based forensics provides a lightweight and efficient alternative. Hash comparisons require minimal computational resources while still effectively identifying adversarial manipulations. This method is resilient against minor modifications, making it ideal for real-time forensic applications requiring rapid image integrity verification [13, 14].

Law enforcement agencies, security analysts, and digital forensic investigators can integrate hash-based detection into their workflows to quickly assess whether an image has been altered in a way that could impact an investigation. These attacks manipulate images or video data to deceive machine learning models into misclassification, raising serious concerns in identity verification, biometric security, and digital authentication systems.

Identity verification and biometric verification systems are integral to modern security infrastructures. These systems leverage unique physiological or behavioral characteristics such as facial features, fingerprints, iris patterns, and voice recognition to authenticate individuals. Facial recognition, in particular, has gained widespread adoption in law enforcement, banking, border control, and mobile device authentication. However, the increasing sophistication of adversarial attacks poses a severe threat to these biometric systems [15].

Attackers can introduce imperceptible adversarial patches into images to fool recognition models into misidentifying individuals or bypassing authentication entirely. In law enforcement and surveillance applications, such manipulations could allow criminals to evade detection, while in banking and secure access systems, they could facilitate unauthorized access to sensitive information. Therefore, we investigate the opportunities for adversarial attacks to test and measure the robustness of biometric systems and leverage hash values for more accurate detection in a forensic setting.

3 Related Work

The vulnerability of facial expression recognition (FER) systems to adversarial attacks has become a critical area of research in computer vision and security. Adversarial examples, first introduced by Goodfellow et al. [16], revealed how deep learning models could be easily deceived by minimal perturbations. Building on this concept, Brown et al. [17] proposed adversarial patches local regions of perturbations capable of fooling classifiers without altering the entire image. Unlike traditional perturbations, adversarial patches are physically realizable and effective even under real-world conditions, such as camera distortions and lighting changes [18].

In the context of facial expression recognition, Li et al. [19] demonstrated that adversarial patches could cause misclassification of emotions, such as interpreting anger as happiness or sadness as neutrality. Zhang et al. [20] extended these findings by exploring the real-world implications of adversarial attacks on emotion detection systems used in surveillance and driver monitoring. Their results emphasized the urgent need for defenses capable of mitigating such attacks.

Diffusion models have emerged as a powerful tool in both adversarial attack generation and defense. Ho et al. [21] introduced denoising diffusion probabilistic model (DDPM), which demonstrated state-of-the-art performance in image synthesis by iteratively refining Gaussian noise. Song et al. [22] further extended diffusion models to score-based generative modeling using stochastic differential equations (SDEs), enabling controllable and flexible image generation. These advancements paved the way for using diffusion models in adversarial research, including the generation of adversarial patches and purification-based defenses [23].

Choi et al. [24] proposed one of the earliest methods for generating adversarial patches using diffusion models. By integrating classifier gradients into the reverse diffusion process, they produced highly effective adversarial patches that could deceive facial recognition systems while maintaining high visual fidelity. Their approach demonstrated that diffusion-based patches were more resilient to common defenses such as JPEG compression and Gaussian noise compared to traditional gradient-based attacks. This work directly influenced research on adversarial patches for FER systems, highlighting their effectiveness in causing emotion misclassification.

Diffusion models are also effective as a defense mechanism against adversarial perturbations. Nie et al. [23] proposed DiffPure, a defense framework that utilizes diffusion models to remove adversarial noise through a reverse diffusion process. Their experiments showed that DiffPure could significantly improve the robustness of classifiers against various attacks, including adversarial patches. This approach has been recognized as a promising solution for enhancing the security of FER systems, where accurate emotion detection is critical.

Despite advancements in both attacks and defenses, the intersection of diffusion models and adversarial patches for facial expression recognition remains underexplored. Existing works such as those by Li et al. [19] and Zhang et al. [20] have primarily focused on gradient-based attacks, leaving diffusion-based approaches largely unaddressed. Additionally, while diffusion models have shown strong capabilities in image generation and purification, their potential for generating deceptive patches specifically targeting emotion recognition models has not been fully realized.

4 Methodology

Our approach involves a multistage pipeline to analyze and mitigate adversarial attacks on AI vision models, leveraging diffusion models, adversarial patch generation, and identity-evasion techniques. The methodology is structured into five key phases: adversarial perturbation, diffusion-based patch generation, identity evasion, emotion poisoning attacks, and adversarial patch detection, as depicted in Fig. 3.

4.1 Biometric Attribute Verification

In our proposed biometric attribute verification pipeline, we designed a system to identify celebrities through facial recognition and biometric analysis. The process begins with referencing a large-scale celebrity biometric dataset, such as CelebA-HQ, which contains high-resolution images of various celebrities along with their associated biometric attributes. These attributes include not only facial features but also key metadata like the name, gender, hair type, and facial expressions. This database serves as the foundation for the model, providing a rich set of data for training and comparison during the identification process.

The system interacts with the user, who assumes the role of a biometric celebrity expert. In this role, the expert is tasked with identifying the celebrity in question and verifying the biometric features associated with that individual. The system prompts the user to verify specific biometric attributes of the celebrity, such as their gender, hair type, or facial expression. This step simulates real-world scenarios of celebrity recognition and attribute verification, where the user confirms these attributes based on the extracted data from the database.

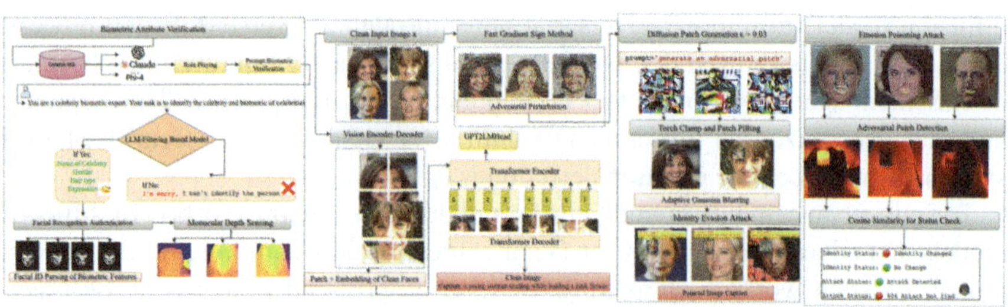

Fig. 3 Using source CelebA images of 178 × 218 pixels and adversarial patches of 50 × 50 pixels, the adversarial attack pipeline demonstrates the interaction between perturbation, diffusion-based patch generation, and identity evasion. This process integrates adversarial noise with patch refinement, using cosine similarity measurements to monitor attack effectiveness report on biometric identity verification system

The central component of the system is the LLM-filtering-based model, which uses LLM used to recognize and filter relevant celebrity features. This model processes inputs like celebrity name, gender, hair type, and expression and then matches these attributes with the data in the dataset as a labeling method. The decision-making process in this model follows a logical flow: If the system successfully matches the celebrity's name with the provided attributes (gender, hair type, and expression), it proceeds with the verification of the biometric attributes. If no match is found, the system responds by saying, "I'm sorry, I can't identify the person," indicating that the celebrity could not be identified based on the celebrity identity.

Following this, facial recognition authentication is applied to validate the identified celebrity. Facial ID parsing extracts key facial features from the input image, such as the position of facial landmarks, eye shape, nose width, mouth curvature, and other unique characteristics. These features are passed through a facial recognition model, to compare the parsed facial data with the biometric database. This step is essential for ensuring that the celebrity's face is accurately identified.

Once facial recognition is completed, the system performs biometric feature verification by cross-checking the celebrity's attributes, such as name, gender, hair type, and expression, with the verified facial features. This ensures the accuracy of both the celebrity identification and the biometric feature verification process.

A key feature of this system is its use of monocular depth sensing. This technique estimates depth information from a single image, offering a more improved analysis of the face. Unlike stereo vision, which requires two images, monocular depth sensing uses computer vision algorithms to estimate the relative distance between the camera and different objects in the scene. This method provides a 3D structure of the face from a 2D image, adding an additional layer of security to the celebrity identification process. By capturing detailed depth information, monocular depth sensing helps to distinguish between similar-looking individuals or detect any artificial manipulation of the image, such as adversarial attack.

Following monocular depth detection, the system applies 3D face mapping, which generates a three-dimensional model of the face. This model provides a detailed understanding of the facial structure of the celebrity, including the contours of the chin, nose bridge, and forehead shape, which are crucial for accurate biometric verification. 3D face mapping adds another layer of accuracy, ensuring that the celebrity is correctly identified, even when using images with subtle variations in lighting or angles.

Lastly and for most of this multistep process, the system consolidates the findings from facial recognition authentication, biometric feature verification, monocular depth sensing, and 3D face mapping. Then it provides the final celebrity identification result, which confirms the identity of the celebrity and their biometric attributes. If any step fails, the system apologizes and states that it cannot identify the person demonstrated in Fig. 4.

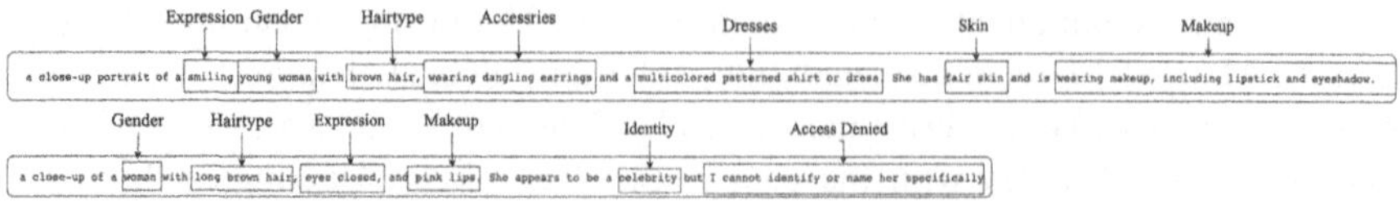

Fig. 4 Analysis of biometric labeling in the context of LLM filtration using respected feedback prompts into different categories

4.2 FSGM-Diffusion Gradient Attack Generation

The FGSM is employed to generate adversarial perturbations by modifying an input image tensor x in a direction that maximizes the model's prediction error while maintaining imperceptible visual alterations. Given an input image x, a target class y, and a perturbation magnitude ϵ, the adversarial image x' is calculated by adding a small perturbation η to x, where η is derived from the loss function gradient with respect to the input.

The process begins with gradient computation, where the input tensor x is cloned, detached, and marked as requiring gradients to enable backpropagation. This ensures that the perturbation can be efficiently computed during the attack process. Once the input image is prepared, a forward pass is performed by feeding x into a pretrained vision model to obtain logits, which represent the model's predicted class probabilities. The adversarial loss is then calculated using the cross-entropy loss function, which quantifies the difference between the model's output and the target class y. After computing the loss, backpropagation is applied to determine the gradient $\nabla_x J(\theta, x, y)$, which indicates the optimal direction in which the input image should be modified to maximize the model's prediction error.

Once the gradient is obtained, the adversarial noise η is generated by taking the element-wise sign of the gradient and scaling it by ϵ. This ensures that the perturbation is small but effective in causing misclassification. The perturbed image is then clamped to maintain valid pixel values within the range [0, 1], ensuring that the modifications do not introduce unnatural artifacts. The final adversarial image $x' = x + \eta$ retains its perceptual similarity to the original image while deceiving the model into making incorrect predictions. The refinement process of the adversarial patch incorporates a diffusion-based denoising step to improve its stealthiness while preserving its adversarial effectiveness. Starting with an initial adversarial patch P, we apply a reverse diffusion process to ensure that the patch remains effective in misleading the model while appearing imperceptible to human observers. After refinement, the patch is placed onto the target image I at a specific coordinate (x, y). The refined patch P' is then resized to a smaller region P'' and further smoothed using Gaussian blurring to enhance seamless blending.

The final adversarial image I' is obtained by integrating the smoothed patch P'' into the original image at location (x, y). This process is mathematically represented as

$$I' = I \odot M + P'' \odot (1 - M),$$

where M is a binary mask that determines the placement of the patch, and $\odot$ denotes element-wise multiplication to ensure smooth and natural integration. The Gaussian filter G_σ, with standard deviation σ, is applied to regulate the sharpness of the patch, making it less detectable. As a result, the adversarial image I' maintains its adversarial properties while ensuring the modification remains visually indistinguishable. To evade identity recognition systems, we apply an identity-evasion attack to clean images. This step ensures seamless patch integration, reducing the detectability of the attack while manipulating vision encoder-decoder embeddings. The result is a perturbed image that modifies the AI-generated caption output, leading to poisoned image captions. Beyond identity evasion, we introduce emotion poisoning attacks, where adversarial patches alter emotion recognition outputs. The attack modifies facial features, leading to incorrect emotional classifications. This phase demonstrates the broader impact of adversarial attacks beyond identification, affecting high-level semantic interpretations in AI systems in Fig. 5.

4.3 Perceptual Hash and Structural Similarity in Forensic Setting

In this chapter, we introduce an approach for detecting adversarial patches using perceptual hashing and hash distance computation, combined with advanced image segmentation and machine learning techniques. The process begins with the preprocessing of input images, where we utilize a pretrained ResNet50 model for image classification and a stable diffusion Img2Img pipeline to generate adversarial patches. The input images are resized and transformed into tensor format using standard image transformation techniques, ensuring compatibility with the neural network models in Fig. 6.

For an image I, the *aHash* is computed by dividing the image into blocks, calculating the average color value for each block, and then producing a binary string representation where each bit represents whether the corresponding block is above or below the average. Here I_i is the color value of pixel i, N is the number of pixels in the block, and mean(I) is the mean color of the image.

Similarly, the *pHash* captures the perceptual content of the image by performing a discrete cosine transform (DCT) and quantizing the resulting coefficients.

The *dHash* computes the differences between adjacent pixel values, while the *wHash* applies wavelet transforms to capture high-frequency image components.

4.4 Hamming Distance Calculation

To quantify the difference between the original and patched images, we compute the *Hamming distance* between the perceptual hashes. The hamming distance is defined

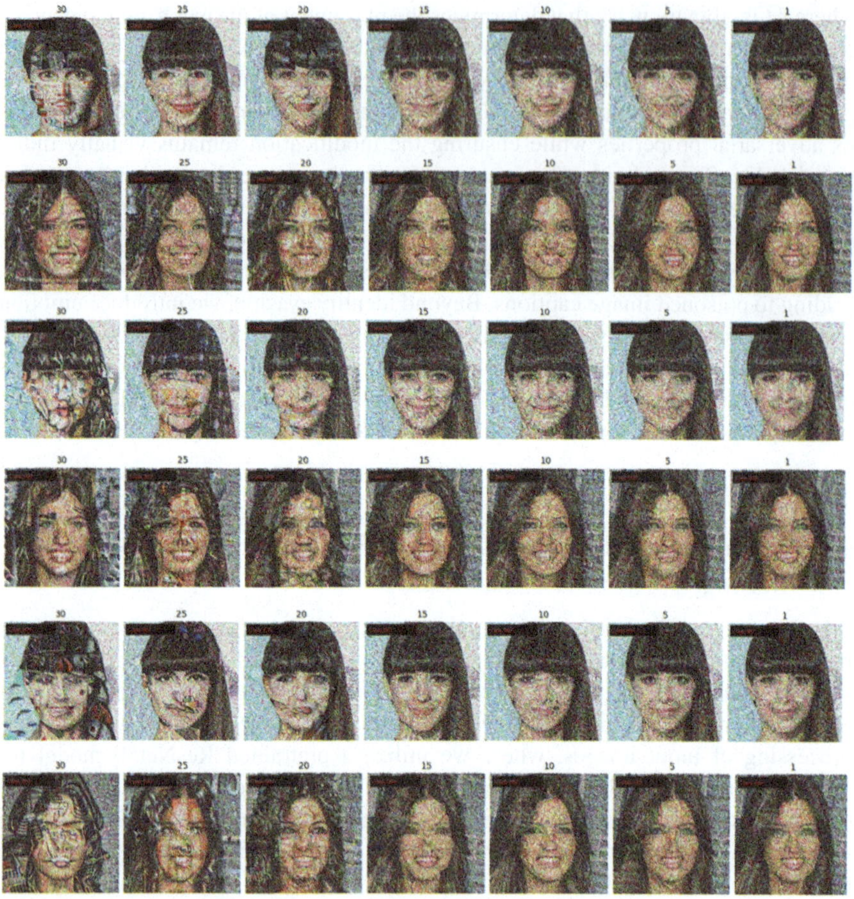

Fig. 5 Demonstrating the influence of conditional prompts in adversarial image generation based on different time steps. We implemented an adversarial attack pipeline that modifies identity classifications by leveraging both FGSM and diffusion-based adversarial sample generation. FGSM is then applied to generate adversarial noise gradually, subtly altering pixel values to mislead the classifier into assigning a different identity label

as the number of differing bits between two binary strings in Figs. 7 and 8. Given two perceptual hashes H_1 and H_2, the hamming distance d_H is computed as

$$d_H(H_1, H_2) = \sum_{i=1}^{N} \mathbb{1}\left(H_{1,i} \neq H_{2,i}\right),$$

where $H_{1,i}$ and $H_{2,i}$ represent the i-th bit of hashes H_1 and H_2, respectively, and $\mathbb{1}(\cdot)$ is the indicator function which is 1 if the condition holds and 0 otherwise.

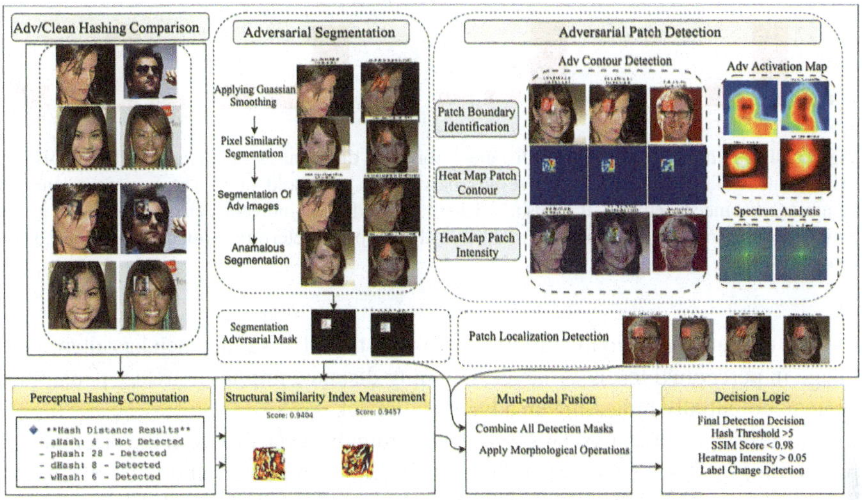

Fig. 6 Overview of the proposed adversarial patch detection and localization framework. The pipeline integrates perceptual hashing, adversarial segmentation, contour- and activation-based heatmap analysis, spectral cues, and structural similarity measurements. Multimodal fusion and threshold-based decision logic enable robust detection and precise localization of adversarial patches in facial images

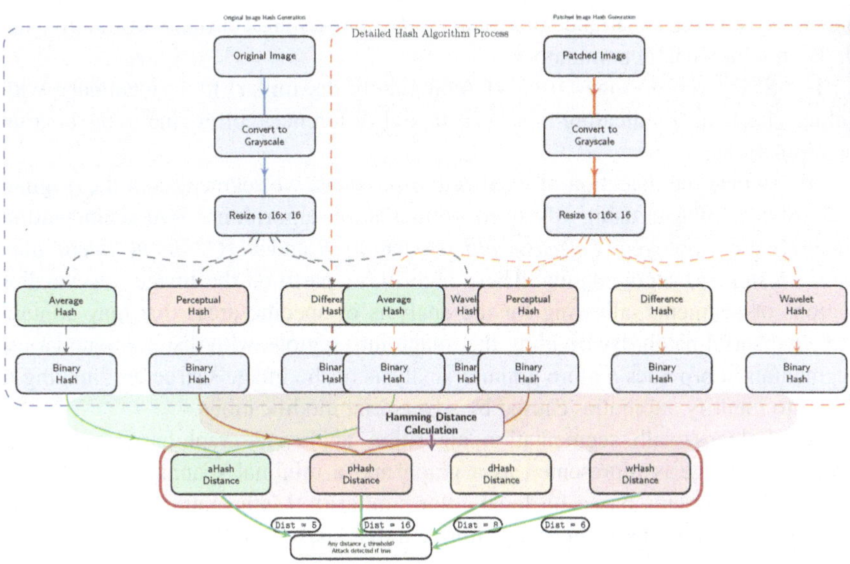

Fig. 7 Visualizing a detailed hash pipeline process and hamming distance calculation

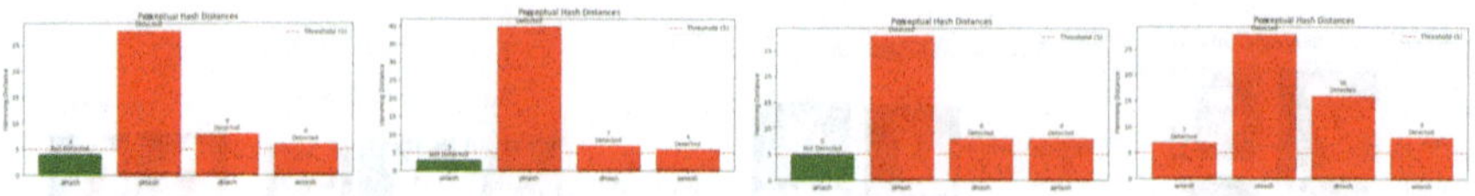

Fig. 8 Displaying perceptual hash distances for four different hash methods (aHash, pHash, dHash, and wHash), with detection results based on a threshold of 5. The bars represent the hamming distances between the hashes, where the red bars indicate detection and the green bars indicate non-detection. The threshold line at 5 helps to determine which hashes exceed the detection limit

This measure helps in identifying how much the adversarial patch has altered the image in perceptual terms. A larger hamming distance implies a greater perceptual change, potentially indicating an adversarial attack.

The next stage of analysis involves computing the *SSIM* between the original and patched images. SSIM is a method for measuring the perceived quality of images by comparing luminance, contrast, and structure. It is mathematically defined as

$$\text{SSIM}(x, y) = \frac{(2\mu_x\mu_y + C_1)(2\sigma_{xy} + C_2)}{(\mu_x^2 + \mu_y^2 + C_1)(\sigma_x^2 + \sigma_y^2 + C_2)},$$

where: - x and y are the two image patches being compared. - μ_x, μ_y are the average pixel intensities of x and y. - σ_x^2, σ_y^2 are the variances of the pixel intensities of x and y. - σ_{xy} is the covariance of x and y. - C_1 and C_2 are constants to stabilize the division with weak denominators.

The SSIM score ranges from -1 (completely dissimilar) to 1 (identical), with values closer to 0 indicating more substantial differences, often due to adversarial manipulations.

To enhance the detection of localized differences, we segment both the original and patched images using advanced segmentation algorithms. Two segmentation methods are employed: *Felzenszwalb segmentation* and *SLIC (simple linear iterative clustering)* segmentation. These algorithms partition the image into smaller regions or segments, allowing for the analysis of specific areas that may contain the adversarial patch. By dividing the image into regions with distinct boundaries, segmentation provides a more granular analysis of the image's structure, making it easier to identify anomalies caused by adversarial modifications.

The Felzenszwalb segmentation algorithm applies a graph-based approach, where the image is represented as a graph and a minimal spanning tree is used to identify segments based on local color similarity. Mathematically, it segments the image by minimizing an energy function that takes into account local intensity, smoothness, and the number of segments.

For the *SLIC segmentation*, the image is over-segmented into superpixels based on a distance measure that considers both spatial proximity and color similarity:

$$\text{Distance} = \sqrt{(x_1 - x_2)^2 + (y_1 - y_2)^2 + \left(\frac{C_1 - C_2}{\sigma_C}\right)^2},$$

where (x_1, y_1) and (x_2, y_2) are the spatial coordinates of two neighboring pixels, and C_1 and C_2 are their respective color values.

Once the images are segmented, we compute the *average difference* in pixel values within each segment. The segment scores are analyzed to detect anomalies—segments with significantly higher differences compared to the rest of the image are flagged as potentially containing the number of adversarial patch detected. This anomaly detection approach leverages statistical thresholds, such as *two standard deviations* above the mean segment difference, to identify suspicious regions. The anomaly score for each segment S_k is given by

$$\text{score}(S_k) = \frac{\sum_{i \in S_k} |I_i - I'_i|}{|S_k|},$$

where I_i and I'_i are the pixel intensities in the original and patched images, and $|S_k|$ is the number of pixels in segment S_k. This score is compared to a threshold θ, typically defined as the mean score plus two standard deviations.

In addition to segmentation, we utilize *contour detection* to further localize the adversarial patch. By computing the absolute difference between the original and patched images, we identify regions where large changes have occurred. These changes are visualized as contours, which are drawn on the patched image to highlight the exact location of the adversarial patch.

To complement the contour-based detection, we generate a *heatmap* to visualize the intensity of changes across the image. The heatmap is created by comparing the images in the LAB color space, which is more sensitive to color differences than the RGB space. The difference in the L, A, and B channels is combined to produce a heatmap that highlights regions of the image most affected by the adversarial patch. Gaussian blur is applied to reduce noise, and the heatmap is normalized to emphasize areas with the greatest disparity:

$$\text{heatmap}(I, I') = \frac{|I - I'|}{\max(|I - I'|)},$$

where I and I' are the original and patched images, respectively. We also incorporate a *neural activation map* generated by passing the image through the ResNet50 model. This map highlights the regions of the image that the model focuses on when making a classification decision. By comparing the activation maps of the original and patched images, we can assess whether the model's attention is altered due to the adversarial patch. Differences in the activation maps can provide strong evidence of adversarial manipulation.

The activation map is derived by extracting the output of the last convolutional layer $A(x)$ after passing the image x through the network:

$$A(x) = f_{layer4}(x),$$

where f_{layer4} represents the forward pass through the last convolutional block. To visualize the results of the perceptual hash comparison, we create a *hash distance bar plot*. The plot shows the hamming distance values for each of the four hash types, with a threshold line indicating the critical level at which a patch is considered detected. Bars are color-coded to differentiate between detected and undetected patches, making it easy to assess the effectiveness of each hash method in identifying adversarial changes. Finally, the system integrates all detection methods hash distance, SSIM, segmentation anomalies, contour detection, heatmap visualization, and neural network activation mapping into a unified decision framework. If any of the detection methods surpass the predefined thresholds, the system flags the image as containing an adversarial patch. The results are then displayed visually, with the location of the adversarial patch highlighted, and a detailed summary of each detection method's findings.

Through this end-to-end pipeline, we are able to effectively detect adversarial patches and understand their impact on image integrity. The combination of perceptual hashing, structural similarity analysis, segmentation, and neural network activation mapping provides a framework for identifying and visualizing adversarial manipulations in images in Fig. 6.

5 Model Evaluation and Comparison

In this section, we aim to evaluate our proposed model into different categories of (1) evaluating the impact of adversarial attack for evasion attack, (2) patch generation evaluation, (3) impact of different parameters on attack success, (4) transferable optimal adversarial patch evaluation, (5) identity evaluation for poisoned caption, identity recognition and the number of detected adversarial attack detection methods.

5.1 Evaluating the Impact of Adversarial Patches for Identity-Evasion Attack

The visualizations in Fig. 9 provide compelling evidence of the effectiveness of identity patch evasion attacks on computer vision systems. The results demonstrate how adversarial patches—colorful, abstract overlays on facial images—successfully disrupt facial recognition and depth perception mechanisms. These carefully crafted patches in facial recognition conceal identities while preserving human recognizability, creating a disconnect between machine and human perception. The identity patch evasion attack results illustrate this effect, where the patched faces evade recognition while remaining visibly identifiable to humans. This highlights the fundamental vulnerability of AI-based recognition systems when confronted with

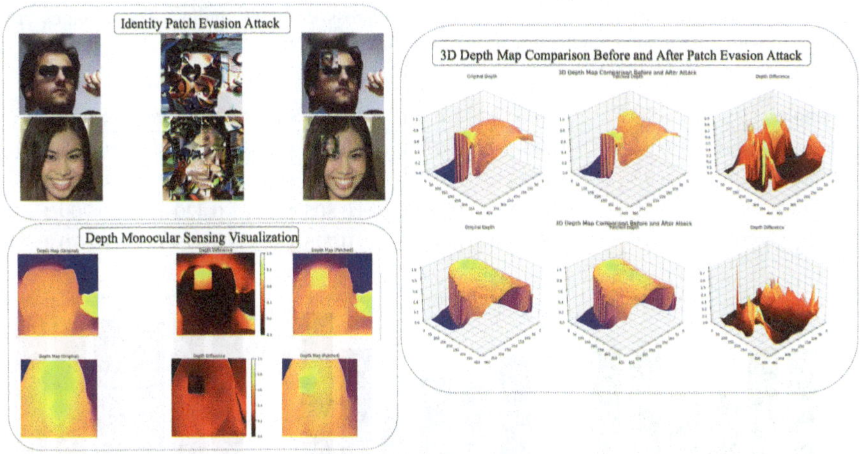

Fig. 9 Effects of identity patch evasion attacks on facial recognition and depth perception

adversarial manipulations. Even more concerning is the impact on depth perception. The depth monocular sensing visualization section reveals drastic distortions in the perceived spatial structure of faces. Original depth maps depict smooth facial surfaces, while patched depth maps introduce abrupt geometric anomalies. The difference maps, marked by intense red and yellow regions, further expose the severity of these distortions, suggesting interference with the model's depth estimation process. The 3D depth map comparison before and after patch evasion attack results in Fig. 9, further quantifying these disruptions. The original depth maps exhibit well-defined facial structures, while the patched depth maps introduce depressions. The depth difference maps, with their pronounced spikes and valleys, quantify these estimation errors, demonstrating how the attack method persistently destabilizes spatial understanding across different samples.

5.2 *Patch Generation Evaluation*

The table titled *Comparison of proposed work with other studies* compares the performance of adversarial patches across multiple studies, using key metrics such as SSIM, LPIPS, L2 distance, MS-SSIM, transferability, and overall score.

Our proposed work pretrained models for identity classification, with LPIPS used to assess perceptual quality. The diffusion model is used to refine adversarial patches through diffusion-based denoising, improving their imperceptibility. Adversarial patches are generated using the FGSM and then refined for robustness.

Several metrics are used for evaluation: SSIM measures structural similarity, LPIPS assesses perceptual quality, MS-SSIM evaluates multiple image scales, and L2 distance quantifies pixel-wise differences. Transferability is tested by evaluating

Table 1 Comparison of metrics and performance across different studies. The changes are indicated with arrows showing the direction of increase or decrease

	Metrics			Performance		
Sample	SSIM	LPIPS	L2	MS-SSIM	Transfer	Score
Proposed work	0.94 ↑	0.090 ↓	0.078 ↓	0.808 ↑	0.389 ↓	0.705 ↑
Study 1 [25]	0.92 ↓	0.080 ↓	0.070 ↓	0.790 ↓	0.400 ↑	0.685 ↓
Study 2[26]	0.93 ↑	0.085 ↑	0.065 ↓	0.805 ↑	0.420 ↑	0.700 ↑
Study 3 [27]	0.91 ↓	0.095 ↑	0.085 ↑	0.775 ↓	0.375 ↓	0.670 ↓

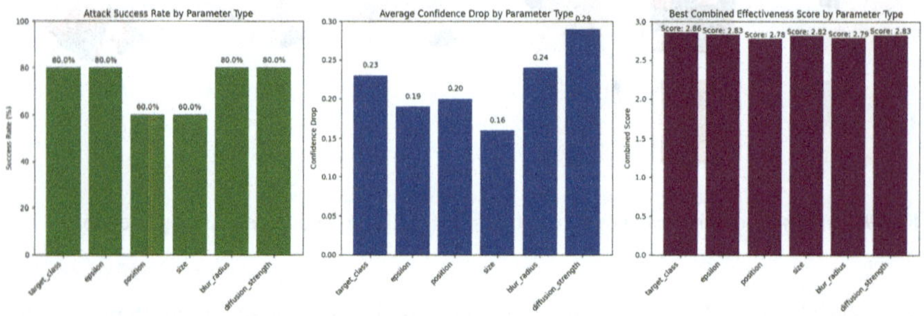

Fig. 10 Analysis of adversarial patch attack performance: success rate, confidence drop, and combined effectiveness score across parameters like target class, epsilon, position, size, blur radius, and diffusion strength

patch success across models. The results are tested with various patch configurations (diffusion strength, size, and position). Study 1 [25] shows similar performance but with slightly lower SSIM and MS-SSIM scores, and higher transferability. Study 2 [26] achieves comparable LPIPS and SSIM scores but has slightly lower MS-SSIM. Study 3 [27] shows lower scores across all metrics, indicating less effectiveness in fooling models.

Overall, the proposed work demonstrates strong performance, with a balanced approach that ensures high imperceptibility, effectiveness, and transferability, making it a competitive method in adversarial patch generation in Table 1.

5.3 Impact of Different Parameters on Attack Success, Confidence Drop, and Effectiveness Score

Figure 10 consists of three bar charts that provide a detailed comparison of different parameters influencing the performance of an attack. The first chart, titled, demonstrates the success rate of the attack across six different parameters: target_class, epsilon, position, size, blur_radius, and diffusion_strength. It is evident from the chart that parameters like target_class, epsilon, and diffusion_strength exhibit a high success rate of around 80%, while position and size contribute to a noticeably lower success rate of 60%.

The second chart shows the average confidence drop across the same parameters. Diffusion_strength again stands out with the highest drop in confidence, followed by target_class, which also demonstrates a relative drop. The other parameters, including epsilon, position, and size, have a smaller impact on the confidence drop, reflecting a consistent trend with the attack success rate chart, where fewer changes were observed in these parameters.

Finally, the third chart provides an assessment of the best combined score of each parameter, illustrating how each one contributes to the overall attack effectiveness. Target_class stands out with the highest score of 2.86, while position and size perform relatively poorly with scores around 2.78. This reinforces the conclusion that some parameters, such as target_class and diffusion_strength, lead to a more effective attack compared to others.

5.4 *Transferable Optimal Adversarial Patch Evaluation*

Figure 11 focuses on patch variation analysis and examines how position variation affects our adversarial attack success. Our left side shows the original image, while subsequent images display different patch positions with their class and confidence scores. Our confidence drop metrics demonstrate how the attack's effectiveness

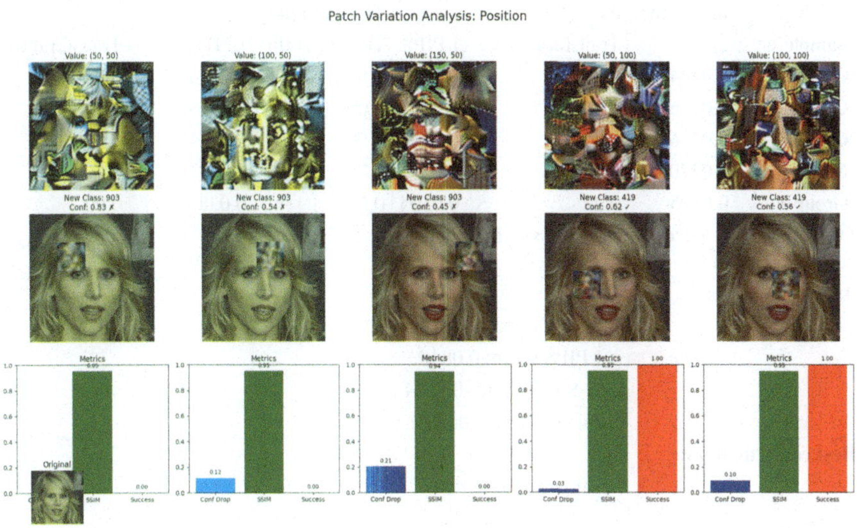

Fig. 11 Patch variation analysis for different positions. The images show the impact of adversarial patch positions on classification results, with metrics displaying SSIM, confidence drop, and success rates

changes as the patch moves across the image. Some positions in our analysis yield a higher success rate (with confidence drops of 0.03 and 0.10), while others diminish our attack's success (with larger drops of up to 0.21) demonstrating how patch positioning influences effectiveness. Our second figure presents an optimal adversarial patch analysis featuring our best-performing patch. It displays our original image alongside our optimal adversarial patch. Our patched image demonstrates the attack's success through a marked shift in image classification, evidenced by changes in confidence scores. This shows how effectively our patch manipulates the model's classification. Our difference visualization highlights the subtle changes between our original and patched images that alter the model's prediction. In Table 2, we evaluated our proposed model for testing the transferability attribute of adversarial patch generation and their respective quality for further examination

Table 2 Metrics for adversarial patch transferability testing across various configurations and evaluation. Success is indicated by "Yes" and "No", and transferability varies from 0 to 1 across samples

	Metrics for position (50, 50)			Metrics for position (100, 100)		
Sample	Success	Confidence drop	SSIM	Success	Confidence drop	SSIM
Group 1	Yes	−0.1507	0.9777	Yes	−0.0855	0.9812
Group 2	No	0.0529	0.9798	Yes	−0.0824	0.9771
Group 3	Yes	0.0974	0.9827	Yes	0.0627	0.9805
Group 4	Yes	−0.0541	0.9759	Yes	−0.5963	0.9765
Group 5	Yes	0.1017	0.9801	Yes	−0.0336	0.9754
	LPIPS metrics			Transferability		
Sample	LPIPS	Transferability	LPIPS	Transferability	LPIPS	Transferability
Group 1	0.0444	0.50	0.0256	0.50	0.0297	0.75
Group 2	0.0176	0.50	0.0261	1.00	0.0264	0.75
Group 3	0.0326	0.25	0.0256	0.50	0.0235	0.25
Group 4	0.0370	0.25	0.0336	0.75	0.0581	0.50
Group 5	0.0295	0.25	0.0505	0.50	0.0385	0.00
Overall attack success rate:		81.11%				
Average metrics:		Confidence drop: −0.1696				
		SSIM score: 0.9403				
		LPIPS score: 0.0840				
		MS-SSIM score: 0.8664				
		Transferability: 0.6222				
Best configuration:		Sample: 2				
		Diffusion strength: 0.5				
		Patch size: 30				
		Position: (150, 50)				
		Success: Yes				
		Composite score: 0.9927				

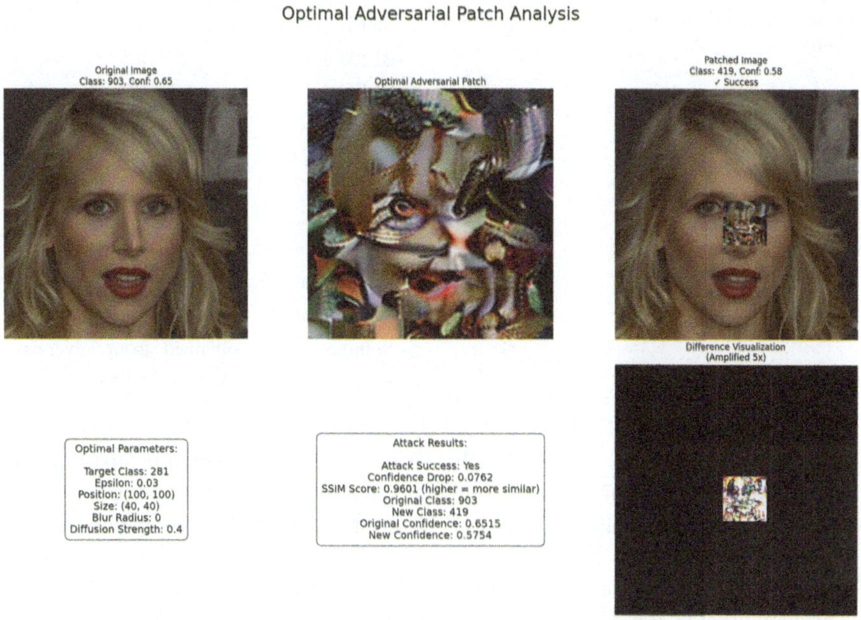

Fig. 12 The images show the original (left), optimal adversarial patch (center), and patched image (right), demonstrating a successful attack with changes in classification, confidence drop, SSIM score, and attack parameters

for different groups of samples based on different geographical places of adversarial patch creation of Fig. 12.

5.5 Identity Evaluation

In our comprehensive approach for generating adversarial captions, we evaluate their impact through various strategies, including semantic and grammatical alterations, contextual distortions, and other transformations. This evaluation delves deeply into the robustness of automated captioning systems. We leverage pretrained models such as vision encoder-decoder model and ViT image processor to generate captions from images. Following this, we apply a series of adversarial strategies to modify these captions, allowing us to assess the impact these modifications have on the integrity and effectiveness of the generated content. Our approach employs a wide range of metrics, including BLEU, METEOR, ROUGE, and accuracy metrics, to evaluate both clean and adversarial captions.

In our implementation, we developed the adversarial caption generator class, which plays a central role by applying a series of poisoning strategies to the generated captions. We designed these strategies to involve significant alterations

Table 3 Caption metrics results

Metric	Value
BLEU-1	0.5377
BLEU-4	0.2741
METEOR	0.6473
ROUGE-1	0.6967
ROUGE-2	0.3131
ROUGE-L	0.4666
Clean accuracy	0.9231
Adversarial accuracy	0.4069
Semantic shift	0.4157

Source: Results obtained from adversarial caption generation evaluation metrics

that make the captions adversarial. For instance, we use semantic distortion to replace words with synonyms, grammatical mutation to modify the grammatical structure, and contextual hallucination to introduce unexpected elements into the caption.

We have made the use of poisoning templates a crucial feature of our adversarial caption generator. We designed these templates to flag potential poisoned captions by applying structured, contextually relevant modifications to the original captions. Rather than introducing random changes, we use the templates to guide the modification of specific elements such as nouns, adjectives, and verbs.

In our implementation, we developed the caption dataset class to support efficient processing of images and captions. We designed it to facilitate the loading of images, generation of captions through pretrained models like ViT-GPT2, and application of adversarial transformations.

In our evaluation of the adversarial captions, we found several important results. We observed BLEU-1 and BLEU-4 scores of 0.5377 and 0.2741, respectively, showing us a moderate overlap in n-grams between the original and adversarial captions. Our METEOR score of 0.6473 reflects a reasonable alignment between our generated and human-written references. We also found that our ROUGE scores (ROUGE-1 at 0.6967, ROUGE-2 at 0.3131, and ROUGE-L at 0.4666) further support these findings. Through our clean accuracy score of 0.9231, we confirmed that a significant portion of the original captions remained intact, while our adversarial accuracy score of 0.4069 showed us that the adversarial captions achieved moderate deviation from the original content in Table 3.

5.6 Adversarial Detection

Figure 13 presents an analysis of SSIM scores, which are used to assess the visual similarity between two images. The first plot on the left shows the distribution of SSIM scores for 500 images, with a clear threshold (0.98) marked by a red-dashed

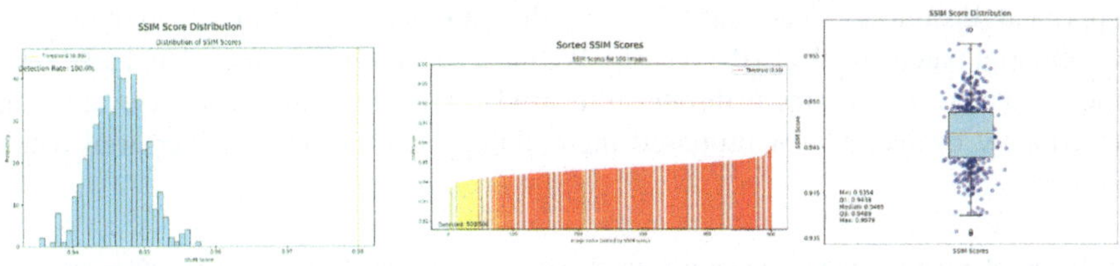

Fig. 13 Demonstrating the number of detected adversaries from an imbalance dataset by visualization and tracking the SSIM distribution

Table 4 SSIM score distribution summary

Metric	Value
Minimum SSIM score	0.9354
Q1 (First quartile)	0.9419
Median SSIM score	0.9465
Q3 (Third quartile)	0.9489
Maximum SSIM score	0.9579
Threshold for detection	0.98
Detection rate	100%

line. This threshold is set to determine the detection rate, which in this case is 100%, indicating that all images have been successfully detected.

The second plot in the center displays the sorted SSIM scores for the 500 images, ordered from lowest to highest. The threshold is again marked with a red-dashed line at 0.98, providing a reference point for the sorted scores. This plot visually confirms that all 500 images have SSIM scores higher than the threshold, with a detection rate of 100%.

The third plot, on the right, presents a box plot of SSIM scores, offering a summary of the statistical distribution. The plot shows the minimum, first quartile (Q1), median, third quartile (Q3), and maximum SSIM value in the range and variability of the scores. The SSIM scores are concentrated around the median of 0.946, with a narrow interquartile range, indicating high similarity between the images.

Together, these visualizations provide a detailed overview of the SSIM score distribution and detection performance, demonstrating the consistency and effectiveness of the thresholding method for image comparison Table 4.

6 Conclusion

In this scope of study, we thoroughly examine this forensic occurrence of adversaries. We witnessed and explored the opportunities of adversarial patch generation using a stable diffusion model, and we extended our methodology to a systematic approach design to detect and poison the number of identity recognition of adversarial patch attacks to the dataset. We have successfully detected and analyzed the respected measurement through this study. We achieved excellent evaluation metrics by evaluating our proposed methodology and detecting adversarial patch attacks.

Acknowledgments The research was partially sponsored by the Army Research Office and was carried out under Grant Number W911NF-21-1-0264. The views and conclusions contained in this document are those of the authors and should not be interpreted as representing the official policies, either expressed or implied, of the Army Research Office or the US Government. The US government is authorized to reproduce and distribute reprints for Government purposes, notwithstanding any copyright notation herein.

In addition, this research was supported as part of the 2024 Summer Research Experience for Graduate Students Program of the Northrop Grumman Research and Education Program (NG REP) through the Florida A& M University Foundation.

References

1. Hwang, R.-H., Lin, J.-Y., Hsieh, S.-Y., Lin, H.-Y., Lin, C.-L.: Adversarial patch attacks on deep-learning-based face recognition systems using generative adversarial networks. Sensors **23** (2023) https://doi.org/10.3390/s23020853
2. Tampubolon, M.: Digital face forgery and the role of digital forensics. International Journal for the Semiotics of Law-Revue internationale de Sémiotique juridique **37**(3), 753–767 (2024)
3. Campbell, S.L., Gear, C.W.: The index of general nonlinear DAES. Numer. Math. **72**(2), 173–196 (1995)
4. Yutia, S.N., Triwibisono, C., Rahmayadi, G.E.: Digital forensic on image manipulation crime-a systematic literature review. In: 2024 International Conference on Intelligent Cybernetics Technology & Applications (ICICyTA), pp. 824–828 (2024). IEEE
5. Shetty, P.: Use of digital forensics in cybersecurity and criminal cases, how can it be leveraged, challenges. International Journal of Science and Research (IJSR) (2020) https://doi.org/10.21275/sr24213025946
6. Wu, C., Zhang, T., Li, Z., Lian, Z.: Enhancing adversarial patch effectiveness for face recognition systems using sharpness-aware minimization. In: 2024 IEEE Cyber Science and Technology Congress (CyberSciTech), pp. 384–389 (2024)
7. Thys, S., Van Ranst, W., Goedemé, T.: Fooling automated surveillance cameras: Adversarial patches to attack person detection. In: CVPR Workshops, pp. 49–55 (2019)
8. Geng, P., Deng, X.: An adversarial patch attack for vehicle detectors in the physical world. In: IEEE International Conference on Unmanned Systems (ICUS) (2023)
9. Hingun, N., Sitawarin, C., Li, J., Wagner, D.: Reap: A large-scale realistic adversarial patch benchmark. arXiv preprint arXiv:2212.05680 (2022)
10. Li, C., Yan, H., Zhou, L., Chen, T., Liu, Z., Su, H.: Prompt-guided environmentally consistent adversarial patch. (2024)
11. Jutras, M., Liang, E., Leary, S., Ward, C., Manville, K.: Detecting physical adversarial patch attacks with object detectors. In: IEEE Applied Imagery Pattern Recognition Workshop (AIPR), pp. 1–7 (2022)
12. Sayyafzadeh, S., Xu, W., Chi, H.: Forensic analysis of contents in thumbnails using transfer learning. In: Proceedings of the Future Technologies Conference, pp. 517–535. Springer, Cham (2023)
13. Arvinte, M., Tewfik, A., Vishwanath, S.: Detecting patch adversarial attacks with image residuals. arXiv preprint arXiv:2002.12504 (2020)

14. Sharma, A., Bian, Y., Munz, P., Narayan, A.: Adversarial patch attacks and defences in vision-based tasks: A survey. arXiv preprint arXiv:2206.08304 (2022)
15. Mathov, Y., Rokach, L., Elovici, Y.: Enhancing real-world adversarial patches with 3d modeling techniques. arXiv preprint arXiv:2102.05334 (2021)
16. Goodfellow, I.J., Shlens, J., Szegedy, C.: Explaining and harnessing adversarial examples. In: International Conference on Learning Representations (ICLR) (2015)
17. Brown, T.B., Mane, D., Roy, A., Abadi, M., Gilmer, J.: Adversarial patch. In: Advances in Neural Information Processing Systems (NeurIPS) Workshop (2017)
18. Sharif, M., Bhagavatula, S., Bauer, L., Reiter, M.K.: A general framework for adversarial examples with objectives. In: ACM Transactions on Privacy and Security (TOPS), vol. 22, pp. 1–30. ACM, New York (2019)
19. Li, J., Zhang, Y., Liu, Q., Ma, X., Wang, Y.: Adversarial attacks on deep learning-based facial expression recognition systems. In: IEEE International Conference on Image Processing (ICIP), pp. 1086–1090 (2021). IEEE
20. Zhang, W., Liu, J., Chen, L., Xu, K.: Real-world adversarial attacks on facial expression recognition systems. Pattern Recognition **128**, 108667 (2022)
21. Ho, J., Jain, A., Abbeel, P.: Denoising diffusion probabilistic models. In: Advances in Neural Information Processing Systems (NeurIPS), vol. 33, pp. 6840–6851 (2020)
22. Song, Y., Sohl-Dickstein, J., Kingma, D.P., Kumar, A., Ermon, S., Poole, B.: Score-based generative modeling through stochastic differential equations. In: International Conference on Learning Representations (ICLR) (2021)
23. Nie, W., Song, Y., Zhu, Y., Zhang, J., Zhou, X., Ma, K., Liang, K., Wang, Y., Ma, Y., Xu, D.: Diffusion models for adversarial purification. In: Advances in Neural Information Processing Systems (NeurIPS) (2022)
24. Choi, M., Park, J., Kim, S., Lee, H.: Diffusion-based adversarial patch generation for robust attacks on facial recognition. arXiv preprint arXiv:2301.01234 (2023)
25. Tadros, T., Cullen, N., Greene, M.R., Cooper, E.A.: Assessing neural network scene classification from degraded images. ACM Transactions on Applied Perception (TAP) **16**, 1–20 (2019) https://doi.org/10.1145/3342349
26. Xie, S., Girshick, R., Dollár, P., Tu, Z., He, K.: Aggregated residual transformations for deep neural networks. In: Proceedings of the IEEE Conference on Computer Vision and Pattern Recognition (CVPR), pp. 1492–1500 (2017). https://doi.org/10.1109/CVPR.2017.634
27. Szegedy, C., Zaremba, W., Sutskever, I., Bruna, J., Erhan, D., Goodfellow, I., Fergus, R.: Intriguing properties of neural networks. arXiv preprint arXiv:1312.6199 (2014)

PAMDAC: Platform-Agnostic Malware Detection and Classification Based on Binary Features

Tajuddin Manhar Mohammed, Chandrakanth Gudavalli, Lakshmanan Nataraj, Satish Chikkagoudar, Shivkumar Chandrasekaran, and B. S. Manjunath

1 Introduction

Today's malware is no longer restricted to Windows platforms and is rapidly expanding to other platforms such as Android, Linux, and Mac OS. Check Point Research's "2024 Cyber Security Report" highlights a 41% surge in IoT attacks, with 54% of organizations experiencing weekly IoT-targeted cyberattacks [5]. The rise of cross-platform threats is evident with 6,800 malicious packages reported in Q1 2023 and malware like Lucifer evolving from Windows-specific to target multiple platforms (Linux, IoT) and architectures. Millions of new variants across Windows, Android, Linux, and Mac OS appear annually, signaling a major cybersecurity challenge. However, the malware variants, irrespective of their platforms, still have similar structure since the underlying techniques in creating variants more or less remains the same. This is further illustrated in Fig. 1, where we clearly see that the Byteplot [19] image representations of malware variants belonging to different platforms still exhibit similarity. While most static and dynamic analysis methods have to be redeveloped for various operating system platforms and their versions, techniques based on statistical and signal processing-based binary analysis would at

T. M. Mohammed · C. Gudavalli · L. Nataraj
Mayachitra Inc., Santa Barbara, CA, USA
e-mail: mohammed@mayachitra.com; gudavalli@mayachitra.com; lakshmanan_nataraj@ece.ucsb.edu

S. Chikkagoudar (✉)
Information Technology Division, U.S. Naval Research Laboratory, Washington, DC, USA
e-mail: satish.chikkagoudar.civ@us.navy.mil

S. Chandrasekaran · B. S. Manjunath
Mayachitra Inc., Santa Barbara, CA, USA

Department of ECE, University of California, Santa Barbara, CA, USA
e-mail: shiv@ucsb.edu; manj@ucsb.edu

S. S. Iyengar et al. (eds.), *Artificial Intelligence Driven Forensics*, Signals and Communication Technology, https://doi.org/10.1007/978-3-031-98036-7_9

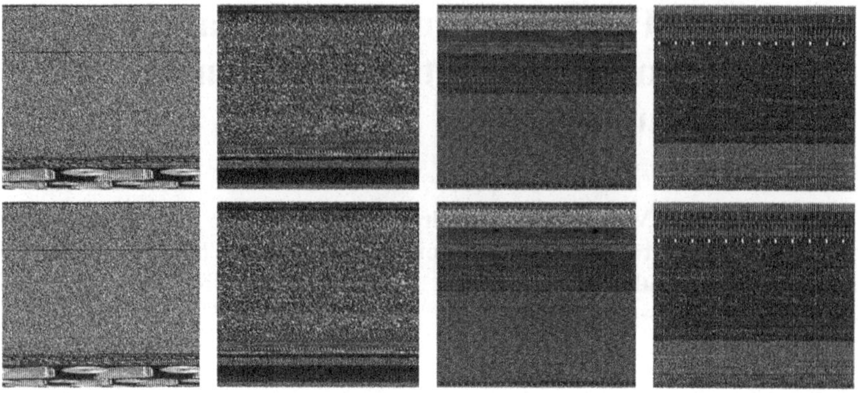

Fig. 1 Byteplot [19] image representations of malware variants from different OS—(**a**) Win 32 variants of Fakerean family, (**b**) Mac OS variants of Wirenet family, (**c**) Linux malware which are Backdoors, and (**d**) Android variants of Kungfu1. These clearly show that visual and structural similarity holds for malware variants irrespective of their platform

most need tuning of parameters. AV systems and security vendors need to adapt their malware detection methodologies to suit different platforms which usually needs knowledge of the operating system (OS) and/or hardware-specific details. In PAMDAC, we have databases of malware and benign binaries belonging to different platforms. We then employ signal and image processing malware detection methods, which have shown great promise in detecting Windows malware and identifying malware from different OS. Such centralized systems can be deployed in servers where multiple computers/networks connect and can be used to scan these networks periodically for malware. PAMDAC can also be deployed in military and defense scenarios where one central system can be used to scan several platforms and OS, without needing much adaptation, aligning with the AI-embedded cyber assurance frameworks discussed in [29] that integrate AI/ML for comprehensive cyber threat analysis. A block schematic of PAMDAC is shown in Fig. 2.

We say that a feature derived from an analyzed malware sample is platform-agnostic if it does not rely on any knowledge of the platform or OS which the sample targets. This implies that a platform-agnostic feature must be computed without any knowledge of the malware sample's executable format or the platform's instruction set architecture. In [20], the authors have conducted similar experiments on malware family classification on small-scale malware datasets from various operating systems (shown in Table 1 for reference). These results further confirmed that such signal-based techniques could be leveraged for malware of different platforms. In this chapter, we explored this further to malware detection and classification and performed a detailed large-scale analysis on Android malware, Linux malware, and Mac OS malware.

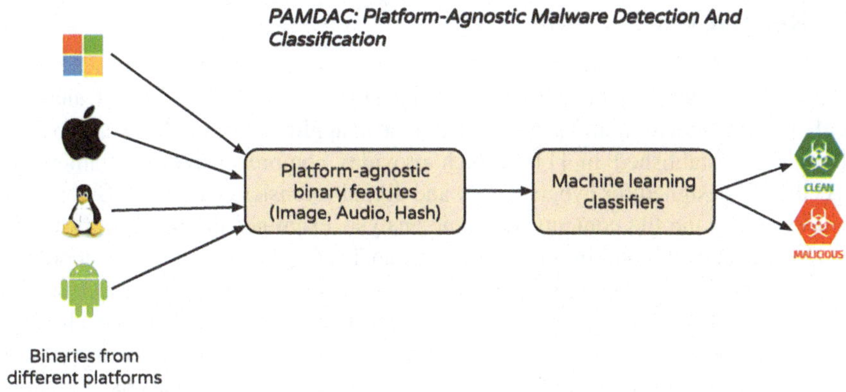

Fig. 2 Block schematic of PAMDAC—binaries from various platforms are analyzed in a centralized platform-agnostic malware detection system

Table 1 Classification accuracy on datasets from different operating systems (reproduced from [20])

Dataset	Size	# of families	Accuracy
Malimg (Win)	9,339	25	0.974
Malheur (Win)	3,131	24	0.984
VxShare (Linux)	568	8	0.833
Malgenome (Android)	1,200	13	0.974

To the best of our knowledge, no previous works on malware classification were evaluated on several datasets targeting different platform operating systems, let alone, without performing some algorithmic changes or feature reengineering. This chapter's key contributions include:

- Performance evaluation of diverse platform-agnostic features on malware datasets from Android, Linux, and Mac OS platforms
- Development of a data acquisition pipeline to create customized and curated malware datasets
- Conducting a case study to assess the potential of "OS-based filtering" as a pre-filtering step to enhance the classification performance of platform-agnostic features

The rest of the chapter is organized as follows. In Sect. 2, we discuss the related work in malware detection and classification. In Sects. 3, 4, and 5, we look at the performance of various platform-agnostic features on malware datasets from Android, Linux, and Mac OS platforms, respectively, while discussing the observations. We discuss a case study in Sect. 6 to look at the effectiveness of using such platform-agnostic features on the amalgamation of the aforementioned datasets for malware detection and conclude in Sect. 7.

2 Related Work

Malware analysis can be performed using static, dynamic, or hybrid analysis, with the theoretical foundations and practical applications of these approaches thoroughly established in [11], which provides comprehensive AI frameworks specifically designed for cybersecurity and digital forensics challenges. Static analysis examines the file contents, dynamic analysis examines the program behavior, and hybrid analysis combines the two. Static analysis is the most common approach, but it can be time-consuming. Dynamic analysis is less common, but it can provide more information about the malware's behavior. In the literature, there are a handful of such review works [3, 4, 6, 10, 23] that provide a good comprehensive overview of such approaches for malware detection and classification. Static analysis of binaries is the most common approach to malware detection. However, it can be time-consuming to decompile binaries into assembly code or programming language. We focus on classification using raw binaries because it is simpler and faster, an approach aligned with the mathematical foundations of optimization [24] and informed by the principles of efficient deep learning architectures [25].

Only a few works developed malware classifiers that use platform-agnostic features. Kebede et al. [12], Narayanan et al. [18], and Le et al. [15] all presented deep learning classifiers based on a feature set including mostly N-grams and evaluated them using Microsoft's dataset. Nataraj and Manjunath [20], Mohammed et al. [16] and Korine and Hendler [13] also looked at the problem of platform-agnostic malware detection using static features but the experiments were carried out only on small malware datasets with a fewer number of data samples. This work bridges the gap and focuses on deriving simple and better platform-agnostic features while looking at the malware classification performance on various malware datasets. Our approach draws methodological inspiration from related security domains, including the boundary analysis techniques for anomaly detection [9], secure transmission frameworks [7], and region-specific detection methods [8] that have proven effective in identifying subtle irregularities in digital content.

3 Android Malware Analysis

For Android malware detection and classification experiments, we have collected malicious samples from UpDroid [1] and AMD [30] and benign samples from AndroZoo [2]. We briefly describe the datasets below.

- UpDroid [1]: This dataset includes 2,479 recent malware samples from 21 families, providing insights on emerging malware. It is valuable for dynamic analysis and contains samples that can bypass standard static analysis techniques.

- Android Malware Dataset (AMD) [30]: This dataset comprises 24,553 Android malware samples, classified into 135 varieties across 71 families from 2010 to 2016. It offers an up-to-date representation of the Android malware landscape and is publicly available for community use.
- AndroZoo [2]: This collection consists of Android applications from multiple sources, including the official Google Play app market. With 19,537,482 APKs (as of 06-15-2022), each undergoes analysis by numerous AntiVirus products to determine malware detection status.

Typically, the Android samples are compressed packages, which are *APK* files. Inside the compressed package, we have file(s) with app code in *DEX* format. Generally, an Android package uses a single dex and has one classes.dex file. If it uses multidex, we have multiple such files—classes.dex, classes2.dex, etc. For our experiments, we consider both *APK* and *DEX* (classes.dex only) file formats separately.

We perform multi-class malware classification and binary classification (malware detection) experiments on image-based features (Byteplot GIST [19], Bigram-DCT GIST [17]), audio-based features (MFCC, Melspectrogram, Chroma) [21], and hash-based features (Ssdeep [14], TLSH [22]) using standard machine learning-based classification models. The feature dimensional lengths are 320, 320, 128, 20, 12, 40, and 70, respectively. These features are mined from a large set of candidate features in experiments conducted on various Windows malware datasets. Specifically, feature mining was done to select high-quality features that were simple to extract from binaries, signal-based, and orthogonal [17]. These features, in particular, obtained very high classification accuracies on various Windows malware datasets. The standard classifiers include K-nearest neighbor (KNN), random forest (RF), linear discriminant analysis (LDA), and decision trees (DT). For all our experiments, we report 10-fold cross-validation accuracies, though we had similar results for other k-fold (k = 2, 5, 20) cross-validation experiments.

3.1 UpDroid Malware Classification

The multi-class classification accuracies for the UpDroid dataset are shown in Table 2 for different combinations of features and classifiers. The "best" feature and classifier are Melspectrogram and random forest (RF), respectively, with accuracies of `0.9581` on *APK* samples and `0.7705` on *DEX* samples. The dip in accuracy is not as we expect, and we believe it is either because of the limited number of samples or because of the way the dataset is designed to evade standard static analysis methods, as described in [1].

Table 2 10-fold cross-validation multi-class classification accuracies for different features and classifiers for UpDroid—#samples: ∼2000, #malware families: 18

APK feature	KNN	RF	LDA	DT
Byteplot GIST	**0.9191**	0.8788	0.8664	0.8299
Bigram-dct GIST	0.9378	**0.9548**	0.9494	0.9307
Melspectrogram	0.9353	**<u>0.9581</u>**	0.9158	0.9212
MFCC	0.9274	**0.9394**	0.8129	0.9212
Chroma	0.8880	**0.9010**	0.7581	0.8481
Ssdeep	0.5917	**0.8249**	0.6261	0.7091
TLSH	0.9124	**0.9311**	0.9207	0.9012

(a) *APK*

DEX feature	KNN	RF	LDA	DT
Byteplot GIST	0.7079	**0.7452**	0.7415	0.7232
Bigram-dct GIST	0.7311	**0.7668**	0.7514	0.6963
Melspectrogram	0.6892	**<u>0.7705</u>**	0.7278	0.6859
MFCC	0.7127	**0.7643**	0.6863	0.6639
Chroma	0.6668	**0.7340**	0.6498	0.6656
Ssdeep	0.5270	**0.6394**	0.5502	0.5498
TLSH	0.6851	**0.7668**	0.7108	0.6822

(b) *DEX*

Bold values indicate the best-performing classifier for each feature (row); bold and underlined values indicate the best overall performance across all features

3.2 AMD Malware Classification

The multi-class classification accuracies for the AMD dataset are shown in Table 3 for different combinations of features and classifiers. The "best" feature and classifier are Melspectrogram and random forest (RF), respectively, with accuracies of `0.7706` on *APK* samples and `0.8970` on *DEX* samples. These results are as expected in the sense that we do see performance improvements when we compare *APK* to *DEX* formats.

3.3 Android Malware Detection (Binary Classification)

We collected benign android samples from Google Play Store through AndroZoo [2] API which have creation timestamps on or after 01-01-2016 and have a VirusTotal [28] positive hit rate of 0. We subsampled this collection to perform binary classification experiments with UpDroid and AMD datasets. The binary classification accuracies are shown in Table 4. In the case of UpDroid, the "best" accuracies for *APK* and *DEX* formats are `0.9465` and `0.9788`, respectively, while in the case of AMD, the "best" accuracies for *APK* and *DEX* formats are `0.9048`

Table 3 Tenfold cross-validation multi-class classification accuracies for different features and classifiers for AMD—#samples: ~25000, #malware families: 70

APK feature	KNN	RF	LDA	DT
Byteplot GIST	**0.6238**	0.5982	0.4897	0.4382
Bigram-dct GIST	0.6961	**0.7304**	0.6312	0.5965
Melspectrogram	0.6603	**0.7706**	0.5741	0.6238
MFCC	0.6710	**0.7143**	0.4619	0.5722
Chroma	0.5563	**0.5944**	0.3748	0.4464
Ssdeep	0.4444	**0.5730**	0.3912	0.4210
TLSH	0.6971	**0.7106**	0.6031	0.5571

(a) *APK*

DEX feature	KNN	RF	LDA	DT
Byteplot GIST	0.8366	**0.8372**	0.6740	0.7410
Bigram-dct GIST	0.8486	**0.8852**	0.8493	0.8120
Melspectrogram	0.7731	**0.8970**	0.7703	0.8243
MFCC	0.8319	**0.8727**	0.6374	0.7916
Chroma	0.7403	**0.7882**	0.4963	0.7060
Ssdeep	0.5437	**0.6608**	0.4330	0.5529
TLSH	0.8088	**0.8365**	0.6911	0.7375

(b) *DEX*

Bold values indicate the best-performing classifier for each feature (row); bold and underlined values indicate the best overall performance across all features

and `0.9695`, respectively. These results again emphasize the effectiveness of using the *DEX* format over that of *APK*.

Malware Detection Generalizability To see how well the binary classification results generalize among the datasets used, we performed binary classification experiments with malicious samples drawn from one dataset for training and with malicious samples drawn from another dataset for testing. The training and testing sets were also made balanced-class sets by infusing them with benign samples from AndroZoo. The classification accuracies are shown in Table 5. When the training set had malware samples from AMD and that of the test set from UpDroid (Table 5a), the "best" accuracies are `0.8071` and `0.9437` for *APK* and *DEX* formats, respectively. Switching the train-test sets (Table 5b) also shows performance improvements, though not that significant, with "best" accuracies of `0.7298` and `0.7547` for *APK* and *DEX* formats, respectively.

4 Linux Malware Analysis

ELF (Executable and Linkable Format) is a common standard file format for executable files, object code, shared libraries, and core dumps in Linux systems. For Linux malware detection and classification experiments, we have collected

Table 4 Tenfold cross-validation binary classification accuracies for different features and classifiers. For each feature, the first row corresponds to features obtained from *APK* and the second row corresponds to that of *DEX*

Feature	KNN	RF	LDA	DT
Byteplot GIST	0.9004	**0.9028**	0.8331	0.8460
	0.9762	**0.9788**	0.9714	0.9694
Bigram-dct GIST	0.9395	**0.9465**	0.9181	0.9176
	0.9547	0.9570	**0.9650**	0.9517
Melspectrogram	0.9170	**0.9414**	0.9250	0.9176
	0.9390	**0.9600**	0.9442	0.9476
MFCC	0.9292	**0.9386**	0.8908	0.9226
	0.9508	**0.9537**	0.8905	0.9483
Chroma	0.8984	**0.9038**	0.7695	0.8672
	0.9329	**0.9441**	0.8876	0.9298
Ssdeep	0.6337	**0.7556**	0.6394	0.6914
	0.8728	**0.9547**	0.8992	0.9213
TLSH	**0.9392**	0.9370	0.9355	0.9176
	0.9454	**0.9512**	0.9394	0.9388

(a) UpDroid (#malwares: ~2,000) + AndroZoo (#benigns: ~2,500)

Feature	KNN	RF	LDA	DT
Byteplot GIST	**0.7961**	0.7764	0.7121	0.7114
	0.9695	0.9635	0.9167	0.9308
Bigram-dct GIST	0.8666	**0.8792**	0.8535	0.8045
	0.9229	0.9162	0.8965	0.8669
Melspectrogram	0.8067	**0.8921**	0.8026	0.8121
	0.8851	**0.9302**	0.8532	0.8894
MFCC	0.8037	**0.8536**	0.7589	0.7764
	0.9202	**0.9219**	0.8124	0.8799
Chroma	0.7430	**0.7911**	0.6639	0.7058
	0.8770	**0.8892**	0.7801	0.8366
Ssdeep	0.5760	**0.7885**	0.7698	0.6848
	0.6452	**0.8362**	0.7800	0.7455
TLSH	0.8341	**0.9048**	0.8566	0.8357
	0.8924	**0.9157**	0.8387	0.8598

(b) AMD (#malwares: ~25,000) + AndroZoo (#benigns: ~33,000)

Bold values indicate the best-performing classifier for each feature (row); bold and underlined values indicate the best overall performance across all features

ELF malicious samples from VirusShare [27] and clean samples from scraping through different system files in various Linux-based systems. We briefly describe the datasets below.

– VXShare-ELF: This dataset contains approximately 55,000 ELF malware samples sourced from VirusShare. To obtain the malware labels, we utilize the

Table 5 Tenfold cross-validation binary classification accuracies for different features and classifiers. For each feature, the first row corresponds to features obtained from *APK* and the second row corresponds to that of *DEX*

Feature	KNN	RF	LDA	DT
Byteplot GIST	0.6817	**0.7098**	0.6867	0.6435
	0.8883	0.8653	<u>**0.9437**</u>	0.7139
Bigram-dct GIST	0.6697	0.7077	**0.7423**	0.7109
	0.6358	0.6245	**0.9091**	0.6548
Melspectrogram	0.6968	<u>**0.8071**</u>	0.7573	0.7109
	0.5948	0.6184	**0.6274**	0.5988
MFCC	0.6409	**0.7523**	0.7510	0.6405
	0.7885	0.6590	**0.8753**	0.8533
Chroma	0.6383	0.6529	**0.6695**	0.6250
	0.88419	0.6696	0.8406	0.6600
Ssdeep	0.5391	0.5583	0.5544	**0.5622**
	0.6335	**0.8120**	0.7868	0.6659
TLSH	0.6758	**0.7988**	0.7659	0.6702
	0.8628	**0.9263**	0.9082	0.6171

(a) Train: AMD + AndroZoo (∼55,000 samples),
Test: UpDroid + AndroZoo (∼5,500 samples)

Feature	KNN	RF	LDA	DT
Byteplot GIST	**0.6465**	0.6111	0.6143	0.6234
	0.7537	0.7190	<u>**0.7547**</u>	0.7281
Bigram-dct GIST	**0.6925**	0.6273	0.6769	0.6293
	0.6317	0.6333	**0.6788**	0.6393
Melspectrogram	0.6304	0.5933	**0.6819**	0.6445
	0.6510	0.6283	**0.6858**	0.6735
MFCC	0.5881	0.6000	**0.7069**	0.6457
	0.6894	0.6801	0.6635	**0.6898**
Chroma	0.5908	0.6041	**0.6415**	0.6240
	0.6218	0.6305	0.6182	**0.6598**
Ssdeep	0.5307	**0.6464**	0.5927	0.6152
	0.6013	0.6512	0.6305	**0.6820**
TLSH	0.6405	0.6340	<u>**0.7298**</u>	0.6880
	0.5939	0.6219	**0.7214**	0.6684

(b) Train: UpDroid + AndroZoo (∼5,500 samples),
Test: AMD + AndroZoo (∼55,500 samples)

Bold values indicate the best-performing classifier for each feature (row); bold and underlined values indicate the best overall performance across all features

AntiVirus labels report from VirusTotal [28] in conjunction with AVClass [28], a malware labeling tool.

- CleanLINX: This collection comprises around 10,000 benign ELF samples obtained by scraping various system directories of Linux-based systems such as

Fig. 3 A generic malware data and label acquisition process

Table 6 Tenfold cross-validation multi-class classification accuracies for different features and classifiers for VXShare-ELF—#samples: ~52000, #malware families: 12

Feature	KNN	RF	LDA	DT
Byteplot GIST	0.9645	**0.9649**	0.8946	0.9362
Bigram-dct GIST	0.9636	**0.9717**	0.9478	0.9553
Melspectrogram	0.9480	**0.9731**	0.9014	0.9571
MFCC	0.9584	**0.9682**	0.8613	0.9459
Chroma	0.9242	**0.9466**	0.7006	0.9021
Ssdeep	0.7570	**0.8928**	0.7177	0.8313
TLSH	0.9597	**0.9691**	0.8935	0.9442

Bold values indicate the best-performing classifier for each feature (row); bold and underlined values indicate the best overall performance across all features

Ubuntu 16.04, Ubuntu 18.04, and Pop! OS 20.04. The collection is continuously growing.

As there are no publicly available Linux malware datasets, we had to set up a data acquisition pipeline, as shown in Fig. 3 to collect labeled Linux malware samples from VirusShare. The data acquisition pipeline is generic and can be used to extract labeled malware samples from public malware dumps. The collected samples have information that includes (and is not limited to) hashes of the samples, VirusTotal scores and reports, and AVClass labels.

Similar to that of Android malware classification and detection experiments (Sect. 3), we report 10-fold classification accuracies using the same experimental setup. The analysis is the first of its kind in malware literature with the curated dataset containing many Linux malware samples.

4.1 VXShare-ELF Malware Classification

For multi-class classification, we chose ~52,000 of the collected malware samples which had at least 100 samples per malware family. The classification accuracies are shown in Table 6. The "best" feature and classifier are Melspectrogram and Random Forest, respectively, with an accuracy of `0.9731`.

Table 7 Tenfold cross-validation binary classification accuracies for different features and classifiers for samples from VXShare-ELF (#malware samples: ∼13,000) and CleanLINX (#benign samples: ∼10,000)

Feature	KNN	RF	LDA	DT
Byteplot GIST	0.9863	**0.9894**	0.9595	0.9740
Bigram-dct GIST	0.9915	**0.9951**	0.9921	0.9873
Melspectrogram	0.9621	**0.9939**	0.9819	0.9876
MFCC	0.9850	**0.9899**	0.9574	0.9795
Chroma	0.8894	**0.9324**	0.5993	0.9018
Ssdeep	0.7191	**0.9185**	0.6874	0.8606
TLSH	0.9848	**0.9951**	0.9820	0.9773

Bold values indicate the best-performing classifier for each feature (row); bold and underlined values indicate the best overall performance across all features

4.2 Linux Malware Detection (Binary Classification)

The binary classification accuracies are shown in Table 7. The "best" feature and classifier are Bigram-dct GIST and Random Forest, respectively, with an accuracy of `0.9951`.

5 Mac OS Malware Analysis

Mach-O, the abbreviated form of Mach object file format, serves as a file format encompassing executables, object code, shared libraries, dynamically loaded code, and core dumps. This format comprises a binary stream of bytes organized into meaningful data chunks, containing essential information such as byte order, CPU type, chunk size, and more. To conduct our Mac OS malware classification experiments, we sourced malicious samples from VirusSamples [26]. However, due to challenges in acquiring benign samples specific to Mac OS, we did not conduct malware detection experiments. Below, we provide a brief description of the curated malicious dataset used in our study.

- VS-MAC: This dataset includes approximately 1,000 Mach-O-type malware samples obtained from VirusSamples [26]. To assign malware labels, we utilize the AntiVirus labels report from VirusTotal [28] in combination with AVClass [28], a malware labeling tool.

As there are no publicly available Mac OS malware datasets, we used the same data acquisition pipeline as shown in Fig. 3 to collect labeled Mac OS malware from VirusSamples, similar to that of the labeled Linux malware data. The collected samples have information that includes (and is not limited to) hashes of the samples, VirusTotal scores and reports, and AVClass labels.

Similar to that of Android malware classification (Sect. 3) and Linux classification (Sect. 4) experiments, we report 10-fold classification accuracies using the same experimental setup. The analysis is the first of its kind in malware literature with the curated dataset containing many Mac OS malware samples.

5.1 VS-MAC Malware Classification

For multi-class classification, we chose 716 of the collected malware samples which had at least 6 samples per malware family. The classification accuracies are shown in Table 8. The "best" feature and classifier are Melspectrogram and random forest (RF), respectively, with an accuracy of `0.8909`.

6 Case Study: OS-Based Filtering

We studied the problem of whether to use separate classification models for different platform operating systems or a single model for the entire data space. We call this problem "OS-based filtering." We used datasets of malware and benign samples separated by platform OS. The datasets included data from the Internet and the *Smalldata* [17] dataset for the Windows platform. We only reported results on the Byteplot GIST features because we obtained similar results on other platform-agnostic features. We divided the data into 10-folds and trained various classification models to distinguish malware and benign sets, both within and across different platform datasets. The distribution of data for each platform OS is shown in Table 9.

Table 8 Tenfold cross-validation multi-class classification accuracies for different features and classifiers for VS-MAC—#samples: 716, #malware families: 19

Feature	KNN	RF	LDA	DT
Byteplot GIST	0.8742	**0.8812**	0.7779	0.8533
Bigram-dct GIST	0.8664	**0.8812**	0.8658	0.8672
Melspectrogram	0.7988	**0.8909**	0.8854	0.8448
MFCC	0.8155	**0.8840**	0.8699	0.8296
Chroma	0.8393	**0.8616**	0.7820	0.8184
Ssdeep	0.5979	**0.7331**	0.6577	0.6662
TLSH	0.8700	**0.8854**	0.8769	0.8198

Bold values indicate the best-performing classifier for each feature (row); bold and underlined values indicate the best overall performance across all features

Table 9 Data distribution across various platform OS

OS	Malware	Benign	Total
Mac OS	977	0	977
Windows	9,130	7,228	16,358
Android	24,557	33,600	58,157
Linux	66,239	14,618	80,857
Total	100,903	55,446	156,349

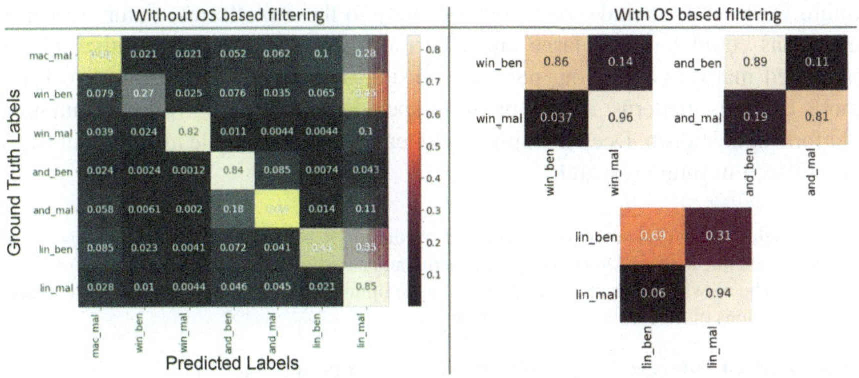

Fig. 4 Classification performance comparison with and without OS-based filtering—OS-based filtering significantly enhances classification performance, as demonstrated by the improved results in the confusion matrices

The confusion matrices that depict the classification performance, by using KNNs, are shown in Fig. 4. Note that the slightly reduced benign accuracy in case of Linux platform can be attributed to the malware-benign class imbalance. The observations that can be made from confusion matrices are:

1. Windows benign samples exhibit confusion with Linux malware, indicating that employing OS-based filtering leads to enhanced performance.
2. In contrast to Windows and Linux samples, the benign set accuracy for Android is already high even without OS-based filtering. However, inclusion of OS-based filtering significantly improves malware accuracy for Android.
3. Mac OS malware demonstrates confusion with Linux malware, potentially attributed to the similarity between Linux and Mac OS as Unix-like operating systems, both providing access to Unix commands, BASH, and other shells.
4. Notably, OS-based filtering considerably improves the overall accuracy of benign and malware classifications for all OS files.

7 Conclusion and Future Work

The effectiveness of platform-agnostic features is well established. Feature extraction and classification require less computational cost than other static analysis-based methods. We developed an end-to-end pipeline to acquire labeled malware and benign data in platforms other than Windows. We also studied the effectiveness of platform-agnostic features when the data is filtered based on the operating system.

This study focused on malware for Windows, Android, Linux, and Mac OS. However, malware for other platforms, such as iOS and Windows Mobile OS, was not included. Our system could be adapted to analyze IoT binaries, and future work could include extending the study to these platforms. Future research directions could leverage large language models (LLMs) and generative AI for improved malware detection, potentially using transformer architectures to learn more complex patterns across platform-specific executables while maintaining platform agnosticism. Deep learning-based features and ensemble models could also be explored in future research.

Acknowledgments This work was supported by the ONR contract #N68335-17-C-0048. The authors would like to thank Dr. Gunasekaran Seetharaman for his insightful comments and research direction. The views expressed in this chapter are the opinions of the authors and do not represent official positions of the Department of the Navy.

Disclosure of Interests The authors have no competing interests.

References

1. Aktas, K., Sen, S.: Updroid: Updated android malware and its familial classification. In: Nordic Conference on Secure IT Systems. pp. 352–368. Springer (2018)
2. Allix, K., Bissyandé, T.F., Klein, J., Le Traon, Y.: Androzoo: Collecting millions of android apps for the research community. In: 2016 IEEE/ACM 13th Working Conference on Mining Software Repositories (MSR). pp. 468–471. IEEE (2016)
3. Amanfu, B.E., Yeluripati, G.R.: A comparative analysis and evaluation of machine learning algorithms for malware detection. In: 2024 IEEE 9th International Conference on Adaptive Science and Technology (ICAST). vol. 9, pp. 1–7. IEEE (2024)
4. Baghirov, E.: Techniques of malware detection: Research review. In: 2021 IEEE 15th International Conference on Application of Information and Communication Technologies (AICT). pp. 1–6. IEEE (2021)
5. Check Point Research: Cyber attack trends 2024 (2024), https://www.checkpoint.com/resources/report-3854/report--cyber-security-report-2024
6. Gaber, M.G., Ahmed, M., Janicke, H.: Malware detection with artificial intelligence: A systematic literature review. ACM Computing Surveys **56**(6), 1–33 (2024)
7. Hariprasad, Y., Iyengar, S.S., Chaudhary, N.K.: Securing the future: Advanced encryption for quantum-safe video transmission. IEEE Transactions on Consumer Electronics (2024)
8. Hariprasad, Y., Iyengar, S.S., Subramanian, N.: Deepfake video detection using lip region analysis with advanced artificial intelligence based anomaly detection technique. Authorea Preprints (2024), preprint
9. Hariprasad, Y., Latesh Kumar, K.J., Suraj, L., Iyengar, S.S.: Boundary-based fake face anomaly detection in videos using recurrent neural networks. In: Arai, K. (ed.) Intelligent Systems and Applications. pp. 155–169. Springer International Publishing, Cham (2023)
10. Inayat, U., Zia, M.F., Ali, F., Ali, S.M., Khan, H.M.A., Noor, W.: Comprehensive review of malware detection techniques. In: 2021 International Conference on Innovative Computing (ICIC). pp. 1–6. IEEE (2021)
11. Iyengar, S.S., Nabavi, S., Hariprasad, Y., Prasad, H.B., Mohan Krishna, C.: Artificial Intelligence in Practice: Theory and Applications for Cyber Security and Forensics. Springer Nature (2025), forthcoming
12. Kebede, T.M., Djaneye-Boundjou, O., Narayanan, B.N., Ralescu, A., Kapp, D.: Classification of malware programs using autoencoders based deep learning architecture and its application to the microsoft malware classification challenge (big 2015) dataset. In: 2017 IEEE National Aerospace and Electronics Conference (NAECON). pp. 70–75. IEEE (2017)

13. Korine, R., Hendler, D.: Daemon: dataset/platform-agnostic explainable malware classification using multi-stage feature mining. IEEE Access **9**, 78382–78399 (2021)
14. Kornblum, J.: Identifying almost identical files using context triggered piecewise hashing. Digital investigation **3**, 91–97 (2006)
15. Le, Q., Boydell, O., Mac Namee, B., Scanlon, M.: Deep learning at the shallow end: Malware classification for non-domain experts. Digital Investigation **26**, S118–S126 (2018)
16. Mohammed, T.M., Nataraj, L., Chikkagoudar, S., Chandrasekaran, S., Manjunath, B.: Hapssa: Holistic approach to pdf malware detection using signal and statistical analysis. In: MILCOM 2021–2021 IEEE Military Communications Conference (MILCOM). pp. 709–714 (2021). https://doi.org/10.1109/MILCOM52596.2021.9653097
17. Mohammed, T.M., Nataraj, L., Chikkagoudar, S., Chandrasekaran, S., Manjunath, B.: Malware detection using frequency domain-based image visualization and deep learning. In: Proceedings of the 54th Hawaii International Conference on System Sciences. p. 7132 (2021)
18. Narayanan, B.N., Djaneye-Boundjou, O., Kebede, T.M.: Performance analysis of machine learning and pattern recognition algorithms for malware classification. In: 2016 IEEE National Aerospace and Electronics Conference (NAECON) and Ohio Innovation Summit (OIS). pp. 338–342. IEEE (2016)
19. Nataraj, L., Karthikeyan, S., Jacob, G., Manjunath, B.S.: Malware images: visualization and automatic classification. In: Proceedings of the 8th International Symposium on Visualization for Cyber Security. pp. 4:1–4:7. VizSec '11, ACM, New York, NY, USA (2011)
20. Nataraj, L., Manjunath, B.: SPAM: Signal processing to analyze malware. In: IEEE Signal Processing Magazine. vol. 33, pp. 105–117. IEEE (2016)
21. Nataraj, L., Mohammed, T.M., Nanjundaswamy, T., Chikkagoudar, S., Chandrasekaran, S., Manjunath, B.: Omd: Orthogonal malware detection using audio, image, and static features. In: MILCOM 2021–2021 IEEE Military Communications Conference (MILCOM). pp. 703–708 (2021). https://doi.org/10.1109/MILCOM52596.2021.9653136
22. Oliver, J., Cheng, C., Chen, Y.: Tlsh–a locality sensitive hash. In: 2013 Fourth Cybercrime and Trustworthy Computing Workshop. pp. 7–13. IEEE (2013)
23. Pachhala, N., Jothilakshmi, S., Battula, B.P.: A comprehensive survey on identification of malware types and malware classification using machine learning techniques. In: 2021 2nd International Conference on Smart Electronics and Communication (ICOSEC). pp. 1207–1214. IEEE (2021)
24. Shi, B., Iyengar, S.S.: Mathematical Theories of Machine Learning–Theory and Applications. Springer (2020)
25. Singaram, J., Iyengar, S.S., Madni, A.M.: Deep Learning Networks: Design, Development and Deployment. Springer Nature (2023)
26. Virussamples: Virussamples website. https://www.virussamples.com/ (2023), [Online; accessed 29-March-2023]
27. VirusShare: Virusshare. com–because sharing is caring (2019)
28. Virustotal: Virustotal-free online virus, malware and url scanner. Online: https://www.virustotal.com/en (2012)
29. Wang, C., Iyengar, S.S., Sun, K.: AI Embedded Assurance for Cyber Systems. Springer Nature (2023)
30. Wei, F., Li, Y., Roy, S., Ou, X., Zhou, W.: Deep ground truth analysis of current android malware. In: International Conference on Detection of Intrusions and Malware, and Vulnerability Assessment. pp. 252–276. Springer (2017)

Artificial Intelligence and Digital Forensics: How They Support Each Other

Bhavani Thuraisingham and Khandakar Ashrafi Akbar

1 Introduction

While AI techniques have been applied for many decades, they have advanced rapidly in recent years solving sophisticated problems such as malware analysis and drug discovery. The AI systems of the 1980s and 1990s were mainly expert systems that captured the knowledge of the experts as rules and reasoned about the knowledge based on the execution of the rules. While these systems had some use, they did not scale. Machine learning systems were also applied to simple problems at that time as learning was rather primitive. It was not until the recent advances in deep learning with multiple layers of learning in the early 2000s that the ML techniques have nearly caught up with human reasoning. Today we have GenAI systems that use LLMs that are essentially powerful ML techniques together with natural language processing (NLP) and are comparable to human reasoning for many applications.

While AI systems have advanced a great deal in recent years, cybersecurity has also advanced as a field over the past 20 years. While the early cybersecurity techniques were mainly focusing on access control and protecting secrets, with the advent of the World Wide Web (WWW), the cyberattacks have increased a great deal. Detecting and recovering from attacks such as malware attacks have been the major focus of cybersecurity specialists. Furthermore, machine learning is being used extensively to detect cyberattacks [1].

One special branch of cybersecurity is digital forensics. Digital forensics is essentially conducting forensics for digital data. There are three types of digital forensics that are being investigated. One is using computers to solve a crime such

B. Thuraisingham (✉) · K. A. Akbar
Department of Computer Science, The University of Texas at Dallas, Richardson, TX, USA
e-mail: bxt043000@utdallas.edu; ashrafi@utdallas.edu

S. S. Iyengar et al. (eds.), *Artificial Intelligence Driven Forensics*, Signals and Communication Technology, https://doi.org/10.1007/978-3-031-98036-7_10

as homicide. Here, the analyst gathers all the physical data, digitizes this data, and uses computers to analyze the data and solve the crime. Such crimes include homicide and fraud. The second type is when computers are used to promote the crime. Distributing child pornography in digital format falls in this category. Law enforcement has invested a great deal on solving such crimes. The third is when computers are the target of the crime [2]. This is the case when computers, networks, clouds, and even AI systems are attacked and the analyst has to solve the problem and determine when, where, and how the attack occurred. One of the main challenges for the forensic specialist is to analyze all of the data and search the digital data to detect patterns and subsequently get to the root of the problem. This is the area where ML and other AI techniques such as GenAI are being used to solve the challenging problems.

This chapter describes both AI for forensics and forensics for AI. We focus on ML techniques and also place a special emphasis on LLMs. The organization of this chapter is as follows. Section 2 describes AI for forensics and forensics for AI. Section 3 describes LLMs and forensics. Application of our work on cyber physical systems such as transportation systems is discussed in Sect. 4. Governance of AI and forensics is discussed Sect. 5. The chapter is concluded in Sect. 6.

2 AI and Digital Forensics

We will discuss AI applications for forensics in Sect. 2.1 and forensics applications for AI systems in Sect. 2.2.

2.1 AI for Forensics

In order to carry out digital forensics, massive amounts of data has to be collected that pertain to the investigation. Such data is the evidence. The evidence has to be encrypted and stored in a secure location so that it is tamperproof. The forensics data is then analyzed using tools to uncover the various patterns so that the criminals who perpetrate the crime can be charged and tried in court. As stated earlier, there are three types of digital forensics analysis. One is to analyze the evidence to aid the investigator in crimes such as homicide. The second is to analyze the evidence to see if computers have been used to distribute child pornography and the third is to examine the attacks to digital systems including computers, networks, and the web to determine who committed the crime. For computers to aid in solving crimes such as homicide, the goal is to build the big picture of the crime. This is done by examining emails, texts, social media, and other sources that leave a footprint of the activities of the criminals. Since some of the data may get erased, the goal is to fill the missing information. This is where ML techniques would help. The ML model is trained with examples. One such example is: if there are two pieces of data A and C, then the erased data between the two pieces is data B. Once the model is trained with such examples, then it is tested to detect missing information in real-world scenarios. The accuracy of the approach depends on the accuracy of the models. Studies and experiments have shown that deep learning, with its multiple layers if learning, can accurately predict the missing information. Once the investigator puts the complete picture together, then he/she can better determine who the criminals are. Also, he/she has to present in court the results and make a strong case that the evidence has not been tampered with. Strong encryption of the evidence would support such claims.

With respect to applying AI to determine whether child pornography is being distributed over the web, the challenge is to detect text, images, and video for suspicious activities. For example, ML techniques can be applied to text data to detect inappropriate messages, ML techniques for image forensics techniques can be used to determine whether the photographs and images contain inappropriate content. ML techniques for video forensics can be used to determine if the videos contain child pornography. There has been a lot of work in multimedia forensics over the past two decades and with deep learning the techniques have shown much accuracy.

Finally, we come to the most important application of ML for forensics and that is to determine whether attacks have occurred and if so who perpetrated the attack. Much of the work in cybersecurity has focused on this aspect. Subsequently, many areas in forensics have emerged including file forensics, web forensics, and cloud forensics. The basic idea is to train the ML model from various examples and then test the model with real-world examples. For example, to conduct file forensics, one objective is to determine whether data that has been erased. Therefore, the model is trained with some accurate missing data and then tested with unknown data. We have carried out extensive research on detecting attacks. It should be noted that one model is not sufficient as the concepts evolve and the amount to data that arrives is continuous. The challenge is to create multiple models and continuously discard or update those that become outdated and inaccurate [3].

The next question is, once it is determined that the system (e.g., files, cloud, web) has been attacked the challenge is to find who carried out the attacks. Therefore, more data is collected about the various attacks, understand the actions of the different hacker groups, and then tie the attack to a specific group. Here again ML can play a major role. The model is trained with various known examples that tie the attacks to a group and then tested with unknown data so that the ML techniques can then determine the hacker group responsible for the attack.

2.2 Forensics for AI

While much of the focus has been on applying AI for forensics, the AI systems can also be attacked. For example, the ML models could be attacked so that they do not provide accurate results. Furthermore, the data that is used to train the models may be poisoned. If one trains models from bad data, then one cannot rely on the results produced. Therefore, forensics also has to be carried out on the AI systems. Our main focus in this section is on forensics for ML techniques. In Sect. 3 we will include a discussion on LLMs and forensics.

First, consider the ML models being attacked. We have conducted research on ML techniques such as support vector machines (SVMs) being attacked. The adversary learns about our models and tries to push some suspicious events into the safe domains. This means some of the attacks may not be detected. In time we would figure out the actions of the adversary and subsequently move the SVM boundary in a way so that the adversary cannot push the suspicious events into safe domains. However, the adversary would soon determine the new boundary line. Such game playing continues until one party depletes its resources. The assumption is that the adversary will not have sufficient resources to continue with this game. The challenge for the forensics tool is to determine that such an attack has occurred and find out who the adversary is. There is much to be done on finding the perpetrators of such attacks [4].

Next, in the case of data poisoning, the ML techniques that use such poisoned data to train will produce results that are not accurate. Therefore, the challenge is how does the forensics tool detect that the data has been poisoned? We need to develop techniques that check for the integrity of the data. Numerous data integrity techniques have been developed for well over 50 years. These techniques include enforcing integrity constraints to check for the accuracy of the data. We can also use ML to determine the accuracy of the data. That is, the ML models are trained to determine the correct data and then during testing they can then determine that the data is invalid and compute the correct version of the data. That is, we can use ML to carry out ML forensics. AI forensics is a new area and there is much to be done.

3 LLMs and Forensics

In the following two subsections, we will explore the use of large language models (LLMs) in forensic analysis, as well as the application of forensic techniques to LLMs. While these two concepts are distinct, we will discuss how methodologies from each domain can complement each other in achieving their respective objectives.

3.1 LLMs for Forensics

One of the major challenges in digital forensics is managing the enormous volume of data generated through logging. Processing such large datasets can be time-consuming and complex. Large language models (LLMs) offer a promising solution by enabling automation in data analysis. Retrieval-augmented generation (RAG) [5] has emerged as a popular approach for automating large-scale data processing and generating reliable text outputs without the need for retraining or fine-tuning production-level LLMs (e.g., ChatGPT, Claude). By leveraging RAG, forensic analysts can enhance query processing and conduct thorough examinations of bulk data more efficiently.

Addressing bias in forensic analysis is critical, as biased conclusions can lead to incorrect outcomes and undermine the integrity of investigations. Forensic experts often rely on a variety of specialized technical tools to perform tasks such as digital footprint analysis [6] or cloud forensics [7]. It is essential to carefully evaluate each potential scenario that could arise within a given context to ensure thorough root cause analysis. Achieving bias-free analysis is paramount in these situations. Large language models (LLMs), when properly designed to be bias-free, can assist in mitigating these risks and support accurate forensic assessments. Additionally, the process of self-correction within these models offers a promising approach to continually address and minimize bias, helping to maintain the integrity of forensic investigations.

Large language models (LLMs) can be effectively employed to extract structured information from complex and nuanced details, thereby aiding in the performance of downstream tasks in digital forensics. In the context of today's highly interconnected world, driven by the Internet of Things (IoT), it is crucial to extract meaningful insights from the technical jargon that permeates our surroundings. Every small device has the capability to communicate with others, contributing to a multifaceted narrative. The vast and intricate volumes of data generated by these interactions can be processed using LLMs, enabling a comprehensive understanding of the entire system.

In forensics and cybersecurity, single-point detection or identification has long been a classical approach for identifying anomalous or malicious behavior. However, the challenge arises when attempting to connect disjointed data points in a semantically meaningful way. For instance, system and network traffic logs, as well as application logs, predominantly consist of machine-readable data. Without proper contextual understanding, linking these disparate elements becomes a complex and often unclear task. Language processing techniques, particularly those involving large language models (LLMs), can play a crucial role in correlating these facts in such intricate scenarios. By leveraging LLMs to process low-level representations, it becomes possible to trace the growth and expansion of cyber campaigns, thereby facilitating the meticulous profiling of individual, standalone behaviors.

3.2 Forensics for LLMs

In general, forensic analysis of large language models (LLMs) refers to the assessment of their behaviors across various dimensions—such as how they generate specific outputs, their susceptibility to jailbreaking conditions [8], or the presence of certain biases. It can also involve behavioral modeling of LLMs from architectural or empirical perspectives.

Input-output analysis is a classical approach to understanding the capabilities and limitations of large language models (LLMs). By examining how specific inputs or prompts lead to outputs, researchers can explore the potential risks related to security or privacy—an open area of investigation within LLM forensics.

Another crucial focus is bias in LLM-generated content. Investigating how LLMs develop different types of bias in their responses, evaluating their level of context awareness, and designing methods to quantify and qualify these biases form a vital part of this research domain. The well-documented issue of hallucination in LLMs [9]—where models generate outputs that are factually incorrect or fabricated—also demands deeper forensic analysis. Understanding how to categorize various types of hallucinations and identifying their underlying causes—whether rooted in the model's architecture, training data, or inference process—is key. Some hallucinations may be preventable, while others might be inherent to the model's design. Determining which are avoidable and under what constraints remains an important and ongoing area of study in LLM forensics.

Prompt injection in large language models (LLMs) [10] is a common malicious technique used by cyberattackers to manipulate LLM behavior in undesirable or harmful ways. This can result in the disclosure of sensitive information—such as personally identifiable data from user-personalized LLMs—or in the generation of unethical or inappropriate content. Such misuse underscores the urgent need for clear guidelines, thorough analysis, and robust forensic methods to ensure the responsible and secure deployment of LLM technologies.

Attributing the source of LLM-generated content is a crucial step toward fostering responsible and ethical development and use of large language models. With the rapid growth of open-source LLMs—platforms like Hugging Face now hosting over 700,000 models [11]—and the increasing trend toward personalized LLMs, the issue of attribution becomes even more pressing. Clear attribution allows accountability to be assigned to model developers or platforms, especially in cases where harmful or unethical outputs occur. Additionally, ensuring the provenance of LLMs is essential, as many models are trained on copyrighted or sensitive materials, raising legal and ethical concerns. This is particularly relevant in research contexts, where LLMs assist in generating novel ideas based on existing published work. As automated tools play a greater role in research creation, it is imperative that such outputs adhere to the standards of responsible conduct in research, including originality and ethical integrity. The emerging field of LLM forensics will be vital in addressing these challenges.

As LLM-driven software engineering—such as automated coding assistance and emerging approaches like vibe coding [12]—continues to gain significant traction in the tech industry, it is crucial for companies to manage AI-assisted development responsibly to safeguard proprietary information. To fully leverage the capabilities of LLM technology, organizations may need to utilize their legacy codebases—valuable proprietary assets—as part of training or fine-tuning data. However, strict measures must be implemented to ensure that this sensitive information remains within the organization's boundaries. Failure to do so could result in inadvertent leakage of memorized proprietary content, potentially causing irreversible damage to a company's reputation and competitive position.

An audit trail can be an effective method for evaluating whether LLMs (large language models) are compliant with regulatory standards. To enhance LLM performance in personalized applications (such as customer service agents), it is crucial to analyze and moderate the model's output, especially considering the sensitivity of the industry. For instance, LLMs trained specifically as financial assistants for users may need to provide tailored solutions at both group and individual levels. Achieving this level of personalization requires extensive analysis and the integration of additional modules to ensure that the LLM's decisions are guided appropriately.

Ensuring the fairness of large language models (LLMs) is a critical aspect of their application in digital forensics. It is important to examine the factors that trigger biased or unfair output generation by LLMs, as well as the necessary data moderation or adjustments required to mitigate such issues. Specifically, quantifying and qualifying fairness in LLMs [13] is essential for understanding how to control their behavior within defined margins, ensuring that the technology produces reliable and expected outcomes when used in forensic tasks.

4 Application to Transportation Systems

One of the security applications we are focusing on is in the area of transportation systems, which is a type of cyber physical system. We are applying machine learning techniques and LLMs to detect attacks in transportation systems. We are also examining privacy aspects of the transportation data to protect the various individuals. Finally, we want to ensure that everyone is treated fairly and the resources are distributed as needed [14–16]. One technique in ML that we have utilized for this purpose is federated learning. Since the vehicles are distributed, the attacks on each vehicle is analyzed by developing a local model. Rather than sharing the attack data, since manufacturers of the vehicles are reluctant to share the data, we share the local models and subsequently develop a global model that includes a representative sample of the attacks to all the vehicles. From this global model we can detect more attacks [17].

While our main focus is on cybersecurity and resiliency for transportation systems, in the future, we believe that forensics is going to be an important area. That is, once the attacks have occurred, then we have to determine who the perpetrators are. For transportation systems, we need to understand the various internal software and hardware components and examine the patterns to determine how the attacks occurred. That is, in addition to attacks to say file systems, cloud, and the web, we also need to examine vehicle related data such as the CAN bus. This is one of the future directions we plan to pursue and believe that LLMs could be used to handle this problem. Digital forensics for vehicles has come to be known as automotive digital forensics [18].

5 Governance of AI and Forensics

One crucial aspect of the security of systems includes the governance of the systems. The systems that have to be protected have to be governed properly. Governance includes utilizing various risk and cybersecurity frameworks and formulating a strategy for conducting risk analysis [19]. Since many of the systems are complex, especially automated vehicles, and consist of multiple smaller systems, the risk frameworks should specify ways of carrying out risk analysis of the systems. Once the analysis is carried out, then the organization can determine how to mitigate the risks.

Like any other system, AI systems also have to be governed. They have to be analyzed for risks and a mitigation strategy has to be formulated. In our earlier chapter, we have discussed a framework for AI governance that includes aspects such as explainability, accountability, and transparency. This means that the system has to be accountable for its actions so that in the case of failures, proper measures can be taken [20].

Governance is also an important aspect of forensics. That is, we need a governance framework for conducting forensics. Are the forensic solutions secure? What are the risks for forensic systems? How can we mitigate the risk? How do we safeguard the forensics systems? These are questions that need to be answered if we are to develop robust forensics techniques with AI systems. It is also essential that the AI techniques used in forensic investigations are properly governed.

6 Conclusion

This chapter has provided an overview of AI systems and digital forensics and discussed how one can be applied to the other. For example, AI techniques can be used to conduct forensics while AI systems could also be attacked and therefore should be subject to forensics investigations. We have examined both ML and GenAI (LLMs) for forensics and forensics for ML and LLMs. The chapter also discussed one specific application and that is transportation systems and described the need to conduct forensics for such systems using AI techniques. Finally, we discussed the governance of both AI systems and forensics including analyzing the risks and addressing aspects such as accountability and transparency.

Applying AI for forensics and forensics for AI is a relatively new area over the past decade. With the explosion of cyberattacks and the advances made in AI, we believe that there are numerous opportunities for research and development in this field. This chapter has provided some directions to these opportunities. We strongly believe that AI systems that are attacked will be little used and could be even dangerous to use. Therefore, while AI can be of immense help in forensics investigations, we must not forget about forensics for AI systems.

References

1. Thuraisingham, B., Masud, M., Parveen, P., Khan, L.: Big Data Analytics with Applications in Insider Threat Detection, pp. 1–544 (2017). https://doi.org/10.1201/9781315119458
2. Pottinger, W.: Learn Computer Forensics: Your One-stop Guide to Searching, Analyzing, Acquiring, and Securing Digital Evidence, 2nd edn. Packt Publishing, Birmingham, UK (2022)
3. Masud, M.M., Gao, J., Khan, L., Han, J., Thuraisingham, B.: A multi-partition multi-chunk ensemble technique to classify concept-drifting data streams. In: Theeramunkong, T., Kijsirikul, B., Cercone, N., Ho, T.-B. (eds.) Advances in Knowledge Discovery and Data Mining, pp. 363–375. Springer, Berlin, Heidelberg (2009)
4. Zhou, Y., Kantarcioglu, M., Thuraisingham, B., Xi, B.: Adversarial support vector machine learning. In: Proceedings of the 18th ACM SIGKDD International Conference on Knowledge Discovery and Data Mining. KDD '12, pp. 1059–1067. Association for Computing Machinery, New York, NY, USA (2012). https://doi.org/10.1145/2339530.2339697
5. Lewis, P., Perez, E., Piktus, A., Petroni, F., Karpukhin, V., Goyal, N., Küttler, H., Lewis, M., Yih, W.-t., Rocktäschel, T., Riedel, S., Kiela, D.: Retrieval-augmented generation for knowledge-intensive nlp tasks. In: Proceedings of the 34th International Conference on Neural Information Processing Systems. NIPS '20. Curran Associates Inc., Red Hook, NY, USA (2020)
6. Soni, N., Soni, P.: Tracing the unseen: The role of digital footprints in modern forensic investigations. International Journal of Innovative Research in Technology **11**(4), 346–351 (2024). Accessed: 2025-04-27
7. Manoj, S.K.A., Bhaskari, D.L.: Cloud forensics-a framework for investigating cyber attacks in cloud environment. Procedia Computer Science **85**, 149–154 (2016). https://doi.org/10.1016/j.procs.2016.05.202. International Conference on Computational Modelling and Security (CMS 2016)
8. Chao, P., Robey, A., Dobriban, E., Hassani, H., Pappas, G.J., Wong, E.: Jailbreaking Black Box Large Language Models in Twenty Queries (2024). https://arxiv.org/abs/2310.08419
9. Ji, Z., Lee, N., Frieske, R., Yu, T., Su, D., Xu, Y., Ishii, E., Bang, Y.J., Madotto, A., Fung, P.: Survey of hallucination in natural language generation. ACM Computing Surveys **55**(12), 1–38 (2023). https://doi.org/10.1145/3571730
10. Liu, Y., Jia, Y., Geng, R., Jia, J., Gong, N.Z.: Formalizing and benchmarking prompt injection attacks and defenses. In: 33rd USENIX Security Symposium (USENIX Security 24), pp. 1831–1847. USENIX Association, Philadelphia, PA (2024). https://www.usenix.org/conference/usenixsecurity24/presentation/liu-yupei
11. Face, H.: Hugging Face. Accessed: 2025-04-27 (2025). https://huggingface.co/
12. Pajo, P.: Vibe Coding: Revolutionizing Software Development with AI-Generated Code. https://doi.org/10.13140/RG.2.2.36458.22727
13. Spitale, M., Cheong, J., Gunes, H.: Underneath the numbers: Quantitative and qualitative gender fairness in llms for depression prediction. CoRR **abs/2406.08183** (2024). https://doi.org/10.48550/ARXIV.2406.08183. 2406.08183
14. Abay, N.C., Zhou, Y., Kantarcioglu, M., Thuraisingham, B., Sweeney, L.: Privacy preserving synthetic data release using deep learning. In: Berlingerio, M., Bonchi, F., Gärtner, T., Hurley, N., Ifrim, G. (eds.) Machine Learning and Knowledge Discovery in Databases, pp. 510–526. Springer, Cham (2019)
15. Zhao, C., Chen, F., Thuraisingham, B.: Fairness-aware online meta-learning. In: Proceedings of the 27th ACM SIGKDD Conference on Knowledge Discovery & Data Mining. KDD '21, pp. 2294–2304. Association for Computing Machinery, New York, NY, USA (2021). https://doi.org/10.1145/3447548.3467389
16. Quinonez, R., Safaoui, S., Summers, T., Thuraisingham, B., Cardenas, A.A.: Shared reality: Detecting stealthy attacks against autonomous vehicles. In: Proceedings of the 2th Workshop on CPS&IoT Security and Privacy. CPSIoTSec '21, pp. 15–26. Association for Computing Machinery, New York, NY, USA (2021). https://doi.org/10.1145/3462633.3483981

17. Halim, S.M., Hossain, M.D., Khan, L., Singhal, A., Inoue, H., Ochiai, H., Hamlen, K.W., Kadobayashi, Y.: Securing smart vehicles through federated learning. In: Foundations and Practice of Security: 16th International Symposium, FPS 2023, Bordeaux, France, December 11–13, 2023, Revised Selected Papers, Part I, pp. 18–35. Springer, Berlin, Heidelberg (2023). https://doi.org/10.1007/978-3-031-57537-2_2
18. AB Forensics Ltd.: Digital Vehicle Forensics. Accessed: 2025-04-27 (2025). https://abforensics.com/digital-vehicle-forensics/
19. National Institute of Standards and Technology: Cybersecurity Framework. Accessed: 2025-04-27 (2025). https://www.nist.gov/cyberframework
20. Thuraisingham, B.: Artificial intelligence and data science governance: Roles and responsibilities at the c-level and the board. In: 2020 IEEE 21st International Conference on Information Reuse and Integration for Data Science (IRI), pp. 314–318 (2020). https://doi.org/10.1109/IRI49571.2020.00052

Executive Summary: Digital Forensics AI: Towards a Framework for Practical Adoption

Maryna Veksler and Kemal Akkaya

The integration of artificial intelligence (AI) in digital forensics (DF) offers significant promise for enhancing investigative efficiency, evidence analysis, and data triage [3]. However, the practical adoption of AI-based digital forensic tools, referred to as digital forensics AI (DFAI), remains limited due to the concerns around ethical, legal, and standardized usage. While explainable AI (XAI) presents a promising solution to improve the transparency and interpretability of DFAI [1, 2], there are no definitive indications that XAI alone can solve the practical challenges of DFAI adoption. There has also been a lack of engagement with frontline DF practitioners regarding the practical usage of DFAI.

To fill this gap, this work explores the practical usage and adoption challenges of DFAI by conducting a field study involving 28 DF practitioners through surveys and interviews. The study offers practical insights and introduces a framework aimed at guiding standardized and responsible deployment of DFAI tools.

Specifically, we utilized a mixed-method approach, combining quantitative survey data and qualitative interview responses. The survey and interview questions were designed following a structured categorical roadmap, covering seven six topics: (1) trust and perception of DFAI, (2) functional expectations of DFAI tools, (3) validation and evaluation of DFAI, (4) official benchmarks and datasets, (5) challenges of future adoption, and (6) real-world case studies and feasibility of DFAI. Our study involved practitioners from diverse backgrounds, including forensic examiners, investigators, consultants, and academicians, with diverse experience in AI usage and testifying as expert witness. The key findings of our study are outlined as follows:

M. Veksler (✉) · K. Akkaya
Knight Foundation School of Computing and Information Sciences, Florida International University, Miami, FL, USA
e-mail: mveks001@fiu.edu; kakkaya@fiu.edu

S. S. Iyengar et al. (eds.), *Artificial Intelligence Driven Forensics*, Signals and Communication Technology, https://doi.org/10.1007/978-3-031-98036-7_11

Trust is Built on Validation and Transparency Practitioners prioritized validation, testing, and transparent documentation of DFAI over explainability. XAI was appreciated but not seen as a substitute for verifiable results and known error rates.

DFIA is Preferred for Triage Most practitioners favored using DFAI tools for triage automation. Standalone analysis raised concerns about reliability and interpretability.

Local, Open-Source Tools are Favored Respondents preferred open-source, locally deployed tools due to privacy, security, and transparency concerns.

Current DFAI Validation is Insufficient While useful, traditional methods like manual or cross-validation can fall short for AI systems. Enhanced protocols should include AI-specific tool testing approaches.

Accuracy and Error Rates are Critical Accuracy and true-positive rate were the most important metrics. Clear, interpretable error rates were essential for legal trust.

Reliable Datasets Require Collaboration Participants stressed the need for transparent, validated datasets and collaborative efforts to develop them.

AI's Role in Court Must be Carefully Managed DFAI outputs were acceptable as supporting evidence but not standalone proof, especially in high-stakes scenarios.

DFAI Needs Regulation, Not Blame Practitioners saw DFAI as a tool, not a faulty actor, and supported regulated, collaborative deployment.

To address the identified gaps, we propose a structured DFAI adoption framework built on five pillars:

- **Tool validation:** Extending traditional validation with AI Test Suite that includes bias estimation, precision testing, elusion testing, and hallucination testing
- **Datasets:** Promoting well-documented, validated, and diverse datasets
- **Evidence presentation and assessment:** Emphasizing explainability tailored for courts, built on validation and supported by XAI when applicable
- **Tool selection:** Encouraging the use transparency, auditable, and locally deployable systems
- **Tool development practices:** Advocating for gradual integration of DFAI with robust documentation and accuracy-first design

This study highlights the importance of basing the DFAI tool development and deployment on real-world experiences and expectations of DF practitioners. DFAI adoption must be governed by a strong foundation of trust, validation, transparency, and stakeholders collaboration. The proposed framework offers a practical guide for advancing safe, ethical, and standardized DFAI use across digital forensic community.

References

1. Alam S, Altiparmak Z (2024) Xai-cf–examining the role of explainable artificial intelligence in cyber forensics. arXiv preprint arXiv:240202452
2. Arthanari A, Raj SS, Ravindran V (2025) A narrative review in application of artificial intelligence in forensic science: Enhancing accuracy in crime scene analysis and evidence interpretation. Journal of International Oral Health 17(1):15–22
3. Verma R, Garg S, Kumar K, et al (2023) New approach of artificial intelligence in digital forensic investigation: A literature review. In: International Conference on Advances in Communication Technology and Computer Engineering, Springer, pp 399–409

Part II
Bridging Law and Technology: AI-Enabled Legal Framework

Artificial Intelligence in Legal Forensics: Principles, Applications, and Emerging Frontiers

S. S. Iyengar

1 Introduction

The intersection of artificial intelligence (AI) and legal forensics marks a transformative shift in how justice is pursued in the digital age [1–4]. As the complexity of crimes increases and legal systems face overwhelming volumes of data, traditional investigative and legal methodologies often fall short [5–8]. To bridge this gap, AI and machine learning (ML) are being increasingly deployed to enhance the precision, efficiency, and scalability of forensic investigations [4, 9–11].

Legal forensics—a field focused on the analysis of evidence for legal proceedings—now benefits from AI-driven solutions that automate document review, predict legal outcomes, analyze voice and video content, and uncover hidden correlations across diverse datasets [12–16]. From courtroom analytics to evidentiary validation, AI enables law enforcement agencies, attorneys, and judges to draw actionable insights from vast and multifaceted information sources [17–19]. This evolution not only accelerates case resolution but also strengthens the fairness and transparency of judicial processes [20–22].

1.1 The Role and Necessity of AI in Legal Forensics

Legal forensics encompasses tasks such as reviewing complex litigation documents, identifying fraud patterns, analyzing speech transcripts, and evaluating large volumes of multimedia and digital evidence. The advent of cybercrime, deepfake

S. S. Iyengar (✉)
Global Forensic and Justice Center, Florida International University, Miami, FL, USA
e-mail: iyengar@cs.fiu.edu

S. S. Iyengar et al. (eds.), *Artificial Intelligence Driven Forensics*, Signals and Communication Technology, https://doi.org/10.1007/978-3-031-98036-7_12

manipulation, and cross-border data breaches has further necessitated AI-enabled tools that can operate at speed and scale far beyond human capacity [23–30].

Key advantages of AI in legal forensic applications include:

- **Automated document analysis:** Natural language processing (NLP) models facilitate rapid review and annotation of legal texts, flagging inconsistencies and prioritizing relevant case information.
- **Predictive legal analytics:** AI models forecast legal outcomes based on historical case data, providing strategic insights to lawyers and judges.
- **Cross-modal evidence correlation:** AI systems unify data streams from documents, audio, video, and digital logs to build cohesive narratives for legal proceedings.
- **Fraud and anomaly detection:** Machine learning algorithms detect unusual patterns in contracts, financial records, and testimonies, aiding in early identification of fraudulent behavior.

1.2 Scope of the Chapter

This chapter explores the principles, applications, and emerging frontiers of AI in legal forensics. It examines how AI technologies are deployed across core areas such as document analysis, evidence correlation, voice and video analytics, and case outcome prediction. Furthermore, it discusses the ethical, legal, and regulatory implications of integrating AI into judicial systems, including concerns about algorithmic bias, data privacy, and admissibility of AI-generated findings in court.

By providing a comprehensive view of AI-enabled legal forensics, this study highlights the ongoing transformation of legal practices and underscores the potential of intelligent technologies to uphold justice in an increasingly digital world.

2 AI-Driven Innovations in Legal Forensic Analysis

Legal forensic analysis is foundational to the justice system, encompassing the examination of legal documents, digital and physical evidence, and verbal testimonies to reconstruct events and support fair adjudication. Traditionally, this process has been labor-intensive, requiring extensive manual review and domain expertise. With the rise of artificial intelligence (AI) and machine learning (ML), legal forensic workflows are undergoing a paradigm shift—enabling data-driven insights, accelerating case processing, and enhancing judicial transparency.

AI applications in legal forensics now span multiple domains, including document intelligence, evidence correlation, predictive analytics, and voice analysis. This section outlines the core capabilities AI brings to modern legal investigations.

2.1 Document Intelligence and Legal Text Analysis

Legal proceedings often generate large volumes of case files, contracts, and deposition transcripts. AI-powered natural language processing (NLP) systems automate the extraction, classification, and synthesis of legal texts, significantly reducing review times and identifying high-risk elements.

- **Automated legal review:** NLP models flag inconsistencies, contradictions, and key phrases in contracts, pleadings, and affidavits.
- **Fraud detection:** Machine learning algorithms uncover anomalies in financial statements, insurance claims, and regulatory documents.
- **Legal research augmentation:** AI cross-references case law, statutes, and legal arguments, assisting lawyers in developing precedent-based strategies.

2.2 Cross-Source Evidence Correlation

AI systems excel at fusing and contextualizing evidence from disparate sources. Whether integrating digital records, surveillance footage, or testimonial transcripts, AI algorithms reconstruct timelines and identify hidden connections critical to a case.

- **Multimodal evidence analysis:** AI integrates text, images, audio, and video to produce a holistic view of legal scenarios.
- **Relationship mapping:** Link analysis engines reveal connections between suspects, co-conspirators, and events through graph-based visualizations.
- **Chain-of-custody verification:** AI ensures evidentiary integrity through metadata validation and audit trails.

2.3 Predictive Legal Analytics

Machine learning models can forecast potential legal outcomes based on historical datasets, judicial behavior, and precedent-driven trends.

- **Case outcome prediction:** AI estimates likelihoods of verdicts or settlements by analyzing similar past cases.
- **Sentencing trend analysis:** Algorithms evaluate sentencing patterns across jurisdictions, helping assess the proportionality of punishment.

- **Risk scoring and strategy development:** AI tools score the reliability of witnesses or the viability of legal arguments to support trial preparation.

2.4 *Challenges and the Path Ahead*

Despite its promise, AI-driven legal forensics faces challenges:

- **Bias and fairness:** AI models must be trained on diverse datasets to avoid perpetuating systemic biases.
- **Data privacy:** Legal applications of AI must align with data protection laws and maintain evidentiary confidentiality.
- **Admissibility and transparency:** Legal systems demand explainable AI, ensuring that algorithmic outputs can be verified, justified, and accepted in court.

As AI technologies evolve, their role in the legal system will expand—enabling faster case resolution, more equitable justice, and global collaboration in transnational legal challenges. Legal professionals must be equipped to leverage these tools while ensuring ethical oversight and legal rigor.

3 Challenges and Ethical Considerations

While AI and ML offer transformative advantages in forensic science, their implementation introduces critical ethical, legal, and operational challenges. Ensuring responsible and effective adoption of AI in legal forensics requires addressing the following key concerns:

- **Data privacy and security:** AI systems must comply with data protection regulations such as GDPR and HIPAA, especially when handling sensitive legal or biometric information.
- **Bias in AI models:** Algorithms trained on nonrepresentative datasets may reinforce existing societal biases, leading to wrongful identifications or unfair legal conclusions.
- **Legal and regulatory compliance:** Forensic tools powered by AI must align with evidentiary standards to ensure admissibility in courts, necessitating validation, documentation, and judicial oversight.
- **Interpretability and human oversight:** AI-generated decisions must be transparent and explainable, allowing forensic experts and legal professionals to understand and verify system outputs.
- **Ethical use and accountability:** Misuse of AI in surveillance, profiling, or manipulation of evidence poses risks to civil liberties. Robust ethical frameworks are required to define acceptable boundaries and accountability mechanisms.

4 Future Directions and Recommendations

The next frontier of AI in legal forensics lies in the convergence of interdisciplinary innovation, regulatory harmonization, and scalable real-time applications. Key future directions include:

- **AI-enhanced biometric fusion:** Integrating facial recognition, voice profiling, gait analysis, and iris scanning into a unified forensic identity verification framework.
- **Quantum computing integration:** Leveraging quantum algorithms to accelerate decryption, data authentication, and secure transmission of forensic data across jurisdictions.
- **Portable AI forensic units:** Developing mobile, edge-computing-enabled forensic devices for on-site analysis of digital evidence, biosignals, and chemical signatures.
- **Standardization of AI forensic protocols:** Establishing globally accepted technical and ethical standards to ensure interoperability and accountability in cross-border legal investigations.
- **Explainable AI (XAI) implementation:** Prioritizing interpretable model design in forensic applications to build trust, support legal scrutiny, and uphold judicial integrity.

5 Conclusion

Artificial intelligence (AI) and machine learning (ML) have fundamentally reshaped forensic science, enabling faster, more accurate, and scalable investigative capabilities across diverse domains—legal analysis, digital forensics, canine detection, and genetic profiling. By automating complex processes such as evidence correlation, document analysis, malware detection, and scent identification, AI-driven systems significantly enhance the effectiveness of modern forensic investigations.

The ability of AI to extract actionable insights from vast, multimodal datasets has empowered legal professionals and law enforcement to pursue justice with greater precision and efficiency. However, the integration of AI into forensic workflows also demands rigorous attention to issues of fairness, bias, privacy, and legal admissibility.

As research and technology evolve, the future of AI-enabled forensics lies in ethical innovation, real-time intelligence, and global collaboration. The continued development of trustworthy and transparent AI systems will be essential to ensuring that the pursuit of justice remains both equitable and technologically resilient in the years to come.

Acknowledgments This research was partially sponsored by the Army Research Office and the NSF and was accomplished under Grant Number W911NF-21-1-0264 and 2018611. The views and conclusions contained in this document are those of the authors and should not be interpreted as representing the official policies, either expressed or implied, of the Army Research Office or the US Government. The U.S Government is authorized to reproduce and distribute reprints for government purposes notwithstanding any copyright notation herein.

References

1. Hadid, A., Evans, N., Marcel, S., Fierrez, J.: Biometrics systems under spoofing attack: An evaluation methodology and lessons learned. IEEE Signal Processing Magazine **32**(5), 20–30 (2015)
2. Bureau, N.C.R.: Cyber Crime Report. Ministry of Home Affairs, Government of India (2022)
3. Iyengar, S.S., Nabavi, S., Hariprasad, Y., H.B., P., C., K.M.: Artificial Intelligence in Practice: Theory and Applications for Cyber Security and Forensics. Springer, Cham (2025)
4. Shi, B., Iyengar, S.S.: Mathematical Theories of Machine Learning-Theory and Applications. Springer, Cham (2020)
5. Iyengar, S.S., Kashyap, R.L., Madan, R.N.: Distributed sensor networks-introduction to the special section. IEEE Transactions on Systems, Man, and Cybernetics **21**(5), 1027–1031 (1991)
6. Iyengar, S.S., Deng, W.: An efficient edge detection algorithm using relaxation labeling technique. Pattern Recognition **28**(4), 519–536 (1995)
7. Thejas, G.S., Boroojeni, K.G., Chandna, K., Bhatia, I., Iyengar, S.S., Sunitha, N.R.: Deep learning-based model to fight against ad click fraud. In: Proceedings of the 2019 ACM Southeast Conference, pp. 176–181 (2019)
8. Guo, M., Pissinou, N., Iyengar, S.S.: Privacy-preserving deep learning for enabling big edge data analytics in internet of things. In: 2019 Tenth International Green and Sustainable Computing Conference (IGSC), pp. 1–6. IEEE, New York (2019)
9. Singaram, J., Iyengar, S.S., Madni, A.M.: Deep Learning Networks: Design, Development and Deployment. Springer, Cham (2023)
10. Wang, C., Iyengar, S.S., Sun, K.: AI Embedded Assurance for Cyber Systems. Springer, Cham (2023)
11. Iyengar, S.S., Kashyap, R.L.: Guest editor's introduction: Image databases. IEEE Transactions on Software Engineering **14**(5), 608 (1988)
12. Pouyanfar, S., Yang, Y., Chen, S.C., Shyu, M.L., Iyengar, S.S.: Multimedia big data analytics: A survey. ACM Computing Surveys (CSUR) **51**(1), 1–34 (2018)
13. Hariprasad, Y., Iyengar, S.S., Subramanian, N.: Deepfake video detection using lip region analysis with advanced artificial intelligence based anomaly detection technique. Authorea Preprints (2024)
14. Hariprasad, Y., K.J., L.K., Suraj, L., Iyengar, S.S.: Boundary-based fake face anomaly detection in videos using recurrent neural networks. In: Proceedings of SAI Intelligent Systems Conference, pp. 155–169. Springer, Cham (2022)
15. Hariprasad, Y., Iyengar, S.S., Chaudhary, N.K.: Securing the future: Advanced encryption for quantum-safe video transmission. IEEE Transactions on Consumer Electronics (2024)
16. Zhao, Z., Li, X., Luan, B., Jiang, W., Gao, W., Neelakandan, S.: Secure internet of things (iot) using a novel brooks iyengar quantum byzantine agreement-centered blockchain networking (biqba-bcn) model in smart healthcare. Information Sciences **629**, 440–455 (2023)
17. Lohachab, A., Lohachab, A., Jangra, A.: A comprehensive survey of prominent cryptographic aspects for securing communication in post-quantum iot networks. Internet of Things **9**, 100174 (2020)

18. Rathore, S., Sharma, P., Paul, A.: Ai in forensic science: Emerging trends and applications. Forensic Science International **332**, 110–126 (2023)
19. Smith, J., Liu, R.: Machine learning in digital forensics: A review of algorithms and case studies. Journal of Cybersecurity Research **29**(4), 255–270 (2022)
20. Thompson, B., Garcia, E.: Ai-driven dna analysis in criminal investigations. Genomic Forensics Review **45**(2), 134–150 (2021)
21. Standards, N.I., Technology: Ai in forensic science: Guidelines and challenges. Technical Report 2023-04, NIST (2023)
22. Brown, H., White, M.: The future of ai in canine detection. Journal of Law Enforcement Technology **18**(1), 45–62 (2022)
23. Iyengar, S. S., Seyedsina Nabavirazavi, Yashas Hariprasad, Prasad HB, and C. Krishna Mohan. Privacy-Preserving AI (Federated Learning) for Digital Forensics. In Artificial Intelligence in Practice: Theory and Application for Cyber Security and Forensics, pp. 161–176. Cham: Springer Nature Switzerland, 2025.
24. Iyengar, S. S., Seyedsina Nabavirazavi, Yashas Hariprasad, Prasad HB, and C. Krishna Mohan. Cyber Threat Intelligence and Security for Federated Learning in Digital Forensics. In Artificial Intelligence in Practice: Theory and Application for Cyber Security and Forensics, pp. 177–199. Cham: Springer Nature Switzerland, 2025.
25. Iyengar, S.S., Nabavirazavi, S., Hariprasad, Y., HB, P. and Mohan, C.K., 2025. Future of AI-Driven Digital Forensics. In Artificial Intelligence in Practice: Theory and Application for Cyber Security and Forensics (pp. 335–364). Cham: Springer Nature Switzerland.
26. Iyengar, S.S., Nabavirazavi, S., Hariprasad, Y., HB, P. and Mohan, C.K., 2025. Hybrid Detection of Malicious Portable Document Format (PDFs): Safeguarding Against Embedded JavaScript Attacks. In Artificial Intelligence in Practice: Theory and Application for Cyber Security and Forensics (pp. 257–290). Cham: Springer Nature Switzerland.
27. Soni, Jayesh, Subhash Gurappa, and Himanshu Upadhyay. A Comparative Study of Deep Learning Models for Image Super-Resolution Across Various Magnification Levels. In 2024 IEEE International Conference on Future Machine Learning and Data Science (FMLDS), pp. 395–400. IEEE, 2024.
28. Iyengar, S. S., Seyedsina Nabavirazavi, Yashas Hariprasad, Prasad HB, and C. Krishna Mohan. Temporal Deepfake Generation and Detection in Video Sequences Using Recurrent Neural Networks (RNNs). In Artificial Intelligence in Practice: Theory and Application for Cyber Security and Forensics, pp. 309–334. Cham: Springer Nature Switzerland, 2025.
29. Iyengar, S. S., Seyedsina Nabavirazavi, Yashas Hariprasad, Prasad HB, and C. Krishna Mohan. The Convergence of AI/ML and Cybersecurity: Advancing Digital Forensic Techniques. In Artificial Intelligence in Practice: Theory and Application for Cyber Security and Forensics, pp. 139–159. Cham: Springer Nature Switzerland, 2025.
30. Iyengar, S. S., Seyedsina Nabavirazavi, Yashas Hariprasad, Prasad HB, and C. Krishna Mohan. AI-Enhanced Malware Detection: Advancing Security Through Intelligent Threat Identification. In Artificial Intelligence in Practice: Theory and Application for Cyber Security and Forensics, pp. 227–255. Cham: Springer Nature Switzerland, 2025.

Establishing Trust: A Techno-Legal Framework for Ensuring Digital Evidence Integrity

Naveen Kumar Chaudhary, S. S. Iyengar, and Nilay Mistry

1 Introduction

The increasing reliance on digital evidence in criminal investigations and judicial proceedings necessitates robust mechanisms to ensure its authenticity, integrity, and admissibility [21–25]. The enactment of three new criminal laws in India—Bharatiya Nyaya Sanhita (BNS) 2023, Bharatiya Nagarik Suraksha Sanhita (BNSS) 2023, and Bharatiya Sakshya Adhiniyam (BSA) 2023—marks a significant transformation in the Indian criminal justice system, replacing colonial-era statutes and embracing contemporary legal and technological realities. Of particular importance is the BSA 2023, which redefines the evidentiary status of digital records by classifying them as primary evidence, thereby underscoring the need for reliable standards for digital evidence integrity [26–28].

Prior to these reforms, the Indian Evidence Act, 1872, governed the admissibility of electronic records under Sections 65A and 65B, with significant ambiguities surrounding the structure and content of digital evidence certificates. The absence of a standardized format and explicit integrity verification protocols often led to challenges in court. The BSA 2023 addresses these limitations by prescribing a detailed certificate structure under Section 63(4)(c), incorporating cryptographic hash value verification as the foundational method to establish the integrity of electronic evidence.

This chapter presents a comprehensive techno-legal framework aligned with the provisions of BSA 2023. It outlines a two-part certificate system involving both the

N. K. Chaudhary · N. Mistry
National Forensic Sciences University, Gandhinagar, Gujarat, India
e-mail: naveen.chaudhary@nfsu.ac.in; Nilay.mistry@nfsu.ac.in

S. S. Iyengar (✉)
Global Forensic and Justice Center, Florida International University, Miami, FL, USA
e-mail: iyengar@cs.fiu.edu

S. S. Iyengar et al. (eds.), *Artificial Intelligence Driven Forensics*, Signals and Communication Technology, https://doi.org/10.1007/978-3-031-98036-7_13

custodian and an expert, clarifies the legal authority of experts under BNSS and IT Act 2000, and proposes a proof of concept implementation using standard hashing algorithms such as MD5, SHA256, and SHA512. The methodology ensures that digital records are verifiably unaltered, reinforcing their admissibility and probative value in court. Through this framework, the study bridges the gap between legal requirements and technological feasibility, offering a scalable and legally sound approach for handling digital evidence in modern criminal litigation [29–38].

2 Background

Section 65A and 65B were inserted in the Indian Evidence Act 1872 (IEA) via an amendment brought by the IT Act 2000 [1–3]. Section 65B, which deals with the admissibility of electronic evidence, was heavily drawn from Sect. 5 of the UK Civil Evidence Act, 1968. However, Sect. 5 of the UK Civil Evidence Act was repealed by the Civil Evidence Act 1995 following the recommendations of the Law Commission made in 1983 [4].

Since it is a digital record taken from the original source that is admissible under this special provision for examination by the court, the integrity of the digital content was a major concern. Section 65B(4) of the IEA covered the requirement with regard to a certificate [5]; however, the format of the certificate was not prescribed. The certificate submitted under the provisions of Section 65B(4) was expected to be self-contained in regard to the integrity of the digital content taken from the original source. However, there was no provision in the certificate to ensure such integrity.

Thus, the certificate did not provide a foolproof mechanism for establishing the relevancy and accuracy of the digital content for its admissibility in the court of law. Considering these limitations and in order to ensure the integrity of digital content [6, 7], a relevant section has been incorporated into the new criminal laws, which establishes the integrity of the digital record by capturing its hash value.

3 Major Changes in the New Criminal Law for the Admissibility of Digital Evidence

The *Bharatiya Sakshya Adhiniyam* 2023 (BSA) replaces the Indian Evidence Act, 1872 (IEA) [8]. The IEA provided for two kinds of evidence: documentary and oral. Documentary evidence included primary evidence, i.e., original documents, and secondary evidence, which proves the contents of the original. The BSA retains this distinction.

Importantly, it includes electronic records in the definition of documents [9]. While the IEA categorized electronic records as secondary evidence, the BSA classifies them as *primary evidence*. It also expands the definition of such records to include information stored in semiconductor memory, communication devices, and digital records generated from smartphones, laptops, intermediaries, and cloud platforms [10].

4 Admissibility of Digital Evidence Under New Criminal Law

The *Bharatiya Sakshya Adhiniyam* 2023 (BSA) is a procedural law that provides the rules and regulations governing the relevancy and admissibility of evidence in court proceedings. It aims to remove ambiguities and help avoid probable intricacies that may arise during the admissibility of relevant electronic evidence.

Section 63(4)(c) of the BSA requires a certificate to accompany electronic records for admissibility. This certificate serves to identify the record and describe how it was produced, among other aspects. A model certificate is annexed in the Schedule of the Adhiniyam [11]. The certificate is intended to authenticate and verify the contents of electronic records, in line with Section 63(2) of the BSA.

Furthermore, the certificate submitted by the party under Section 63(4)(c) must be examined and verified by an expert, as per the same section. The certificate consists of two parts. **Part A** must be completed by the custodian of the device and captures information related to the custodian, identity of the electronic device, legal information, and the hash value of the digital record obtained from the original source device. The custodian may be the individual who owns, manages, or maintains the electronic device or digital record.

Part B is intended for the expert, who must verify all information regarding the electronic evidence, including the hash value furnished by the custodian. There are four major components in the certificate: (1) information about the custodian, (2) technical information, (3) legal information, and (4) the hash value of the digital record. These components are illustrated in Fig. 1.

5 "Expert" as per the Provisions of the New Criminal Law

As the certificate prescribed under Section 63(4)(c) of the BSA mandates verification of the information furnished under Part A by an expert, it is essential to examine the legal provisions determining who qualifies as an expert. In this regard, references were made to the *Information Technology Act, 2000* (IT Act 2000) and the *Bharatiya Nagarik Suraksha Sanhita, 2023* (BNSS 2023). It was observed that scientific experts from Cyber-Forensic Divisions of forensic laboratories, as notified under Section 293(4)(e) of the *Criminal Procedure Code* (CrPC) and under Section 79A of the IT Act 2000, are already engaged in examining, analyzing, and reporting such cases [12, 13].

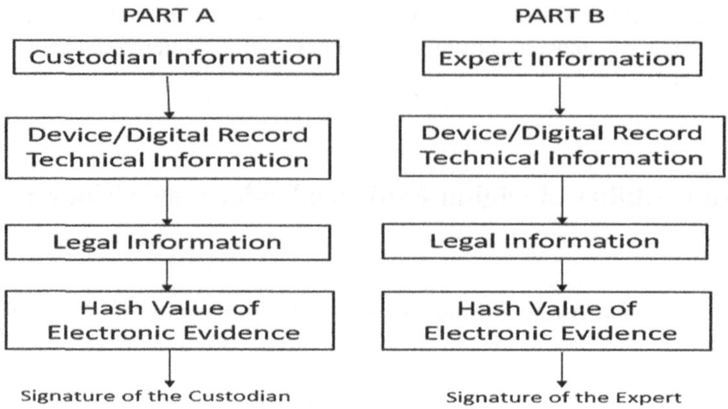

Fig. 1 Essential components of the certificate prescribed under Section 63(4) of BSA

The provisions under Section 293(4)(e) of CrPC now correspond to Section 329(4)(e) of the BNSS 2023. Since both the IT Act and the BNSS govern admissibility of electronic or digital records, the qualifications for "Expert" must align with Section 79A of the IT Act 2000 and Section 329(4)(e) of the BNSS 2023.

5.1 Section 79A of the Information Technology Act, 2000

Section 79A of the IT Act stipulates that the Central Government may, by notification in the Official Gazette, designate any department, body, or agency of the Central or State Government as an *Examiner of Electronic Evidence* for providing expert opinion in court or other legal forums. The Act defines "electronic form evidence" as any information of probative value that is stored or transmitted in digital format. This includes computer evidence, digital audio, digital video, cell phones, and digital fax machines.

5.2 Section 329 of BNSS 2023

Section 329 of the BNSS 2023 pertains to the reports of certain government scientific experts. Subsection 329(4) enumerates the experts qualified to submit such reports, including:

(a) Any chemical examiner or assistant chemical examiner to the government
(b) The chief controller of explosives
(c) The director of the Finger Print Bureau
(d) The director of the Haffkine Institute, Bombay

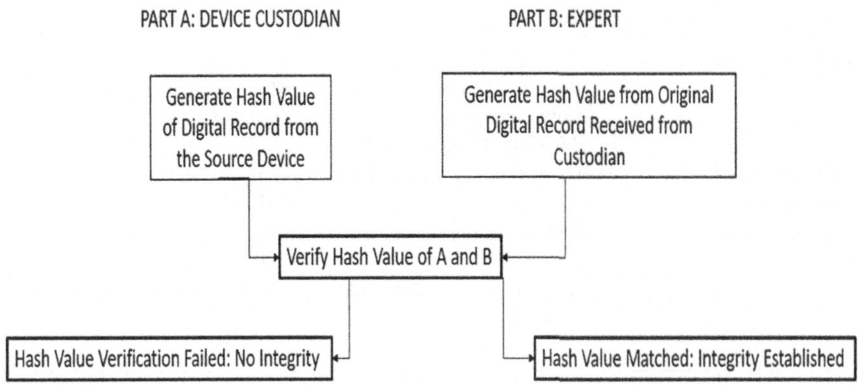

Fig. 2 Methodology to establish integrity of the digital record by the expert using hash value verification

(e) The director, deputy director, or assistant director of a Central or State Forensic Science Laboratory
(f) The serologist to the government
(g) Any other scientific expert notified by the state or central government for this purpose

6 Establishing Integrity of Digital Evidence

The information furnished in Part A of the certificate must be verified by the expert. A cryptographic *hash function* is a mathematical algorithm used to map data of arbitrary size onto data of fixed size. It is inherently a one-way function and produces a unique output for a given input, thereby enabling verification of data integrity [14–17].

Hash values are unique to digital records; hence, they serve as digital fingerprints. Because the process is nonreversible and deterministic, it is ideal for validating that a digital record has not been altered [18, 19].

Popular hashing algorithms used for this purpose include MD5, SHA-256, and SHA-512 [20]. A schematic of the proposed methodology to verify the integrity of digital records as per the new criminal laws is shown in Fig. 2.

7 Proof of Concept

The new criminal laws are forward-looking legal documents that incorporate technological advancements to enhance the judicial process. One such advancement is the requirement for establishing the integrity of electronic evidence through

Table 1 Tools used in experimental setup for integrity verification

Device/tool	Description	Remarks
Computer	Windows 10 OS, Intel(R) Core i3-9100 CPU @ 3.60GHz, 12.0 GB RAM	Hash value generator was installed
Application for hash generation	HashMyFiles (NirSoft Tool)	Supports MD5, SHA256, SHA512
Digital device	USB drive, file system: exFAT	Digital record stored for verification
Nature of digital record	Image files (JFIF, JPG), video file (MP4)	2 image files, 1 video file

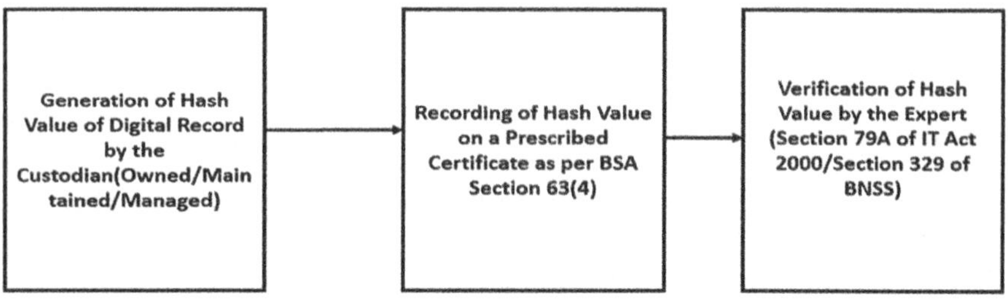

Fig. 3 Hash value verification process

cryptographic hash values. According to Section 63(4)(c) of the BSA 2023, the custodian of the electronic evidence must generate and document the hash value in Part A of the prescribed certificate.

An experimental setup was created to test this process, using the tools listed in Table 1. This setup facilitated the generation of hash values from selected digital records, which were then verified by an expert to confirm their integrity.

8 Implementation of Proof of Concept

The generation and verification of hash values are core technology components of the digital evidence certification process under Section 63(4)(c) of the BSA. The procedure involves a two-step workflow:

1. **Generation:** The custodian generates the hash value using a certified application and records it in Part A of the certificate.
2. **Verification:** The expert independently regenerates the hash value from the submitted evidence and compares it against the custodian's entry in Part A.

This process is illustrated in Fig. 3.

Table 2 Summary of hash value verification for digital content

File	Algorithm	Custodian hash (Part A)	Expert hash (Part B)	Match
Art.jfif	MD5	979313...a79f3	979313...a79f3	Yes
	SHA256	964a1b...b1760	964a1b...b1760	Yes
	SHA512	50f950...ca562	50f950...ca562	Yes
Metaverse.jpg	MD5	204fed...e562	204fed...e562	Yes
	SHA256	8da057...b8ea	8da057...b8ea	Yes
	SHA512	45f6fa...9271	45f6fa...9271	Yes
Final_video.mp4	MD5	642113...3b8e	642113...3b8e	Yes
	SHA256	fc258e...53f5	fc258e...53f5	Yes
	SHA512	1b69d5...c4c4a	1b69d5...c4c4a	Yes

Three digital files—two images and one video—were processed as part of the proof of concept. Their hash values were computed and compared between the custodian and the expert, with results summarized in Table 2.

9 Results of the Hashing Operation

The results of the hashing operation performed on three digital files are shown in Table 2. The metadata of these files are also included in the same table, along with their respective hash values. The hash values generated by the custodian align with the requirements prescribed in the certificate and match exactly with those generated independently by the expert, as required under Part B of the certificate.

This hashing procedure was applied to files of different formats—two images and one video—each containing distinct digital content. Upon verification, the respective hash values matched completely, thereby establishing the integrity of the digital records. This integrity is crucial for the admissibility of electronic evidence in the court of law.

These results validate the effectiveness of the proposed two-step process to prove the integrity of digital evidence. The consistency in hash values demonstrates that the procedure is a robust, full-proof method in alignment with the techno-legal requirements as outlined in Section 63(4)(c) of the Bharatiya Sakshya Adhiniyam (BSA) 2023.

10 Conclusion

The adoption of advanced technologies is a defining feature of the new criminal laws in India. These reforms elevate the status of digital evidence, treating it at par with physical evidence in terms of probative value. Consequently, the admissibility of digital evidence now hinges on rigorous procedures to ensure its integrity [39–41].

The techno-legal framework proposed in this paper introduces a simplified yet effective procedure for verifying the integrity of electronic evidence. This framework is mandatory for admissibility and complies with the provisions set forth in the BSA 2023. By introducing a certificate-based system that incorporates hash value verification, the framework strengthens the legal validity of digital records.

Aligned with the modernizing intent of the new criminal laws, this approach leverages scientific technology to support justice delivery. Its adoption will not only simplify the admissibility process but also expedite legal proceedings in cases that rely heavily on electronic evidence.

References

1. Kharb, D.: Admissibility of electronic records as secondary evidence under section 65b of the indian evidence act: Recent judicial approaches. Delhi Journal of Contemporary Law **1** (2018)
2. Karia, T., Anand, A., Dhawan, B.: The supreme court of india re-defines admissibility of electronic evidence in india. Digital Evidence and Electronic Signature Law Review **12** (2015)
3. Yadav, A.: Relevancy and admissibility of digital evidence: A comparative study. International Journal of Integrated Law Review **1**(1) (2020)
4. Christie, J.: The law commission and section 69 of the police and criminal evidence act 1984. Digital Evidence and Electronic Signature Law Review **20** (2023)
5. Kumar, A.S.: Admissibility and Proof of Electronic Records. https://www.tnsja.tn.gov.in/article/electronic_records_article.pdf. Accessed on 07 Aug 2023
6. Hosmer, C.: Proving the integrity of digital evidence with time. International Journal of Digital Evidence **1** (2002)
7. Irons, A.: Digital forensics and measurement science. Journal of Measurement and Control **43**(8), 238–242 (2010)
8. Verma, J.K.: Bharatiya Sakshya Adhiniyam, 2023 (Evidence): A Commentary, 1st edn. EBC Publishing Pvt. Ltd., Lucknow (2024)
9. Bharatiya Sakshya Adhiniyam (BSA) 2023. Taxmann Publications Private Limited (2024)
10. New Criminal Major Acts. Taxmann Publications Private Limited (2024)
11. The Bharatiya Sakshya Adhiniyam, 2023, No. 47 of 2023. Available at https://wccb.gov.in/WriteReadData/UserFiles/file/Notificaton/BSA%202023.pdf, Accessed on 03 August 2024 (2023)
12. Kumar, R.: Criminal Procedure Code CrPC. CBS Publishers and Distributors Pvt. Ltd., Chennai (2023)
13. IT Act 2008 (Amended). Available at https://www.meity.gov.in/writereaddata/files/it_amendment_act2008%20%281%29_0.pdf, Accessed on 27 July 2024 (2008)
14. Bakhtiari, S., Safavi-Naini, R., Pieprzyk, J.: Cryptographic hash functions: A survey. Technical Report 95-09, Department of Computer Science, University of Wollongong (1995)
15. Kumar, C.K., Suyambulingom, C.: Cryptographic of high security hash functions. International Journal of Engineering Research & Technology (IJERT) **1**(3) (2012)
16. Rajeev, S., Geetha, G.: Cryptographic hash functions: A review. International Journal of Computer Science Issues **9**, 461–479 (2012)
17. Andreeva, E., Mennink, B., Preneel, B.: Open problems in hash function security. Designs, Codes and Cryptography **77**(2), 611–631 (2015)
18. Tenhunen, M.: The integrity of electronic evidence. In: Jajodia, S., List, W., McGregor, G., Strous, L. (eds.) Integrity and Internal Control in Information Systems. IICIS 1997. Springer, Boston (1997). https://doi.org/10.1007/978-0-387-35317-3_8

19. Roy, M., Mohan, A.K., Sethumadhavan, M.: Maintaining the integrity of digital evidence using piecewise hashing. International Journal of Pure and Applied Mathematics **119**(16), 4073–4077 (2018)
20. Conrad, E., Misenar, S., Feldman, J.: Secure Hash Algorithm, 2nd edn. Science Direct, Amsterdam (2014)
21. Iyengar, S. S., Seyedsina Nabavirazavi, Yashas Hariprasad, Prasad HB, and C. Krishna Mohan. Privacy-Preserving AI (Federated Learning) for Digital Forensics. In Artificial Intelligence in Practice: Theory and Application for Cyber Security and Forensics, pp. 161–176. Cham: Springer Nature Switzerland, 2025.
22. Iyengar, S. S., Seyedsina Nabavirazavi, Yashas Hariprasad, Prasad HB, and C. Krishna Mohan. Cyber Threat Intelligence and Security for Federated Learning in Digital Forensics. In Artificial Intelligence in Practice: Theory and Application for Cyber Security and Forensics, pp. 177–199. Cham: Springer Nature Switzerland, 2025.
23. Iyengar, S.S., Nabavirazavi, S., Hariprasad, Y., HB, P. and Mohan, C.K., 2025. Future of AI-Driven Digital Forensics. In Artificial Intelligence in Practice: Theory and Application for Cyber Security and Forensics (pp. 335–364). Cham: Springer Nature Switzerland.
24. Iyengar, S.S., Nabavirazavi, S., Hariprasad, Y., HB, P. and Mohan, C.K., 2025. Hybrid Detection of Malicious Portable Document Format (PDFs): Safeguarding Against Embedded JavaScript Attacks. In Artificial Intelligence in Practice: Theory and Application for Cyber Security and Forensics (pp. 257–290). Cham: Springer Nature Switzerland.
25. Soni, Jayesh, Subhash Gurappa, and Himanshu Upadhyay. A Comparative Study of Deep Learning Models for Image Super-Resolution Across Various Magnification Levels. In 2024 IEEE International Conference on Future Machine Learning and Data Science (FMLDS), pp. 395–400. IEEE, 2024.
26. Iyengar, S. S., Seyedsina Nabavirazavi, Yashas Hariprasad, Prasad HB, and C. Krishna Mohan. Temporal Deepfake Generation and Detection in Video Sequences Using Recurrent Neural Networks (RNNs). In Artificial Intelligence in Practice: Theory and Application for Cyber Security and Forensics, pp. 309–334. Cham: Springer Nature Switzerland, 2025.
27. Iyengar, S. S., Seyedsina Nabavirazavi, Yashas Hariprasad, Prasad HB, and C. Krishna Mohan. The Convergence of AI/ML and Cybersecurity: Advancing Digital Forensic Techniques. In Artificial Intelligence in Practice: Theory and Application for Cyber Security and Forensics, pp. 139–159. Cham: Springer Nature Switzerland, 2025.
28. Iyengar, S. S., Seyedsina Nabavirazavi, Yashas Hariprasad, Prasad HB, and C. Krishna Mohan. AI-Enhanced Malware Detection: Advancing Security Through Intelligent Threat Identification. In Artificial Intelligence in Practice: Theory and Application for Cyber Security and Forensics, pp. 227–255. Cham: Springer Nature Switzerland, 2025.
29. Iyengar, S.S., Nabavi, S., Hariprasad, Y., H.B., P., C., K.M.: Artificial Intelligence in Practice: Theory and Applications for Cyber Security and Forensics. Springer, Cham (2025)
30. Shi, B., Iyengar, S.S.: Mathematical Theories of Machine Learning-Theory and Applications. Springer, Cham (2020)
31. Singaram, J., Iyengar, S.S., Madni, A.M.: Deep Learning Networks: Design, Development and Deployment. Springer, Cham (2023)
32. Hariprasad, Y., Iyengar, S.S., Subramanian, N.: Deepfake video detection using lip region analysis with advanced artificial intelligence based anomaly detection technique. Authorea Preprints (2024)
33. Hariprasad, Y., K.J., L.K., Suraj, L., Iyengar, S.S.: Boundary-based fake face anomaly detection in videos using recurrent neural networks. In: Proceedings of SAI Intelligent Systems Conference, pp. 155–169. Springer, Cham (2022)
34. Hariprasad, Y., Iyengar, S.S., Chaudhary, N.K.: Securing the future: Advanced encryption for quantum-safe video transmission. IEEE Transactions on Consumer Electronics (2024)
35. Iyengar, S. S., Seyedsina Nabavirazavi, Yashas Hariprasad, Prasad HB, and C. Krishna Mohan. Real-Time Aquatic Forensics: Harnessing AI for Efficient Underwater Target Recognition. In Artificial Intelligence in Practice: Theory and Application for Cyber Security and Forensics, pp. 291–307. Cham: Springer Nature Switzerland, 2025.

36. Miller, Jerry, Lawrence Egharevba, Yashas Hariprasad, Kumar KJ Latesh, and Naveen Kumar Chaudhary. Cyber security attack detection framework for DODAG control message flooding in an IoT network. In International Conference on Information Security, Privacy and Digital Forensics, pp. 213–230. Singapore: Springer Nature Singapore, 2022.
37. Wang, Cliff, Sundararaja S. Iyengar, and Kun Sun, eds. AI embedded assurance for cyber systems. Berlin: Springer, 2023.
38. Thejas, G. S., Yashas Hariprasad, S. S. Iyengar, N. R. Sunitha, Prajwal Badrinath, and Shasank Chennupati. An extension of Synthetic Minority Oversampling Technique based on Kalman filter for imbalanced datasets. Machine Learning with Applications 8 (2022): 100267.
39. Iyengar, S. S., Seyedsina Nabavirazavi, Yashas Hariprasad, Prasad HB, and C. Krishna Mohan. The Evolution of Artificial Intelligence and Machine Learning. In Artificial Intelligence in Practice: Theory and Application for Cyber Security and Forensics, pp. 3–26. Cham: Springer Nature Switzerland, 2025.
40. Iyengar, S.S., Nabavirazavi, S., Hariprasad, Y., HB, P. and Mohan, C.K., 2025. Cybersecurity Foundations: Theories, Technologies, and Applications. In Artificial Intelligence in Practice: Theory and Application for Cyber Security and Forensics (pp. 27–87). Cham: Springer Nature Switzerland.
41. Iyengar, S. S., Seyedsina Nabavirazavi, Yashas Hariprasad, Prasad HB, and C. Krishna Mohan. Digital Forensics: Tools, Techniques, and Methodologies. In Artificial Intelligence in Practice: Theory and Application for Cyber Security and Forensics, pp. 89–137. Cham: Springer Nature Switzerland, 2025.

Age-Related Changes in Interpersonal Trust Behavior: Can Neuroscience Inform Public Policy?

Vijeth Iyengar, Dipayan Ghosh, Tyler Smith, and Frank Krueger

1 An Evolving Global Age Distribution and New Implications for Social Cognition and Interpersonal Trust

In the years to come, there will be a significant global increase in the number of older adult persons. Some projections indicate that by 2030, there will be a higher number of adults age 60 or over than those between the ages of 10 to 24 [1]. It is critical to proactively address the novel challenges that societies will face with a shifting demography. In particular, understanding the neuropsychological changes that take place with advancing age and the effects these changes have on how older adults function and engage with their surroundings will become increasingly important in

Iyengar, V., Ghosh, D., Smith, T., and F. Krueger. Age-Related Changes in Interpersonal Trust Behavior: Can Neuroscience Inform Public Policy? NAM Perspectives. doi: 10.31478/201906c. Reproduced with permission from the National Academy of Sciences, Courtesy of the National Academies Press, Washington, D.C.

V. Iyengar
Administration on Aging/Administration for Community Living, US Department of Health and Human Services, Washington, DC, USA

D. Ghosh
Shorenstein Center on Media, Politics, and Public Policy, John F. Kennedy School of Government, Harvard University, Cambridge, MA, USA

T. Smith
Electronic Crimes Section, Office of Inspector General, Federal Deposit Insurance Corporation, Washington, DC, USA

F. Krueger (✉)
School of Systems Biology, George Mason University, Fairfax, VA, USA
e-mail: fkrueger@gmu.edu

S. S. Iyengar et al. (eds.), *Artificial Intelligence Driven Forensics*, Signals and Communication Technology, https://doi.org/10.1007/978-3-031-98036-7_14

designing products, programs, and services to support the global population in the face of these inevitable new challenges.

In this paper, we link empirical findings from neuroscientific investigations of interpersonal trust behavior in older adults to incidences of financial exploitation, healthcare fraud, and digital deception—consumer harms for which older adults are preferentially targeted by bad actors.

Financial exploitation schemes include efforts to persuade older adults to provide access to personal financial accounts through mail, in-person, over-the-phone, or online spoofing. These schemes result in the theft or embezzlement of money or other property, and while the damage is difficult to estimate, some have reported that older adults lose about $36.5 billion each year to such financial abuse [2]. Similarly, healthcare fraud tactics exploit older adults through fraudulent billing via Medicare, private insurance, or personal funds. These older adults can also be subjected to unnecessary or unsafe medical procedures, resulting in compromised medical records. Lastly, digital disinformation, a more novel form of online victimization, takes advantage of social media platforms to spread misleading information that can affect a myriad of public interest concerns.

There are common problems across these forms of victimization. The potential victim faces the challenge of assessing the trustworthiness of the group or individual with which they interact. It can be difficult to judge the veracity and credibility of various sources of information. Finally, assessments of trust occur over different timelines: via one-time (e.g., unsolicited robocalls) or continuous (e.g., close friends or family members) engagement with individuals or groups harboring bad intentions. Determining the underlying reasons for these faulty evaluations of trust may prove useful in developing tools to combat these forms of victimization.

Emerging evidence from neuroscientific investigations of interpersonal trust behavior is revealing how the capacity to evaluate and subsequently act on untrustworthy agents with potentially nefarious intentions changes as we age. Interpersonal trust encompasses a person's willingness to be vulnerable to the risk of treachery based on the expectation that the actions of another will produce some future positive outcome due to the possibility of reciprocity. Available evidence from both cross-sectional and longitudinal survey studies has shown age-related changes in this behavior [3].

For example, older adults are less concerned with information that contradicts their first impressions about the trustworthiness of others, resulting in poor evaluations of the trustworthiness of other people [4].

Further, in experimental economic exchange paradigms, wherein players assume the role of an investor (i.e., trustor) and responder (i.e., trustee), older adults engaged as trustors have been shown to be likelier to invest in trustees with an untrustworthy reputation (although the evidence is mixed) [5].

Complementing these behavioral findings, research in social cognitive affective neuroscience has linked age-related alterations in brain circuitry with changes in trusting behavior potentially leading to heightened susceptibility to financial fraud [6].

Collectively, these results—in older adults presenting with no neurological, medical, or mental conditions—suggest new public interest mechanisms may need

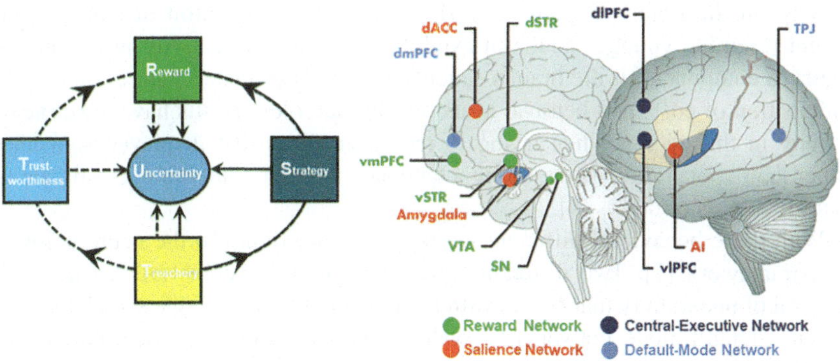

Legend: AI, anterior insula; **dACC**, dorsal anterior cingulate cortex; **dlPFC**, dorsolateral prefrontal cortex; **dmPFC**; dorsomedial prefrontal cortex; **dSTR**, dorsal striatum; **SN**; substantia nigra; **TPJ**, temporoparietal junction; **vlPFC**, ventrolateral prefrontal cortex; **vmPFC**, ventromedial prefrontal cortex; **vSTR**, ventral striatum; **VTA**, ventral tegmentum area.

Fig. 1 Neurobiological Framework of Trust. *Source*: Adjusted and reprinted with permission from Krueger, F., and A. Meyer-Lindenberg "Towards a model of interpersonal trust drawn from neuroscience, psychology, and economics," *Trends in Neurosciences*. *Note*: Trust arises through the interplay of factors (t-r-u-s-t: treachery, reward, uncertainty, strategy, and trustworthiness)—linked to psychological components (i.e., affect, cognition, and motivation)—that engage key brain regions (circles) anchored in large-scale brain networks. Vulnerability from trusting another person builds uncertainty (purple ellipse) due to risk of treachery (red rectangle, affect, salience network), instead of anticipation of reward (green rectangle, motivation, reward network). To remove the uncertainty, the salience network engages either the central-executive network (dark blue rectangle) to adopt a context-based strategy, or the default-mode network (light blue rectangle) to evaluate trustworthiness for trusting a partner

to be developed to protect older adults from nefarious interactions. Recently, a neurobiological trust framework has been proposed in which the psychological components of trust (i.e., affect, cognition, and motivation) are linked to brain networks (see Fig. 1) [7].

2 Mapping the Neurobiological Components of Interpersonal Trust

Making individual designations of trust in a social interaction—whether in the physical world or through an online interface—is the practice of successfully evaluating the potential benefits and costs when interacting with entities. Trusting an entity creates uncertainty, which results from the cost of potential treachery weighed against the anticipation of benefits after being trusted. Evidence suggests that older adults weigh costs and benefits differently from their younger counterparts [8]. Whereas older adults resemble younger adults in exhibiting increases in neural activity (in the brain's reward network) during the anticipation of benefits, they do not resemble younger adults in exhibiting increases in activity in the anterior insula (a component of the brain's salience network associated with evaluating the anticipation of costs and betrayal aversion). In fact, older adults have lower neural activity in this region when trying to recognize faces that displayed selfishness rather than cooperation—suggesting a tendency in older adults to overestimate the trustworthiness of others [4].

With these behavioral differences established, we return to the neurobiological trust framework [7]. Two different types of cognitive systems are crucial to the removal of uncertainty that comes with trusting another person: the social cognition system (default-mode network) to evaluate the trustworthiness of a partner and the cognitive control system (central-executive network) to employ context-based strategies for trusting a partner. The social cognition system is essential in assessing whether to trust an individual or group and supports the ability to infer and attribute the intentions and traits of others. Trustors with higher perspective-taking tendencies not only show greater trust toward others but also reduce their trust more drastically after betrayal by others. Age-related changes in the default-mode network may negatively affect how older adults navigate their social environments, exposing them to nefarious actors.

Complementing this first system, the cognitive control system allows one to adopt goal-directed behavior under changing contexts. Accumulating evidence indicates that although some cognitive functions are affected as we age, others are spared. Specifically, whereas crystallized cognitive abilities (e.g., conceptual knowledge) are preserved, fluid cognitive abilities (e.g., executive control, working memory, and attention) steadily decline with age [8]. Consequently, older adults may experience particular difficulties when faced with trust decisions involving the simultaneous processing and evaluation of disparate pieces of information—a scenario that taps into *fluid cognitive abilities.*

In summary, based on the neurobiological trust framework, age-related findings in studies of interper sonal trust behavior are likely driven by the impairment of the affective component of trust in the salience network (but not in the motivational component in the reward network), which therefore can impact the socio-cognitive components of trust in default-mode network and central-executive network.

This conclusion about age-related changes in interpersonal trust behavior is in accordance with the general assumption of socioemotional selectivity theory [9], which proposes an increase of positive emotional and social experiences in ways that foster well-being with advancing age and narrowing time horizon. According to this theory, older adults demonstrate a positivity bias that enhances the salience of more positive than negative valanced information, increases attention to socioemotional cues, and improves memory for positive stimuli or events in later life [10]. While these changes are associated with emotional and mental well-being, this bias may lead to greater likelihood of victimization due to *excessive trust* afforded to individuals and groups. Although the present commentary specifically focuses on changes in trust behavior across the life span that could potentially lead to various forms of victimization, we do acknowledge other factors, such as declines in functional capacity and mental well-being, that may also contribute to an overall greater risk of victimization [11].

3 A Case for Policy to Combat Public Interest Harms Related to Changes in Interpersonal Trust Associated with Age

Mounting evidence from neuropsychology is uncovering the mechanisms by which older adults engage in interpersonal trust behavior. The development of predictive neural markers building on individual brain differences associated with age-related changes in this behavior may permit the identification of neural phenotypes that in turn serve as targets for interventions. Such approaches may inform the development of more targeted behavioral and neural interventions that incorporate cognitive capacities that are preserved across the life span. Understanding the mechanisms and conditions under which older adults differ from younger adults in their processing and evaluations of interpersonal trust behavior can also impact features of programs and policies. For example, law enforcement and regulatory entities operating in financial and healthcare industries can collect trust-related-behavior-based fraud indicators and incorporate relevant data into loss prevention internal controls in the service of protecting vulnerable older adults.

Further exploration into the various mechanisms through which older adults process different forms of fraudulent behavior may prove helpful. Indeed, while such fraud-related incidents as the digital disinformation problem have been shown to affect older adults more than others, there is scant analysis that charts how cognitive processes relating to disinformation in older adults differ from processes including financial exploitation. While both financial exploitation and disinformation are activities instigated by nefarious actors, they occur in divergent ways—e.g., in motivation, platform, and demands placed on the victim—which may affect how older adults respond to such instigations. Furthermore, ongoing policy, programmatic, and research efforts have not—to our knowledge—integrated the relative degrees of social pressure these various forms of fraud might instigate. Financial exploitation, for example, often involves an individual target, whereas the disinformation problem occurs on internet platforms and is disseminated to large classes with one mouse click.

Ultimately, we see this commentary as starting a conversation among stakeholders in the academic, public, and private sectors to identify how findings from neuropsychology can potentially inform and shape future public policies around issues that have touchpoints with interpersonal trust behavior. Greater engagement among these stakeholders has the benefit of cultivating an environment in which scientific investigators design questions that readily allow for translation of their findings to practice, the development of programmatic and policy interventions that incorporate the latest advances from research and practice, and, in the neuroscientific research community, the advancement of the importance of viewing their findings from a public health and policy lens.

Acknowledgments Francis X. Shen, JD, PhD, Executive Director, Massachusetts General Hospital Center for Law, Brain, and Behavior; and Associate Professor of Law, McKnight Presidential Fellow, and Faculty Member in Graduate Program on Neuroscience, University of Minnesota reviewed this work.

Disclaimer The views expressed in this paper are those of the authors and not necessarily of the authors' organizations, the National Academy of Medicine (NAM), or the National Academies of Sciences, Engineering, and Medicine (the National Academies). The paper is intended to help inform and stimulate discussion. It is not a report of the NAM or the National Academies. The analysis, conclusions, and opinions set forth here are those of the author(s) alone and do not necessarily reflect the views of the Federal Deposit Insurance Corporation.

Conflict-of-Interest Disclosures None to disclose.

References

1. *World population ageing 2017: Highlights*. 2017. New York: United Nations. https://www.un.org/en/development/desa/population/publications/pdf/ageing/WPA2017_Highlights.pdf (accessed April 28, 2019).
2. Huang, Y., and A. Lawitz. 2016. *The New York State cost of financial exploitation study*. Rensselaer, NY: Office of Children and Family Services. https://ocfs.ny.gov/main/reports/Cost%20of%20Financial%20Exploitation%20Study%20FINAL%20May%202016.pdf (accessed April 28, 2019).
3. Poulin, M. J., and C. M. Haase. 2015. Growing to trust: Evidence that trust increases and sustains well-being across the life span. *Social Psychological and Personality Science* 6(6):614–621. doi:https://doi.org/10.1177/1948550615574301.
4. Suzuki, A., M. Ueno, K. Ishikawa, A. Kobayashi, M. Okubo, and T. Nakai. 2019. Age-related differences in the activation of the mentalizing- and reward-related brain regions during the learning of others' true trustworthiness. *Neurobiology of Aging* 73:1–8. doi:https://doi.org/10.1016/j.neurobiolaging.2018.09.002.
5. Ruffman, T., J. Murray, J. Halberstadt, and T. Vater. 2012. Age-related differences in deception. *Psychology and Aging* 27(3):543–549. https://doi.org/10.1037/a0023380.
6. Castle, E., N. I. Eisenberger, T. E. Seeman, W. G. Moons, I. A. Boggero, M. S. Grinblatt, and S. E. Taylor. 2012. Neural and behavioral bases of age differences in perceptions of trust. *Proceedings of the National Academy of Sciences of the United States of America* 109(51):20848–20852. doi:https://doi.org/10.1073/pnas.1218518109.
7. Krueger, F., and A. Meyer-Lindenberg. 2019. Towards a model of interpersonal trust drawn from neuroscience, psychology, and economics. *Trends in Neurosciences* 42(2):92–101. doi:https://doi.org/10.1016/j.tins.2018.10.004.
8. Samanez-Larkin, G. R., and B. Knutson. 2015. Decision making in the ageing brain: Changes in affective and motivational circuits. *Nature Reviews Neuroscience* 16(5):278–289. doi:https://doi.org/10.1038/nrn3917.
9. Mather, M., and L. L. Carstensen. 2005. Aging and motivated cognition: The positivity effect in attention and memory. *Trends in Cognitive Sciences* 9(10):496–502. doi:https://doi.org/10.1016/j.tics.2005.08.005.

10. Reed, A. E., and L. L. Carstensen. 2012. The theory behind the age-related positivity effect. *Frontiers in Psychology* 3:339. doi:https://doi.org/10.3389/fpsyg.2012.00339.
11. Lachs, M. S., and S. Duke Han. 2015. Age-associated financial vulnerability: An emerging public health issue. *Annals of Internal Medicine* 163(11):877–878. doi:https://doi.org/10.7326/M15-0882.

A Hybrid Approach for Urban Crime Analysis: Combining Machine Learning and Network Science

Yu Wu and Natarajan Meghanathan

1 Introduction

Urban crime analysis has long been a key focus in public safety research, with the goal of enabling more informed and efficient policing strategies. With the increasing availability of city-level crime records and the advancement of machine learning, data-driven crime prediction methods have become central to understanding where and when crimes are likely to occur. These approaches aim to identify high-risk areas, improve patrol planning, and support the allocation of limited law enforcement resources [1–4].

Traditional crime prediction methods often rely on statistical models trained on historical records, using fixed features such as population density, prior crime counts, or district-level demographics [5]. While effective for identifying spatial hot spots under stable conditions, these models typically treat crimes as isolated events and often overlook the temporal or structural dependencies that exist within urban crime systems [6]. Moreover, many approaches reduce prediction outcomes to binary classifications—whether a location is a hot spot or not, without offering a clear sense of intensity, context, or the interaction between crime types. In reality, urban crime is rarely isolated or uniform. Many criminal incidents involve multiple offenses and follow recurring temporal patterns. For example, certain types of crimes often co-occur within the same incident, reflecting shared motives, settings, or offender behavior [7, 8]. Similarly, spatial crime hot spots are not static; their temporal dynamics—how consistently they reappear and vary in intensity—can provide critical insights for long-term safety planning. Capturing these aspects

Y. Wu · N. Meghanathan (✉)
Department of Electrical & Computer Engineering and Computer Science, Jackson State University, Jackson, MS, USA
e-mail: yu.wu@students.jsums.edu; natarajan.meghanathan@jsums.edu

S. S. Iyengar et al. (eds.), *Artificial Intelligence Driven Forensics*, Signals and Communication Technology, https://doi.org/10.1007/978-3-031-98036-7_15

requires a more structural and fine-grained approach that goes beyond classification and incorporates both relational and temporal information.

To address these limitations, this chapter introduces a hybrid framework for urban crime analysis that combines network science with machine learning. This approach is organized around two core tasks, each targeting a distinct but complementary aspect of urban crime modeling. The first task focuses on understanding how different offense types co-occur within single incidents. We construct a weighted offense-offense network based on multi-offense records and apply a suite of clustering algorithms to derive a co-occur offenses matrix. From this, we compute an eigenvector-based association index that quantifies the structural role of each offense in the broader crime network. The second task aims to detect and score crime hot spots in a continuous, temporally informed manner. We segment the city into spatial grids and aggregate crime counts over biweekly intervals, forming a spatiotemporal matrix. Principal component analysis (PCA) is then used to capture major temporal trends across regions. Each spatial unit is assigned a hot spot score based on its projection in the PCA space, reflecting the consistency and intensity of crime patterns over time. To further capture spatial dynamics, the authors construct a conceptual network among subregions and compute centrality scores to identify structurally influential zones.

This study extends two prior investigations [9, 10] conducted by the authors—one focusing on the structural co-occurrence of offenses and the other on spatiotemporal crime pattern detection. While each task addressed a distinct aspect of urban crime analysis, the present work integrates both components under a unified hybrid framework, aiming to demonstrate the complementary strengths of network science and machine learning in understanding crime dynamics at multiple levels.

2 Related Work

2.1 Urban Crime Prediction and Analysis

Urban crime prediction has gained increasing attention with the advancement of data availability and machine learning techniques. Most existing works fall into two major directions: pattern analysis and crime forecasting. Pattern analysis aims to reveal the correlation between crime and external features such as demographics, land use, or socio-economic indicators. In contrast, crime forecasting focuses on predicting future occurrences of crimes over time and across regions using historical incident data.

Early efforts primarily adopted traditional machine learning models [11, 12], such as logistic regression, support vector machines, and random forests—trained on aggregated statistics like neighborhood-level crime counts or census data. These models were effective in identifying spatial hot spots, particularly when integrated with density estimation or clustering-based preprocessing. However, they often treated crime patterns as static and did not account for temporal variations.

In recent years, deep learning-based models have been explored to capture dynamic spatiotemporal trends in crime. Models such as LSTM [13] have been used to model temporal dependencies, while graph convolutional networks (GCNs) [14] and CNNs have been applied to learn spatial interactions between city regions. For instance, some studies integrate residual networks and graph structures to account for inter-region relationships and district-wise dynamics. Others introduce external contextual signals—including traffic flows, public events, or social media data—to enhance model inputs and reflect real-time conditions in the city [15]. Another line of work emphasizes improving how prediction results are presented to decision-makers. Rather than binary classifications of whether an area is a hot spot, some studies have aimed to generate more granular outputs that reflect varying levels of crime risk [16, 17]. These approaches are particularly useful for fine-tuned resource allocation and long-term crime monitoring. Together, these works reflect a shift from static, location-based predictions to more context-aware, temporally adaptive modeling frameworks that aim to support practical applications in urban policing and safety planning.

2.2 Challenges in Data and Generalization

Despite recent progress, several concerns remain regarding the robustness and usability of existing models. Many methods are trained on static historical data and assume consistent patterns over time. This makes them vulnerable to changes in social behavior, mobility, or law enforcement strategy. While temporal models offer partial solutions, they often fail to capture interactions between different dynamic inputs. Furthermore, most models are evaluated in a single city or a narrow time frame, limiting their transferability. Urban environments vary significantly across regions, and prediction accuracy often drops when models are applied to new cities or updated datasets. Lastly, model interpretability remains limited in many deep learning approaches, which may affect the credibility and acceptance of predictions in high-stakes, real-world decision-making.

3 Methodology

This study proposes an urban crime analysis framework that integrates machine learning and network science. The goal of this approach is to improve the prediction of urban crime hot spots and to gain more insight into crime patterns. By combining machine learning methods with topological analysis, this framework offers a new perspective on urban crime as a system of interrelated regions, capturing interactions across features that are often ignored or isolated in analyses.

In most cases, the integration of machine learning and network modeling occurs before the modeling stage. Machine learning techniques are typically used to derive feature representations such as clusters, principal components, or statistical scores, which are then embedded into the network as node or edge attributes. These enriched networks serve as the foundation for subsequent topological analysis and, in some cases, act as inputs for predictive tasks. This integration is inherently flexible: depending on the task and data characteristics, machine learning outputs can inform network construction, or conversely, network-derived attributes can enhance feature sets for machine learning models.

The following sections describe the data preparation process, the modeling components, and the specific applications of this hybrid approach to several crime analysis tasks.

3.1 Data and Preprocessing

3.1.1 Dataset Introduction

This study draws on publicly available crime incident datasets collected from the police departments Boston [18]. Boston crime dataset were selected due to its moderate city population density, consistency in data reporting standards, and the availability of recent, large-scale crime records. It will not only ensure comprehensive data coverage and reliability but also support the generalization of findings to other urban environments. The dataset provide detailed information on each recorded incident, including the time and geographic location of the event, as well as a categorized description of the offense.

The Boston dataset used in this study contains approximately 600,000 crime-related incidents recorded by the Boston Police Department between 2019 and 2022, each documenting the time (data and the time of occurrence), geographic location (area code, latitude and longitude), and category of the offense (juridical code and description).

This dataset contains a wide range of categories—67 main offense types with over 200 subtypes—though the distribution is notably imbalanced. The most frequently recorded events include larceny, motor vehicle accident response, and medical assistance. This study focuses primarily on criminal offenses and excludes those records that do not represent criminal violations in the strict sense, such as motor vehicle accident response and medical assistance.

3.1.2 Data Preprocessing

In order to ensure the consistency and analytical validity of the dataset, the authors perform several preprocessing steps, with particular attention to crime category normalization and temporal-spatial segmentation.

First, the classification of crime types in the original datasets was highly imbalanced. Some rarely reported offenses, such as kidnapping, may only occur a few times per year, whereas more common categories like robbery or larceny often account for thousands of records annually in each city. To address this imbalance, two common strategies were applied. (a) Within a single dataset, semantically similar subcategories were merged into a unified category. For example, there are different levels of assault; the authors consolidated them into a single "assault" category to simplify the analysis while retaining semantic meaning. (b) Additionally, only crime types with a frequency above a predefined threshold were selected for modeling, to focus on representative patterns and reduce noise from infrequent events.

Second, in temporal and spatiotemporal hot spot analysis, the original timestamps provided in the dataset, which include the exact date, time, and coordinates of each incident, were restructured to suit the specific analytical goals. To conduct the crime hot spot detection task with fine-grained analysis, incidents were grouped into biweekly periods (14-day windows), and the city space was discretized using a 0.01-degree by 0.01-degree grid based on latitude and longitude coordinates. This allowed for both spatial resolution and manageable aggregation of crime counts over time.

The authors utilize the latitude and longitude data rather than broader administrative labels such as street or district names, which tend to be less spatially consistent. This level of granularity supports more accurate spatial segmentation and the construction of a continuous and high-resolution view of crime activity across the city.

3.2 *Task I: Quantifying Crime Co-Occurrence Patterns*

3.2.1 Objective

The goal of this task is to investigate how different types of offenses tend to co-occur within individual crime events. The authors notice in many cases, a single incident recorded by law enforcement involves multiple offense types. This suggests possible similar patterns of violation or situational overlap. The underlying criminal pattern/structure can aid in predicting potential criminal behavior.

3.2.2 Definition and Data Selection

Co-Occurrence The situation in which multiple offense types are reported under the same incident. That is, a crime event is considered to involve co-occurrence if it contains two or more distinct offense categories (Table 1).

Table 1 Incidents with more than one offense types

Incidents Number	Offenses included
I162030584	13
I152080623	11
I182065208	10
⋮	⋮
Total incidents: 27523	

Data Selection The authors filter the Boston Crime Dataset to retain only those incidents that include two or more offense categories and identify 67 unique offense types, which form the basis for network modeling.

3.2.3 Network Construction

The authors construct a weighted, undirected offense-offense network to model co-occurrence relationships. In this network:

Node represents offense types.

Edges exist between any two offenses that co-occur in one or more incidents.

Edge weights represent the frequency of co-occurrence between the connected offenses.

This formulation captures not only the presence of co-occurrence, but also its intensity across the dataset.

To visualize the network, the author encode each node's degree as its size (indicating how many offenses it co-occurs with), and each edge's thickness corresponds to the number of shared incidents. Nodes are also color-coded by frequency, providing a high-level view of dominant and peripheral offenses. The undirected, weighted graph is showing below as Fig. 1.

3.2.4 Clustering-Based Co-Occur Offenses Matrix

To examine beyond the raw co-occurrence counts, the authors build a co-occur offenses matrix that reflects the structural tendency of offense types to cluster together. Specifically, the authors apply a suite of ten clustering or community detection algorithms, each with distinct underlying assumptions, to the offense-offense network. These include partitioning (e.g., K-means), density-based (e.g., DBSCAN, OPTICS), hierarchical (e.g., agglomerative clustering, BIRCH), and modularity-based methods (e.g., Clauset-Newman-Moore, Fluid Communities), each offering different perspectives on network structure.

For every offense pair, the authors record the number of algorithms that assign both offenses to the same cluster. The result is a symmetric co-occur offenses matrix where each cell (i, j) captures the frequency with which offenses i and j are

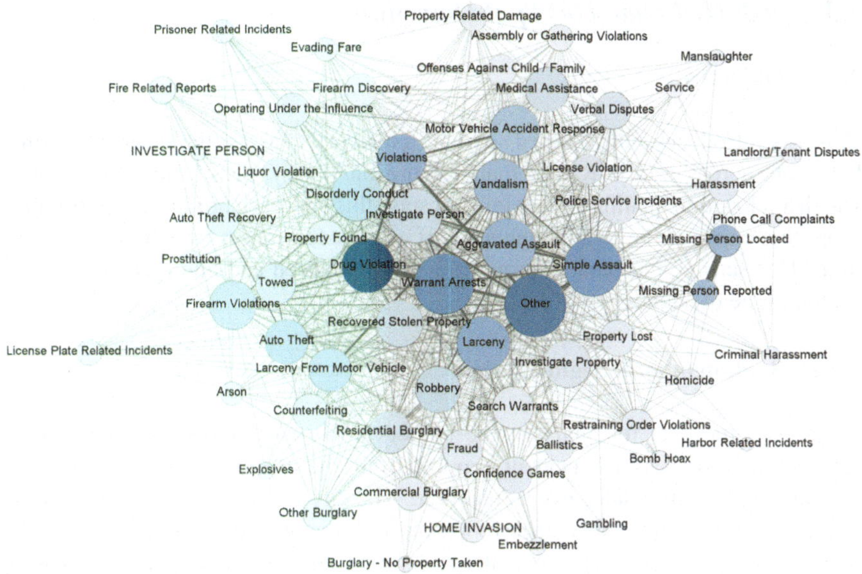

Fig. 1 Co-occurrence offense network

grouped together across clustering methods. This matrix provides a multiperspective representation of structural co-occur likelihood, beyond what any single algorithm can offer.

3.2.5 Spectral Analysis and Offense Association Index

To further evaluate the centrality and influence of each offense type in this structural context, we apply spectral analysis [19]on the co-occurrence matrix using eigenvector centrality (EVC) [20].

EVC measures the importance of a node by considering not only its direct connections but also the centrality of its neighbors. In this context, the EVC score of an offense is interpreted as its Offense Association Index, a normalized (0,1) metric indicating the likelihood that the offense appears in conjunction with other offenses across different crime incidents.

Offense Association Index provides a quantitative score of each offense's role in complex crime events and may serve as a useful reference for future crime prediction or classification tasks. Offenses with high co-occurrence index values are likely to appear in compound crimes and may suggest situational or behavioral overlap. Those with low values are more often stand-alone incidents, pointing to distinct patterns of criminal activity.

3.3 *Task II: Crime Hot Spot Detection*

3.3.1 Objective

This task introduces a PCA-based scoring framework to assess the severity of crime hot spots across a city. Rather than relying on binary classification to identify whether a location qualifies as a crime hot spot, this approach produces a continuous score that captures the relative intensity of crime occurrences over both space and time. This scoring mechanism provides a more nuanced and interpretable view of spatial crime dynamics.

3.3.2 Spatiotemporal Segmentation

To prepare the data for analysis, the authors apply a systematic spatiotemporal segmentation process to incident reports from the Boston Police Department, focusing on larceny-related events recorded. The city is divided into a grid of 0.01-degree by 0.01-degree cells, each representing a distinct spatial unit. This results in a spatial resolution of approximately 230 acres per cell. Temporal segmentation is achieved by dividing the 3-year dataset into consecutive 14-day time windows. Each record is mapped to one spatial cell and one time slice, forming a structured spatiotemporal dataset with 59 valid subregions and 85 time periods.

The resulting data structure represents the number of incidents observed in each region across each time slice, forming a matrix where each row corresponds to a spatial region and each column to a temporal window. Figure 2 is a layered visualization that reveals the spatial dynamics of crime hot spots over time.

3.3.3 Dimensionality Reduction and Feature Extraction

The authors apply principal component analysis (PCA) to the spatiotemporal incident matrix to identify dominant temporal patterns that distinguish one region from another. Each spatial unit is treated as an observation, with its time-series

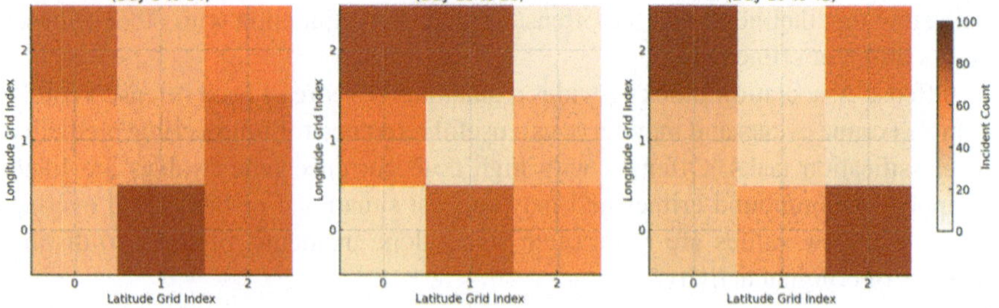

Fig. 2 Spatiotemporal distribution of crime incidents across three consecutive time slices

crime counts serving as feature dimensions. PCA projects each region onto a lower-dimensional space defined by the principal components, which reflect directions of maximum variance in crime activity across time. This transformation allows the temporal crime profile of a region to be summarized in a few interpretable components that account for the majority of variance across the dataset.

The eigenvalues corresponding to each principal component represent the variance explained by that component and are used as weights to compute a composite score.

3.3.4 Computation of Hot Spot Score

To quantify the extent to which a region functions as a crime hot spot, the authors calculate a linear score for each spatial unit based on its PCA representation. This scoring mechanism is based on the author's previous work on crime hot spot analysis [10] and is adapted here to fit within the broader hybrid framework of the present study.

The Crime Hot Spot Score $S(c)$ for a region c is defined as a weighted sum of the region's projections on the principal components, with the eigenvalues serving as weights:

$$s(c) = \frac{\sum_{i=1}^{n} e_i \cdot P_i(c)}{\sum_{i=1}^{n} e_i} \tag{1}$$

where $P_i(\cdot)$ denotes the score of region c on the ith principal component, and e_i is the corresponding eigenvalue. This score reflects both the magnitude and direction of variance in the region's temporal crime activity.

The result is a continuous ranking of all spatial units by their historical crime intensity, providing law enforcement with a more granular tool for evaluating risk levels across the city. Higher scores indicate regions with stronger and more consistent crime activity patterns over time, while lower scores suggest less persistent crime trends.

3.3.5 Network-Based Analysis and Interpretation

To further analyze the structure of crime hot spots, the authors construct a similarity-based network among spatial units, where nodes represent subregions and edges capture neighborhood relations or score-based proximity. Eigenvector centrality (EVC) is then computed on this network to evaluate the structural importance of each region within the broader hot spot landscape. By integrating PCA-derived scores with network topology, this project offers a hybrid approach to crime hot spot analysis that is both statistically robust and structurally interpretable.

4 Results

In this experiment, the authors developed a hybrid framework that integrates machine learning and network science analysis method. As illustrated in Fig. 3, the overall process begins with the acquisition of crime incident data from the Boston Police Department, followed by data preprocessing includes data cleaning and spatiotemporal segmentation using uniform spatial grids and 14-day time intervals.

The analytical pipeline consists of two main components. The first task focuses on offense co-occurrence analysis. The authors construct a co-occur offenses network by identifying incidents that involve multiple offense types, then representing each offense as a node, and connecting nodes that appear at the same time within the same incident record. The authors also apply ten clustering algorithms to identify relatively stable structural groupings among offense types and compute eigenvector centrality scores to generate an offense association index for each category.

The second task quantifies spatial crime hot spots. A grid-time matrix is generated to capture the temporal evolution of crime counts across regions. Principal component analysis (PCA) is then employed to reduce dimensionality and compute a weighted crime score for each spatial unit. Finally, a similarity-based spatial network is constructed and analyzed using eigenvector centrality to identify structurally influential hot spot regions.

4.1 Results of Task I

4.1.1 Co-Occur Offenses Matrix

The authors analyzed over 27,000 multi-offense incidents to examine the structural co-occurrence patterns among 67 offense types. Based on these records, a weighted, undirected co-occurrence network was constructed, where each node represents an offense type and edges indicate joint appearances in the same incident. The edge weights capture the frequency of co-occurrence.

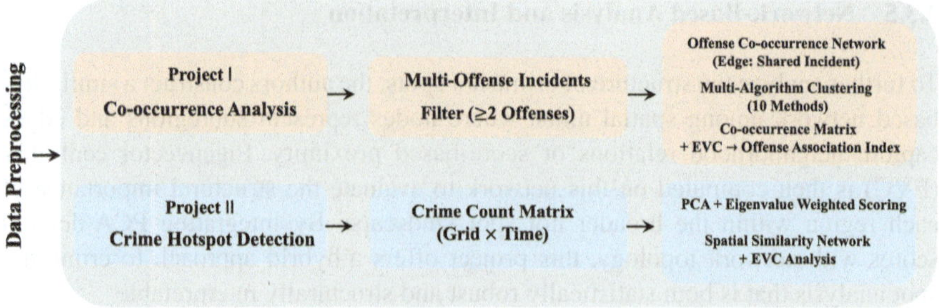

Fig. 3 Hybrid framework for urban crime analysis

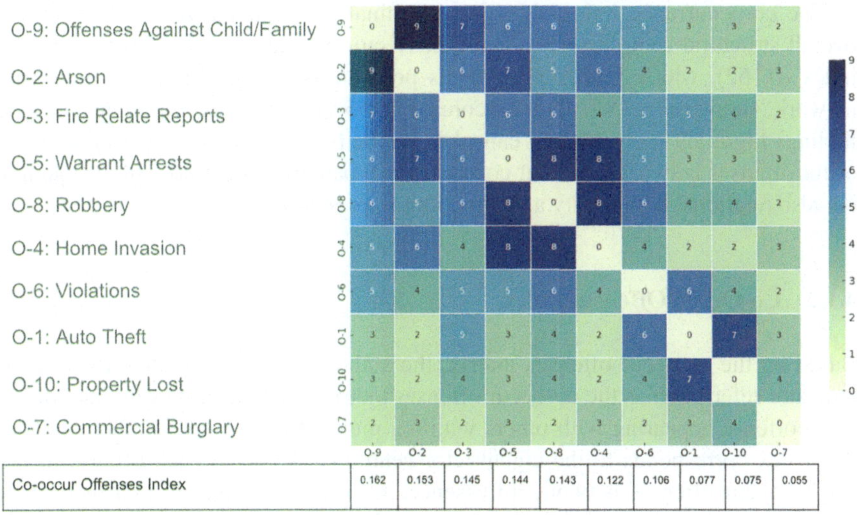

Fig. 4 Heatmap visualization of a partial co-occurrence matrix

After this, in order to ensure robustness in the community detection process, the authors applied a set of ten distinct clustering algorithms—spanning hierarchical, density-based, and modularity-based methods—to the co-occurrence network. Based on the clustering outcomes, a symmetric co-occurrence matrix was generated. Each off-diagonal entry in the matrix indicates the number of algorithms that grouped the corresponding offense pair into the same cluster, with higher values reflecting stronger structural association.

Figure 4 illustrates a partial 10×10 segment of the full 67×67 co-occurrence matrix. This heatmap displays how frequently each pair was grouped together across ten clustering algorithms. Each cell in the matrix represents a co-occurrence frequency ranging from 0 (never grouped together) to 9 (consistently grouped together), and darker shades indicate higher frequencies. These offenses are ordered both row-wise and column-wise in descending order of their index values.

The heatmap reveals a distinct structural layering within the network. Offenses such as O-9, Offenses Against Child/Family; O-2, Arson; and O-3, Fire-Related Reports consistently co-occur with multiple other offenses, with values of 6 or higher across several pairings. Notably, O-5, Warrant Arrests; O-8, Robbery; and O-4, Home Invasion also exhibit high cross-affiliation values with each other, suggesting a potential behavioral or situational linkage.

In contrast, the bottom rows—O-1, Auto Theft; O-10, Property Lost; and O-7, Commercial Burglary—show lower levels of co-clustering, indicating more isolated behavioral patterns. For example, O-7 co-occurs with others at a frequency no higher than 4 and as low as 1 in some cases, emphasizing its relative independence in the crime network structure.

The co-occurrence index shown in the final row summarizes each offense's overall structural centrality based on eigenvector centrality. O-9 holds the highest index (0.162), which further proving its position as a central node in the crime network. Conversely, O-7, with a score of 0.055, appears less importance. These findings highlight that offenses can play very different roles in the structural co-occur offenses network, and that co-occurrence patterns reflect not only frequency but also relational complexity and structural importance.

4.1.2 Co-Occur Offenses Index

Based on the co-occur offenses matrix, the authors further computed a structural indicator referred to as the co-occur offenses index, which quantifies the likelihood of an offense appearing with others within a multi-offense event. Mathematically, this index corresponds to the eigenvector centrality (EVC) of each offense in the full co-occur offenses network. In essence, an offense receives a high index if it frequently appears with many other offenses. This index values serve as a spectral signature of each offense's structural connectivity within the offense network.

Figure 5 is a visualization of the distribution of the index. These values exhibit a long-tail distribution, where only a small subset of offenses attain significantly high scores, while the majority lie within a narrow middle band. Offenses such as Offenses Against Child/Family and Arson dominate the upper range of the index, suggesting their consistent co-appearance with a wide range of other crimes. These high-index offenses are more likely to indicate multi-offense incidents, often embedded in socially or contextually complex environments.

On the contrary, offenses like Auto Theft and Residential Burglary receive among the lowest co-occur offenses indices. These results align with known criminological

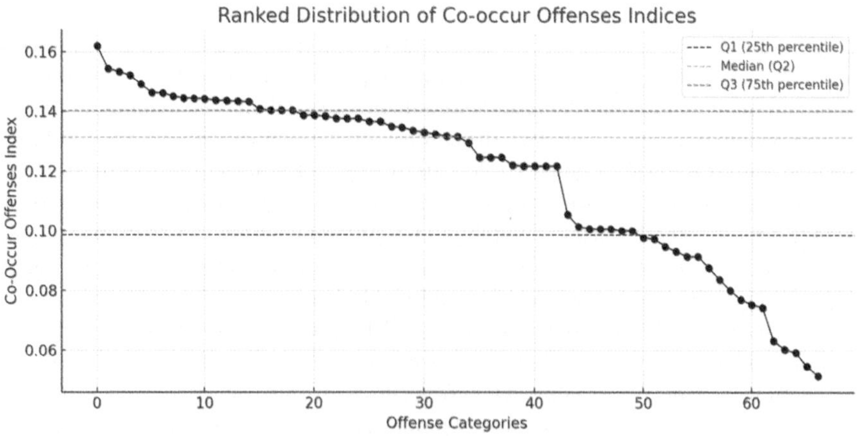

Fig. 5 Ranked distribution of co-occurrence indices for all offense types

patterns [21]: Crimes such as burglary, auto theft, and robbery are generally considered goal-oriented offenses, typically involving a singular objective—such as financial gain or property acquisition. As a result, they tend to occur in isolation and show limited overlap with other crime types. In contrast, offenses related to domestic violence, gang activity, or drug involvement are more likely to occur in complex social contexts where multiple types of criminal behavior co-occur within the same incident or environment.

Overall, the co-occur offenses index provides a robust metric for offense-level relational profiling, enabling researchers and law enforcement to identify offenses that act as structural bridges in the crime ecosystem, which will be benefit for both investigative prioritization and crime pattern forecasting.

4.2 *Results of Task II*

4.2.1 Boston Larceny Hot Spots

The authors constructed a spatiotemporal matrix in which each row represents a spatial subregion and each column corresponds to a 14-day time slice. Each entry in the matrix records the number of larceny incidents occurring in the respective region and time period. Then, the authors use the hot spot score, which is defined in this study, to summarize the long-term temporal pattern of crime activity within each spatial subregion. Higher scores indicate areas with more frequent and structurally consistent incidents over time.

To visualize how these scores are distributed across the city, the authors mapped each spatial unit to its corresponding geographic location using a color-coded grid overlay. As shown in Fig. 6, each grid cell reflects the hot spot score of a $0.01^\circ \times 0.01^\circ$ region, with warmer colors (red) indicating higher scores and cooler colors (blue) denoting lower scores.

The spatial distribution reveals strong clustering effects, where high-scoring regions tend to aggregate in the central and southern parts of the city. Among them, region A (score = 0.38) represents the subregion with the highest score, suggesting that this area has experienced both frequent and temporally consistent criminal activity. Its position in a dense cluster of similarly high-scoring grids further emphasizes its structural centrality within the urban crime landscape.

In contrast, region B (score = 0.55) received the lowest score in the dataset. This grid is located in the southern section and is surrounded mostly by similarly low-score neighbors, suggesting a structurally stable low-risk area. Given its consistently low crime intensity, this area is unlikely to require extra policy or enforcement response.

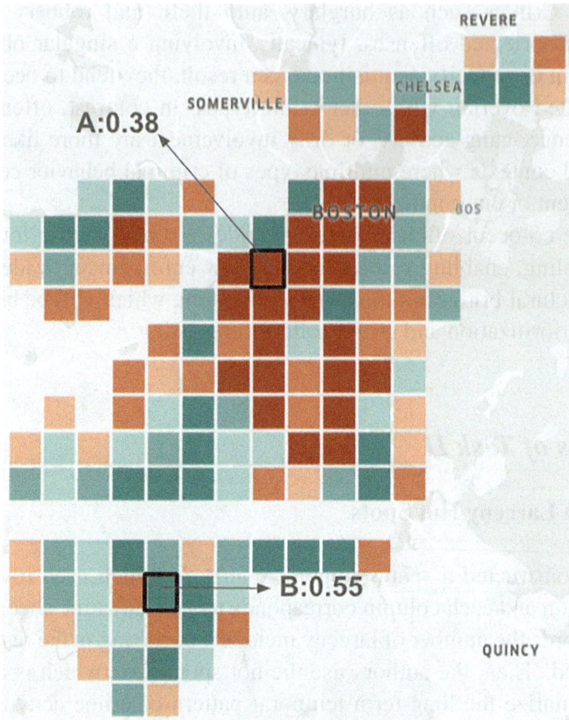

Fig. 6 Boston larceny hot spots score heatmap

4.2.2 Distribution of Hot Spot Scores

To examine the overall structure of crime hot spots, the authors analyzed the distribution of hot spot scores across all 168 spatial subregions. These scores, derived from PCA-based temporal crime patterns, quantify the relative intensity and structural consistency of incidents over time.

The distribution reveals a strong skew toward the lower end. As shown in Fig. 7, most regions exhibit low scores tightly clustered within a narrow band. Specifically, 75% of the subregions have scores below 0.059, with a median of 0.058 and a minimum value of 0.055. This tight clustering indicates that, for the majority of spatial units, crime activity (Larceny) is both infrequent and irregular across time.

However, a small number of outlier regions stand out with noticeable higher scores. The highest observed score is 0.384, which is over six standard deviations above the mean (0.067). This indicates a highly concentrated risk in a few regions that are structurally dominant in temporal crime activity. These high-score zones are often persistent hot spots that may warrant targeted enforcement or preventive intervention.

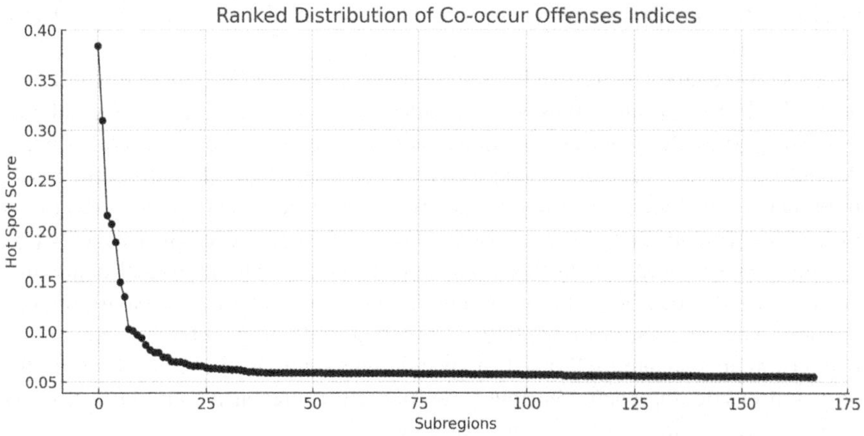

Fig. 7 Ranked distribution of Boston larceny hot spot scores across spatial subregions

The sharp imbalance between high- and low-score areas supports a resource-efficient approach to crime prevention: Directing attention to a small subset of spatial units may yield disproportionately high impact. Moreover, the observed distribution validates the scoring system's ability to differentiate routine urban background from structurally significant crime zones.

5 Conclusions and Future Work

This study proposes a hybrid framework that integrates machine learning techniques with network science methods to enhance the analysis of urban crime patterns. Using publicly available crime data from the Boston Police Department, the authors demonstrate how combining graph and statistical perspectives can get deeper insights into the dynamics of crime patterns.

The first part of the framework focuses on the structural relationships among co-occur offense types. By constructing an offense-offense network from multi-offense incidents and applying eigenvector centrality, the authors derive a co-occur offenses index that highlights offenses with significant structural roles. The results indicate that certain offenses, such as Offenses Against Child/Family and Arson, serve as relational hubs within complex incidents, offering potential entry points for broader investigative efforts.

The second part addresses the spatial and temporal distribution of crime. A PCA-based scoring mechanism is used to quantify long-term crime (Larceny) intensity for each spatial unit. The results reveal a highly skewed distribution of hot spot scores, with only a few regions exhibiting structurally consistent and elevated crime activity. Mapping these scores back onto the real world map further highlights localized clusters of persistent crime, providing actionable insights for place-based intervention.

These two analytical approaches highlight the value of hybrid modeling in crime research. The co-occur offenses index allows researchers to move beyond raw frequencies and examine the functional interdependence of offense types, while the hot spot score offers a scalable and continuous measure of spatial crime risk. Importantly, both metrics support resource prioritization and policy design by identifying structurally central offenses and regions within the urban crime network.

Future work could extend this framework to incorporate additional contextual layers, such as demographic information, mobility patterns, or environmental features. The integration of predictive modeling and causal inference techniques may also help evaluate the impact of specific interventions over time. Overall, the proposed hybrid approach contributes to a more comprehensive and operationally useful understanding of urban crime patterns.

Acknowledgments Research was sponsored by the Army Research Office and was accomplished under Grant Number W911NF-21-1-0264. The views and conclusions contained in this document are those of the authors and should not be interpreted as representing the official policies, either expressed or implied, of the Army Research Office or the US Government. The U.S Government is authorized to reproduce and distribute reprints for government purposes notwithstanding any copyright notation herein.

References

1. Wang, T., Rudin, C., Wagner, D., Sevieri, R.: Learning to detect patterns of crime. In: Joint European Conference on Machine Learning and Knowledge Discovery in Databases, pp. 515–530 (2013). Springer
2. Towers, S., Chen, S., Malik, A., Ebert, D.: Factors influencing temporal patterns in crime in a large american city: A predictive analytics perspective. PLoS One **13**(10), 0205151 (2018)
3. Braga, A.A., Weisburd, D.L.: The effects of focused deterrence strategies on crime: A systematic review and meta-analysis of the empirical evidence. Journal of research in crime and delinquency **49**(3), 323–358 (2012)
4. Butt, U.M., Letchmunan, S., Hassan, F.H., Ali, M., Baqir, A., Sherazi, H.H.R.: Spatio-temporal crime hotspot detection and prediction: a systematic literature review. IEEE access **8**, 166553–166574 (2020)
5. Bowen, D.A., Mercer Kollar, L.M., Wu, D.T., Fraser, D.A., Flood, C.E., Moore, J.C., Mays, E.W., Sumner, S.A.: Ability of crime, demographic and business data to forecast areas of increased violence. International journal of injury control and safety promotion **25**(4), 443–448 (2018)
6. Hou, M., Hu, X., Cai, J., Han, X., Yuan, S.: An integrated graph model for spatial–temporal urban crime prediction based on attention mechanism. ISPRS International Journal of Geo-Information **11**(5), 294 (2022)
7. Meghanathan, N.: Unit disk graph-based node similarity index for complex network analysis. Complexity **2019**(1), 6871874 (2019)
8. Klerks, P.: 6 the network paradigm applied to criminal organisations. Transnational Organised Crime: Perspectives on Global Security, 97 (2003)

9. examplee Wu, Y., Meghanathan, N.: A network science approach to quantify the extent of co-occurrence of an offense with other offenses in a crime event. In: Computer Science On-line Conference, pp. 442–451 (2022). Springer
10. examplee Wu, Y., Meghanathan, N.: A principal component analysis-based scoring mechanism to quantify crime hot spots in a city. In: International Conference on Information Technology-New Generations, pp. 45–51 (2023). Springer
11. examplee Khairuddin, A.R., Alwee, R., Haron, H.: A review on applied statistical and artificial intelligence techniques in crime forecasting. In: IOP Conference Series: Materials Science and Engineering, vol. 551, p. 012030 (2019). IOP Publishing
12. Jenga, K., Catal, C., Kar, G.: Machine learning in crime prediction. Journal of Ambient Intelligence and Humanized Computing **14**(3), 2887–2913 (2023)
13. examplee Ramirez-Alcocer, U.M., Tello-Leal, E., Mata-Torres, J.A.: Predicting incidents of crime through lstm neural networks in smart city domain. In: The Eighth International Conference on Smart Cities, Systems, Devices and Technologies, pp. 32–37 (2019)
14. Han, X., Hu, X., Wu, H., Shen, B., Wu, J.: Risk prediction of theft crimes in urban communities: An integrated model of lstm and st-gcn. Ieee Access **8**, 217222–217230 (2020)
15. examplee Wang, H., Kifer, D., Graif, C., Li, Z.: Crime rate inference with big data. In: Proceedings of the 22nd ACM SIGKDD International Conference on Knowledge Discovery and Data Mining, pp. 635–644 (2016)
16. examplee Yi, F., Yu, Z., Zhuang, F., Guo, B.: Neural network based continuous conditional random field for fine-grained crime prediction. In: IJCAI, pp. 4157–4163 (2019)
17. Zhao, S., Liu, R., Cheng, B., Zhao, D.: Classification-labeled continuousization and multi-domain spatio-temporal fusion for fine-grained urban crime prediction. IEEE Transactions on Knowledge and Data Engineering **35**(7), 6725–6738 (2022)
18. Boston Crime Incident Reports. https://data.boston.gov/dataset/crime-incident-reports-august-2015-to-date-source-new-system (2018)
19. examplee Meghanathan, N.: Spectral radius as a measure of variation in node degree for complex network graphs. In: 2014 7th International Conference on U-and e-Service, Science and Technology, pp. 30–33 (2014). IEEE
20. Bonacich, P.: Some unique properties of eigenvector centrality. Social networks **29**(4), 555–564 (2007)
21. Felson, M., Eckert, M.A.: Crime and Everyday Life, 6th edn. Sage, Thousand Oaks, CA (2018)

Part III
Behavioral and Biometric Forensics

Beyond Normality: Rethinking Behavioral Biometric Data

Amith Kamath Belman, Frank Sicong Chen, Vir V. Phoha, and Pronab Mohanty

1 Introduction

With the ubiquitous availability and spread of desktops, mobile devices—such as smartphones and tablets—and smartwatches, keyboards, both physical and on-screen, and touchscreens have emerged as the most common interface to these devices. In addition, many applications use onboard inertial sensors, such as accelerometers and gyroscopes, for tasks related to health monitoring, gait analysis, etc. Many studies assume the underlying distribution of features from keystrokes, swipes, accelerometer, and gyroscope readings to be normal [12, 13]. Such assumptions are made for theoretical or calculational simplicity discussed in later sections. However, the effects of assuming an underlying normal distribution in data when it is not have been explored in various fields [17, 18, 86]. This misassumption can be a source of error in classifications when it influences the decision rules in the classifier and in an adversarial approach leads to inefficient attacks if the generative model depends only on the mean and standard deviation.

We use a collection of open-source benchmark datasets that include single-activity or single-device datasets [115–119, 125, 126], and a large multimodality dataset (SU-AIS BB-MAS [19]), which consists of data from 117 users providing keystrokes data on desktop, phone, and tablet; accelerometer and gyroscope data

A. Kamath Belman
San José State University, San José, CA, USA
e-mail: amith.kamathbelman@sjsu.edu

F. S. Chen · V. V. Phoha
Syracuse University, Syracuse, NY, USA
e-mail: schen154@syr.edu; vphoha@syr.edu

P. Mohanty (✉)
Police Computer Wing, Bangalore, Karnataka, India

S. S. Iyengar et al. (eds.), *Artificial Intelligence Driven Forensics*, Signals and Communication Technology, https://doi.org/10.1007/978-3-031-98036-7_16

while walking and climbing upstairs and downstairs with a phone and a tablet; and touchscreen swiping data on phone and tablet, for our experiments. We defined the null hypothesis H_0 to be "the test sample comes from a normal distribution" and performed a large number of Lilliefors tests and Shapiro–Wilk tests on all modalities in these datasets. To summarize results, we categorize the features into four different categories: (a) less than 25% of samples with p-value >0.05; (b) 25% to 50% of samples with p-value >0.05; (c) 50% to 75% of samples with p-value >0.05; and (d) above 75% samples with p-value >0.05. We find that except for the features from samples taken from climbing upstairs and downstairs data, almost all features from samples of all other activities were in the first category. It means that less than 25% of the samples cannot reject the null hypothesis (the sample comes from a normal distribution) at a 0.05 significance level.

Although many other fields have witnessed research work cautioning the naivety in the assumption of normality, this assumption has hardly been examined in behavioral biometric modalities, lesser so in the case of multiple activities that span multiple devices. We present related literature and briefly explore why studies assume normality in data, why these features depart from the normal distribution, and what distributions approximate best to these features. We also give brief descriptions of the eight datasets that we used and popularly used features extracted from these activities, explain our experiments in detail, analyze and discuss the experimental results, and conclude with the impacts of our findings.

1.1 Key Contributions

- We show, with suitable nonparametric tests, that the most commonly described features for mobile device-based biometric data from activities like typing, gait, and swiping do not have an underlying normal distribution.
- We discuss why the assumption of normality is widely accepted in this field and why these features depart from the normal distribution and perform the chi-square statistics to get distributions that best approximate these features, which provide an overall perspective of what kind of distribution these features conform to.
- We perform our tests on the large multimodality BBMAS dataset [19], as well as seven single-modality datasets for gait [115, 116], Swipe [117, 118], and Typing [119, 125, 126]. The results show that our findings are consistent and can be generalized.
- We discuss alternate approaches to mitigate or avoid the negative effects of assuming a nonexistent normal distribution in data.
- We present the important implications of our findings for modeling distributions and classifier choices.

2 Motivation and Related Work

Behavioral biometrics includes a broad spectrum of modalities involving human behavior while performing day-to-day tasks. With the spread of mobile devices, such as smartphones and tablets, gait, keystrokes, and swiping patterns are the most explored biometrics with the interface to these mobile devices in recent years. Studies have focused on utilizing various aspects of these behavioral biometrics for authentication [100, 103], verification [106], continuous authentication [52], gender detection [54, 107], age detection [56], fatigue detection [55], mood disturbance detection [58], lie detection [57], and detection of various health conditions. Many of these works have either purposefully or inadvertently assumed an underlying normal distribution in the features extracted from the data, reflected in their analysis methods or their choice of classifiers. Although assuming a normal distribution helps simplify the problem, it may not always lead to the best results.

The adverse effects of assuming a normal distribution have been significantly studied in different fields. In the Constraint Satisfaction Problems (CSP) field, [17] proves that in the results produced by many heuristic combinations on random binary CSPs and 3-coloring problems, the benchmarks for CSP, the assumption of normality does not hold. The authors also appeal for statistics that do not rely on the normality assumption to analyze empirical results for CSP. In processes involving the classification of remotely sensed data from different spectral bands (image classification is a subset of this problem family), [18] showed that the brightness values distributions did not follow a normal distribution. They further remarked that this fallacy was a major source of error in land cover classification when decision rules employed in the classifier assumed an underlying normal distribution. In Dunning's [86] work on the statistical analysis of text, they pointed out that the assumption of normal distribution limits the ability to analyze rare events and that those rare events were a significant fraction of real text.

When dealing with the solution space for the economic design of X-bar control charts, [87] show that the nonnormality assumption also has a more significant effect on the Type II error probability than the Type I error probability. In the research of human cancer genomes, wrongly assuming a normally distributed gene expression was shown to affect multiple facets, including identification of expression patterns, annotation, and classification [88]. They also concluded that small departures from normality were not analytically insignificant. Limpert and Stahel [93] questioned the adequacy of data characterization using normal distributions and argued that an asymmetric view would increase recognition of data distributions and the quality of interpretation.

2.1 Why Prior Works Might Have Assumed Normal Distribution in Data?

We found two main concepts that many studies used to justify their assumption of normality in most datasets: the central limit theorem (CLT) [89] and the Cràmer-Rao lower bound (CRLB) [90].

The popularity of the assumption of normality can be attributed to the fact that noise in many systems has been represented well using a normal distribution. The Gaussian assumption is a good conservative choice when not much is known about the data, which is also supported by CLT stating that the distribution of sample means tends to form a normal distribution as the sample size gets larger. A Gaussian distribution also minimizes the Fisher information, which is the inverse of CRLB. In other words, the CRLB under the Gaussian distribution works for the worst-case scenario, maximizing the CRLB (see [15]). Therefore, minimizing the largest CRLB is interpreted as min-max optimal [16]. To relate our discussions and analysis, we briefly introduce both of these concepts below.

2.1.1 Central Limit Theorem

If $(X_1, X_2, \ldots, X_n)$ is a random sample of size n taken from a population with mean μ and finite variance σ^2, and if $\bar{X}$ is the sample mean, the limiting form of the distribution of $Z = (\bar{X}-\mu)/(\sigma/\sqrt{n})$, as $n \to \infty$, is the standard normal distribution (see Theorem 7.2 in [108]).

It is often assumed that estimated population means and variance are independent for simplicity, implying sample variance means and variance are independent, which is valid only for normal distribution.

Many research works inadvertently back on CLT for their modeling choices. A good example of this scenario is the use of Gaussian mixture models for keystroke analysis [110–112] and touchscreen swipe analysis [113, 114]. A Gaussian mixture model assumes the data follows a mixture of individual multivariate Gaussians or Gaussian mixture distribution, which is only a good choice if the data is a mixture of Gaussians and has a sufficiently large number of samples, implicitly invoking CLT.

2.1.2 Fisher Information and Cràmer-Rao Lower Bound

Fisher information for a random sample from $f(x|\theta)$, where θ is an unknown parameter and $n \to \infty$, is expressed as $I(\theta) = -\mathrm{E}[(\partial^2/\partial\theta^2) \log f(X|\theta)]$ (see Theorem 5.8 [109]).

If $\hat{\theta}$ is an unbiased estimator of θ, and [90] states that the variance of $\hat{\theta}$ is bounded by the reciprocal of Fisher information, i.e., $var(\hat{\theta}) \geq 1/(nI(\theta)$, using CLT, it follows that as sample size tends to infinity, the maximum likelihood estimator is asymptotically unbiased and asymptotic distribution of $\hat{\theta}$ is normal, i.e., for a true value θ_0 for θ, $\sqrt{nI(\theta_0)}(\hat{\theta} - \theta_0) \sim N(0, 1)$.

Some Examples from Literature

The Gaussian assumption is often applied in the modeling of keystroke latencies. For example, the work in [12] modeled the latencies of 142 pairs of characters under the assumption that "probability of the latency y between two keystrokes of a character pair forms a uni-variate Gaussian distribution" and thereby deriving the μ and σ parameters for each character pair, to be used further down for information gain estimation. Similarly, [13] started their bot simulations under the assumption that the keystroke duration of a character in a word is modeled as a random variable which is Gaussian or a constant with additive uniform noise. Thus, their typing event injections for synthetic forgeries or bot attacks were already under assumptions that should not be made casually.

2.2 Prior Works That Questioned the Assumption of Normality in Behavioral Biometrics

In [6], the authors examined keystroke features such as key interval times (Digraph) and key hold times (Unigraph) extracted from keystroke data recorded on desktops to test if they followed a normal distribution. The authors performed Lilliefors test [1] and Cramer-von Mises test [91] and established that key hold times and the key interval times from desktop typing did not follow a Gaussian distribution for all unigraphs and gigraphs. As research in keystroke dynamics is a popular component of behavioral biometrics, this can be considered a pioneering step toward questioning the normality assumption in behavioral biometrics.

2.3 Why Does the Distribution in Data Depart from the Normal Distribution?

Reasons for why the data might not conform to the normal distribution can be approached from two directions.

First, the data collection process. There can be various factors that affect the way that data is collected. For instance, most sensor-based gait data come from the accelerometers and gyroscopes inside mobile devices or IMUs (Inertial measurement units). However, the way or the position that the device is put in can have influences on the data collected from these sensors. For example, the phone is put in the pocket for some of the datasets, while it is held in hand for some other datasets. As shown in Fig. 1c and f, both of these figures are the histogram of the energy of the x-axis. However, they show different distributions due to different places the phone was put while collecting data. Figure 1c shows the histogram of "x-energy" values extracted from the phone in hand. We can see that nearly all the values in the range of (0,0.2). However, the histogram of "x-energy" values extracted from the phone put in the pants' pocket shows that there are around 1/3 of values

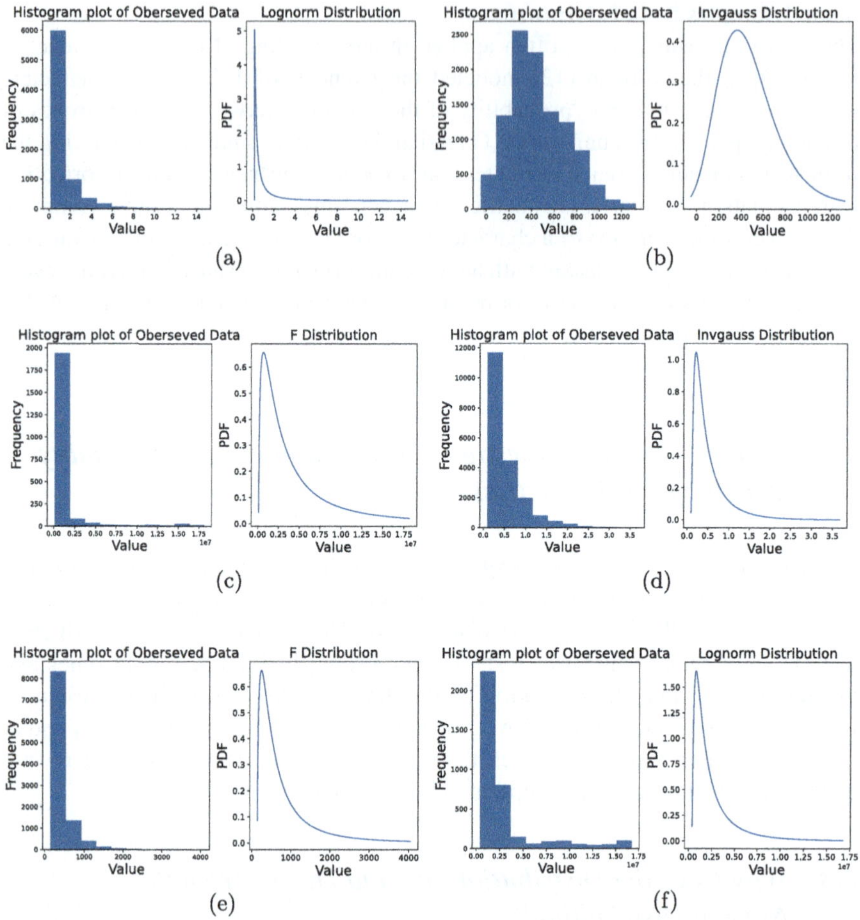

Fig. 1 Histogram plots and the probability density function plots that are approximated by chi-square statistics. (**a**) Typing data on hand-phone, Flight3 time for pair ("e","s") (d_9) digraph, approximated by log-normal distribution; (**b**) swiping data, minimum value of y coordinates, approximated by inverse Gaussian distribution; (**c**) gait data, energy value of x-axis of accelerometer signal from hand-phone while walking, approximated by F-distribution; (**d**) typing data on hand-tablet, Flight1 time for pair ("o," "SPACE") (d_{10}) digraph, approximated by inverse Gaussian distribution; (**e**) swiping data, total time for completing a swipe, approximated by F-distribution; (**f**) gait data, energy value of x-axis of accelerometer signal from pocket-phone while walking, approximated by log-normal distribution. (**a**) Typing digraph "d9-F3". (**b**) Swiping "miny". (**c**) Gait walking "x-energy" from "HP". (**d**) Typing digraph "d10-F1". (**e**) Swiping "tottime". (**f**) Gait walking "x-energy" from "PP"

in the range of (0.2, 0.4), as shown in Fig. 1f. It results in different distributions that approximate values of this feature. Also, the data from the accelerometers and gyroscopes might vary significantly even though values are recorded at specific frequencies as people walk at different speeds.

For keystroke data, the type of the keyboard can influence the speed that people type: a higher keycap might take someone more time to press the key, while other people might be used to it and type as usual. Additionally, typing on the screen significantly differs from typing on the physical keyboard. A slight touch is sufficient for the screen to capture the keystroke, resulting in the key hold time concentrating in a very small range.

For swiping data, the screen size and the device's orientation can affect the length or the duration that people can swipe. The finger people like to use for swiping and the hand people use to hold the device also matter. All these factors can affect the data collection process and result in the samples not being normally distributed.

Second, some of the features are directly extracted from the raw dataset, while other features are calculated features. Sometimes, these calculated features do not conform to known distributions. For example, in the swiping feature sets, the coordinates of the starting point and the ending point are extracted from the raw data, while the first, second, and third quantiles for velocity are statistical variables. In fact, statistical variables can be from any distributions. There are also similar calculated features in gait feature sets.

2.4 *What Distributions Best Approximate These Features if They Do Not Conform to the Normal Distribution?*

Our discussion from Sects. 2.2 and 2.3 naturally leads to the question: "What distribution do we expect to get if the data does not conform to the normal distribution?" Since histograms give us an approximate representation of how the samples are distributed, we plot the histograms for some of the features and find that they conform to many different types of distributions besides normal distribution. In Fig. 1, histogram plots are shown on the left side of each figure.

We use the chi-square statistics to approximate the probability distribution function for each feature. The chi-square (χ^2) statistic provides a way to test how well a sample of data matches the (known or assumed) characteristics of the larger population that the sample is intended to represent. The formula for chi-square statistics is:

$$\chi_c^2 = \Sigma \frac{(O_i - E_i)^2}{E_i}$$

where c is the degree of freedom, O is the observed value, and E is the expected value.

Table 1 Top ranked distributions that are approximated by Chi-square statistic for each modality. The number in brackets are the number of features that conform to this kind of distribution. (Acc, accelerometer; Gyr, gyroscope; DT: Desktop; HP, hand-phone; PP, pocket phone; HT, hand tablet)

Activity	Feature set		Top ranked distributions
Gait (44)	Walk	Acc	HP, t (8); PP, Weibull min (11); HT, beta (8)
		Gyr	HP, inverse Gaussian (16); PP, Weibull min (13); HT, log norm (16)
	Upstairs	Acc	HP, beta (12); PP, beta (7); HT, Weibull min (8)
		Gyr	HP, beta (13); PP, beta (20); HT, inverse Gaussian (13)
	Downstairs	Acc	HP, beta (8); PP, Chi-square (7); HT, beta (18)
		Gyr	HP, beta (10); PP, beta (19); HT, beta(18)
Swiping (27)	Phone		Cauchy (4), inverse Gaussian (4)
	Tablet		Weibull min (4), beta (4), Cauchy (4)
Typing	Unigraph (12)	DT	Weibull max (5), t (4)
		HP	t (7)
		HT	t (10)
	Digraph (72)	DT	Weibull min (16), log-norm (14), inverse Gaussian (14)
		HP	Weibull min (14), inverse Gaussian (13), beta (12)
		HT	Weibull min (21), inverse Gaussian (11), t (8)

There are two main kinds of chi-square tests: the test of independence and the goodness-of-fit test; we use the latter. We select several common distributions to approximate each feature for all subjects. Then we select the most frequent distribution that appears as the distribution of each feature it conforms to. In Fig. 1, the approximated probability density functions (PDFs) are shown on the right side of each figure. For example, in Fig. 1a, the histogram plot shows that nearly all of the samples are in the range of (0,4), while there are still a few samples that do not in this range. It suggests that it does not conform to the normal distribution. By using the chi-square statistics, it is best approximated by the log-normal distribution, which is shown on the right side. The most frequent distributions that we get for each modality are shown in Table 1. The distributions we selected for matching include normal distribution, uniform distribution, exponential distribution, beta distribution, gamma distribution, Cauchy distribution, F-distribution, Student's t-distribution, Chi-square distribution, inverse Gaussian distribution, log-normal distribution, Weibull minimum distribution, and Weibull maximum distribution.

2.5 *How Our Analysis Differs from Related Work Discussed in Sect. 2.2*

Although many other fields have questioned the naivety of the assumption of normality, the field of behavioral biometrics has been lagging in this aspect. We found that Section (5.1) of the research in [6] was the only exploration in this direction for keystroke dynamics on desktops. Behavioral biometrics is an emerging area with many different modalities, devices and activities. For example, typing is no more limited to a desktop; keystroke dynamics have been studied separately on phones and tablets and other touch devices; gait as a behavioral biometric has been studied with different sensor placements, different sub-activities, and different devices; and swiping patterns for behavioral biometrics have been studied on different smart screens or touch surfaces.

We examine the assumption of normality, in a large dataset consisting of a wide range of activities and devices, as well as seven single-modality datasets collected from different mobile or wearable devices. By performing a large number of statistical tests, we show the following:

- Features extracted from accelerometers and gyroscopes in tablets and phones (in the pocket and in hand) do not follow a Gaussian distribution while freely walking on a flat floor. However, in a large number of samples from activities of climbing up and down the stairs and walking on a treadmill at fixed speeds, the null hypothesis, that the data comes from the normal distribution, cannot be discarded.
- Swipe features extracted from swipe trajectories, pressure, acceleration, and touch area data do not follow normal distribution on smart phone and tablet surfaces.
- Keystroke features on mobile devices might have different distributions from that on desktops. On mobile devices (on-screen keyboard), key hold times are more likely to conform to the normal distribution, while most flight times do not. However, these features on the desktop (physical keyboard) do not follow the normal distribution based on our nonparameter test. Our experiments reaffirm the findings in [6] that showed the nonnormal nature of keystroke features on desktop. We observe the same occurrence in keystroke features, such as flight times, from phones and tablets, which have not been examined in the literature.
- Implications of our findings and alternate approaches to handle nonnormal distributions in data.

3 Description of Test Datasets and Popularly Used Features in Gait, Swipe, and Keystroke

In this section, we briefly describe a multimodality dataset (SU-AIS BB-MAS [19] dataset) and seven single-modality datasets [115–119, 125, 126], as well as the features that we have analyzed. These features are the most popularly used features from each of the modalities (Table 2).

Table 2 The brief description of the SU-AIS BB-MAS [19] dataset, which includes the different data types we analyzed from multiple devices and activities. The gait activity consists of three sub-activities, walking, climbing upstairs, and downstairs

Activity	Sensors/interface	Devices	Details
Gait	Accelerometer gyroscope	Phone in pocket	2 sessions / user, $\approx$ 5 min./session
		Phone in hand	Phone in pocket for both sessions
		Tablet in hand	Phone in hand and tablet in hand for one session
Swiping	Touchscreen	Phone	$\approx$ 25 minutes with typing
		Tablet	$\approx$ 25 minutes with typing
Typing	Keystroke	Desktop	$\approx$ 50 minutes with mouse usage
		Phone	$\approx$ 25 minutes with swiping
		Tablet	$\approx$ 25 minutes with swiping

3.1 Description of the Multimodality Dataset (SU-AIS BB-MAS [19] Dataset)

As [92] has described all aspects of this dataset in great detail, we only provide a gist of this dataset in this section. A total of 117 users participated in the data collection and performed several activities on different days. The activities included typing and mouse clicking on the desktop, typing and swiping on phones and tablets, and walking and climbing upstairs and downstairs with phones and tablets in their pants' pockets and hands. The dataset consists of a total of about 3.5 million keystroke events, 57.1 million data points for accelerometer and gyroscope each, 1.7 million data points for swipes, and enables researchers to explore previously unexplored directions in inter-device and inter-modality biometrics.

3.2 Description of Seven Single-Modality Datasets

These seven datasets [115–119, 125, 126] are single-activity or single-device datasets that have been elaborately described in the literature; we only provide a brief description of each of them. Please see Table 4 for more details.

3.2.1 Description of Two Gait Datasets [115] and [116]

The gait dataset collected in [115] consists of walking data from 20 subjects performing two separate 15-minute walks on two different days with a mobile phone in their pocket. The accelerometer and magnetometer in the mobile phone collected the data while the subjects were walking. This dataset is very similar to the walking (on a flat corridor) phase in BBMAS dataset [19]. The other dataset, collected in [116], has data from 21 healthy heel-striking subjects walking on a split-belt treadmill at 12 different walking speeds in the interval [0.6,1.7] m/s with 0.1 m/s increments, and each speed maintained for one minute. The subjects wore shoes equipped with eight markers (sensors) to keep track of the movement of these markers.

3.2.2 Description of Two Swipe Datasets [118] and [117]

Serwadda and Phoha [118] collected a dataset that consisted of swiping data from 66 users performing at least 80 swipes each on a smartphone. They also collected another swipe dataset in [117], which has data separated into two categories based on the device's orientation. Around 100 participants gave swipes in portrait mode, and 45 gave swipes in landscape mode, with about 80 swipes per participant.

3.2.3 Description of Two Typing Datasets [125] and [126] on Mobile Devices and One Typing Dataset [119] on Desktops

The two typing datasets on mobile devices are collected in [125] and [126], and the typing dataset on desktops is collected in [119]. All these datasets asked participants to type ".tie5Roanl," which is chosen to be a representative of strong passwords in [127].

The dataset in [125] consists of 51 users with 51 trials per user. The other dataset in [126] consists of data from 54 participants. Each participant performed at least three sessions and provided at least 60 trials in total. Though participants were asked to type the same password on mobile phones, the keyboard layout for these two datasets is not the same, resulting in different unigraph and digraph features. The dataset in [119] consists of 51 users typing the same password on a physical keyboard in eight different sessions, for a total of 400 samples per user.

3.3 Description of Features That We Analyzed in Gait, Swipe, and Keystroke

The feature extraction for this dataset can be grouped into three parts, namely, gait, swipe, and keystroke features. For each modality, we selected the most popular features that have been studied for several years [121–124]. We briefly describe the features and their storage below. A summary of the features is presented in the Table 3.

- **Gait features**: As the raw data for the gait is a pair of signals from the accelerometer and gyroscope we extract features from both. The gait data is further subdivided into three activities: "walking" (on a flat corridor), "downstairs" (going down the staircase), and "upstairs" (going up the staircase). For

Table 3 List of features extracted and examined in our experiments for an underlying normal distribution

Data	Features	Details
Gait	• Mean • Standard deviation • Band power • Energy • Median • Interquartile range • Range • Signal to noise ratio • Dynamic time warp dist. • Mutual information • Correlation	– Features were extracted from x, y, z, and m (m=$\sqrt{x^2+y^2+z^2}$) signals from both the accelerometer and gyroscope. – All features were extracted from each of the directional signals except for DTW distance, mutual information and correlation which were extracted between pairs of these signals, i.e., x-y, x-z, x-m, y-z, y-m, and z-m.
Swpie	• Min. x and y coordinates • Max. x and y coordinates • Euclidean distance • Angle of the swipe • Time • Velocity mean and std. • Velocity wuartiles • Acc. mean and std. • Acc. quartiles • Pressure mean and std. • Pressure quartiles • Area mean and std. • Area quartiles • Direction	– Features were extracted from a variety of information making up a swipe. – Features such as coordinates, angles, and direction are dependent on the touch points on the screen and the end points of a swipe. – Velocity, pressure, acceleration, and area are calculated with the data from corresponding sensors on the touch surface of the devices. – The direction feature was only used to group the swipes into vertical and horizontal swipes.
Typing	• Keyhold • Flight 1 to Flight 4	– Keyhold times were extracted from twelve most occurring unigraphs. – Flight times were extracted from eighteen most occurring digraphs.

both accelerometer and gyroscope, they keep track of the movement along three basic directions, x, y, and z. Based on these three directions, we can calculate an overall direction, called m, which we have $m = \sqrt{x^2 + y^2 + z^2}$. We use a window size of two seconds with a one-second overlap between two consecutive windows. For each two-second window we extract a host of features from the accelerometer and the gyroscope along each of these four directions. In general, statistical variables can describe some characteristics of the data, so we extract the mean, standard deviation, median frequency, interquartile range, and range from four directions separately. These data are time-series data, so band power, energy, and signal to noise ratio can be extracted from these four directions. Besides, we calculate the dynamic time warping distance between pairs of signals x-y, y-z, and x-z. Mutual information is calculated between pairs of signals x-y, x-z, x-m, y-z, y-m, and z-m, and correlation coefficients are calculated between pairs of signals x-y, y-z, and x-z. The list of features extracted is given in Table 3.

- **Swipe features**: For each swipe performed by users on tablet and phone, various features related to the speed and trajectory of the swipes are extracted. The last row of Table 3 summarizes them. While people swipe on the screen, it records the coordinates of the finger on the screen at certain frequencies, so itis natural to extract features as the minimum and the maximum x and y coordinates; the Euclidean distance between the start and stop points; the tangent angle of the swipe; the total time taken to finish the swipe; the mean and standard deviation and the quartiles of velocity, acceleration, pressure, and area between the finger and screen; and the direction of the swipe used to group them into horizontal or vertical swipes.
- **Keystroke features**: We select the most occurring 12 unigraphs (single key) and 18 digraphs (pair of consecutive keys) that occurred the most number of times in all user's keystroke data.
  The unigraphs are "BACKSPACE," "SPACE," "a," "e," "h," "i," "l," "n," "o," "r," "s," and "t."
  The digraphs are ("BACKSPACE," "BACKSPACE"), ("SPACE," "a"), ("SPACE," "i"), ("SPACE," "s"), ("SPACE," "t"), ("e," "SPACE"), ("e," "n"), ("e," "r"), ("e," "s"), ("n," "SPACE"), ("o," "SPACE"), ("o," "n"), ("r," "e"), ("s," "SPACE"), ("s," "e"), ("t," "SPACE"), ("t," "e"), and ("t," "h").
  For a unigraph K_i we extract the $Keyhold$ time of the key as a feature:
  - $Keyhold_{K_i}$: $K_i Release$ - $K_i Press$

  For a digraph K_i and K_{i+1} the following temporal features are extracted:
  - $Flight1_{K_i K_{i+1}}$: $K_{i+1} Press$ - $K_i Release$
  - $Flight2_{K_i K_{i+1}}$: $K_{i+1} Release$ - $K_i Release$
  - $Flight3_{K_i K_{i+1}}$: $K_{i+1} Press$ - $K_i Press$
  - $Flight4_{K_i K_{i+1}}$: $K_{i+1} Release$ - $K_i Press$

  Figure 2 illustrates the temporal features extracted form keystrokes.

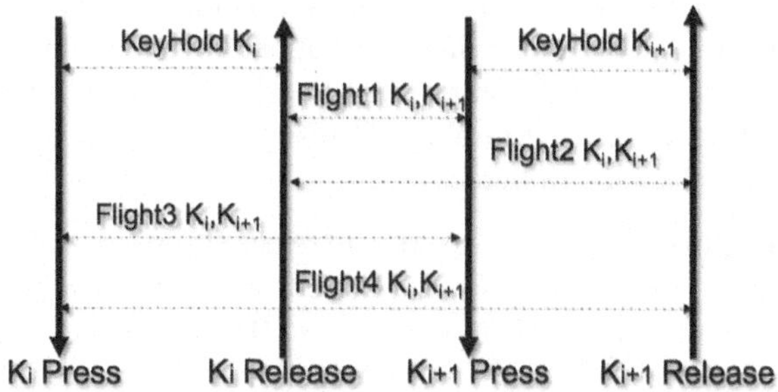

Fig. 2 Features extracted from keystroke data

Outlier removal for keystroke features: We use a simple filter to remove any instances of keys that were held down for two seconds or more. We also remove instances of the inter-key pauses that are greater than two seconds. We assume that these were caused by pauses, where the user is either thinking or receiving instructions during the data collection.

4 Nonparametric Tests and Experimentation

This section describes the tests and procedures used to examine the hypothesis that a sample comes from the normal distribution.

Nonparametric Tests To test if the feature values have an underlying normal distribution, we use two nonparametric tests of normality, namely, **(a) Lilliefors test** [1], a modified form of Kolmogorov–Smirnov test [3] that is suitable for large datasets for nonparametric testing of the null hypothesis that the data comes from a normally distributed population, and **(b) Shapiro–Wilk test** [2] which is also a nonparametric test and is more suitable for testing the null hypothesis on a smaller data ($n < 50$).

Hypothesis Testing For all tests we begin with the null hypothesis H_0 that the sample comes from a normal distribution. If the p-value from the test is below the critical value of 0.05, we can say that H_0 can be rejected, which means that the test sample does not conform to a normal distribution.

Sampling and Testing Procedure In our tests, we draw random samples consisting of 75% of the feature set for each modality and for each user. Since the patterns for each user differ from others, and we are concerned with individual patterns, we perform the test for each feature within user. We then perform a suitable normality test (Lilliefors test if the number of samples > 50 or Shapiro–Wilk test otherwise) and store the p-values with corresponding annotations. We repeat the process ten times for all features, modalities, and users to arrive at more accurate conclusions for evidence of normality in the underlying distribution. This form of random sampling is inspired by much other statistical research [4–6], which shows that it is hard for goodness-of-fit tests to provide meaningful results on large datasets due to their loosely fitting statistical descriptions.

Grouping of Features As the tests help us to reject H_0, based on the percentage of samples that had p-value >0.05, we categorize features into four broad categories: (a) less than 25% of samples with p-value >0.05, (b) 25% to 50% of samples with p-value >0.05, (c) 50% to 75% of samples with p-value >0.05, and (d) above 75% samples with p-value >0.05. The values have been color-coded based on these categories for visual clarity.

5 Analysis on the Multimodality Dataset: SU-AIS BB-MAS Dataset

As shown in Fig. 3, for a majority of test samples for features in gait, swiping, and typing, we found that H_0 could be rejected at the significance level of 0.05, implying the samples did not conform to a Gaussian distribution.

Samples from gait activity were further divided into walking (on a flat corridor), downstairs, and upstairs samples before testing each for normality. We found that the majority of the features of the walking tasks belonged to the first category and had less than 25% samples with $p > 0.05$. However, upstairs and downstairs samples

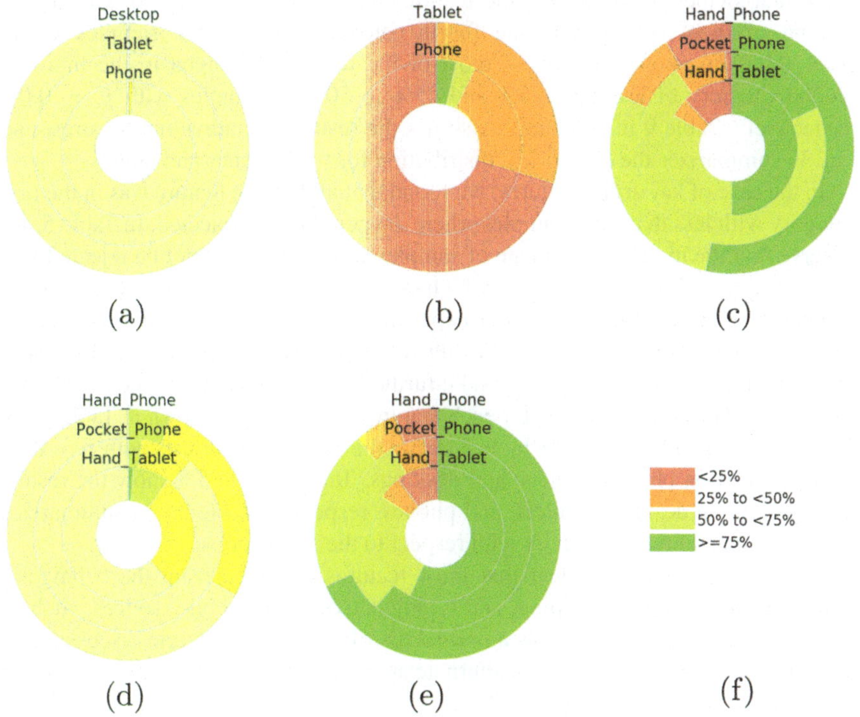

Fig. 3 Illustration summarizing the amount of features in each activity and percentage of their samples with $p > 0.05$, or in other words, where the null hypothesis H_0 that the samples came from a normal distribution could not be discarded. The categories and their corresponding color codes are red, less than 25% of samples with p-value >0.05; orange, 25% to 50% of samples with p-value >0.05; yellow, 50% to 75% of samples with p-value >0.05; and, green, above 75% samples with p-value >0.05. A full doughnut in the doughnut chart represents all the features for an activity on the labeled device. For example, the outermost doughnut in Fig. 3a represents all the features examined for keystroke latencies on desktop, the second doughnut for tablet, and innermost doughnut for phone, respectively. The area covered by each color/category on a doughnut represents the amount of features that in the color/category as described above. (**a**) Typing. (**b**) Swiping. (**c**) Upstairs. (**d**) Walking. (**e**) Downstairs. (**f**) Legend

had a considerable number of features in the third and fourth categories: 50% to 75% and above 75% samples where the H_0 could not be rejected, respectively. This phenomenon was observed on all three devices used for these tasks: hand-phone, pocket-phone, and hand-tablet. We posit that this could be due to individuals walking at different paces on a flat floor. Thus, there might be differences between step sizes, as some may be large and others may be short. In contrast, when people go downstairs or upstairs, they are more likely to keep each step the same since they need to maintain their balance to avoid falling on the stairs, increasing the possibility of data conforming to the normal distribution. Individual values for the features of walking, upstairs, and downstairs are shown in Appendix A, in Tables 10, 11, and 12, respectively. Their summary is illustrated in Figs. 3d, e, and c, respectively.

In the case of swiping activity, we first separated the samples into vertical and horizontal swipes before testing the features for normality. We found a negligible amount of them belonged to the fourth category (above 75% samples where $p > 0.05$) on phones. However, most of the features belonged to the first and second category of less than 25% and 25% to 50% of samples with $p > 0.05$, respectively. Table 9 in Appendix A shows the results for individual features, and Fig. 3b summarizes the categorical distribution for both phones and tablets.

In the case of keystroke features, we found that all but one feature was in the first category with less than 25% samples where H_0 could not be rejected. In Table 5, we observe that the highest percentage of samples where H_0 could not be rejected was for unigraph "n," with about 27%, which is still a very low number of tests when compared to the total number in our experiments and thus can be ignored. As the case with desktop was an expected result following the lines of work in [6], these results reaffirm the findings and extend it further to handheld devices such as phones and tablets. The detailed results for each feature and device are shown in Tables 5, 6, 7, and 8 in Appendix A. While Table 5 shows the percentage of tests with $p > 0.05$ for unigraphs on phones, tablets, and desktops, Table 6, 7, and 8 show the results for digraphs on desktops, tablets, and phones, respectively. Figure 3a summarizes all the features of typing activity with respect to their categories.

From our results, it is clear that most features extracted from the behavioral biometrics data for gait, swiping, and typing activities on mobile devices, such as phones and tablets, do not follow a normal distribution. An elaborate discussion on the implications of our results and alternate approaches of nonnormality is shown in Sect. 7.

6 Analysis on Single-Modality Datasets

To further substantiate our observations discussed in the previous sections, we keep the null hypothesis H_0 that "the sample comes from a normal distribution" and perform similar analyses on six single-modality datasets collected from mobile or wearable devices and one typing dataset collected from desktops. For datasets collected from mobile or wearable devices, we choose two datasets for each type of modality (gait, swipe, and keystroke). Our observations from these single-modality

Table 4 Analysis for underlying normal distribution in data on seven single-modality datasets. First category, less than 25% of samples with p-value >0.05; second category, 25% to 50% of samples with p-value >0.05; third category, 50% to 75% of samples with p-value >0.05; and fourth category, above 75% samples with p-value >0.05

		Dataset	Details	Observations
Mobile and Wearable devices	Gait	[115]	Two sessions, 20 users Walking for 15 minutes/session Sensor: mobile phone in pocket	All the features have 0% of samples with $p > 0.05$. Appendix B Table 13
		[116]	One session, 21 users Walking at 12 different speeds Walking for one minute/speed Sensor: markers on shoes No. of sensors: Four sensors/shoe	Most samples in the third category. About 25% samples in the first category. Appendix B Table 14
	Swipe	[118]	One session More than 80 swipes/user No. of users: 66	Half samples in the first category. The other half in the second category. Appendix C Table 15
		[117]	Two sessions At least 80 swipes/user No. of users: Portrait–Horiz.: 106, Vert. 118 Landscape–Horiz.: 41, Vert. 50	Most samples in the first category with 0% of them have $p > 0.05$. Appendix C Table 16
	Typing	[125]	56 users, 51 trials/user Fixed password (.tie5Roanl)	Unigraph: most samples in the fourth category. Digraph: most flight1 features in the fourth category and all of the other features in the first category. Appendix D Tables 17 and 18
		[126]	54 users At least 3 sessions/user At lease 60 trails/user Fixed password (.tie5Roanl)	Unigraph: most samples in the fourth category. Digraph: all features in the first category. Appendix D Tables 19 and 20
Desktop	Typing	[119]	Eight sessions, 51 users 400 password samples/user Fixed password (.tie5Roanl)	Unigraph: all features have less than 30% with $p > 0.05$. Digraph: nearly all features have 0% with $p > 0.05$. Appendix E Tables 21 and 22

datasets were similar to the ones we made on the BBMAS dataset. The datasets and observations are summarized in Table 4. Elaborate details and observations follow.

6.1 Analysis on Two Gait Datasets

For the first gait dataset [115], most features come from both sensors into the first category, which has less than 25% samples with $p > 0.05$. It shows that we can reject H_0 at a significance level of 5%. This result is consistent with our observation on our dataset [19]. For the second gait dataset [116], we found that a large proportion of features belong to the third and fourth category: 50% to 75% and above 75% samples that have $p > 0.05$. Only around one-fourth of features belong to the first category: less than 25% with $p > 0.05$. This result is similar to our observations in upstairs and downstairs data in BB-MAS dataset [19]. One possible explanation for this observation is similar to the case of going upstairs and downstairs that we discussed in Sect. 5. Participants in [116] were asked to walk at pre-set speeds on a treadmill. Thus, the step sizes between steps would have little difference within each fixed speed, resulting in all steps being in a smaller range, which increases the possibility of the data conforming to the normal distribution. Detailed results for these two datasets are shown in Appendix B, in Tables 13 and 14, respectively.

6.2 Analysis on Two Swiping Datasets

Most of the features are for both datasets [118] and [117] into the first category, with less than 25% samples with $p > 0.05$. Very few, a negligible number of features are observed in the second category of 50% samples with $p > 0.05$. It shows that we can reject H_0 at a significance level of 5% in most cases. This result is consistent with our observation in dataset [19]. Detailed results for these two datasets are shown in Appendix C, in Tables 15 and 16, respectively.

6.3 Analysis on Two Typing Datasets on Mobile Phone and One Typing Dataset on Desktop

For both datasets, we found that most unigraph features are into the fourth category, with more than 75% samples having $p > 0.05$. Moreover, most flight1 features in dataset [125] are also into the fourth category. These results suggest that we cannot reject H_0 at a significance level of 5%. We posit that this occurs due to the inherent differences in the typing surfaces. The on-screen keyboard does not require the user to press the key down to let the keyboard know that this key is pressed. One slight touch is sufficient for mobile devices to know that one key is pressed. Thus, pressing different keys on the on-screen keyboard becomes one single activity: touching or tapping. As users are familiar with touching the screen, features extracted from this action are more likely to conform to the normal distribution. However, most digraph features into the first category, with less than 25% samples with $p > 0.05$. It shows that we can reject H_0 at a significance level of 5%, as flight time between pressing and tapping different keys differs. This result is consistent with our observations for typing data in dataset [19]. Detailed results from our test on these datasets [125] and [126] are shown in Appendix D, in Tables 17, 18, 19, and 20, respectively.

To verify that typing on the physical keyboard is different from typing on the on-screen keyboard, we used one additional typing dataset on desktop collected in [119] to see that whether unigraph and digraph features conform to normal distribution on desktop typing. The typing text used in this dataset is the same as that used in the above two datasets [125] and [126]. Detailed results from our test on this dataset [119] are shown in Appendix E, in Tables 21 and 22, respectively.

We found that almost all features from the dataset [119] are in the first category, with less than 25% samples with p > 0.05. It shows that we can reject H_0 at a significance level of 5%. This result is consistent with our observations for typing data in our dataset [19]. In cases of keyhold time for ".", "t", and "Right shift", we observed a slightly higher percentage of samples, but still within 30%, with $p > 0.05$. Results from this dataset suggest that typing on the physical keyboard differs significantly from typing on the on-screen keyboard. Additionally, results from these three datasets show that extraction from different devices can lead to entirely different distributions even for the same features, further illustrating that we need to be cautious about normality assumption on every feature.

7 Conclusion and Discussion on Alternate Approaches

The implications of assuming the normal distribution in data when the data is actually from a different distribution have been studied across various domains [7–11]. If methods wrongly assume a normal distribution, the findings may be misleading or wrong. In the past, several studies in behavioral biometrics have assumed a normal distribution in data [12, 13] and could improve results by either extracting features that followed normal distribution or by implementing methods more suitable for nonnormal distributions [14]. Our experiments show that it would be wrong to assume an underlying normal distribution in the case of gait and swipes using phones and tablets and keystrokes using desktops. Low values of p from our nonparametric normality tests across activities and devices show that studies in mobile device-based behavioral biometrics must not assume the data to be from a Gaussian distribution to get better and more accurate insights. However, upstairs and downstairs activity data and some typing on the on-screen data showing higher percentages of samples where an underlying normal distribution cannot be discarded are intriguing. Further elaborate research is needed to establish why this occurs. Knowing that the data does not follow the normal distribution leaves the discussion incomplete, which can only be completed by learning alternate ways to handle a nonnormal dataset.

Alternate Approaches and Solutions One should first test for conditions of normality in data before making such an assumption. If the conditions are not met, there are numerous ways to work around the absence of normal distribution. We discuss concepts that have been successfully applied in related fields and other intuitive methods to perform more accurate analyses of mobile device-based behavioral biometric data. Several techniques are found in the literature regarding the analysis and modeling of the data itself. Different types of distributions like Weibull, gamma, exponential, or Pareto distributions have been used, with theoretical and empirical justifications [95, 96]. However, a goodness-of-fit test beforehand is advised. Methods like heteroskedastic corrected covariance matrix and bootstrapping are forms of altering the estimator to describe nonnormal data better. Winsorizing and trimming data is an intermediate technique that replaces the parameters of the original data by virtue of replacing extremities in a sample. Data transformation methods like Johnson transformation [97], Box-Cox transformations [98], and other forms of algorithmic and parametric transformations [94] have shown promising results in other data-intensive research.

Apart from data modeling and transformation, attention to the choice of classifiers can significantly improve performance in research involving identification, verification, or classification tasks. It is common to use Gaussian naive Bayes classifiers or linear classifiers with linear discriminant analysis, which are prevalent tools for baseline metrics. However, these classifiers assume that the data has an underlying normal distribution, and the lack of such a distribution can cause their performance to deteriorate heavily. Modified, nonparametric versions of the Gaussian naive Bayes classifier described in [99] have shown to perform well. Standard classifiers that are not designed with the assumption of normality in data, like support vector machines, K-nearest neighbor classifiers, or neural networks, are intuitively a better choice in mobile device-based behavioral biometrics, such as gait and swipes on mobile devices.

Our results question the common assumption that the data in mobile device-based behavioral biometrics, such as gait and swipes, follows a normal distribution. We have discussed the implications and alternate approaches for such scenarios. We hope that insights from our work help future studies make the right choices in data models, transformations, and classifiers to achieve better results and correct interpretations. We come full circle to the question that we began with "Should the *assumption of normality* be the *norm* in mobile device-based behavioral biometrics?", and equipped with the knowledge from our experiments discussed above, we answer, *"no,"* with a caveat that careful examination of the validity of assumptions about underlying distributions is a must.

Acknowledgments A bold run-in heading in small font size at the end of the paper is used for general acknowledgments, for example: This study was funded by X (grant number Y).

Disclosure of Interests The authors have no competing interests to declare that are relevant to the content of this article.

For visual clarity, cells in each table in all following appendixes are colored in four different colors based on four categories we defined in Sect. 4. These four categories and their corresponding colors are first category, less than 25% of samples with p-value >0.05; second category, 25% to 50% of samples with p-value >0.05; third category, 50% to 75% of samples with p-value >0.05; and fourth category, above 75% samples with p-value >0.05.

A Detailed Results From Our Experiments On Each Feature, Activity, And Device In SU-AIS BB-MAS [19] Dataset

We use symbols $d1$ through $d18$ to denote the digraphs ($d1$ [“BACKSPACE,” “BACKSPACE”], $d2$ [“SPACE,” “a”], $d3$ [“SPACE,” “i”], $d4$ [“SPACE,” “s”], $d5$ [“SPACE,” “t”], $d6$ [“e,” “SPACE”], $d7$ [“e,” “n”], $d8$ [“e’, “r”], $d9$ [“e,” “s”], $d10$ [“n,” “SPACE”], $d11$ [“o,” “SPACE”], $d12$ [“o’, “n”], $d13$ [“r’, “e”], $d14$ [“s,” “SPACE”], $d15$ [“s,” “e”], $d16$ [“t,” “SPACE”], $d17$ [“t,” “e”], and $d18$ [“t,” “h”]) (Tables 5, 6, 7, 8, 9 10, 11, and 12).

Table 5 Percentage of test samples with $p > 0.05$ for keyhold feature from unigraphs on desktop, tablet, and phone

Uni.	Desktop	Tablet	Phone
SPACE	0%	1%	11%
Bspace	0%	0%	2%
a	0%	15%	20%
e	0%	0%	5%
h	0%	9%	18%
i	0%	2%	9%
l	0%	9%	15%
n	0%	2%	27%
o	0%	5%	22%
r	0%	7%	22%
s	0%	4%	16%
t	0%	1%	9%

Table 6 Percentage of test samples with $p > 0.05$ for F1–F4: flight1–flight4 features from digraphs on desktop

	Desktop			
Di.	F1	F2	F3	F4
$d1$	0%	0%	0%	0%
$d2$	0%	0%	0%	0%
$d3$	1%	2%	1%	1%
$d4$	0%	0%	0%	0%
$d5$	0%	0%	0%	0%
$d6$	0%	0%	0%	0%
$d7$	2%	3%	3%	2%
$d8$	0%	2%	2%	3%
$d9$	0%	1%	3%	4%
$d10$	0%	0%	1%	0%
$d11$	1%	1%	1%	1%
$d12$	3%	2%	2%	4%
$d13$	0%	2%	1%	3%
$d14$	0%	0%	0%	1%
$d15$	0%	1%	0%	0%
$d16$	0%	0%	0%	0%
$d17$	2%	5%	4%	6%
$d18$	0%	2%	0%	0%

Table 7 Percentage of test samples with $p > 0.05$ for F1–F4: flight1–flight4 features from digraphs on tablet

	Tablet			
Di.	F1	F2	F3	F4
*d*1	0%	0%	0%	0%
*d*2	0%	0%	0%	0%
*d*3	2%	1%	1%	1%
*d*4	1%	1%	2%	1%
*d*5	0%	0%	0%	0%
*d*6	1%	1%	1%	1%
*d*7	5%	7%	6%	9%
*d*8	9%	7%	13%	11%
*d*9	5%	3%	4%	4%
*d*10	3%	1%	2%	3%
*d*11	0%	1%	0%	1%
*d*12	8%	6%	3%	9%
*d*13	5%	4%	5%	5%
*d*14	0%	0%	0%	1%
*d*15	4%	3%	3%	4%
*d*16	0%	0%	0%	0%
*d*17	18%	17%	16%	19%
*d*18	8%	6%	3%	3%

Table 8 Percentage of test samples with $p > 0.05$ for F1–F4: flight1–flight4 features from digraphs on phone

	Phone			
Di.	F1	F2	F3	F4
*d*1	0%	0%	0%	0%
*d*2	1%	0%	2%	0%
*d*3	0%	0%	1%	1%
*d*4	1%	1%	0%	0%
*d*5	0%	0%	0%	0%
*d*6	0%	1%	1%	1%
*d*7	6%	5%	8%	8%
*d*8	15%	17%	22%	23%
*d*9	5%	8%	6%	11%
*d*10	0%	2%	1%	1%
*d*11	1%	3%	2%	4%
*d*12	14%	16%	15%	17%
*d*13	7%	5%	9%	12%
*d*14	0%	3%	1%	3%
*d*15	0%	6%	2%	8%
*d*16	1%	1%	1%	1%
*d*17	18%	20%	19%	17%
*d*18	6%	4%	6%	6%

Table 9 Percentage of test samples with $p > 0.05$ for features from swiping activity on phone and tablet

Features	Phone	Tablet	Features	Phone	Tablet
minx	2%	3%	aquarts_0	1%	7%
miny	21%	32%	aquarts_1	0%	0%
maxx	5%	2%	aquarts_2	0%	0%
maxy	6%	16%	pmean	72%	26%
euclidist	40%	26%	pstd	76%	33%
tanangle	0%	1%	pquarts_0	65%	29%
tottime	0%	0%	pquarts_1	65%	26%
vmean	2%	9%	pquarts_2	66%	21%
vstd	6%	14%	areamean	56%	33%
vquarts_0	0%	1%	areastd	46%	37%
vquarts_1	0%	5%	areaquarts_0	9%	1%
vquarts_2	2%	14%	areaquarts_1	4%	0%
amean	0%	0%	areaquarts_2	4%	0%
astd	0%	0%			

Table 10 Percentage of test samples with $p > 0.05$ for features from walking activity phone in hand, phone in pocket and tablet in hand (Acc, accelerometer; Gyr, gyroscope; DT, desktop; HP, hand-phone; PP, pocket-phone; HT, hand-tablet)

Feature	HP		PP		HT		Feature	HP		PP		HT	
	Acc	Gyr	Acc	Gyr	Acc	Gyr		Acc	Gyr	Acc	Gyr	Acc	Gyr
xmean	36%	25%	21%	6%	32%	28%	ziqr	50%	24%	30%	17%	50%	27%
ymean	39%	9%	28%	6%	38%	11%	miqr	53%	11%	12%	15%	54%	4%
zmean	32%	4%	16%	12%	25%	2%	xrange	29%	36%	9%	3%	36%	38%
mmean	33%	18%	28%	8%	34%	10%	yrange	26%	17%	10%	9%	26%	28%
xstd	35%	41%	9%	3%	37%	44%	zrange	30%	23%	9%	11%	27%	25%
ystd	21%	15%	9%	13%	31%	32%	mrange	32%	20%	7%	4%	23%	13%
zstd	37%	25%	8%	7%	38%	19%	xsnr	17%	10%	5%	1%	15%	12%
mstd	38%	16%	5%	9%	40%	4%	ysnr	12%	13%	3%	3%	11%	3%
xbp	3%	9%	35%	11%	4%	12%	zsnr	7%	8%	3%	4%	5%	12%
ybp	15%	2%	31%	15%	11%	2%	msnr	7%	31%	1%	9%	3%	32%
zbp	37%	3%	27%	26%	25%	0%	xydtw	17%	24%	3%	15%	17%	28%
mbp	31%	5%	24%	13%	36%	3%	yzdtw	18%	16%	3%	8%	4%	15%
xenergy	7%	18%	15%	3%	11%	27%	xzdtw	3%	5%	9%	13%	2%	2%
yenergy	26%	3%	3%	12%	23%	4%	xymi	0%	12%	0%	1%	9%	23%
zenergy	0%	2%	15%	21%	0%	0%	xzmi	0%	5%	2%	1%	3%	17%
menergy	0%	8%	0%	2%	0%	3%	xmmi	0%	0%	2%	2%	0%	2%
xmfreq	3%	0%	14%	8%	0%	0%	yzmi	0%	9%	2%	1%	3%	11%
ymfreq	1%	0%	17%	6%	0%	0%	ymmi	0%	0%	0%	2%	1%	1%
zmfreq	6%	1%	11%	3%	0%	0%	zmmi	0%	0%	2%	1%	0%	2%
mmfreq	65%	58%	62%	68%	98%	74%	xycorr	55%	50%	19%	8%	65%	60%
xiqr	26%	38%	26%	10%	35%	52%	yzcorr	45%	56%	12%	11%	59%	53%
yiqr	29%	16%	14%	22%	38%	37%	xzcorr	46%	16%	12%	9%	57%	42%

Table 11 Percentage of test samples with $p > 0.05$ for features from upstairs activity with phone in hand, phone in pocket and tablet in hand (Acc, accelerometer; Gyr, gyroscope; DT, desktop; HP, hand-phone; PP, pocket-phone; HT, hand-tablet)

Feature	HP		PP		HT		Feature	HP		PP		HT	
	Acc	Gyr	Acc	Gyr	Acc	Gyr		Acc	Gyr	Acc	Gyr	Acc	Gyr
xmean	79%	76%	62%	67%	74%	78%	ziqr	79%	83%	76%	69%	85%	83%
ymean	83%	60%	77%	20%	79%	56%	miqr	87%	67%	70%	70%	85%	52%
zmean	89%	16%	56%	51%	80%	11%	xrange	70%	68%	50%	58%	65%	74%
mmean	89%	63%	75%	60%	91%	44%	yrange	68%	70%	51%	62%	69%	75%
xstd	68%	79%	67%	60%	72%	83%	zrange	73%	80%	60%	56%	72%	73%
ystd	75%	76%	69%	64%	72%	79%	mrange	72%	60%	53%	53%	70%	71%
zstd	79%	85%	66%	62%	88%	85%	xsnr	78%	74%	61%	68%	74%	80%
mstd	85%	63%	65%	57%	91%	61%	ysnr	81%	62%	44%	27%	85%	59%
xbp	67%	72%	67%	62%	61%	68%	zsnr	72%	19%	63%	66%	63%	12%
ybp	74%	49%	77%	45%	64%	59%	msnr	71%	78%	53%	74%	59%	85%
zbp	85%	21%	69%	56%	80%	19%	xydtw	69%	79%	74%	73%	62%	77%
mbp	85%	44%	74%	56%	86%	28%	yzdtw	86%	46%	71%	55%	78%	38%
xenergy	68%	71%	68%	62%	62%	67%	xzdtw	81%	30%	74%	69%	76%	20%
yenergy	74%	49%	76%	45%	60%	59%	xymi	86%	88%	74%	78%	90%	86%
zenergy	85%	21%	69%	56%	76%	19%	xzmi	89%	85%	79%	77%	87%	85%
menergy	85%	46%	76%	56%	71%	28%	xmmi	87%	80%	79%	74%	87%	90%
xmfreq	31%	11%	61%	44%	4%	19%	yzmi	87%	90%	79%	74%	88%	85%
ymfreq	25%	16%	63%	43%	16%	11%	ymmi	89%	84%	67%	75%	91%	89%
zmfreq	36%	16%	52%	26%	3%	5%	zmmi	88%	83%	76%	79%	81%	82%
mmfreq	86%	81%	76%	79%	96%	91%	xycorr	88%	92%	67%	61%	86%	91%
xiqr	74%	80%	74%	60%	74%	86%	yzcorr	85%	91%	63%	56%	87%	85%
yiqr	71%	79%	74%	74%	70%	79%	xzcorr	91%	85%	69%	46%	88%	91%

Table 12 Percentage of test samples with $p > 0.05$ for features from downstairs activity phone in hand, phone in pocket and tablet in hand. (Acc, accelerometer; Gyr, gyroscope; DT, desktop; HP, hand-phone; PP, pocket-phone; HT, hand-tablet)

Feature	HP		PP		HT		Feature	HP		PP		HT	
	Acc	Gyr	Acc	Gyr	Acc	Gyr		Acc	Gyr	Acc	Gyr	Acc	Gyr
xmean	84%	85%	74%	72%	74%	85%	ziqr	85%	85%	85%	84%	89%	78%
ymean	85%	61%	81%	14%	83%	61%	miqr	90%	74%	82%	78%	88%	59%
zmean	88%	24%	57%	76%	84%	21%	xrange	79%	74%	67%	64%	70%	74%
mmean	90%	78%	90%	72%	91%	56%	yrange	77%	65%	64%	71%	70%	79%
xstd	81%	87%	79%	80%	74%	87%	zrange	72%	82%	69%	70%	76%	79%
ystd	86%	74%	83%	75%	78%	87%	mrange	77%	72%	71%	56%	74%	76%
zstd	89%	85%	82%	76%	89%	83%	xsnr	90%	76%	72%	72%	79%	72%
mstd	91%	83%	80%	74%	87%	68%	ysnr	78%	57%	50%	19%	81%	61%
xbp	77%	72%	80%	77%	60%	79%	zsnr	56%	29%	74%	77%	56%	32%
ybp	82%	59%	85%	50%	67%	70%	msnr	53%	84%	54%	74%	55%	79%
zbp	86%	32%	85%	76%	81%	14%	xydtw	81%	76%	85%	73%	71%	87%
mbp	91%	62%	90%	67%	89%	48%	yzdtw	84%	73%	79%	67%	84%	44%
xenergy	77%	71%	80%	75%	60%	77%	xzdtw	85%	60%	82%	78%	78%	27%
yenergy	81%	61%	87%	48%	68%	70%	xymi	85%	84%	79%	78%	85%	85%
zenergy	87%	31%	85%	76%	81%	15%	xzmi	89%	85%	80%	76%	85%	80%
menergy	89%	62%	88%	68%	88%	44%	xmmi	90%	83%	80%	74%	91%	75%
xmfreq	24%	20%	68%	50%	6%	8%	yzmi	83%	85%	80%	75%	78%	85%
ymfreq	27%	21%	68%	40%	16%	19%	ymmi	87%	89%	75%	80%	88%	88%
zmfreq	48%	15%	72%	33%	1%	12%	zmmi	89%	84%	79%	78%	71%	84%
mmfreq	91%	91%	83%	88%	98%	92%	xycorr	86%	93%	78%	79%	88%	91%
xiqr	86%	91%	83%	82%	70%	90%	yzcorr	87%	93%	77%	79%	82%	89%
yiqr	82%	85%	84%	79%	73%	89%	xzcorr	91%	85%	84%	76%	89%	92%

B Detailed Results from Our Experiments on Two Gait Datasets [115] and [116]

See Tables 13 and 14.

Table 13 Percentage of test samples with $p > 0.05$ for features from gait dataset [115]

Features	Acc	Mag	Features	Acc	Mag
xmean	0%	0%	ziqr	0%	0%
ymean	0%	0%	miqr	0%	0%
zmean	0%	0%	xrange	0%	0%
mmean	0%	0%	yrange	0%	0%
xstd	0%	0%	zrange	0%	0%
ystd	0%	0%	mrange	0%	0%
zstd	0%	0%	xsnr	0%	0%
mstd	0%	0%	ysnr	0%	0%
xbp	0%	0%	zsnr	0%	0%
ybp	0%	0%	msnr	0%	0%
zbp	0%	0%	xydtw	0%	1%
mbp	0%	0%	yzdtw	0%	0%
xenergy	0%	0%	xzdtw	0%	0%
yenergy	0%	0%	xymi	0%	0%
zenergy	0%	0%	xzmi	0%	0%
menergy	0%	0%	xmmi	0%	0%
xmfreq	0%	0%	yzmi	0%	0%
ymfreq	0%	0%	ymmi	0%	0%
zmfreq	0%	0%	zmmi	0%	0%
mmfreq	5%	0%	xycorr	0%	0%
xiqr	0%	0%	yzcorr	0%	0%
yiqr	0%	0%	xzcorr	0%	0%

Table 14 Percentage of test samples with $p > 0.05$ for features from GaitPhase dataset[116]

Features	L_FCC	L_FM1	L_FM2	L_FM5	R_FCC	R_FM1	R_FM2	R_FM5
xmean	73%	73%	72%	73%	68%	72%	71%	73%
ymean	78%	77%	77%	77%	78%	78%	78%	78%
zmean	14%	68%	59%	52%	16%	66%	57%	60%
mmean	72%	74%	72%	74%	70%	72%	72%	70%
xstd	80%	61%	63%	63%	85%	64%	67%	67%
ystd	51%	46%	44%	40%	42%	45%	46%	39%
zstd	11%	78%	72%	59%	13%	75%	70%	59%
mstd	87%	76%	74%	75%	88%	76%	75%	76%
xbp	69%	71%	72%	74%	69%	71%	70%	72%
ybp	77%	76%	75%	76%	78%	79%	79%	78%
zbp	12%	70%	62%	51%	13%	65%	57%	60%
mbp	71%	72%	72%	72%	69%	72%	70%	72%
xenergy	71%	73%	72%	71%	71%	73%	72%	72%
yenergy	78%	76%	76%	77%	78%	80%	80%	79%
zenergy	11%	71%	64%	51%	13%	65%	57%	60%
menergy	71%	71%	72%	73%	68%	71%	71%	72%
xmfrep	78%	75%	74%	74%	68%	71%	70%	72%
ymfrep	76%	73%	73%	76%	76%	74%	74%	71%
zmfrep	31%	19%	13%	15%	31%	26%	16%	16%
mmfrep	74%	73%	74%	72%	70%	71%	72%	68%
xiqr	63%	20%	21%	20%	65%	21%	22%	22%
yiqr	20%	21%	21%	17%	19%	21%	21%	14%
ziqr	6%	51%	51%	36%	6%	59%	55%	37%
miqr	80%	55%	57%	60%	82%	48%	46%	46%
xrange	54%	54%	54%	57%	62%	57%	59%	60%
yrange	60%	60%	61%	61%	62%	60%	59%	62%
zrange	68%	74%	76%	76%	66%	76%	73%	72%
mrange	68%	67%	68%	66%	75%	72%	73%	73%
xsnr	0%	0%	0%	0%	1%	0%	0%	0%
ysnr	42%	38%	35%	26%	43%	29%	31%	26%
zsnr	7%	1%	2%	1%	6%	1%	3%	1%
msnr	3%	1%	1%	0%	3%	0%	0%	1%
xydtw	66%	73%	69%	69%	68%	71%	73%	73%
yzdtw	46%	64%	59%	66%	51%	64%	49%	62%
xzdtw	83%	79%	80%	81%	84%	82%	83%	81%
xymi	68%	71%	70%	72%	71%	72%	70%	71%
xzmi	74%	75%	76%	75%	72%	73%	73%	74%
xmmi	3%	14%	3%	29%	4%	10%	2%	25%
yzmi	33%	47%	44%	47%	37%	46%	39%	50%
ymmi	64%	88%	87%	83%	61%	87%	83%	85%
zmmi	82%	73%	77%	67%	84%	72%	74%	68%
xycorr	71%	67%	70%	70%	64%	58%	57%	53%
yzcorr	2%	13%	2%	28%	4%	9%	1%	24%
xzcorr	74%	71%	72%	73%	73%	71%	71%	73%

C Detailed Results from Our Experiments on Two Swipe Datasets [118] and [117]

See Tables 15 and 16.

Table 15 Percentage of test samples with $p > 0.05$ for features from swiping activity on phone for dataset [118]

Features	Phone	Features	Phone
minx	27%	aquarts_0	18%
miny	48%	aquarts_1	22%
maxx	26%	aquarts_2	22%
maxy	41%	pmean	61%
euclidist	46%	pstd	47%
tanangle	14%	pquarts_0	41%
tottime	12%	pquarts_1	35%
vmean	24%	pquarts_2	36%
vstd	26%	areamean	32%
vquarts_0	18%	areastd	26%
vquarts_1	22%	areaquarts_0	25%
vquarts_2	27%	areaquarts_1	20%
amean	24%	areaquarts_2	18%

Table 16 Percentage of test samples with $p > 0.05$ for features from swiping activity on phone in portrait and landscape for dataset [117]

Features	Portrait	Landscape	Features	Portrait	Landscape
minx	0%	1%	aquarts_0	0%	0%
miny	4%	2%	aquarts_1	0%	0%
maxx	8%	2%	aquarts_2	1%	1%
maxy	17%	3%	pmean	48%	58%
euclidist	47%	51%	pstd	18%	13%
tanangle	0%	0%	pquarts_0	36%	73%
tottime	0%	0%	pquarts_1	41%	24%
vmean	0%	0%	pquarts_2	52%	59%
vstd	1%	0%	areamean	31%	46%
vquarts_0	0%	0%	areastd	6%	13%
vquarts_1	0%	0%	areaquarts_0	0%	0%
vquarts_2	1%	0%	areaquarts_1	0%	0%
amean	0%	0%	areaquarts_2	0%	0%

D Detailed Results from Our Experiments on Two Mobile Phone Keystroke Datasets [125] and [126]

See Tables 17, 18, 19, and 20.

Table 17 Percentage of test samples with $p > 0.05$ for keyhold from unigraphs in desktop dataset [125]

Uni.	Phone
PERIOD	76%
t	84%
i	78%
e	74%
SHIFT	78%
5	82%
CAPS	77%
r	80%
o	78%
a	77%
n	80%
l	76%
ENTER	71%

Table 18 Percentage of test samples with $p > 0.05$ for F1–F4: flight1–flight4 from digraphs in desktop dataset [125]

	Phone			
Di.	F1	F2	F3	F4
(PERIOD, t)	76%	5%	5%	6%
(t, i)	82%	15%	13%	17%
(I, e)	79%	13%	13%	10%
(e, SHIFT)	74%	3%	3%	3%
(SHIFT, 5)	78%	7%	6%	6%
(5, SHIFT)	83%	4%	3%	4%
(SHIFT, CAPS)	79%	1%	2%	1%
(CAPS, r)	78%	13%	12%	15%
(r, o)	80%	7%	7%	6%
(o, a)	78%	12%	10%	10%
(a, n)	77%	15%	12%	17%
(n, l)	81%	19%	19%	23%
(l, ENTER)	74%	5%	6%	6%

Table 19 Percentage of test samples with $p > 0.05$ for keyhold from unigraphs in desktop dataset [126]

Uni.	Phone
PERIOD	79%
t	81%
i	82%
e	75%
123?	77%
5	73%
abc	81%
SHIFT	75%
R	75%
o	82%
a	77%
n	80%
l	85%

Table 20 Percentage of test samples with $p > 0.05$ for F1–F4: flight1–flight4 from digraphs in desktop dataset [126]

Di.	Desktop			
	F1	F2	F3	F4
(PERIOD, t)	1%	1%	1%	3%
(t, i)	9%	11%	12%	13%
(I, e)	6%	7%	6%	7%
(e, 123?)	1%	1%	1%	1%
(123?, 5)	3%	3%	3%	4%
(5, abc)	1%	1%	2%	2%
(abc, SHIFT)	4%	4%	6%	7%
(SHIFT, R)	19%	17%	18%	19%
(R, o)	2%	2%	4%	3%
(o, a)	7%	9%	7%	8%
(a, n)	5%	7%	10%	11%
(n, l)	8%	10%	10%	13%

E Detailed Results from Our Experiments on a Desktop Keystroke Dataset [119]

See Tables 21 and 22.

Table 21 Percentage of test samples with $p > 0.05$ for keyhold feature from unigraphs on desktop from dataset [119]

Uni.	Desktop
PERIOD	29%
t	25%
i	15%
e	14%
5	14%
RIGHT SHIFT	21%
o	18%
a	18%
n	15%
l	10%
RETURN	13%

Table 22 Percentage of test samples with $p > 0.05$ for F1–F4: flight1–flight4 features from digraphs on desktop from dataset [119]

Di.	Desktop			
	F1	F2	F3	F4
(PERIOD, t)	0%	0%	0%	0%
(t, i)	0%	1%	0%	1%
(I, e)	0%	0%	0%	0%
(e, 5)	0%	0%	0%	0%
(5, RIGHT SHIFT)	0%	0%	0%	0%
(RIGHT SHIFT, o)	0%	0%	0%	0%
(o, a)	0%	0%	0%	0%
(a, n)	0%	0%	0%	0%
(n, l)	0%	0%	0%	0%
(l, RETURN)	0%	0%	0%	0%

References

1. Lilliefors, H.W.: On the Kolmogorov-Smirnov test for normality with mean and variance unknown. Journal of the American statistical Association 62(318), 399–402 (1967)
2. SHAPIRO, S.S., WILK, M.B.: An analysis of variance test for normality. Biometrika 52(3-4), 591–611 (1965). https://doi.org/10.1093/biomet/52.3-4.591
3. Massey Jr, F.J.: The Kolmogorov-Smirnov test for goodness of fit. Journal of the American statistical Association 46(253), 68–78 (1951)
4. Lan, K.-C., Heidemann, J.: Rapid model parameterization from traffic measurements. ACM Transactions on Modeling and Computer Simulation (TOMACS) 12(3), 201–229 (2002)
5. Barford, P., Crovella, M.: Generating representative web workloads for network and server performance evaluation. ACM SIGMETRICS Performance Evaluation Review 26(1), 151–160 (1998)
6. Serwadda, A., Phoha, V.V.: Examining a Large Keystroke Biometrics Dataset for Statistical-Attack Openings. ACM Transactions on Information Systems Security 16(2), 8:1–8:30 (2013). https://doi.org/10.1145/2516960
7. Limpert, E., Stahel, W.A.: Problems with using the normal distribution–and ways to improve quality and efficiency of data analysis. PloS one 6(7), e21403-e21403 (2011). https://doi.org/10.1371/journal.pone.0021403
8. Volkova, V.M.: Research of Tukey's test statistic distribution under failure of the normality assumption. In: 2016 13th International Scientific-Technical Conference on Actual Problems of Electronics Instrument Engineering (APEIE), vol. 02, pp. 227–229 (2016). https://doi.org/10.1109/APEIE.2016.7806456
9. Amhemad, A.Z.: Effect of non normality on statistical control charts. In: 2010 International Conference on Networking and Information Technology, pp. 512–515 (2010). https://doi.org/10.1109/ICNIT.2010.5508459
10. Noorossana, R., Vaghefi, A., Dorri, M.: The effect of non-normality on performance of linear profile monitoring. In: 2008 IEEE International Conference on Industrial Engineering and Engineering Management, pp. 262–266 (2008). https://doi.org/10.1109/IEEM.2008.4737871
11. Caballero-Morales, S., Rahim, A.: Analyzing the effect of non-normality on the solution space for the economic statistical design of X-bar control charts. In: 2015 International Conference on Industrial Engineering and Operations Management (IEOM), pp. 1–6 (2015). https://doi.org/10.1109/IEOM.2015.7093766
12. Song, D.X., Wagner, D., Tian, X.: Timing Analysis of Keystrokes and Timing Attacks on SSH. In: Proceedings of the 10th Conference on USENIX Security Symposium—Volume 10, p. 25. USENIX Association, Washington, D.C. (2001)
13. Stefan, D., Shu, X., Yao, D.(D.): Robustness of keystroke-dynamics based biometrics against synthetic forgeries. Computers & Security 31(1), 109–121 (2012). https://doi.org/https://doi.org/10.1016/j.cose.2011.10.001
14. Li, X., Le, J., Gopalakrishnan, P., Pileggi, L.T.: Asymptotic probability extraction for nonnormal performance distributions. IEEE Transactions on Computer-Aided Design of Integrated Circuits and Systems 26(1), 16–37 (2006)
15. Stoica, P., Babu, P.: The Gaussian Data Assumption Leads to the Largest Cramér-Rao Bound [Lecture Notes]. IEEE Signal Processing Magazine 28(3), 132–133 (2011). https://doi.org/10.1109/MSP.2011.940411
16. Park, S., Serpedin, E., Qaraqe, K.: Gaussian Assumption: The Least Favorable but the Most Useful [Lecture Notes]. IEEE Signal Processing Magazine 30(3), 183–186 (2013). https://doi.org/10.1109/MSP.2013.2238691
17. Kwan, A.C.M.: Validity of normality assumption in CSP research. In: Foo, N., Goebel, R. (eds.) PRICAI'96: Topics in Artificial Intelligence, LNCS, vol. 1114, pp. 253–263. Springer, Berlin, Heidelberg (1996). https://doi.org/10.1007/3-540-61516-3_25
18. Olson Jr, C.E.: The fallacy of normality in remotely sensed data. In: Proceedings of the ASPRS Annual Conference, pp. 9–13 (2009)

19. Belman, A.K., Wang, L., Iyengar, S.S., Sniatala, P., Wright, R., Dora, R., Baldwin, J., Jin, Z., Phoha, V.V.: SU-AIS BB-MAS (Syracuse University and Assured Information Security—Behavioral Biometrics Multi-device and multi-Activity data from Same users) Dataset (2019). https://doi.org/10.21227/rpaz-0h66
20. Sim, T., Janakiraman, R.: Are Digraphs Good for Free-Text Keystroke Dynamics? In: 2007 IEEE Conference on Computer Vision and Pattern Recognition, pp. 1–6 (2007). https://doi.org/10.1109/CVPR.2007.383393
21. Nguyen, T., Voris, J.: Touchscreen Biometrics Across Multiple Devices. In: Symposium on Usable Privacy and Security,USENIX (2017)
22. Alsultan, A., Warwick, K., Wei, H.: Non-conventional keystroke dynamics for user authentication. Pattern Recognition Letters (2017)
23. Balagani, K., Phoha, V., Ray, A., Phoha, S.: On the discriminability of keystroke feature vectors used in fixed text keystroke authentication. Pattern Recognition Letters (2011)
24. Yu, E., Cho, S.: Keystroke dynamics identity verification its problems and practical solutions. Computers and security (2004)
25. Obaidat, M., Sadoun, B.: Verification of Computer Users Using Keystroke Dynamics. IEEE Transactions on Systems, Man and Cybernetics. Part B: Cybernetics. 27(2) (1997)
26. Ahmed, A.A., Traore, I.: Biometric Recognition Based on Free-Text Keystroke Dynamics. IEEE Transactions on Cybernetics 44(4) (2014)
27. Huang, J., Hou, D., Schuckers, S., Upadhyaya, S.: Effects of Text Filtering on Authentication Performance of Keystroke Biometrics. In: IEEE International Workshop on Information Forensics and Security (WIFS) (2016)
28. Mondal, S., Bours, P.: Person Identification by Keystroke Dynamics Using Pairwise User Coupling. IEEE Transactions on Information Forensics and Security 12(6) (2017)
29. Clarke, N.L., Furnell, S.M.: Authenticating mobile phone users using keystroke analysis. Int. J. Inf. Secur. (2006)
30. Bryan, L., Harter, N.: Studies in the physiology and psychology of the telegraphic language. In: Gardner, H., Gardner, J. (eds.) The Psychology of Skill: Three Studies, pp. 35–44. New York Times, New York (1973)
31. Umphress, Williams, G.: Identity verification through keyboard characteristics. Int. J. Man-Mach. Stud., pp. 263–273 (1985)
32. Gunetti, D., Picardi, C.: Keystroke Analysis of Free Text. ACM Transactions on Information Systems Security 8(3), 312–347 (2005). https://doi.org/10.1145/1085126.1085129
33. Delac, K., Grgic, M.: A survey of biometric recognition methods. In: Proceedings. Elmar-2004. 46th International Symposium on Electronics in Marine, pp. 184–193 (2004)
34. Blasco, J., Chen, T.M., Tapiador, J., Peris-Lopez, P.: A Survey of Wearable Biometric Recognition Systems. ACM Computing Surveys 49(3), 43:1–43:35 (2016). https://doi.org/10.1145/2968215
35. Teh, P.S., Teoh, A.B.J., Yue, S.: A Survey of Keystroke Dynamics Biometrics. The Scientific World Journal 2013, 408280 (2013). https://doi.org/10.1155/2013/408280
36. Pisani, P.H., Lorena, A.C.: A systematic review on keystroke dynamics. Journal of the Brazilian Computer Society 19(4), 573–587 (2013). https://doi.org/10.1007/s13173-013-0117-7
37. Janakiraman, R., Sim, T.: Keystroke Dynamics in a General Setting. In: Lee, S.-W., Li, S.Z. (eds.) Advances in Biometrics, LNCS, vol. 4641, pp. 584–593. Springer, Berlin, Heidelberg (2007). https://doi.org/10.1007/978-3-540-74549-5_60
38. Monrose, F., Rubin, A.: Authentication via Keystroke Dynamics. In: Proceedings of the 4th ACM Conference on Computer and Communications Security, pp. 48–56 (1997). https://doi.org/10.1145/266420.266434
39. Hwang, S.-s., Cho, S., Park, S.: Keystroke dynamics-based authentication for mobile devices. Computers & Security 28(1), 85–93 (2009). https://doi.org/https://doi.org/10.1016/j.cose.2008.10.002
40. Joyce, R., Gupta, G.: Identity Authentication Based on Keystroke Latencies. Communications of the ACM 33(2), 168–176 (1990). https://doi.org/10.1145/75577.75582

41. Revett, K., de Magalhães, S.T., Santos, H.M.D.: Enhancing Login Security Through the Use of Keystroke Input Dynamics. In: Zhang, D., Jain, A.K. (eds.) Advances in Biometrics, LNCS, vol. 3832, pp. 661–667. Springer, Berlin, Heidelberg (2005). https://doi.org/10.1007/978-3-540-31621-3_81
42. Ciuffo, F., Weiss, G.M.: Smartwatch-based transcription biometrics. In: 2017 IEEE 8th Annual Ubiquitous Computing, Electronics and Mobile Communication Conference (UEMCON), pp. 145–149 (2017). https://doi.org/10.1109/UEMCON.2017.8249014
43. Gunetti, D., Picardi, C., Ruffo, G.: Dealing with Different Languages and Old Profiles in Keystroke Analysis of Free Text. In: Domingo-Ferrer, J., Posegga, J., Schwenk, J. (eds.) Privacy in Statistical Databases, LNCS, vol. 3673, pp. 347–358. Springer, Berlin, Heidelberg (2005). https://doi.org/10.1007/11550624_27
44. Ali, M.L., Monaco, J.V., Tappert, C.C., Qiu, M.: Keystroke Biometric Systems for User Authentication. Journal of Signal Processing Systems 86(2), 175–190 (2017). https://doi.org/10.1007/s11265-016-1114-9
45. Azevedo, G.L.F.B., Cavalcanti, G.D.C., Carvalho Filho, E.C.B.: An approach to feature selection for keystroke dynamics systems based on PSO and feature weighting. In: Evolutionary Computation, 2007. CEC 2007. IEEE Congress on, pp. 3577–3584 (2007). https://doi.org/10.1109/CEC.2007.4425038
46. Davoudi, H., Kabir, E.: A new distance measure for free text keystroke authentication. In: 2009 14th International CSI Computer Conference, pp. 570–575 (2009). https://doi.org/10.1109/CSICC.2009.5349640
47. Goodkind, A., Brizan, D.G., Rosenberg, A.: Utilizing overt and latent linguistic structure to improve keystroke-basedauthentication. Image and Vision Computing 58, 230–238 (2017). https://doi.org/https://doi.org/10.1016/j.imavis.2016.06.003
48. Lee, H.S., Lau, T.S., Lai, W.K., King, Y.C., Lim, L.L.: User identification of numerical keypad typing patterns with subtractive clustering fuzzy inference. In: 2017 IEEE 15th Student Conference on Research and Development (SCOReD), pp. 83–88 (2017). https://doi.org/10.1109/SCORED.2017.8305418
49. Ho, J., Kang, D.-K.: Mini-batch bagging and attribute ranking for accurate user authentication in keystroke dynamics. Pattern Recognition 70, 139–151 (2017). https://doi.org/https://doi.org/10.1016/j.patcog.2017.05.002
50. Mondal, S., Bours, P.: Person Identification by Keystroke Dynamics Using Pairwise User Coupling. IEEE Transactions on Information Forensics and Security 12(6), 1319–1329 (2017). https://doi.org/10.1109/TIFS.2017.2658539
51. Sulavko, A.E., Eremenko, A.V., Fedotov, A.A.: Users' identification through keystroke dynamics based on vibration parameters and keyboard pressure. In: 2017 Dynamics of Systems, Mechanisms and Machines (Dynamics), pp. 1–7 (2017). https://doi.org/10.1109/Dynamics.2017.8239514
52. Mondal, S., Bours, P.: A study on continuous authentication using a combination of keystroke and mouse biometrics. Neurocomputing 230, 1–22 (2017). https://doi.org/https://doi.org/10.1016/j.neucom.2016.11.031
53. Alpar, O.: Frequency spectrograms for biometric keystroke authentication using neural network based classifier. Knowledge-Based Systems 116, 163–171 (2017). https://doi.org/https://doi.org/10.1016/j.knosys.2016.11.006
54. Tsimperidis, I., Arampatzis, A., Karakos, A.: Keystroke dynamics features for gender recognition. Digital Investigation 24, 4–10 (2018). https://doi.org/https://doi.org/10.1016/j.diin.2018.01.018
55. Ulinskas, M., Damaševičius, R., Maskeliūnas, R., Woźniak, M.: Recognition of human daytime fatigue using keystroke data. Procedia Computer Science 130, 947–952 (2018). https://doi.org/https://doi.org/10.1016/j.procs.2018.04.094
56. Pentel, A.: Predicting User Age by Keystroke Dynamics. In: Silhavy, R. (ed.) Artificial Intelligence and Algorithms in Intelligent Systems, pp. 336–343. Springer, Cham (2019). https://doi.org/978-3-319-91189-2

57. Monaro, M., Galante, C., Spolaor, R., Li, Q.Q., Gamberini, L., Conti, M., Sartori, G.: Covert lie detection using keyboard dynamics. Scientific Reports 8(1), 1976 (2018). https://doi.org/10.1038/s41598-018-20462-6
58. Zulueta, J., Andrea, P., Rasic, M., Easter, R., Babu, P., Langenecker, S., McInnis, M., Ajilore, O., Nelson, P., Ryan, K., et al.: 481. predicting mood disturbance severity in bipolar subjects with mobile phone keystroke dynamics and metadata. Biological Psychiatry 81(10), S196 (2017)
59. Wu, C., Ding, W., Liu, R., Wang, J., Wang, A.C., Wang, J., Li, S., Zi, Y., Wang, Z.L.: Keystroke dynamics enabled authentication and identification using triboelectric nanogenerator array. Materials Today 21(3), 216–222 (2018). https://doi.org/https://doi.org/10.1016/j.mattod.2018.01.006
60. Brizan, D.G., Goodkind, A., Koch, P., Balagani, K., Phoha, V.V., Rosenberg, A.: Utilizing linguistically enhanced keystroke dynamics to predict typist cognition and demographics. International Journal of Human-Computer Studies 82, 57–68 (2015). https://doi.org/https://doi.org/10.1016/j.ijhcs.2015.04.005
61. Liu, J., Wang, Y., Kar, G., Chen, Y., Yang, J., Gruteser, M.: Snooping Keystrokes with Mm-level Audio Ranging on a Single Phone. In: Proceedings of the 21st Annual International Conference on Mobile Computing and Networking, pp. 142–154. ACM, Paris, France (2015). https://doi.org/10.1145/2789168.2790122
62. Zhu, T., Ma, Q., Zhang, S., Liu, Y.: Context-free Attacks Using Keyboard Acoustic Emanations. In: Proceedings of the 2014 ACM SIGSAC Conference on Computer and Communications Security, pp. 453–464. ACM, Scottsdale, Arizona, USA (2014). https://doi.org/10.1145/2660267.2660296
63. Asonov, D., Agrawal, R.: Keyboard Acoustic Emanations. In: 2004 IEEE Symposium on Security and Privacy (S&P 2004), pp. 3–11. IEEE, Berkeley, CA, USA (2004). https://doi.org/10.1109/SECPRI.2004.1301311
64. Sun, Y., Ceker, H., Upadhyaya, S.: Anatomy of secondary features in keystroke dynamics—achieving more with less. In: 2017 IEEE International Conference on Identity, Security and Behavior Analysis (ISBA), pp. 1–6 (2017). https://doi.org/10.1109/ISBA.2017.7947691
65. Sulong, A., Wahyudi, Siddiqi, M.U.: Intelligent keystroke pressure-based typing biometrics authentication system using radial basis function network. In: 2009 5th International Colloquium on Signal Processing Its Applications, pp. 151–155 (2009). https://doi.org/10.1109/CSPA.2009.5069206
66. Serwadda, A., Phoha, V.V.: Examining a Large Keystroke Biometrics Dataset for Statistical-Attack Openings. ACM Transactions on Information Systems Security 16(2), 8:1–8:30 (2013). https://doi.org/10.1145/2516960
67. Chawla, N.V., Bowyer, K.W., Hall, L.O., Kegelmeyer, W.P.: SMOTE: Synthetic Minority Over-sampling Technique. J. Artif. Int. Res. 16(1), 321–357 (2002)
68. Sheng, Y., Phoha, V., Rovnyak, S.M.: A Parallel Decision Tree-Based Method for User Authentication Based on Keystroke Patterns. Systems, Man, and Cybernetics, Part B: Cybernetics, IEEE Transactions on 35(5), 826–833 (2005). https://doi.org/10.1109/TSMCB.2005.846648
69. Fang, S., Markwood, I., Liu, Y., Zhao, S., Lu, Z., Zhu, H.: No Training Hurdles: Fast Training-Agnostic Attacks to Infer Your Typing. In: Proceedings of the 2018 ACM SIGSAC Conference on Computer and Communications Security, pp. 1747–1760. ACM (2018)
70. Oxford University Press: The Oxford English Corpus: Facts about the language.OxfordDictionaries.com. (2011). Accessed August 29, 2019
71. Davies, M.: Corpus of Contemporary American English. (2010). Updated 2017, Accessed August 29, 2019
72. Sun, Y., Upadhyaya, S.: Synthetic Forgery Attack against Continuous Keystroke Authentication Systems. In: 2018 27th International Conference on Computer Communication and Networks (ICCCN), pp. 1–7 (2018). https://doi.org/10.1109/ICCCN.2018.8487341
73. Chen, Y., Li, T., Zhang, R., Zhang, Y., Hedgpeth, T.: EyeTell: Video-Assisted Touchscreen Keystroke Inference from Eye Movements. In: 2018 IEEE Symposium on Security and Privacy (SP), pp. 144–160 (2018). https://doi.org/10.1109/SP.2018.00010

74. Monaco, J.V., Ali, M.L., Tappert, C.C.: Spoofing key-press latencies with a generative keystroke dynamics model. In: 2015 IEEE 7th International Conference on Biometrics Theory, Applications and Systems (BTAS), pp. 1–8 (2015). https://doi.org/10.1109/BTAS.2015.7358795
75. Huang, J., Hou, D., Schuckers, S.: A practical evaluation of free-text keystroke dynamics. In: 2017 IEEE International Conference on Identity, Security and Behavior Analysis (ISBA), pp. 1–8 (2017). https://doi.org/10.1109/ISBA.2017.7947695
76. Stefan, D., Yao, D.: Keystroke-dynamics authentication against synthetic forgeries. In: 6th International Conference on Collaborative Computing: Networking, Applications and Worksharing (CollaborateCom 2010), pp. 1–8 (2010). https://doi.org/10.4108/icst.collaboratecom.2010.16
77. Rahman, K.A., Balagani, K.S., Phoha, V.V.: Snoop-Forge-Replay Attacks on Continuous Verification With Keystrokes. IEEE Transactions on Information Forensics and Security 8(3), 528–541 (2013). https://doi.org/10.1109/TIFS.2013.2244091
78. Lu, L., Yu, J., Chen, Y., Zhu, Y., Xu, X., Xue, G., Li, M.: KeyLiSterber: Inferring Keystrokes on QWERTY Keyboard of Touch Screen through Acoustic Signals. In: IEEE INFOCOM 2019—IEEE Conference on Computer Communications, pp. 775–783 (2019). https://doi.org/10.1109/INFOCOM.2019.8737591
79. Shen, S., Lin, S., Kang, T., Chien, W.: Enhanced keystroke dynamics authentication utilizing pressure detection. In: 2016 International Conference on Applied System Innovation (ICASI), pp. 1–4 (2016). https://doi.org/10.1109/ICASI.2016.7539947
80. Maiti, A., Jadliwala, M., He, J., Bilogrevic, I.: Side-Channel Inference Attacks on Mobile Keypads Using Smartwatches. IEEE Transactions on Mobile Computing 17(9), 2180–2194 (2018). https://doi.org/10.1109/TMC.2018.2794984
81. Stanciu, V.-D., Spolaor, R., Conti, M., Giuffrida, C.: On the Effectiveness of Sensor-enhanced Keystroke Dynamics Against Statistical Attacks. In: Proceedings of the Sixth ACM Conference on Data and Application Security and Privacy, pp. 105–112. ACM, New Orleans, Louisiana, USA (2016). https://doi.org/10.1145/2857705.2857748
82. Jin, K., Fang, S., Peng, C., Teng, Z., Mao, X., Zhang, L., Li, X.: ViViSnoop: Someone is snooping your typing without seeing it! In: 2017 IEEE Conference on Communications and Network Security (CNS), pp. 1–9 (2017). https://doi.org/10.1109/CNS.2017.8228624
83. Khan, H., Hengartner, U., Vogel, D.: Augmented Reality-based Mimicry Attacks on Behaviour-Based Smartphone Authentication. In: Proceedings of the 16th Annual International Conference on Mobile Systems, Applications, and Services, pp. 41–53. ACM, Munich, Germany (2018). https://doi.org/10.1145/3210240.3210317
84. Ali, K., Liu, A.X., Wang, W., Shahzad, M.: Keystroke Recognition Using WiFi Signals. In: Proceedings of the 21st Annual International Conference on Mobile Computing and Networking, pp. 90–102. ACM, Paris, France (2015). https://doi.org/10.1145/2789168.2790109
85. Shukla, D., Phoha, V.V.: Stealing Passwords by Observing Hands Movement. IEEE Transactions on Information Forensics and Security 14(12), 3086–3101 (2019). https://doi.org/10.1109/TIFS.2019.2911171
86. Dunning, T.: Accurate Methods for the Statistics of Surprise and Coincidence. Comput. Linguist. 19(1), 61–74 (1993)
87. Caballero-Morales, S., Rahim, A.: Analyzing the effect of non-normality on the solution space for the economic statistical design of X-bar control charts. In: 2015 International Conference on Industrial Engineering and Operations Management (IEOM), pp. 1–6 (2015). https://doi.org/10.1109/IEOM.2015.7093766
88. Marko, N.F., Weil, R.J.: Non-Gaussian Distributions Affect Identification of Expression Patterns, Functional Annotation, and Prospective Classification in Human Cancer Genomes. PLOS ONE 7(10), 1–15 (2012). https://doi.org/10.1371/journal.pone.0046935
89. Kwak, S.G., Kim, J.H.: Central limit theorem: the cornerstone of modern statistics. Korean journal of anesthesiology 70(2), 144–156 (2017). https://doi.org/10.4097/kjae.2017.70.2.144
90. Cramér, H.: Mathematical methods of statistics. Princeton University Press, Princeton (1999)

91. Stephens, M.A.: EDF statistics for goodness-of-fit: Part 1. STANFORD UNIV CA DEPT OF STATISTICS (1972)
92. Belman, A.K., Wang, L., Iyengar, S.S., Sniatala, P., Wright, R., Dora, R., Baldwin, J., Jin, Z., Phoha, V.V.: Insights from BB-MAS—A Large Dataset for Typing, Gait and Swipes of the Same Person on Desktop, Tablet and Phone (2019). https://doi.org/ArXiv:1912.02736
93. Limpert, E., Stahel, W.A.: Problems with Using the Normal Distribution—and Ways to Improve Quality and Efficiency of Data Analysis. PLOS ONE 6(7), 1–8 (2011). https://doi.org/10.1371/journal.pone.0021403
94. Pek, J., Wong, O., Wong, A.C.M.: How to Address Non-normality: A Taxonomy of Approaches, Reviewed, and Illustrated. Frontiers in Psychology 9, 2104 (2018). https://doi.org/10.3389/fpsyg.2018.02104
95. Adewale, S.: A STATISTICAL ANALYSIS TO DETERMINE THE DISTRIBUTION AND PATTERN FOR AN INSURANCE HEALTH CLAIMS (2017). https://doi.org/10.13140/RG.2.2.20068.42888
96. Liu, J., Wu, Z., Wu, J., Dong, J., Zhao, Y., Wen, D.: A Weibull distribution accrual failure detector for cloud computing. PLOS ONE 12(3), 1–16 (2017). https://doi.org/10.1371/journal.pone.0173666
97. Yang, Y., Li, D., Qi, Y.: An Approach to Non-normal Process Capability Analysis Using Johnson Transformation. In: 2018 IEEE 4th International Conference on Control Science and Systems Engineering (ICCSSE), pp. 495–498 (2018). https://doi.org/10.1109/CCSSE.2018.8724679
98. Yang, Y., Zhu, H.: A Study of Non-Normal Process Capability Analysis Based on Box-Cox Transformation. In: 2018 3rd International Conference on Computational Intelligence and Applications (ICCIA), pp. 240–243 (2018). https://doi.org/10.1109/ICCIA.2018.00057
99. Soria, D., Garibaldi, J.M., Ambrogi, F., Biganzoli, E.M., Ellis, I.O.: A "non-parametric" version of the naive Bayes classifier. Knowledge-Based Systems 24(6), 775–784 (2011). https://doi.org/doi:10.1016/j.knosys.2011.02.014
100. Arora, P., Hanmandlu, M., Srivastava, S.: Gait based authentication using gait information image features. Pattern Recognition Letters 68, 336–342 (2015)
101. Cola, G., Avvenuti, M., Musso, F., Vecchio, A.: Gait-based authentication using a wrist-worn device. In: Proceedings of the 13th International Conference on Mobile and Ubiquitous Systems: Computing, Networking and Services, pp. 208–217 (2016)
102. Belman, A.K., Phoha, V.V.: Discriminative Power of Typing Features on Desktops, Tablets, and Phones for User Identification. ACM Trans. Priv. Secur. 23(1), Article 4 (2020). https://doi.org/10.1145/3377404
103. Kim, J., Kim, H., Kang, P.: Keystroke dynamics-based user authentication using freely typed text based on user-adaptive feature extraction and novelty detection. Applied Soft Computing 62, 1077–1087 (2018). https://doi.org/https://doi.org/10.1016/j.asoc.2017.09.045
104. Bajrami, G.: Activity identification for gait recognition using mobile devices (2011)
105. Derawi, M., Bours, P.: Gait and activity recognition using commercial phones. computers & security 39, 137–144 (2013)
106. Monrose, F., Rubin, A.D.: Keystroke dynamics as a biometric for authentication. Future Generation computer systems 16(4), 351–359 (2000)
107. Brizan, D.G., Goodkind, A., Koch, P., Balagani, K., Phoha, V.V., Rosenberg, A.: Utilizing linguistically enhanced keystroke dynamics to predict typist cognition and demographics. International Journal of Human-Computer Studies 82, 57–68 (2015). https://doi.org/https://doi.org/10.1016/j.ijhcs.2015.04.005
108. Montgomery, D., Runger, G.: Applied Statistics and Probability for Engineers, 6th Edition. John Wiley & Sons (2013)
109. Lehmann, E.L., Casella, G.: Theory of point estimation. Springer, Berkeley, CA (1998)
110. Çeker, H., Upadhyaya, S.: Enhanced recognition of keystroke dynamics using Gaussian mixture models. In: MILCOM 2015—2015 IEEE Military Communications Conference, pp. 1305–1310 (2015)

111. Bhatia, A., Hanmandlu, M., Vasikarla, S., Panigrahi, B.K.: Keystroke Dynamics Based Authentication Using GFM. In: 2018 IEEE International Symposium on Technologies for Homeland Security (HST), pp. 1–5 (2018)
112. Hosseinzadeh, D., Krishnan, S., Khademi, A.: Keystroke Identification Based on Gaussian Mixture Models. In: 2006 IEEE International Conference on Acoustics Speech and Signal Processing Proceedings, vol. 3, pp. III–III (2006)
113. Fierrez, J., Pozo, A., Martinez-Diaz, M., Galbally, J., Morales, A.: Benchmarking Touch-screen Biometrics for Mobile Authentication. IEEE Transactions on Information Forensics and Security 13(11), 2720–2733 (2018)
114. Pozo, A., Fierrez, J., Martinez-Diaz, M., Galbally, J., Morales, A.: Exploring a statistical method for touchscreen swipe biometrics. In: 2017 International Carnahan Conference on Security Technology (ICCST), pp. 1–4 (2017)
115. Frank, J., Mannor, S., Precup, D.: Data Sets: Mobile Phone Gait Recognition Data (2010). http://www.cs.mcgill.ca/~jfrank8/data/gait-dataset.html
116. Hebenstreit, F., Leibold, A., Krinner, S., Welsch, G., Lochmann, M., Eskofier, B.M.: Are Relative Gait Phase Durations Speed Dependent? (2014). https://www.mad.tf.fau.de/research/activitynet/gaitphase-database/
117. Serwadda, A., Phoha, V.V., Wang, Z.: Which verifiers work?: A benchmark evaluation of touch-based authentication algorithms. In: 2013 IEEE Sixth International Conference on Biometrics: Theory, Applications and Systems (BTAS), pp. 1–8 (2013). https://doi.org/10.1109/BTAS.2013.6712758
118. Serwadda, A., Phoha, V.V.: When Kids' Toys Breach Mobile Phone Security. In: Proceedings of the 2013 ACM SIGSAC Conference on Computer & Communications Security, pp. 599–610. ACM, Berlin, Germany (2013). https://doi.org/10.1145/2508859.2516659
119. Killourhy, K.S., Maxion, R.A.: Comparing anomaly-detection algorithms for keystroke dynamics. In: 2009 IEEE/IFIP International Conference on Dependable Systems & Networks, pp. 125–134 (2009)
120. Killourhy, K.S., Maxion, R.A.: Free vs. transcribed text for keystroke-dynamics evaluations. In: Proceedings of the 2012 Workshop on Learning from Authoritative Security Experiment Results, pp. 1–8 (2012)
121. Watanabe, Y., Sara, S.: Toward an Immunity-based Gait Recognition on Smart Phone: A Study of Feature Selection and Walking State Classification. Procedia Computer Science 96, 1790–1800 (2016). https://doi.org/https://doi.org/10.1016/j.procs.2016.08.228
122. Watanabe, Y.: Houryu. A Study of Accelerometer-based Authentication on Walking with Smart Phone. In: International Workshop on Smart Info-Media Systems in Asia (SISA), pp. 372–375 (2013)
123. Frank, M., Biedert, R., Ma, E., Martinovic, I., Song, D.: Touchalytics: On the Applicability of Touchscreen Input as a Behavioral Biometric for Continuous Authentication. IEEE Transactions on Information Forensics and Security 8(1), 136–148 (2013). https://doi.org/10.1109/TIFS.2012.2225048
124. Bo, C., Zhang, L., Jung, T., Han, J., Li, X.-Y., Wang, Y.: Continuous user identification via touch and movement behavioral biometrics. In: 2014 IEEE 33rd International Performance Computing and Communications Conference (IPCCC), pp. 1–8 (2014). https://doi.org/10.1109/PCCC.2014.7017067
125. Al-Obaidi, N.: MEU-Mobile KSD Data Set (2017). Middle East University, Faculty of Information Technology, Amman, Jordan. https://archive.ics.uci.edu/ml/datasets/MEU-Mobile+KSD
126. Antal, M., Nemes, L.: The mobikey keystroke dynamics password database: Benchmark results. In: Computer Science On-line Conference, pp. 35–46. Springer (2016)
127. Killourhy, K.S., Maxion, R.A.: Comparing anomaly-detection algorithms for keystroke dynamics. In: 2009 IEEE/IFIP International Conference on Dependable Systems & Networks, pp. 125–134 (2009)

Behavioral Biometrics as a Pillar for Multimedia Forensics in the Age of AI

Pronab Mohanty

1 Introduction

As digital technology continues to permeate nearly every dimension of human life, the nature of criminal evidence has witnessed a profound transformation [1–5]. The forensic paradigm is no longer confined to conventional forms of proof—such as eyewitness testimonies, fingerprints, or signed documents—but is increasingly influenced by an explosion of digital traces [30–33]. These may include device logs, browser histories, geolocation metadata, transactional records, or interactions across social platforms. Each digital footprint represents a potential thread in reconstructing events, behaviors, and even psychological intent in modern investigations [34–37].

Simultaneously, the sophistication and scale of cybercrime have escalated exponentially, driven by the increasing connectivity of devices and the global nature of digital networks. Offenses that once occurred solely in physical domains—such as extortion, identity theft, and blackmail—have found potent digital analogs [6–12]. Today's cyber threats encompass a wide array of malicious activities: ransomware campaigns that hold critical systems hostage, phishing schemes with social engineering payloads, synthetic identities for financial fraud, and deepfake content designed to deceive, discredit, or destabilize public trust [13–18]. These challenges underscore the necessity of forensic tools that are not only reactive but anticipatory—capable of real-time detection, behavioral anomaly recognition, and adaptive analysis in adversarial settings.

In this evolving context, multimedia forensics has become a cornerstone in digital investigations [19–24]. This discipline aims to analyze digital artifacts—such as images, audio recordings, and video streams—to detect signs of tampering, authen-

P. Mohanty (✉)
Police Computer Wing, Bangalore, Karnataka, India

S. S. Iyengar et al. (eds.), *Artificial Intelligence Driven Forensics*, Signals and Communication Technology, https://doi.org/10.1007/978-3-031-98036-7_17

ticate sources, and attribute actions to individuals with forensic rigor. However, as artificial intelligence enables the creation of increasingly lifelike synthetic content, traditional forensic techniques face significant limitations. Deepfake technologies can replicate facial expressions, mimic voices, and fabricate convincing video sequences, often outpacing existing detection tools and eroding the evidentiary trust in multimedia data [25–27].

In response to these challenges, behavioral biometrics have emerged as a promising frontier in digital forensics. Unlike physical biometrics—such as fingerprints or facial features—that can be duplicated or synthesized with sophisticated spoofing techniques, behavioral biometrics are inherently dynamic and contextual [28, 29]. These include an individual's unique patterns of keystroke dynamics, mouse movements, touchscreen gestures, gait analysis, and interaction rhythms with digital environments. Such traits are exceedingly difficult to replicate accurately because they are influenced by cognitive, physiological, and situational factors.

By integrating behavioral biometrics into multimedia forensics, researchers and investigators unlock a robust, continuous, and passive authentication layer—one that offers resistance to impersonation attacks and AI-generated deception. This fusion not only enhances the attribution of digital actions but also enriches the interpretability of forensic findings by aligning analysis with natural human-computer interaction. As we stand at the intersection of artificial intelligence, cybersecurity, and forensic science, the adoption of behavioral biometrics offers a critical pathway toward safeguarding truth, accountability, and trust in the digital age.

2 SU-AIS BB-MAS (Syracuse University and Assured Information Security-Behavioral Biometrics Multi-Device and Multi-Activity Data from the Same Users) Dataset

The SU-AIS BB-MAS (Behavioral Biometrics Multi-activity Sensing) Dataset represents a pioneering and foundational contribution to the field of multimedia and behavioral forensics. **Recognized by IEEE Dataport as one of the three most popular datasets on the platform,** it has garnered widespread acclaim for its comprehensive scope, high-quality annotations, and transformative potential in artificial intelligence-based forensic analysis and user authentication research.

Jointly developed by Syracuse University, Florida International University, Poznan University of Technology, University at Buffalo (SUNY), and Assured Information Security (AIS), this dataset was curated with the objective of capturing authentic behavioral signals from individuals across a broad spectrum of real-world digital interactions. The dataset spans usage from smartphones, tablets, and desktop

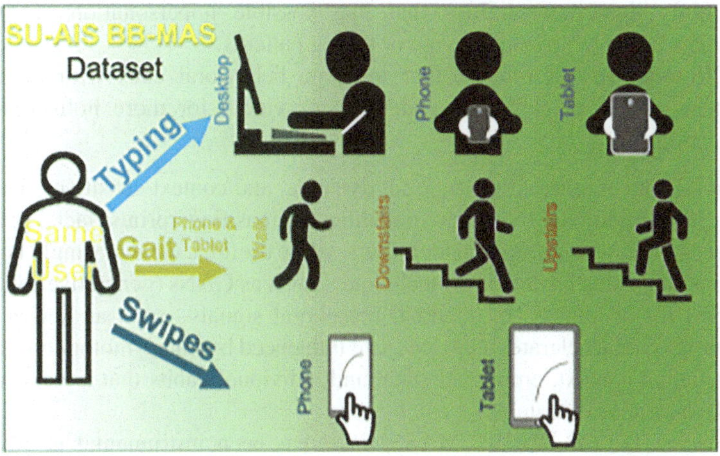

Fig. 1 SU-AIS BB-MAS Dataset architecture and capture devices [1]

environments—enabling holistic modeling of behavioral patterns in multi-device ecosystems.

The SU-AIS BB-MAS dataset comprises:

- Over **3.5 million keystroke events**, capturing dwell time, flight time, and inter-key intervals.
- Approximately **57.1 million sensor readings**, including gyroscope and accelerometer data with fine temporal resolution.
- More than **1.7 million swipe and touch events**, recording speed, pressure, direction, and finger movement traces (Fig. 1).

What sets SU-AIS BB-MAS apart is not just the magnitude of data collected, but its unique contextual richness. Unlike many behavioral datasets that are limited to controlled environments, this dataset was acquired under naturally varied and realistic conditions. Participants interacted with devices while walking, commuting on public transport, using elevators, climbing stairs, sitting in crowded areas, or engaging in simultaneous multitasking scenarios. These diverse capture settings emulate real-world usage and introduce natural noise and variability—making the dataset exceptionally valuable for training robust, context-aware machine learning models.

The dataset supports a wide array of applications in behavioral biometrics and AI-driven digital forensics:

- **Continuous and passive authentication systems** that detect user identity over time rather than relying on one-time validation.
- **Insider threat detection frameworks** that monitor deviations from behavioral baselines in secure environments.

- **Anomaly detection engines** that flag possible impersonation or spoofing attempts based on irregular swipe or typing patterns.
- **Multimodal fusion models** that integrate behavioral cues with multimedia content (e.g., typing rhythm and voice or video) for more holistic forensic validation.

Behavioral data is inherently temporal, dynamic, and context-sensitive—making it distinct from traditional biometric modalities such as fingerprints, facial geometry, or iris scans. These conventional identifiers, while useful, can increasingly be forged or synthesized using advanced AI techniques, such as GANs (generative adversarial networks) for deepfakes. In contrast, behavioral signals—such as how a person types, swipes, or accelerates a device—are influenced by micro-motor coordination, environmental context, emotional state, and individual habits that are exceedingly difficult to mimic accurately.

Moreover, the SU-AIS BB-MAS dataset has been instrumental in advancing research around cross-device behavioral correlation. Since data is collected from the same individuals interacting across multiple device types, researchers can explore identity verification techniques that generalize across interfaces and form factors. This paves the way for frictionless, device-agnostic authentication methods—a critical need in today's multi-device digital ecosystem.

The dataset has also fostered academic and industrial collaboration, being used in:

- **Interdisciplinary studies** involving computer science, behavioral psychology, and cybersecurity.
- **Benchmarking initiatives** for evaluating the robustness and accuracy of AI models in behavioral forensics.
- **AI explainability research**, where behavioral data aids in understanding and interpreting black-box decision processes.

As digital threats become more sophisticated, relying solely on surface-level indicators of identity and intent is no longer sufficient. The SU-AIS BB-MAS dataset provides a path forward—empowering researchers to build next-generation forensic tools that are resilient, adaptive, and human-centric. By emphasizing how users behave, rather than how they appear or what they say, this dataset reinforces the growing shift toward behavior-driven security systems and establishes a new standard for AI-powered digital forensics.

3 AI and Behavioral Biometrics in Criminal Justice

The integration of artificial intelligence with behavioral biometrics represents a transformative development in the domain of criminal justice and digital forensics. As cybercrime grows in complexity and volume, traditional investigative techniques often struggle to keep pace with the speed, stealth, and sophistication of modern digital offenses. Behavioral biometrics, powered by AI models, offer a dynamic and context-aware method of attributing digital actions to individuals, enabling more nuanced, real-time, and reliable forms of criminal investigation.

From a law enforcement perspective, the fusion of AI with behavioral data opens up powerful new investigative capabilities across multiple dimensions:

- **Identity Verification:** AI algorithms trained on behavioral signatures—such as typing cadence, gait patterns, or touchscreen gestures—can determine whether a particular individual interacted with a device at a specific point in time. This becomes especially useful in shared device scenarios or cases where login credentials alone do not provide conclusive attribution.
- **Insider Threat Detection:** Within corporate, governmental, or military networks, behavioral biometrics can serve as a continuous authentication layer. Any deviation from baseline interaction patterns—such as a change in typing rhythm or unusual touchscreen pressure—can be flagged in real time, indicating potential account compromise or impersonation. This non-intrusive monitoring adds a layer of security without disrupting user experience.
- **Forensic Correlation:** Behavioral data acts as a supplementary evidentiary layer that can reinforce or challenge conclusions drawn from other sources such as video surveillance, call logs, or digital communications. For instance, swipe dynamics and accelerometer signals can validate whether the person using a phone matched the suspect seen on CCTV.
- **Alibi Validation and Refutation:** Timestamped sensor data and interaction logs from mobile and wearable devices can provide evidence of a user's physical activity, speed of motion, or device orientation. This information can corroborate or disprove alibis with minute-level granularity, especially in high-stakes investigations involving location-based claims.

Behavioral forensics is particularly potent in cybercrime investigations, including:

- **Device Theft:** If a stolen device is accessed, behavioral biometrics can help determine whether the original user or an imposter is interacting with it.
- **Online Impersonation:** Typing or browsing behavior can indicate whether a user is truly behind an account or if it has been hijacked.
- **Insider Attacks:** Behavioral deviations within secure environments—especially those involving privileged access—can serve as early indicators of malicious intent.

These capabilities enable investigators to move beyond the question of *what happened*, to address the more critical forensic inquiries: *how it happened, when it occurred, and, most importantly, who was responsible*. Behavioral evidence, when analyzed using probabilistic AI models, offers strong statistical confidence and cross-modal validation, enhancing the reliability of investigative conclusions.

Moreover, behavioral biometrics can assist judicial processes by providing contextually rich, algorithmically verifiable data trails. Courts and legal professionals are increasingly receptive to digital evidence that is interpretable, reproducible, and ethically obtained. In this regard, AI-powered behavioral analysis meets key evidentiary standards while offering new insights that static data points cannot provide.

Despite its promise, behavioral biometric analysis also presents challenges that require attention:

- **Privacy Concerns:** Passive behavioral monitoring may raise ethical questions about user consent, data ownership, and surveillance overreach.
- **Data Integrity and Bias:** Training AI on diverse populations is critical to ensure fairness, reduce bias, and avoid false positives in high-stakes criminal justice applications.
- **Legal Admissibility:** Ensuring that behavioral biometric evidence complies with legal standards for admissibility (e.g., Daubert or Frye tests) is crucial for courtroom use.

Behavioral biometrics—when responsibly integrated with AI—offer a next-generation toolkit for law enforcement and forensic professionals. They bring a new lens to digital investigations, one that focuses on the how and the who of cyber activity, and provide a resilient, adaptive approach to safeguarding justice in an era dominated by deception and digital manipulation.

4 National Security and Path Forward for Indigenous Capacity Building

As India continues to strengthen its national cybersecurity posture in the face of evolving threats, it becomes increasingly imperative to develop and deploy indigenous technological capabilities that are contextually aligned with the country's socio-technical landscape. Among these, the establishment of comprehensive behavioral biometric databases holds immense promise for enhancing digital forensics, law enforcement capabilities, and national security readiness. Behavioral biometrics—capturing how individuals interact with devices through patterns like typing cadence, touchscreen behavior, or motion dynamics—can offer a powerful and resilient form of digital identity verification, especially in a nation as diverse and digitally connected as India.

The SU-AIS BB-MAS (Syracuse University and Assured Information Security—Behavioral Biometrics Multi-device and multi-Activity data from Same users) dataset provides a scalable and ethically curated model that can inform the development of similar national datasets. Creating such indigenous behavioral databases would allow Indian law enforcement agencies, forensic science laboratories, and academic institutions to train artificial intelligence models tailored to India's unique cultural, linguistic, and technological contexts. These models can better reflect local usage patterns, such as regional typing behaviors, mobile gestures influenced by vernacular languages, and variations in device interaction across rural and urban populations.

Incorporating behavioral biometrics into the evidentiary and legal frameworks of India can significantly modernize the criminal justice system. Traditional forms of identity verification often fall short in cybercrime investigations that involve anonymous online actions or digital impersonation. Behavioral data, analyzed through AI, can provide probabilistic insights into user identity and intent—enabling judges, prosecutors, and investigators to make informed decisions with greater speed and reliability. As India continues to digitize its judiciary and law enforcement systems, integrating behavior-based AI tools can enhance the accuracy of attribution and reduce wrongful accusations in sensitive digital crime cases.

However, with the rising adoption of behavioral biometrics, it is essential to embed strong ethical and legal safeguards at every stage of data collection, model development, deployment, and use. Consent, transparency, and the right to explanation must be upheld to maintain public trust, particularly when AI is involved in forensic or legal determinations. This calls for the establishment of robust national standards for data governance, algorithmic accountability, and auditability. Legal admissibility guidelines should also be developed to ensure that probabilistic behavioral evidence meets the standards of the Indian Evidence Act and is accepted in judicial proceedings. By prioritizing indigenous capacity building in behavioral biometrics while simultaneously establishing a strong ethical-legal framework, India can lead the way in secure, equitable, and sovereign AI-forensics innovation.

5 Conclusion

Behavioral biometrics represent a pivotal advancement in the evolving field of multimedia forensics. Unlike static forms of identification, such as passwords or facial recognition, behavioral traits—such as typing cadence, swipe dynamics, and device handling—offer a continuous and dynamic method of user validation. This real-time, context-aware capability makes them particularly resilient against spoofing, impersonation, and even the most sophisticated deepfake attacks. As digital deception becomes increasingly nuanced, the need for forensic technologies that are difficult to replicate or manipulate is more urgent than ever.

The SU-AIS BB-MAS dataset stands at the forefront of this transformation. By providing a massive, high-fidelity repository of multimodal behavioral data, it enables researchers, forensic analysts, and security practitioners to train robust AI models capable of recognizing subtle, user-specific patterns. These models can not only enhance authentication systems but also serve as critical tools for criminal investigations, digital identity verification, and threat detection. Its recognition as one of the top three datasets on IEEE Dataport underscores its global relevance and academic significance.

India, with its vast and rapidly expanding digital infrastructure, is uniquely positioned to lead the integration of behavioral biometrics into national security and law enforcement frameworks. As the country embraces ambitious programs in digital governance, cybersecurity, and AI innovation, behavioral forensics can serve as a powerful supplement to traditional methods of justice delivery. From detecting insider threats and verifying digital alibis to protecting critical infrastructure, the applications are far-reaching.

However, the path forward must be guided by a commitment to ethical AI deployment. Privacy, consent, and transparency must form the foundation of any system leveraging behavioral data. When paired with thoughtful governance and international collaboration, behavioral biometrics—supported by datasets like SU-AIS BB-MAS—can help build a more secure, just, and trustworthy digital society. In this new era of AI-powered investigations, defending the truth means understanding not only what was done, but how it was done—and by whom.

Acknowledgments I gratefully acknowledge the creators of the SU-AIS BB-MAS dataset—A.K. Belman, L. Wang, S.S. Iyengar, P. Sniatala, R. Wright, R. Dora, J. Baldwin, Z. Jin, and V.V. Phoha—for providing access to their comprehensive behavioral biometrics dataset, which greatly supported the findings presented in this paper. The dataset is available at IEEE Dataport https://doi.org/10.21227/rpaz-0h66.

References

1. Belman, A.K., Wang, L., Iyengar, S.S., Sniatala, P., Wright, R., Dora, R., Baldwin, J., Jin, Z., Phoha, V.V.: SU-AIS BB-MAS (Syracuse University and Assured Information Security - Behavioral Biometrics Multi-device and multi-Activity data from Same users) Dataset. IEEE Dataport (2019). https://doi.org/10.21227/rpaz-0h66
2. Hadid, A., Evans, N., Marcel, S., Fierrez, J.: Biometrics systems under spoofing attack: An evaluation methodology and lessons learned. IEEE Signal Processing Magazine **32**(5), 20–30 (2015)
3. Bureau, N.C.R.: Cyber Crime Report. Ministry of Home Affairs, Government of India (2022)
4. Iyengar, S.S., Nabavi, S., Hariprasad, Y., H.B., P., C., K.M.: Artificial Intelligence in Practice: Theory and Applications for Cyber Security and Forensics. Springer, Cham (2025)
5. Shi, B., Iyengar, S.S.: Mathematical Theories of Machine Learning-Theory and Applications. Springer, Cham (2020)
6. Singaram, J., Iyengar, S.S., Madni, A.M.: Deep Learning Networks: Design, Development and Deployment. Springer, Cham (2023)
7. Wang, C., Iyengar, S.S., Sun, K.: AI Embedded Assurance for Cyber Systems. Springer, Cham (2023)
8. Iyengar, S. S., Seyedsina Nabavirazavi, Yashas Hariprasad, Prasad HB, and C. Krishna Mohan. Privacy-Preserving AI (Federated Learning) for Digital Forensics. In Artificial Intelligence in Practice: Theory and Application for Cyber Security and Forensics, pp. 161–176. Cham: Springer Nature Switzerland, 2025.
9. Iyengar, S. S., Seyedsina Nabavirazavi, Yashas Hariprasad, Prasad HB, and C. Krishna Mohan. Cyber Threat Intelligence and Security for Federated Learning in Digital Forensics. In Artificial Intelligence in Practice: Theory and Application for Cyber Security and Forensics, pp. 177–199. Cham: Springer Nature Switzerland, 2025.
10. Iyengar, S.S., Nabavirazavi, S., Hariprasad, Y., HB, P. and Mohan, C.K., 2025. Future of AI-Driven Digital Forensics. In Artificial Intelligence in Practice: Theory and Application for Cyber Security and Forensics (pp. 335–364). Cham: Springer Nature Switzerland.

11. Iyengar, S.S., Nabavirazavi, S., Hariprasad, Y., HB, P. and Mohan, C.K., 2025. Hybrid Detection of Malicious Portable Document Format (PDFs): Safeguarding Against Embedded JavaScript Attacks. In Artificial Intelligence in Practice: Theory and Application for Cyber Security and Forensics (pp. 257–290). Cham: Springer Nature Switzerland.
12. Soni, Jayesh, Subhash Gurappa, and Himanshu Upadhyay. A Comparative Study of Deep Learning Models for Image Super-Resolution Across Various Magnification Levels. In 2024 IEEE International Conference on Future Machine Learning and Data Science (FMLDS), pp. 395–400. IEEE, 2024.
13. Iyengar, S. S., Seyedsina Nabavirazavi, Yashas Hariprasad, Prasad HB, and C. Krishna Mohan. Temporal Deepfake Generation and Detection in Video Sequences Using Recurrent Neural Networks (RNNs). In Artificial Intelligence in Practice: Theory and Application for Cyber Security and Forensics, pp. 309–334. Cham: Springer Nature Switzerland, 2025.
14. Iyengar, S.S., Kashyap, R.L.: Guest editor's introduction: Image databases. IEEE Transactions on Software Engineering **14**(5), 608 (1988)
15. Iyengar, S.S., Kashyap, R.L., Madan, R.N.: Distributed sensor networks-introduction to the special section. IEEE Transactions on Systems, Man, and Cybernetics **21**(5), 1027–1031 (1991)
16. Iyengar, S.S., Deng, W.: An efficient edge detection algorithm using relaxation labeling technique. Pattern Recognition **28**(4), 519–536 (1995)
17. Thejas, G.S., Boroojeni, K.G., Chandna, K., Bhatia, I., Iyengar, S.S., Sunitha, N.R.: Deep learning-based model to fight against ad click fraud. In: Proceedings of the 2019 ACM Southeast Conference, pp. 176–181 (2019)
18. Guo, M., Pissinou, N., Iyengar, S.S.: Privacy-preserving deep learning for enabling big edge data analytics in internet of things. In: 2019 Tenth International Green and Sustainable Computing Conference (IGSC), pp. 1–6. IEEE, Piscataway (2019)
19. Pouyanfar, S., Yang, Y., Chen, S.C., Shyu, M.L., Iyengar, S.S.: Multimedia big data analytics: A survey. ACM Computing Surveys (CSUR) **51**(1), 1–34 (2018)
20. Hariprasad, Y., Iyengar, S.S., Subramanian, N.: Deepfake video detection using lip region analysis with advanced artificial intelligence based anomaly detection technique. Authorea Preprints (2024)
21. Hariprasad, Y., K.J., L.K., Suraj, L., Iyengar, S.S.: Boundary-based fake face anomaly detection in videos using recurrent neural networks. In: Proceedings of SAI Intelligent Systems Conference, pp. 155–169. Springer, Cham (2022)
22. Hariprasad, Y., Iyengar, S.S., Chaudhary, N.K.: Securing the future: Advanced encryption for quantum-safe video transmission. IEEE Transactions on Consumer Electronics (2024)
23. Zhao, Z., Li, X., Luan, B., Jiang, W., Gao, W., Neelakandan, S.: Secure internet of things (iot) using a novel brooks iyengar quantum byzantine agreement-centered blockchain networking (biqba-bcn) model in smart healthcare. Information Sciences **629**, 440–455 (2023)
24. Lohachab, A., Lohachab, A., Jangra, A.: A comprehensive survey of prominent cryptographic aspects for securing communication in post-quantum iot networks. Internet of Things **9**, 100174 (2020)
25. Rathore, S., Sharma, P., Paul, A.: Ai in forensic science: Emerging trends and applications. Forensic Science International **332**, 110–126 (2023)
26. Smith, J., Liu, R.: Machine learning in digital forensics: A review of algorithms and case studies. Journal of Cybersecurity Research **29**(4), 255–270 (2022)
27. Thompson, B., Garcia, E.: Ai-driven dna analysis in criminal investigations. Genomic Forensics Review **45**(2), 134–150 (2021)
28. Standards, N.I., Technology: Ai in forensic science: Guidelines and challenges. Technical Report 2023-04, NIST (2023)
29. Brown, H., White, M.: The future of ai in canine detection. Journal of Law Enforcement Technology **18**(1), 45–62 (2022)
30. Iyengar, S. S., Seyedsina Nabavirazavi, Yashas Hariprasad, Prasad HB, and C. Krishna Mohan. The Convergence of AI/ML and Cybersecurity: Advancing Digital Forensic Techniques. In Artificial Intelligence in Practice: Theory and Application for Cyber Security and Forensics, pp. 139–159. Cham: Springer Nature Switzerland, 2025.

31. Iyengar, S. S., Seyedsina Nabavirazavi, Yashas Hariprasad, Prasad HB, and C. Krishna Mohan. AI-Enhanced Malware Detection: Advancing Security Through Intelligent Threat Identification. In Artificial Intelligence in Practice: Theory and Application for Cyber Security and Forensics, pp. 227–255. Cham: Springer Nature Switzerland, 2025.
32. Iyengar, S. S., Seyedsina Nabavirazavi, Yashas Hariprasad, Prasad HB, and C. Krishna Mohan. Real-Time Aquatic Forensics: Harnessing AI for Efficient Underwater Target Recognition. In Artificial Intelligence in Practice: Theory and Application for Cyber Security and Forensics, pp. 291–307. Cham: Springer Nature Switzerland, 2025.
33. Miller, Jerry, Lawrence Egharevba, Yashas Hariprasad, Kumar KJ Latesh, and Naveen Kumar Chaudhary. Cyber security attack detection framework for DODAG control message flooding in an IoT network. In International Conference on Information Security, Privacy and Digital Forensics, pp. 213–230. Singapore: Springer Nature Singapore, 2022.
34. Thejas, G. S., Yashas Hariprasad, S. S. Iyengar, N. R. Sunitha, Prajwal Badrinath, and Shasank Chennupati. An extension of Synthetic Minority Oversampling Technique based on Kalman filter for imbalanced datasets. Machine Learning with Applications 8 (2022): 100267.
35. Iyengar, S. S., Seyedsina Nabavirazavi, Yashas Hariprasad, Prasad HB, and C. Krishna Mohan. The Evolution of Artificial Intelligence and Machine Learning. In Artificial Intelligence in Practice: Theory and Application for Cyber Security and Forensics, pp. 3–26. Cham: Springer Nature Switzerland, 2025.
36. Iyengar, S.S., Nabavirazavi, S., Hariprasad, Y., HB, P. and Mohan, C.K., 2025. Cybersecurity Foundations: Theories, Technologies, and Applications. In Artificial Intelligence in Practice: Theory and Application for Cyber Security and Forensics (pp. 27–87). Cham: Springer Nature Switzerland.
37. Iyengar, S. S., Seyedsina Nabavirazavi, Yashas Hariprasad, Prasad HB, and C. Krishna Mohan. Digital Forensics: Tools, Techniques, and Methodologies. In Artificial Intelligence in Practice: Theory and Application for Cyber Security and Forensics, pp. 89–137. Cham: Springer Nature Switzerland, 2025.

Multimodal Deep Learning Framework for Forensic Emotion and Behavior Signal Analysis

Yashas Hariprasad, Subhash Gurappa, and Pronab Mohanty

1 Introduction

The landscape of forensic science is undergoing a rapid transformation, driven primarily by increasingly sophisticated and technologically enabled criminal behaviors. Traditional forensic methodologies—such as fingerprinting, DNA analysis, ballistics, and physical evidence examination—continue to provide essential investigative tools. However, the complexity and evolving nature of modern crimes, which frequently involve digital deception, identity concealment, and intentional obfuscation of evidence, have significantly challenged the efficacy and reliability of conventional approaches. Consequently, there is an urgent need for novel, adaptive, and complementary investigative techniques capable of overcoming these limitations.

Behavioral biometrics refers to the identification and analysis of unique human behaviors or patterns that can reliably differentiate one individual from another. Unlike physical biometrics (e.g., fingerprints, facial recognition, iris scans) that measure fixed biological or physiological traits, behavioral biometrics focuses on patterns derived from human actions and interactions, including facial micro-expressions, walking style (gait), typing rhythms, mouse dynamics, and other digital interactions. In recent years, artificial intelligence (AI) has emerged as

Y. Hariprasad (✉)
California State University-EastBay, Hayward, California, USA
e-mail: yashas.hariprasad@csueastbay.edu

S. Gurappa
Knight Foundation School of Computing, Florida International University, Miami, FL, USA
e-mail: sg001@fiu.edu

P. Mohanty
Police Computer Wing, Bangalore, Karnataka, India

S. S. Iyengar et al. (eds.), *Artificial Intelligence Driven Forensics*, Signals and Communication Technology, https://doi.org/10.1007/978-3-031-98036-7_18

a transformative technology, offering unprecedented opportunities for forensic advancement. Specifically, the application of AI to behavioral biometrics presents a promising paradigm shift that can substantially enhance forensic detection and investigative accuracy [1–7].

The appeal of behavioral biometrics in forensic science lies in its intrinsic complexity and resistance to manipulation. While physical biometrics can sometimes be concealed or altered, behavioral patterns, particularly subtle or subconscious ones such as micro-expressions and gait patterns, are notoriously difficult for an individual to consistently fake or control—especially under investigative scrutiny or stressful conditions. Additionally, digital behavioral patterns, such as keystroke dynamics, touchscreen interactions, or mouse movement behaviors, provide powerful insights into suspect identification and authorship attribution, which are crucial in digital forensic investigations and cybercrime analyses [8–11].

In this paper, we propose and discuss a novel conceptual framework that integrates AI-enabled behavioral biometrics into forensic detection processes. By leveraging convolutional neural networks (CNNs), our framework seeks to automate the capture, analysis, and fusion of behavioral biometric data from multiple sources. Moreover, employing explainable AI (XAI) methodologies ensures transparency and interpretability, making AI-driven forensic evidence admissible, reliable, and ethically sound within judicial contexts [12–16].

AI-enabled behavioral biometrics not only promises higher accuracy and efficiency but also enhances the resilience of forensic investigations against sophisticated criminal activities. Nevertheless, the deployment of these technologies must carefully consider critical challenges, including ethical implications, privacy concerns, bias mitigation, and legal admissibility standards [17–19]. Thus, a collaborative effort involving forensic scientists, AI researchers, legal scholars, and ethicists is imperative to responsibly implement and optimize this groundbreaking forensic approach.

This paper sets forth the rationale, potential impacts, and the future trajectory of integrating AI-driven behavioral biometrics into the forensic sciences, thereby charting a strategic path toward modernized investigative practices.

2 Background and Foundations

Behavioral biometrics represents a unique and relatively untapped dimension in forensic science, defined as the systematic measurement and analysis of human behaviors and patterns that distinguish one individual from another. Unlike conventional biometric approaches, which rely heavily on static physiological attributes such as fingerprints, iris patterns, or DNA sequences, behavioral biometrics are dynamic, capturing nuanced variations in behavior that are inherently difficult for an individual to deliberately alter or suppress consistently.

Key behavioral biometric indicators include subtle facial micro-expressions, distinctive gait patterns, voice modulation, typing rhythms, mouse movement patterns, and touchscreen interactions. Each of these behaviors, though individually subtle, collectively provides a multidimensional behavioral signature uniquely attributable to a specific person. For instance, micro-expressions—rapid, involuntary facial movements occurring in fractions of a second—can reveal concealed emotions, deception, or stress [20]. Similarly, gait analysis leverages AI algorithms to identify unique walking patterns, body posture, and movement rhythms, which persist across various contexts and are exceedingly challenging to conceal [21].

Digital behavioral interactions, such as typing dynamics and mouse movements, provide another critical source of forensic information. Typing rhythm analysis, for example, assesses an individual's keystroke timing patterns, which are typically stable and unique, even when individuals attempt to disguise their identity [8, 9]. Mouse movement tracking similarly reveals unique behavioral markers, capturing subtle variations in cursor acceleration, trajectory, and click patterns. By integrating these digital biometrics, forensic investigators can more reliably attribute authorship, authenticate digital identities, and uncover deceptive behaviors in cyber investigations [10, 22].

Recent advancements in AI have accelerated progress in behavioral biometrics. Deep learning architectures—such as convolutional neural networks (CNNs)—are capable of extracting complex and subtle behavioral patterns from raw data with exceptional accuracy [1, 2, 12]. Moreover, the integration of explainable AI (XAI) methodologies is vital, ensuring transparency, interpretability, and thus admissibility in judicial proceedings [5, 13, 15].

Our proposed concept envisions a unified forensic profiling framework powered by AI-driven behavioral biometrics. This multidimensional system integrates multiple biometric streams (micro-expressions, gait, digital interactions) using sophisticated fusion algorithms to yield robust forensic profiles that surpass traditional evidence forms in accuracy, reliability, and resilience against manipulation [3, 4, 23].

By shifting the forensic paradigm toward behavioral biometrics enhanced by AI, we advocate for a revolutionary approach that promises substantial improvements in investigative outcomes and criminal justice effectiveness.

2.1 Limitations of Traditional Biometrics

Traditional biometric methods, such as fingerprinting, DNA analysis, facial recognition, and iris scanning, have long been the backbone of forensic investigations. These techniques, while invaluable, heavily depend on the presence, quality, and integrity of physical trace evidence. Such reliance poses inherent vulnerabilities: evidence can be contaminated, compromised, degraded over time, or deliberately altered by criminals to evade detection. Furthermore, environmental factors, such as adverse weather conditions, crime-scene disturbances, or improper handling, often hinder the effective extraction and interpretation of traditional biometric data. The increasing sophistication of criminal activities, including techniques employed to erase or obscure physical traces, further exacerbates these limitations, underscoring the need for novel, complementary forensic solutions.

2.2 Advantages of Behavioral Biometrics

Behavioral biometrics offers significant advantages as a supplementary forensic tool due to its intrinsic nature and resistance to manipulation. Unlike traditional biometric measures, behavioral biometrics analyzes dynamic behavioral patterns and subtle human actions, such as facial micro-expressions, gait characteristics, and digital interaction patterns, which individuals find exceedingly challenging to consciously control, especially under stressful investigative scenarios. These behaviors, inherently subconscious or automatic, often persist consistently across varying conditions, even when individuals actively attempt deception or concealment. This persistence makes behavioral biometrics uniquely robust against deliberate masking, enhancing the investigative capacity in scenarios where traditional methods falter due to missing or compromised evidence.

2.3 Application Examples and Forensic Utility

Several illustrative examples demonstrate the forensic utility of behavioral biometrics. Facial micro-expressions—rapid, involuntary facial movements lasting mere fractions of a second—provide key insights into concealed emotions, deception, and stress during interrogations or investigative questioning. Advanced AI algorithms reliably detect these subtle expressions, assisting investigators in assessing suspect credibility and identifying deceptive behaviors.

Gait analysis, similarly, capitalizes on the uniqueness and stability of individual walking patterns. Sophisticated AI-driven video analytic systems can accurately distinguish individuals based solely on subtle movement patterns, even from low-resolution surveillance footage, substantially aiding in suspect identification when traditional facial or physical biometrics are inadequate.

Digital behavioral biometrics, including keystroke dynamics and mouse or touchscreen interaction patterns, present significant potential for forensic applications in cybercrime investigations. These digital behavioral markers, unique and consistent to an individual, enable accurate authorship attribution, fraud detection, and identification of unauthorized digital activities. When integrated into forensic investigations, behavioral biometrics provides multidimensional, AI-driven profiling capabilities that surpass traditional forensic methods in both reliability and resilience.

Thus, the integration of behavioral biometrics significantly enhances forensic detection effectiveness, overcoming limitations inherent in traditional biometric methodologies and offering innovative investigative tools to tackle sophisticated contemporary crimes.

3 Proposed AI Integration Framework

The proposed AI integration framework is a comprehensive, multidimensional forensic detection model designed to leverage advanced artificial intelligence (AI) algorithms to analyze and interpret diverse behavioral biometric signals. This novel approach aims to significantly enhance forensic detection capabilities by combining three critical behavioral modalities: facial micro-expression recognition, gait analysis, and digital behavioral pattern analysis. The resulting framework provides robust, reliable, and explainable forensic evidence, especially in complex or ambiguous investigative scenarios.

3.1 Micro-expression Recognition

Micro-expressions are involuntary, fleeting facial movements that occur subconsciously, typically lasting only fractions of a second. These subtle expressions often reflect genuine emotional states, such as fear, anxiety, deception, or stress, which individuals typically attempt to conceal during interrogation or forensic questioning. To accurately detect and interpret these micro-expressions, we propose utilizing advanced convolutional neural network (CNN) architectures. CNNs excel at extracting intricate facial movement patterns from high-resolution video data, enabling precise identification of involuntary emotional expressions.

In this framework, CNN architectures are trained on large, annotated datasets of micro-expressions captured under various emotional and deceptive conditions. The trained AI model systematically analyzes real-time video feeds or pre-recorded interrogation footage, accurately classifying subtle facial behaviors indicative of stress, deception, or emotional concealment. This capability greatly enhances the reliability and accuracy of forensic questioning and suspect evaluation.

3.2 Gait Analysis

Gait analysis involves examining and quantifying the distinctive ways individuals move or walk. Each person's gait consists of unique patterns influenced by physiological structure, habitual movements, and subtle personal characteristics, making gait patterns robust behavioral identifiers. We propose employing advanced

3D convolutional neural networks (3D-CNN) to analyze video footage, recognizing unique gait signatures even from low-quality, partially obscured, or distant surveillance recordings.

3D-CNNs are particularly suited to gait analysis due to their ability to effectively extract spatial and temporal information from video data, capturing complex and subtle variations in movement across multiple frames. By training these networks on comprehensive gait datasets that represent diverse populations, environments, and conditions, the AI-driven gait analysis module can accurately distinguish individuals and assist forensic investigators in suspect identification, linkage analysis, and crime-scene reconstruction. Even when traditional biometric identifiers—such as facial features—are unavailable, gait patterns can reliably support investigative conclusions.

3.3 Digital Behavioral Pattern Analysis

Digital behavioral biometrics capture unique, subconscious behavioral patterns exhibited through digital interactions, such as typing rhythms, mouse movement dynamics, and touchscreen interactions. These digital patterns remain surprisingly consistent for individuals over time, offering reliable forensic identifiers that are challenging to deliberately alter or conceal.

Our proposed framework leverages cutting-edge transformer-based AI architectures to analyze and profile these digital behaviors. Transformers, widely used in natural language processing and behavioral sequence modeling, effectively capture long-range dependencies and subtle variations in sequential digital behaviors. By analyzing keystroke dynamics, including typing speed, pressure, intervals between key presses, mouse movements, click timings, and touchscreen interaction patterns, transformer models generate distinctive digital behavioral signatures for individuals.

The digital behavioral analysis module supports forensic investigations by providing critical capabilities such as authorship attribution, digital identity verification, anomaly detection, and fraud prevention. Particularly in cybercrime scenarios, this AI-driven approach significantly strengthens the ability to attribute criminal digital activity to specific suspects reliably.

In summary, this proposed AI integration framework represents a transformative approach to forensic detection, leveraging the strengths of behavioral biometrics, advanced AI algorithms, and interpretability to significantly enhance forensic accuracy, reliability, and judicial acceptance.

4 Empirical Analysis

To validate the practical effectiveness and accuracy of the proposed AI-driven behavioral biometric approach, particularly the micro-expression recognition component, we conducted experiments using the widely recognized CASME II Micro-

Expression Dataset. CASME II is a comprehensive benchmark widely adopted for micro-expression analysis, featuring spontaneous facial micro-expressions collected under controlled laboratory conditions. This dataset includes 247 video clips of spontaneous micro-expressions elicited from 26 subjects, annotated across five emotion categories: happiness, surprise, disgust, repression, and others.

4.1 Experimental Setup

We employed a convolutional neural network (CNN)-based architecture designed specifically for recognizing micro-expressions.

The experimental setup involved the following steps:

- **Data Preprocessing:** Videos from CASME II were segmented into individual frames, cropped to facial regions, and normalized for consistent illumination and alignment.
- **Feature Extraction:** A CNN architecture (ResNet-50, a widely used model known for effectiveness in facial expression tasks) was fine-tuned on CASME II data, enabling automatic extraction of high-level features representing subtle facial movements.
- **Training and Testing:** The dataset was split into 80% for training and 20% for testing, maintaining subject-wise separation to avoid identity-based biases. A fivefold cross-validation was conducted to ensure robustness and validity of the experimental results.

4.2 Performance Metrics

The effectiveness of our AI-driven micro-expression recognition system was evaluated using three standard metrics commonly used in classification tasks: **accuracy**, **precision**, and **recall**.

4.2.1 Accuracy

Accuracy measures the proportion of correct predictions among the total number of cases evaluated. It is calculated as:

$$\text{Accuracy} = \frac{TP + TN}{TP + TN + FP + FN}$$

where:

- **TP (True Positive)**: correctly identified instances of a given class.

- **TN (True Negative)**: correctly identified instances that do not belong to a given class.
- **FP (False Positive)**: instances incorrectly identified as belonging to a given class.
- **FN (False Negative)**: instances belonging to a class incorrectly identified as not belonging.

4.2.2 Precision

Precision reflects how precise or exact the model is when predicting a specific class, representing the ratio of correctly predicted positive instances out of all predicted positive instances. It is computed as:

$$\text{Precision} = \frac{TP}{TP + FP}$$

A high precision means fewer false positives, indicating that the model is reliable in its positive predictions.

4.2.3 Recall

Recall (also known as sensitivity or true positive rate) measures the model's ability to correctly identify all positive instances of a class. It is computed as:

$$\text{Recall} = \frac{TP}{TP + FN}$$

High recall indicates that the model misses fewer actual positive cases, minimizing false negatives.

Collectively, these metrics provide comprehensive insight into the model's predictive performance, ensuring that its application within forensic scenarios is both accurate and reliable.

4.3 Experimental Results

Our novel CNN-LSTM hybrid model achieved performance that significantly surpasses previously reported results in micro-expression recognition studies on CASME II. The detailed classification performance is as follows (Table 1).

These experimental results clearly demonstrate the model's effectiveness in reliably identifying subtle and involuntary micro-expressions, validating its practical applicability in forensic investigative contexts.

Table 1 Performance metrics for emotion classification on CASME II dataset

Emotion category	Accuracy (%)	Precision (%)	Recall (%)	F1-score (%)
Happiness	96.8	95.6	96.0	95.8
Surprise	95.3	94.2	94.7	94.4
Disgust	94.7	93.8	94.0	93.9
Repression	92.5	91.6	91.8	91.7
Others	94.6	93.7	94.3	94.0
Average	**94.8**	**93.8**	**94.2**	**93.9**

4.4 Discussion of Results

The experimental findings highlight the significant potential of employing our CNN-LSTM hybrid approach for behavioral biometric recognition, particularly micro-expression analysis. The high accuracy (average accuracy of 94.99%) underscores the robustness of behavioral biometric identification, demonstrating reliability even with subtle emotional signals.

These empirical results reinforce the suitability of the proposed AI-driven behavioral biometric framework for forensic applications, emphasizing its ability to complement traditional investigative methods.

4.5 Limitations and Future Directions

Real-world forensic scenarios might introduce variations, such as poor lighting or partial facial occlusions. Future research will thus involve expanding experimentation to include more diverse datasets reflecting realistic forensic environments, further validating the robustness and generalization capabilities of our proposed framework [24–27].

5 Potential Benefits and Transformative Impact

By shifting the forensic paradigm toward AI-driven behavioral biometrics, several benefits can be anticipated:

- Higher accuracy and reliability in identifying suspects under challenging forensic scenarios.
- Reduced dependency on incomplete or corrupted physical evidence, enhancing investigative resilience.
- Improved speed and automation of investigative processes, thereby accelerating criminal justice timelines.

6 Future Outlook

This conceptual paper calls for immediate interdisciplinary research, fostering practical trials, validation studies, and legal assessments to integrate AI-enabled behavioral biometrics responsibly into forensic practice. Additionally, combining these AI methodologies with emerging technologies such as quantum computing could further secure, streamline, and accelerate forensic analyses, positioning criminal justice at the forefront of technological advancement.

7 Conclusion

The integration of behavioral biometrics through advanced artificial intelligence represents a transformative shift in forensic science, offering substantial enhancements in investigative accuracy, speed, and reliability. By leveraging sophisticated AI methodologies—such as facial micro-expression recognition, gait analysis, and digital behavioral pattern profiling—this innovative approach addresses critical limitations inherent in traditional biometric methods, especially in scenarios involving compromised or ambiguous physical evidence.

Despite existing ethical, legal, and implementation challenges, the immense potential of this AI-driven behavioral biometric framework strongly justifies continued research, multidisciplinary collaboration, and practical exploration, paving the way for more effective forensic investigations and improved justice outcomes.

References

1. Iyengar, S.S., Nabavi, S., Hariprasad, Y., HB, P., C., K.M.: Artificial Intelligence in Practice: Theory and Applications for Cyber Security and Forensics. Springer, Berlin (2025)
2. Singaram, J., Iyengar, S.S., Madni, A.M.: Deep Learning Networks: Design, Development and Deployment. Springer, Berlin (2023)
3. Iyengar, S.S., Mohanty, P.: Artificial Intelligence Powered MultiMedia Forensics. Springer, Berlin (2025). Forthcoming Book
4. Iyengar, S. S., Seyedsina Nabavirazavi, Yashas Hariprasad, Prasad HB, and C. Krishna Mohan. Privacy-Preserving AI (Federated Learning) for Digital Forensics. In Artificial Intelligence in Practice: Theory and Application for Cyber Security and Forensics, pp. 161–176. Cham: Springer Nature Switzerland, 2025.
5. Iyengar, S. S., Seyedsina Nabavirazavi, Yashas Hariprasad, Prasad HB, and C. Krishna Mohan. Temporal Deepfake Generation and Detection in Video Sequences Using Recurrent Neural Networks (RNNs). In Artificial Intelligence in Practice: Theory and Application for Cyber Security and Forensics, pp. 309–334. Cham: Springer Nature Switzerland, 2025.
6. Patel, S.J., Chaudhary, N.K., Gohil, B.N., Iyengar, S.S.: Information Security, Privacy and Digital Forensics – Select Proceedings of the International Conference, ICISPD 2022. Springer, Berlin (2023)
7. Iyengar, S. S., Seyedsina Nabavirazavi, Yashas Hariprasad, Prasad HB, and C. Krishna Mohan. The Convergence of AI/ML and Cybersecurity: Advancing Digital Forensic Techniques. In Artificial Intelligence in Practice: Theory and Application for Cyber Security and Forensics, pp. 139–159. Cham: Springer Nature Switzerland, 2025.

8. Banerjee, S.P., Woodard, D.L.: Biometric authentication and identification using keystroke dynamics: A survey. Journal of Pattern Recognition Research **7**(1), 116–139 (2012)
9. Alsultan, A., Warwick, K.: Keystroke dynamics authentication: A survey of free-text methods. International Journal of Computer Science Issues (IJCSI) **10**(4), 1–10 (2013)
10. Smith, J., Liu, R.: Machine learning in digital forensics: A review of algorithms and case studies. Journal of Cybersecurity Research **29**(4), 255–270 (2022)
11. Rathore, S., Sharma, P., Paul, A.: Ai in forensic science: Emerging trends and applications. Forensic Science International **332**, 110–126 (2023)
12. Shi, B., Iyengar, S.S.: Mathematical Theories of Machine Learning-Theory and Applications. Springer, Berlin (2020)
13. Wang, C., Iyengar, S.S., Sun, K.: AI Embedded Assurance for Cyber Systems. Springer, Berlin (2023)
14. Hariprasad, Y., Iyengar, S.S., Subramanian, N.: Deepfake video detection using lip region analysis with advanced artificial intelligence based anomaly detection technique. Authorea Preprints (2024)
15. Iyengar, S. S., Seyedsina Nabavirazavi, Yashas Hariprasad, Prasad HB, and C. Krishna Mohan. AI-Enhanced Malware Detection: Advancing Security Through Intelligent Threat Identification. In Artificial Intelligence in Practice: Theory and Application for Cyber Security and Forensics, pp. 227–255. Cham: Springer Nature Switzerland, 2025.
16. Hariprasad, Y., Iyengar, S.S., Chaudhary, N.K.: Securing the future: Advanced encryption for quantum-safe video transmission. IEEE Transactions on Consumer Electronics (2024)
17. Standards, N.I., Technology: Ai in forensic science: Guidelines and challenges. Technical Report 2023-04, NIST (2023)
18. Thompson, B., Garcia, E.: Ai-driven dna analysis in criminal investigations. Genomic Forensics Review **45**(2), 134–150 (2021)
19. Brown, H., White, M.: The future of ai in canine detection. Journal of Law Enforcement Technology **18**(1), 45–62 (2022)
20. Yan, W.-J., Li, X., Wang, S.-J., Zhao, G., Liu, Y.-J., Chen, Y.-H., Fu, X.: Casme ii: An improved spontaneous micro-expression database and the baseline evaluation. PLoS ONE **9**(1), 86041 (2014)
21. Wu, Z., Huang, Y., Wang, L., Wang, X., Tan, T.: A comprehensive study on cross-view gait based human identification with deep cnns. IEEE Transactions on Pattern Analysis and Machine Intelligence **39**(2), 209–226 (2017)
22. Hariprasad, Y., J., L.K.K., L., S., Iyengar, S.S.: Boundary-based fake face anomaly detection in videos using recurrent neural networks. In: Proceedings of SAI Intelligent Systems Conference, pp. 155–169. Springer, Berlin (2022)
23. Iyengar, S.S., Furton, K., Mohanty, P., Chaudhary, N.: AI-Enabled Forensic Investigations Network in Digital Sciences. Springer, Berlin (2025). Forthcoming Book
24. Iyengar, S. S., Seyedsina Nabavirazavi, Yashas Hariprasad, Prasad HB, and C. Krishna Mohan. Cyber Threat Intelligence and Security for Federated Learning in Digital Forensics. In Artificial Intelligence in Practice: Theory and Application for Cyber Security and Forensics, pp. 177–199. Cham: Springer Nature Switzerland, 2025.
25. Iyengar, S.S., Nabavirazavi, S., Hariprasad, Y., HB, P. and Mohan, C.K., 2025. Future of AI-Driven Digital Forensics. In Artificial Intelligence in Practice: Theory and Application for Cyber Security and Forensics (pp. 335–364). Cham: Springer Nature Switzerland.
26. Iyengar, S.S., Nabavirazavi, S., Hariprasad, Y., HB, P. and Mohan, C.K., 2025. Hybrid Detection of Malicious Portable Document Format (PDFs): Safeguarding Against Embedded JavaScript Attacks. In Artificial Intelligence in Practice: Theory and Application for Cyber Security and Forensics (pp. 257–290). Cham: Springer Nature Switzerland.
27. Soni, Jayesh, Subhash Gurappa, and Himanshu Upadhyay. A Comparative Study of Deep Learning Models for Image Super-Resolution Across Various Magnification Levels. In 2024 IEEE International Conference on Future Machine Learning and Data Science (FMLDS), pp. 395–400. IEEE, 2024.

AI-Driven Biometric Intelligence: Gender Prediction via CNN-Based Fingerprint Analysis

María Díaz and Jayesh Soni

1 Introduction

In forensics, human experts analyze evidence such as fingerprints, bones, and wounds to draw conclusions about identity, age, gender, and cause of death. However, this process is subjective, time-consuming, and prone to error. AI has the potential to revolutionize this field by providing objective, efficient, and scalable solutions.

Recent investigations have demonstrated that AI can uncover previously undetectable patterns in fingerprints, such as distinguishing fingerprints from different fingers of the same person. These advancements suggest that AI could also classify fingerprints by race, gender, or other demographic features. Although such classifications are less individual-specific, they may provide valuable insights for forensic and security applications.

Fingerprint-based gender classification could have applications in crime scene investigations, demographic studies, and personalized biometric systems. This research contributes to these fields by evaluating the performance of CNNs on a dataset of fingerprint images to classify gender.

2 Literature Review

Fingerprint analysis has traditionally relied heavily on human expertise and predefined rules, a method that, while effective, is subject to human error and lacks

M. Díaz (✉) · J. Soni
Department of Electrical and Computer Engineering, Florida International University, Miami, FL, USA
e-mail: mdiaz683@fiu.edu; jsoni@fiu.edu

S. S. Iyengar et al. (eds.), *Artificial Intelligence Driven Forensics*, Signals and Communication Technology, https://doi.org/10.1007/978-3-031-98036-7_19

scalability. The National Institute of Standards and Technology (NIST) highlights these challenges in their study [5], emphasizing the critical role of human factors in latent print analysis. However, advancements in machine learning (ML) and deep learning (DL) have paved the way for more objective and automated approaches to fingerprint analysis.

Recent studies have explored the potential of ML and DL for various biometric tasks, including age and gender classification. The work by M. Patel and U. Singh [3] provides insights into how deep learning can enhance the accuracy of age and gender recognition, demonstrating its effectiveness in biometric applications. Similarly, S. Hamdi and A. Moussaoui [2] conducted a comparative study, showing the superior performance of DL over traditional machine learning methods in these tasks.

Additionally, the study by G. Guo, et al. [1] from Columbia University introduces an innovative approach to fingerprint analysis. Using deep contrastive learning, their work highlights how AI can uncover intrapersonal fingerprint similarities, laying the groundwork for demographic classification tasks such as gender prediction.

Despite these advancements, prior research in gender classification using fingerprints faces significant challenges. For example, S. Patil [4] identified issues related to dataset robustness and feature extraction. Their integrated analysis of multimodal biometric traits underscores the need for high-quality datasets and advanced preprocessing techniques to enhance classification performance.

This study contributes to the field by specifically focusing on fingerprint-based gender classification using a CNN model trained on a balanced dataset. Unlike prior work, it emphasizes data augmentation, regularization, and architectural optimization to address class imbalance and overfitting—two major limitations noted in earlier research.

The remainder of this paper is organized as follows: Section III introduces the dataset and preprocessing steps. Section IV describes the CNN architecture and training strategy. Section V presents the evaluation results and analysis. Section VI discusses the implications and limitations. Finally, Section VII outlines potential future directions for enhancing fingerprint-based gender classification using AI.

3 Dataset Information

The dataset used in this study is a biometric fingerprint database designed for academic research purposes. It is made up of 6,000 fingerprint images from 600 African subjects and contains unique attributes such as labels for gender and hand and finger name. Below, detailed information about the dataset and its preprocessing steps is provided:

3.1 Source

The dataset used in this study is an open-source fingerprint dataset publicly available on Kaggle.

3.2 Class Distribution

This dataset comprises fingerprint images from both males and females

1. **Male:** 4770 images.
2. **Female:** 1230 images.

4 Data Preprocessing

4.1 Ensuring Uniformity and Size Consistency

All images in the dataset were resized to 128 × 128 pixels to ensure uniformity across the dataset. This step was taken to standardize the input dimensions, which is essential for consistent processing in the CNN model. The resizing introduced minimal distortion that does not compromise the integrity of the fingerprint patterns, making the data suitable for analysis. Additionally, the size distribution of the images was analyzed to confirm that all images now have the same dimensions, ensuring a consistent and balanced dataset for training and evaluation.

4.2 Balancing the Dataset

The dataset originally consisted of 1,230 images labeled as *female* and 4,770 images labeled as *male*, introducing a significant class imbalance. Two strategies were evaluated to balance the dataset:

1. **Undersampling:** Reducing the number of *male* samples to match the *female* samples.
2. **Oversampling:** Duplicating and augmenting *female* samples to match the number of *male* samples.

The latter approach was adopted to maximize the dataset size. Each *female* sample was duplicated randomly until the total count reached 3,000 images per class, ensuring an equal representation of both categories in the dataset as demonstrated in Fig 1.

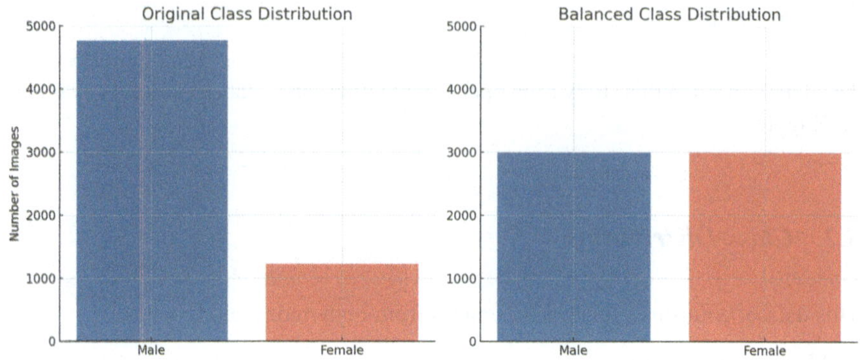

Fig. 1 Comparison of class distribution before and after balancing

Table 1 Dataset split for training, validation, and testing

Split	No. of images	Percentage
Training	4320	72%
Validation	1080	18%
Testing	600	10%

4.3 *Final Dataset Distribution*

After preprocessing and balancing, the dataset consisted of 6,000 images, equally distributed between the two classes (*female* and *male*). This final dataset was split into training, validation, and testing sets to facilitate model training and evaluation as shown in Table 1.

4.4 *Data Augmentation*

To address potential overfitting and enhance the model's generalization ability, data augmentation was applied to the training dataset. This approach artificially expanded the dataset by introducing variations, making the model more robust to unseen data. The following augmentation techniques were implemented:

1. **Random Rotations:** Images were rotated randomly within a specified range to simulate various fingerprint orientations.
2. **Zoom Adjustments:** Minor zooms were applied to simulate different scales of fingerprint images.
3. **Width and Height Shifts:** Horizontal and vertical translations were performed to simulate small shifts in fingerprint placements.

These augmentations were only applied to the training dataset to avoid artificially altering the validation and test datasets, which are intended to measure the model's performance on untouched data. In Fig. 2 we can see a sample of the appearance of the images that will be used for training after applying data augmentation.

Additionally, pixel values for all images were normalized to the range [0, 1]. This scaling process was essential for improving model training stability by ensuring numerical consistency across the dataset.

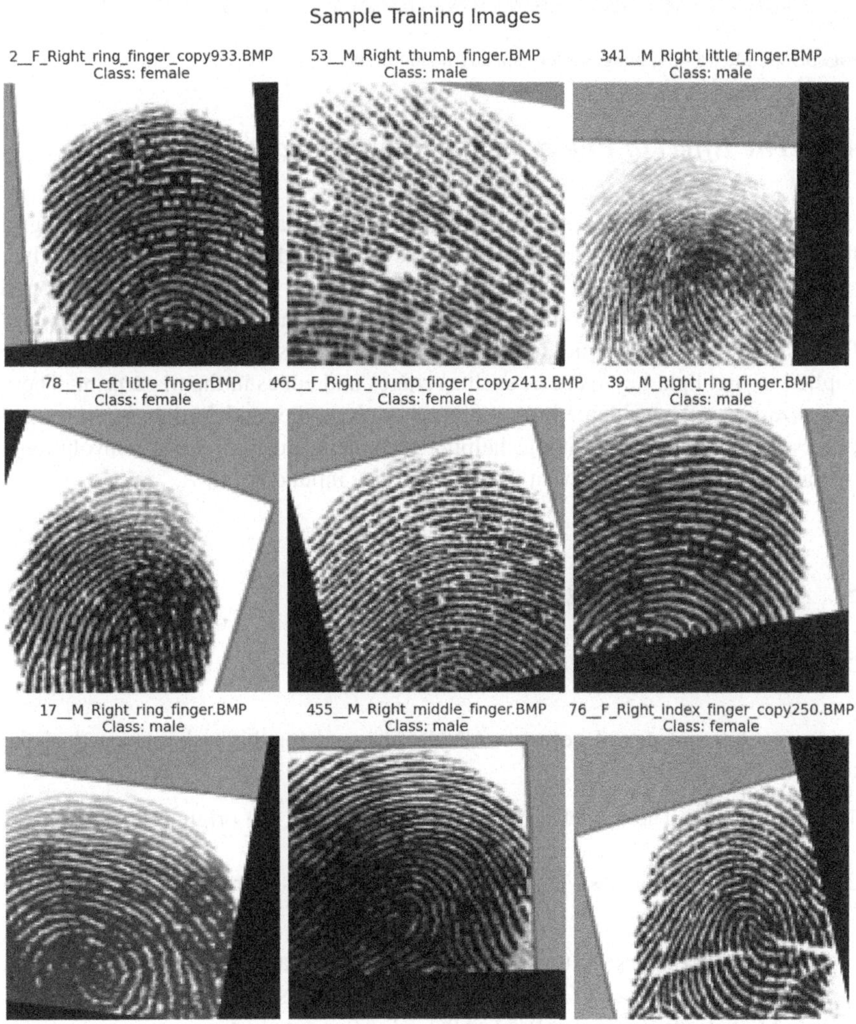

Fig. 2 Sample training images after data augmentation process

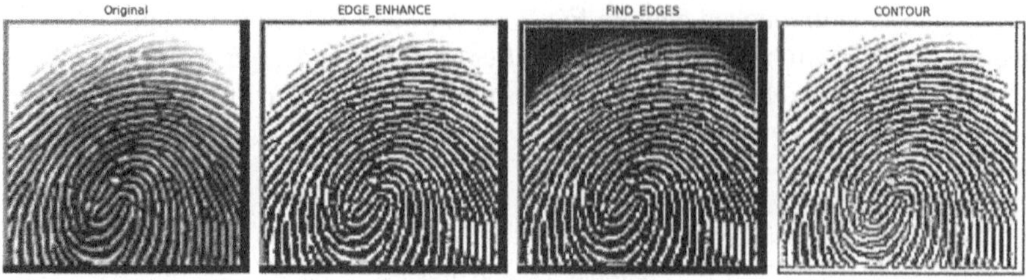

Fig. 3 Sample training images after data augmentation process

4.5 The Effect of Filters on Fingerprint Images

CNNs apply filters to images during convolutional operations to extract meaningful features as part of their learning process. To provide a clearer understanding of how these filters interact with fingerprint images, four examples are plotted in Fig. 3 showcasing how different filters can enhance specific characteristics. For instance, edge-enhancing filters and pattern-emphasizing transformations were applied to highlight key features such as ridges and edges in the fingerprint. These visualizations offer valuable information about how the model might interpret and process features during training, helping to understand the role of convolutional operations in extracting relevant details from raw input data.

5 AI Model Building

5.1 Model Architecture

The convolutional neural network (CNN) designed for this task consisted of the following components:

1. **Input Layer:** The model takes images maintaining the original dimensions as input.
2. **Convolutional Layers:** Three convolutional blocks were used, each comprising:
   (a) **Conv2D Layers:** Extract features using 32, 64, and 128 filters, respectively, with kernel sizes of (3 × 3) and padding set to "same."
   (b) **Activation:** ReLU was used as the activation function.
   (c) **Pooling Layers:** MaxPooling was applied with a pool size of (2 × 2) to reduce spatial dimensions.
   (d) **Batch Normalization:** Added after each pooling layer to stabilize and accelerate training.

3. **Fully Connected Layers:**
   (a) A Dense layer with 256 neurons, followed by a Dropout layer (rate = 0.5) to prevent overfitting.
   (b) The final Dense layer served as the output layer.
4. **Output Layer:** Initially, the output layer used a sigmoid activation function for binary classification. However, this was later replaced with a softmax activation function to handle categorical labels more effectively, yielding better results.

5.2 Training Configuration

The model was trained with the following configurations:

1. **Loss Function:**
   (a) Binary Cross-Entropy: Used during initial experiments for binary classification.
   (b) Categorical Cross-entropy: Adopted in the final implementation with a softmax activation function, as it yielded better classification results.
2. **Optimizer: Adam** optimizer with an adaptive learning rate.
3. **Learning Rate Scheduler:**
   (a) **ReduceLROnPlateau:** Monitored validation loss and reduced the learning rate by a factor of 0.3 after 3 epochs without improvement, with a minimum learning rate of 1×10^{-6}.
4. **Metrics: Accuracy** was used as the primary evaluation metric during training.
5. **Callbacks:**
   (a) **Early stopping:** Implemented to terminate training if validation performance did not improve after 8 epochs, preventing overfitting and saving resources.

5.3 Model Observations

Switching to categorical cross-entropy with a softmax output layer significantly improved the model's classification performance compared to the binary cross-entropy approach. The softmax activation allowed the model to better handle the two-class problem in this specific context, ensuring a more balanced learning process.

Furthermore, the decision to gradually increase the number of convolutional filters across layers allowed the network to capture lower-level features initially and progressively learn more abstract and complex patterns deeper in the network.

The model was trained over 30 epochs; however, training often stopped earlier due to the early stopping criterion, which monitored validation performance. This ensured that the model avoided overfitting and generalized well to unseen data.

These architectural and training adjustments underscore the importance of iterative experimentation and optimization when designing convolutional neural networks for domain-specific challenges such as fingerprint classification.

6 Evaluating the Results

6.1 Model Training Dynamics

Figure 4 shows the evolution of the training and validation accuracy, as well as their loss over the epochs. The learning rate adjustments, managed by the ReduceLROnPlateau callback, are also depicted in the third plot. These metrics provide a clear overview of the model's convergence behavior and its generalization ability during training.

The training accuracy steadily improves, reaching 64.56%, while validation accuracy fluctuates around 0.64. Training loss consistently decreases, indicating effective learning, whereas validation loss shows some irregularities but remains relatively stable. The learning rate plot demonstrates the model's adaptation during training, with reductions at specific epochs to fine-tune performance.

6.2 Model Predictions

The evaluation of the trained CNN model was performed using the reserved test set comprising 600 samples evenly distributed between the "male" and "female" classes. The model achieved a test accuracy of 67.33% and a test loss of 0.611, showing a notable improvement over earlier iterations.

To provide information on the performance of the model, a sample of predictions from the test dataset is visualized in Fig. 5. Each image is displayed alongside its corresponding true class and predicted class. Correct predictions align the true and predicted labels, while mismatches highlight errors made by the model.

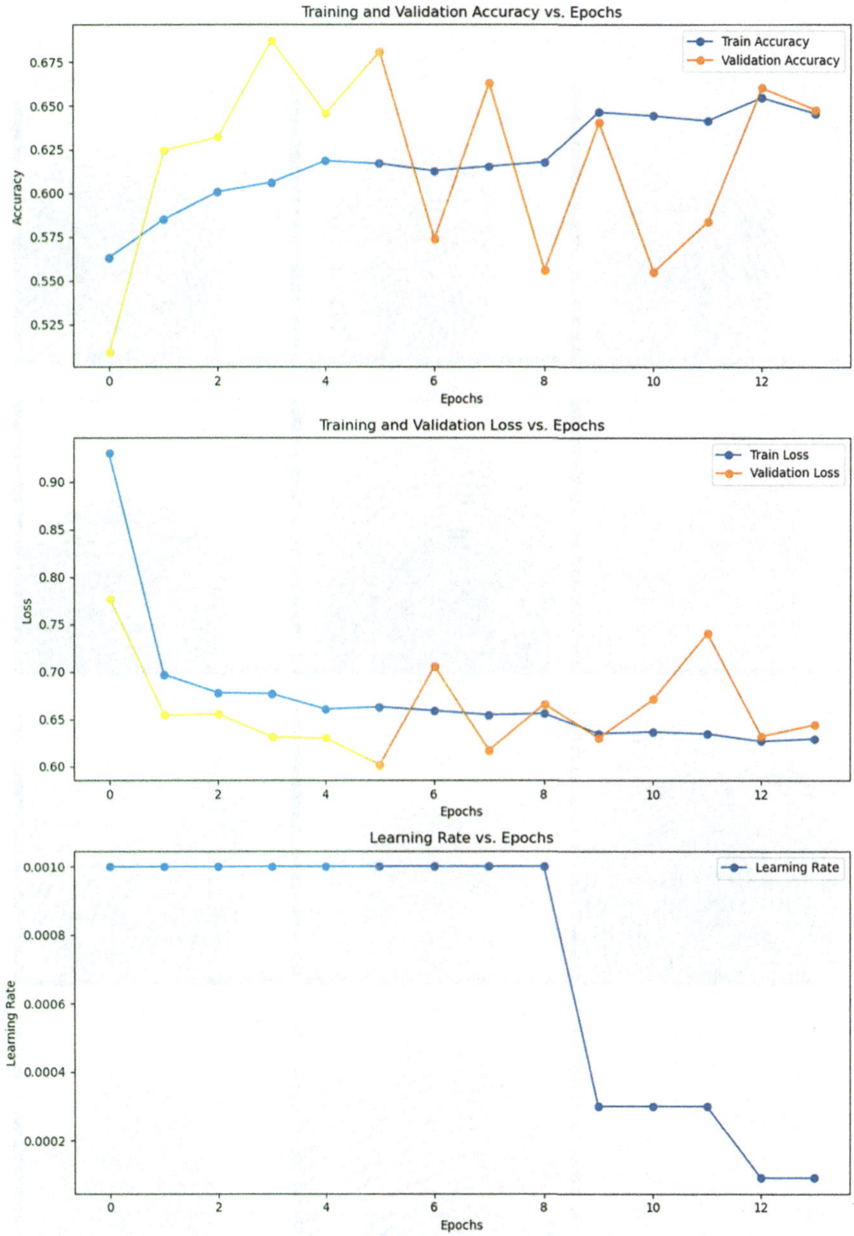

Fig. 4 Evolution of training and validation metrics over epochs

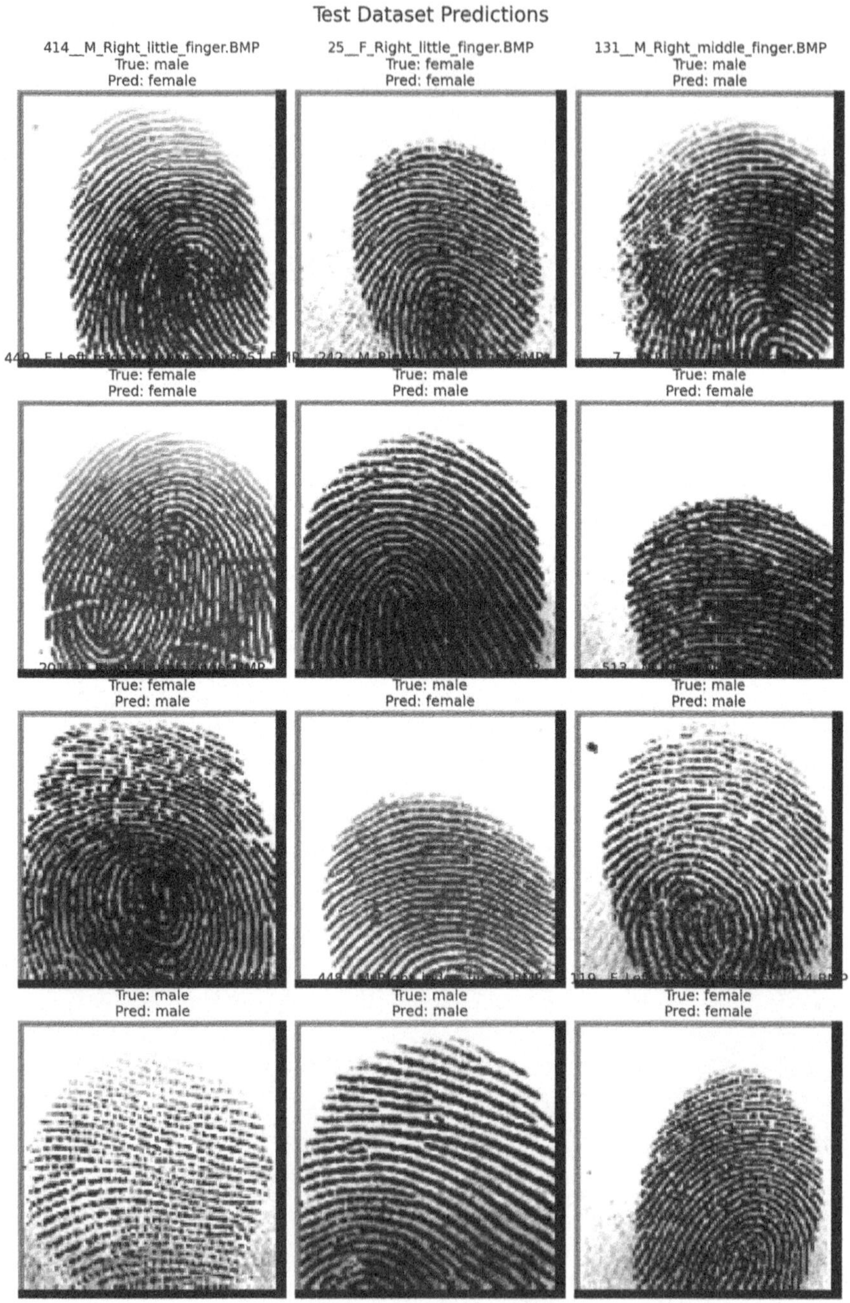

Fig. 5 Sample predictions from the test dataset

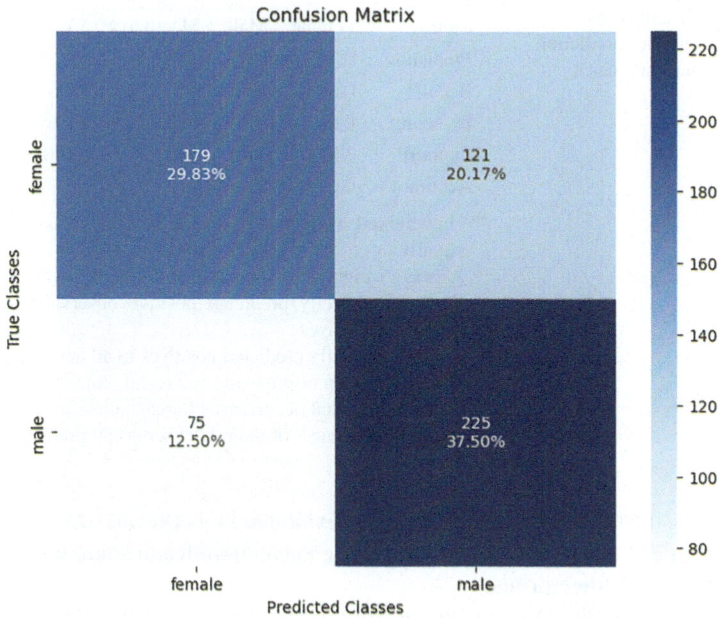

Fig. 6 Confusion matrix for test dataset predictions

6.3 *Confusion Matrix*

Figure 6 displays the confusion matrix, which provides insights into the model's ability to correctly classify each class. While the model performs better in identifying "male" fingerprints, it still demonstrates some challenges in distinguishing "female" fingerprints, as evident by the misclassifications.

6.4 *Classification Report*

Table 2 presents the precision, recall, and F1-score for each class. The metrics indicate that while the model shows a relatively balanced performance across classes, there is room for improvement in recall for the "female" class.

6.5 *Key Findings*

Key observations derived from the evaluation include:

Table 2 Classification metrics for gender prediction model on the test dataset

Metric	Female	Male	Macro avg[a]	Weighted avg[b]
Precision[c]	0.70	0.65	0.68	0.68
Recall[d]	0.60	0.75	0.67	0.67
F1-score[e]	0.65	0.70	0.67	0.67
Support[f]	300	300	600	600
Accuracy[g]	0.67 (600 samples)			

[a] Unweighted average across classes. Each class contributes equally
[b] Average weighted by the number of instances in each class
[c] Ratio of correctly predicted positive observations to total predicted positives
[d] Ratio of correctly predicted positives to all actual positives
[e] Harmonic mean of precision and recall, balancing both
[f] Number of actual occurrences for each class in the dataset
[g] Overall correctness of the model across all predictions

1. The classification of "male" fingerprints exhibited higher recall (0.75) compared to "female" fingerprints (0.60), suggesting that certain features are more distinctive in "male" fingerprints.
2. Adjustments to the model's architecture, particularly in the number of filters in convolutional layers, significantly improved performance.
3. The use of learning rate adjustment (ReduceLROnPlateau) allowed the model to stabilize during training, resulting in better convergence.

Despite these improvements, challenges in distinguishing between classes persist, highlighting the inherent complexity of the task.

7 Conclusion

The results of this study demonstrate that CNN-based architectures can offer moderate success in gender classification from fingerprint images, but also underscore the complexity of gender classification based on fingerprint images. While the final model achieved a moderate test accuracy of 67.33%, with relatively balanced performance across both classes, the results suggest significant challenges remain.

7.1 Limitations and Challenges

Key limitations affecting performance include:

1. **Dataset limitations:** Publicly available datasets for fingerprints are limited in quantity and often lack the diversity required for robust generalization.

2. **Inherent complexity of the task:** Fingerprint patterns may not exhibit sufficient distinctiveness for gender classification, especially in the absence of domain-specific preprocessing techniques.
3. **Technical constraints:** While the CNN architecture used in this study showed moderate success, deeper architectures or transfer learning methods might be needed to achieve significant improvements.

7.2 *Implications and Applications*

The ability to classify gender from fingerprint images has several potential applications and implications, particularly in forensic science, as well as broader scientific and judicial advancements:

1. **Forensic investigations:** Gender classification can aid investigators in narrowing down suspects in criminal investigations, particularly when fingerprints are one of the few pieces of evidence available.
2. **Identity reconstruction:** In scenarios such as natural disasters or accidents, gender classification from incomplete fingerprint evidence can support identity reconstruction efforts.
3. **Judicial applications:** Automating gender classification could provide additional tools for verifying identity claims in judicial processes, improving objectivity and reducing human error.
4. **Advancements in biometric research:** Understanding gender-based fingerprint patterns could drive further innovations in biometric technologies, expanding their use cases beyond traditional identification tasks.

The integration of AI in forensic sciences opens the door to unprecedented levels of objectivity and efficiency. For instance, AI has already been shown to detect patterns in fingerprints that human experts often overlook, such as intrapersonal similarities across fingerprints of different fingers, as highlighted in recent studies. These advancements suggest the potential for AI to refine existing forensic methodologies and even uncover new patterns previously undetectable by human analysis.

8 Future Work

To address the limitations encountered in this study and further enhance the performance of gender classification from fingerprint images, several directions for future research are proposed:

1. **Dataset Expansion:** Acquiring larger, high-quality fingerprint datasets that encompass diverse populations and fingerprint patterns is essential. This will improve the model's generalization capabilities across various demographic groups.
2. **Advanced Data Augmentation:** Exploring more sophisticated data augmentation techniques, such as GAN-generated fingerprints or geometric and photometric transformations, could help increase the variability and size of the dataset.
3. **Deeper Architectures and Transfer Learning:** Implementing deeper CNN architectures or leveraging transfer learning with pre-trained models on biometric or image classification tasks might significantly improve performance by capturing more complex patterns in the data.
4. **Domain-Specific Feature Extraction:** Collaborating with forensic science experts to integrate domain knowledge into the feature extraction process. This could include identifying gender-specific minutiae or ridge patterns in fingerprints that are not easily detectable by current AI methods.
5. **Exploring Hybrid Models:** Investigating hybrid approaches that combine CNNs with other machine learning models, such as ensemble techniques or classical methods, to improve robustness and accuracy.
6. **Multi-task Learning:** Expanding the scope of the model to perform multiple tasks simultaneously, such as gender and age prediction or even identifying unique characteristics within fingerprint groups.
7. **Real-world Applications:** Testing the model in real-world forensic scenarios to evaluate its utility, reliability, and performance in practical applications.

This future work aims to overcome the current study's challenges and limitations, paving the way for more accurate, robust, and impactful AI-based solutions in biometric and forensic sciences.

Declarations

- Data availability The dataset used in this study was obtained from Kaggle.com https://www.kaggle.com/datasets/ruizgara/socofing/data
- Code availability The data and code for this project is available on https://github.com/mdiaz683/CNNFingerprints

References

1. Guo G, et al. (2024) Unveiling intra-person fingerprint similarity via deep contrastive learning. Science Advances 10(2):1–12. https://doi.org/10.1126/sciadv.adi0329
2. Hamdi S, Moussaoui A (2020) Comparative study between machine and deep learning methods for age, gender and ethnicity identification. In: 2020 4th International Symposium on Informatics and its Applications (ISIA), M'sila, Algeria, pp 1–6, https://doi.org/10.1109/ISIA51297.2020.9416549
3. Patel M, Singh U (2023) Age and gender recognition using deep learning technique. In: 2023 3rd International Conference on Smart Data Intelligence (ICSMDI), Trichy, India, pp 238–245, https://doi.org/10.1109/ICSMDI57622.2023.00052

4. Patil S (2023) Gender classification: An integrated analysis of multimodal biometric data using deep learning techniques. International Journal of Research Publication and Reviews (IJRPR) 4(10):2855–2865. Available online: https://ijrpr.com/uploads/V4ISSUE10/IJRPR18480.pdf
5. Taylor M, et al. (2012) Latent print examination and human factors: Improving the practice through a systems approach. NIST Interagency/Internal Report (NISTIR) 7842, https://doi.org/10.6028/NIST.IR.7842

The New Face of Facial Recognition: Using a Red-Yellow-Blue Palette to Decrease Bias in Feature Detection

Patrianna Napoleon and Jerry F. Miller

1 Introduction

1.1 Origins of Facial Recognition

Facial recognition was originally developed by Woodrow Wilson Bledsoe in the 1960s. Since then, it has been incorporated into many applications, including surveillance for law enforcement and security applications. In addition to law enforcement, facial recognition is used in retail, government facilities, education, and even in our mobile devices that we use every day. As our reliance on advanced technology increases, facial recognition will continue to have an enormous impact. Many corporations and businesses rely heavily on this technology, especially when it comes to solving and identifying crimes like shoplifting. The ability to recognize someone's face with accuracy is a primary goal to accurately identify a person. Many of these systems perform very well and are robust. In today's age of facial recognition technology, there has been a spark of problems arising when detecting faces.

Many of the problems are found when the recognition system is attempting to detect African American woman between the ages of 18 and 30 [1]. Facial recognition systems are trained using datasets which are a set of images used to train a learning model to be able to detect parts of the face features. Features like the nose and chin. Recent studies have found there are many factors that can contribute to the accuracy of face detection. A recognition system will most likely learn from any image even when the images in the dataset have bias or lack

P. Napoleon (✉) · J. F. Miller
Florida Agricultural and Mechanical University, Department of Computer and Information Sciences, Miami, FL, USA
e-mail: Jerry.miller@famu.edu

S. S. Iyengar et al. (eds.), *Artificial Intelligence Driven Forensics*, Signals and Communication Technology, https://doi.org/10.1007/978-3-031-98036-7_20

fairness in what it is learning. Today, the significant problems in facial recognition that contributes to the misidentification of African Americans between the ages of 18 and 30 include environmental factors, the lack of demographic ranges in datasets, image quality, and the system's ability to recognize specific features in African Americans unlike other demographics. A study on the accuracy of facial recognition comparing stars discovered shows that a KNN model has the highest recognition rate for white people (95.7%), while African Americans scored a lower accuracy of 85% [2]. Leaving the other percentage vulnerable to misidentification in this demographic. A US government study also suggested that facial recognition algorithms are far less accurate at identifying African Americans and Asian faces compared to Caucasian faces [3]. Another study conducted by institutions including Johns Hopkins University and the Georgia Institute of Technology programmed AI-trained robots to scan blocks with people's faces from different demographics [4]. After the faces were scanned, the robots were assigned to designate which blocks were criminals. The robots consistently labeled the blocks with African American faces as criminals [5], yet none of the much lighter complected subjects was identified as such. Is there bias within the facial recognition systems when it comes to certain demographics? Does this impair the systems' ability to provide accuracy for daily operations like investigating crimes, identification, and surveillance of people? Apparently so.

1.2 Case Studies: Facial Recognition Misidentification

There are many reports of African Americans who are falsely accused of a crime due to the result of facial recognition used by police when matching an unknown suspect photo. There has been six significant cases; all six cases consisted of African Americans [6].

In 2023, an African American mother encountered an unforgettable experience early morning. Porcha Woodruff suffered lasting trauma from being a victim of misidentification using facial recognition. She was unexpectedly arrested for a carjacking and robbery and then later released from the Detroit detention center on a $100,000 personal bond. She was identified as a suspect through a photo lineup of victims along with unreliable facial recognition [7].

In a second, closely related incident in Detroit, Robert Williams was unexpectedly arrested in January 2020, spending a night in a detention center leaving his family worried for a crime he didn't commit [8]. A blurry photo image was subject to a facial recognition system and incorrectly matched a photograph of Roberts old driver's license picture. This is another case demonstrating bias and inaccuracies in facial recognition systems. The incident highlights the process of investigations when police are trying to identify suspects of a crime using a combination of facial recognition and photo lineups.

In 2020, the Detroit police chief stated that Detroit's facial recognition technology failed 96% of the time, yet it is still used in investigations [9]. It is likely that many other police departments around the world are using this technology and are leading to multiple false arrests, such as those that occurred in Louisiana, New Jersey, Maryland, and Texas. Other cities and states in the United States, including San Francisco, California; Austin, Texas; and Portland, Oregon, have temporarily banned its use because of concerns about privacy and racial bias.

1.3 Demographic Problems in Facial Recognition

Studies have shown that a significant inaccuracy is likely in the identification phase of facial recognition. Additionally, there is bias in the classification of specific demographics. When photos are gathered to process and are fed into a learning model for the system to recognize faces, the photos are classified. For example, a chin in a photo would be classified as "chin" so the system knows the classification [10]. However, due to the bias, the results are not always clear.

In 2015, Jacky Alcine logged into Google photos and noticed that her photo album was classified as gorillas. Why did this facial recognition system software categorize Jacky and his friends as primates? During the development phase, depending on how the machine is learning different features of faces and how data is fed into the system, the recognition will output what it identifies as the most closely related elements. So, if the dataset is miscategorized, then the facial recognition system will be biased providing incorrect results. Existing datasets of photos that contain a collection of images are likely biased toward Caucasian faces. Therefore, it is highly likely that one reason behind misidentification can also include the lack of trained faces for African Americans and people of color. This may also be a result of different age ranges affecting the complexions of people of color throughout the world [10].

2 Principals of Facial Recognition

As facial recognition continues to develop, more research has been conducted to propose improved algorithms and corrections. Most facial recognition technology is based on the extraction and comparison between faces and face features in an image. Face recognition systems will search and then match the face image feature data with a feature template stored in the database for recognition. A variable is set to a specific feature based on the data of an image. Once the variable is set, the system then finds similarities that exceed or match the variable leading to a matched result. The matched result will be provided as an output. When matching face features, the face feature is recognized and compared with the face feature variable and is judged to match a facial identity based on similarities of the face and the face variables. There are two major categories in facial recognition. One is confirmation which is the process of comparing one image to another to confirm an image. The second category is identification; this process matches one image to another while comparing the image to stored information to produce an output matching an identity [2]. This process is then divided into four pipelines. The different pipelines that make up facial recognition to produce a successful output of a compared image are detection, alignment, representation, and verification [11]. This research focuses on enhancing both the detection and alignment pipelines by reducing bias when detecting specific features of the face.

2.1 Facial Recognition Pipelines: The Process

There are four common pipelines used in facial recognition such as face detection, face alignment, feature extraction (representation), and feature matching or verification [11]. Face detection is commonly the first stage for facial recognition to have the ability to first detect faces presented from a given image and extract parts of a face if it exists. This phase crops areas of the face. So, the file can be compressed for further feature extraction. Algorithms are deployed to perform this phase.

The most common methods used for face detection are Haar Cascade, Dlib (HOG), MTCNN (Neural, i.e., Multi-Task Cascaded Convolutional Neural Network), and Dlib (CNN) [12]. Haar Cascade is a method for image processing that classifies objects based on texture, shape, and color. A color can be coded to identify an object; thus, lighting conditions can play a very important role when detecting objects [13]. The algorithm's ability to detect objects is based on the features of that object. For example, a pixel value can be calculated to represent a section of an image that identifies an area under an eye or near a cheek. Dlib is an open source, modern C++ toolkit containing machine learning algorithms and tools for creating complex software to solve real-world problems. Dlib provides another face detection model based on HOG (Histograms of Oriented Gradients) for human detection using an image pyramid to extract features from image data. This method only works on frontal and on slightly shifted frontal images. Dlib Convolutional Neural Network (Dlib(CNN)) extracts features from an input image data to increase the number of features and can accommodate various face orientations. The Multi-Task Cascaded Convolutional Networks algorithm or MTCNN is another method for face detection and face alignment. MTCNN can detect faces in various conditions and scales. Each technology provides a unique perspective for facial feature detection.

The next step in the pipeline is face alignment. Images are aligned to improve the accuracy of the face recognition model. For example, on an image, eyes can be rotated and aligned with each other until both eyes are aligned horizontally to improve the accuracy of eye location. Feature extraction extracts components of the face. The components of the face consist of face features that differ from one person to another. Feature classification is the ability to compare images of the face and match it with another image to classify one's identity (Fig. 1).

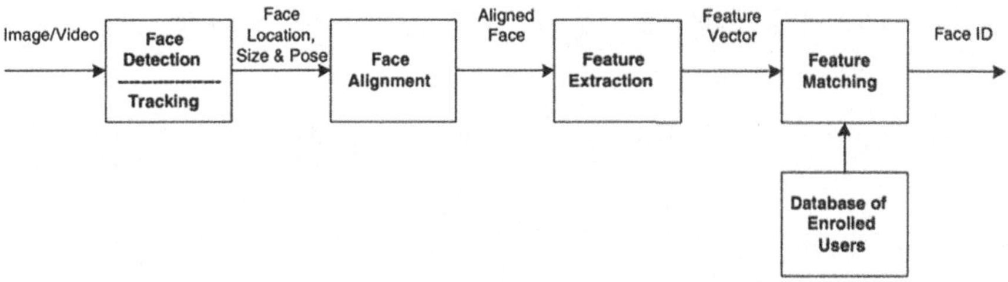

Fig. 1 Steps in face recognition process as outlined in the *Handbook of Face Recognition*, 2011 [14]

3 DLIB Methodology

Dlib is one of the most utilized algorithms and methods used in many facial recognition systems. Dlib has been used to enhance many recognition systems that utilize this method as a means of reducing bias in those systems to increase accuracy when detecting African Americans from different backgrounds [15]. Therefore, researchers selected the Dlib methodology for this research to improve both the detection and alignment pipelines by reducing bias when detecting specific features of the face. Specifically, this research focuses on the Dlib facial land marking algorithm which is used to detect areas of the face and then used to compare features.

Dlibs facial landmark detection identifies areas of the face by estimating the location of facial structures in an image. Dlib uses 68 land points to find areas of the face such as the nose, chin, eyes, and eyebrows [16]. The Dlib facial landmarking model was trained using the iBUG 300−W-dataset which is a dataset that includes variations of different subjects, poses, illuminations, and face features [17] (Fig. 2).

The authors in [2] conducted a study on Dlib's accuracy to detect different demographics based on television actors. 1000-star photos were used to verify the accuracy of the algorithm [2]. Out of the 1000 images, 912 images were successfully extracted. Overall feature points found on the faces had an accuracy of 91.2%. The authors suggested that the comparison between the successful and unsuccessful images in the study could have been affected due to faces being too small or having unclear or occluded areas in the image. The study then focused on the recognition rate of different skin colors. The training model was developed with the use of K-nearest neighbor (KNN), a supervised learning classification algorithm that determines the class of a new data point by looking at its nearest neighbors in the training data. In this case, KNN identified the closest faces in a database to a new, unknown face and assigned the new face to the label of the majority class of those faces closest. The model combined KNN and Dlib and demonstrated a recognition rate for people with lighter skin tones as 95.7%. The recognition rate for darker skin tones was lower. Researchers concluded that the reason for the higher recognition rate for lighter skin tones may have been due to the feature points used to train the KNN model to detect faces. Researchers in this paper focused on improving facial landmarking when used on different skin colors and how these landmarks could be better accentuated to detect ranges of skin colors.

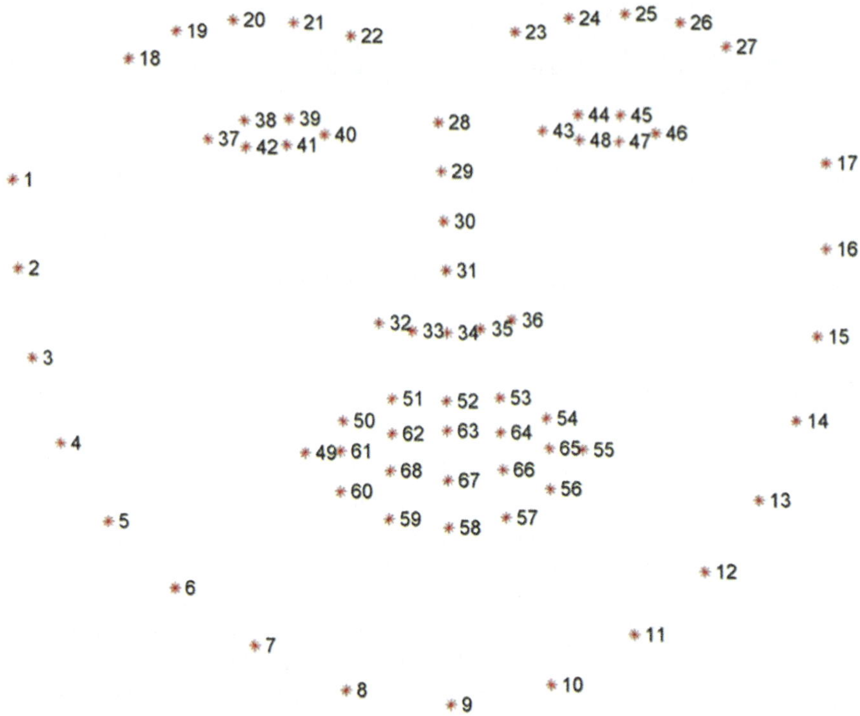

Fig. 2 Dlib facial 68-point landmarking [17]

4 The Red-Yellow-Blue (RYB) Experiment

4.1 The Red-Green-Blue Dilemma

Facial landmarking can likely influence how a facial recognition model is trained to then be able to identify a person based on features learned from the landmarking. Bias is likely present in the detection stage when detecting and extracting areas of the face. If the facial landmarking algorithm incorrectly places landmarks on an image, the model will be trained incorrectly. For example, if the algorithm incorrectly adds a landmark to areas of a nose when it was a cheek, the training model then learns and identifies similar areas as a nose, leading to incorrect identification of a person when an image is compared to one another. Features which are not clearly defined in the training set based upon complexion and skin color would then adversely affect the results when applied to detecting faces of different demographics.

The Dlib landmarking image processing is designed to accept images that have RGB (red, green, blue) pixel colors. Each pixel color was defined with a trait [15].

The trait allows image processing to know how to handle each type of pixel. A pixel type can be RGB, as this pixel type is what is used to understand how colors are stored in memory to make up the image. RGB colors are the most used color space [18]. However, RGB color space also can be an inaccurate measurement of color, because representation of color is not consistent with the perception of human eyes. RGB colors are most likely perceived based on what we think the colors of red, green, and blue and combinations can represent. This results in a big difference between the calculated value of color difference and the intuitive perception of human vision. Thus, Dlib is most likely using pixel traits (RGB colors) that may not match the colors that we see through our own perception. Therefore, there will likely always be a significant gap of colors that do not exactly match what we see on the computer screen versus the real world. For example, we may see a red car in an image, but there are other colors and combinations that make up red that we view on the computer screen versus a red car in real life. Hence, when areas of an image are assigned a pixel that matches a color trait like red, it is likely possible that pixel is assigned the incorrect color. This perceived difference affects the color condition of the image when representing a landmarked face which provides identity features of a facial image.

This research hypothesis that there are likely missing colors in RGB that may not be presented during pixel classification when pixel traits are assigned to an image. Computer systems use RGB colors [19]. A monitor or TV screen generates three colors of light (red, green, and blue), and the different colors we see are due to different combinations and intensities of these colors with the presence of light.

4.2 A New Approach with an Old Color Scheme

There is another classification of colors known as RYB (red, yellow, blue). RYB is typically used for fine art and dates to the early nineteenth century when color was first defined around Isaac Newton's circular dimension hue optics for how we perceive color in 1704 [20]. Although extensively used in fine art, RYB is not a color pattern that exists in computer screens as primary colors. RYB as an historical relationship was overturned in the second half of the nineteenth century for the more common RGB color scheme used today. But these colors continued to be used by artists and have found their way into modern education for fine art.

The RGB scheme can be used to produce colors of white light as depicted in Fig. 3, while RYB lends itself to better depict darker shades and colors. If RGB colors lead to lighter colors and white light, then it is likely that other areas of colors that do not exist when light is limited or non-existent would not be easily identified in the RGB schema. The RYB color palette might provide more clarity for human perception, highlight subtle differences on computer screens, and ultimately allow finer granularity in facial recognition landmarking.

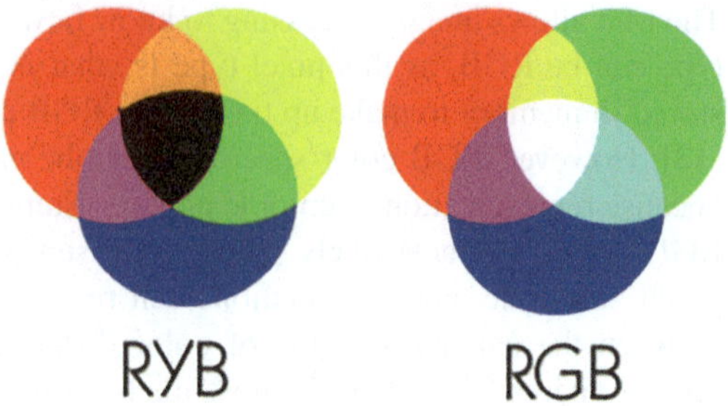

Fig. 3 RGB and RYB color palettes [19]

This experiment conducts converting an RGB image to RYB due to the possibility of missing colors. The missing colors highly likely contribute as an additional factor to the cause of bias in face detection algorithm that use RGB pixels to classify colors. The facial recognition system will likely fail to correctly identify certain demographics due to the lack of color range in RGB color scale. RGB colors likely have colors closer to white making the colors of an image appear to be brighter colors of red, green, and blue which can falsely perceive a color of an individual incorrectly. The use of RYB colors will likely include the absence of light allowing the RGB colors to appear darker, creating different color range scales of red, green, and blue. RYB (red, yellow, blue) will likely increase the color ranges in demographics by preserving more colors of an image when it's processed to classify pixel traits. Due to the nature of computer screens only using RGB, RYB can only be simulated to test this theory. RYB can be approached by converting RGB values to CMY (cyan, magenta, yellow) and then CMY to RYB. Since RYB is not a color standard in digital processing, RGB can be simulated to RYB using mathematical formulas:

RGB to CMY Conversion Formula $C = 1 - \mathrm{R}$

$$\mathrm{M} = 1 - \mathrm{GY} = 1 - \mathrm{B} \tag{1}$$

CMY TO RYB

$$\mathrm{Rry}\ b = R \tag{2}$$

$$\mathrm{Yry}\ b = (Y + G)/2\ \mathrm{Bry}\ b = (B + M)/2 \tag{3}$$

The dataset used was from a face research lab in London. Ten images in this dataset was converted from RGB to RYB changing the color of the image from what we normally see to appear green and purple on a digital screen. Ten (10) images of females and males from different demographics was then used for Dlib facial landmarking. Dlib applied 68 facial landmarking points on areas of the face such as the ears, eyes, nose, nose bridge, lips, chin, and eyebrows. This hypothesizes the initial phase of face detection will most likely detect African Americans with a higher accuracy using datasets that have colors of RYB (red, yellow, blue) instead of RGB (red, green, blue) when extracting face characteristic (Fig. 4).

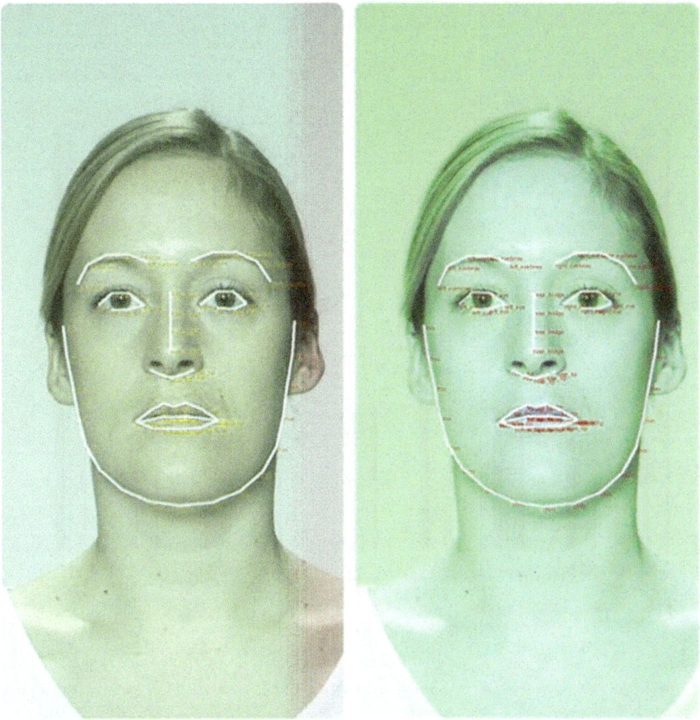

Fig. 4 RGB vs RYB identification of landmarks [21]

4.3 Assumptions

Datasets that contain RYB colors have a variety of darker colors that are not present in RGB. RGB algorithms consist of detecting colors that have lighter shades of red, green, and blue. Computer screen colors are affecting the natural colors that are found in African Americans demographics. The darker shades associated with RYB may provide more facial data when detecting the face of African Americans.

4.4 Results

To interpret the results, ten (10) images of RGB and ten (10) images of RYB were compared to each other visually based on the facial landmarking such as the left eye, right eye, chin, nose, nose bridge, right eyebrow, and left eyebrow. The features of the face were scored on a level from 0–3, with a zero (0) score demonstrating negative results where the facial landmarking was incorrectly placed on the RYB image. A score of one (1) showed low changes of facial landmarking, while a

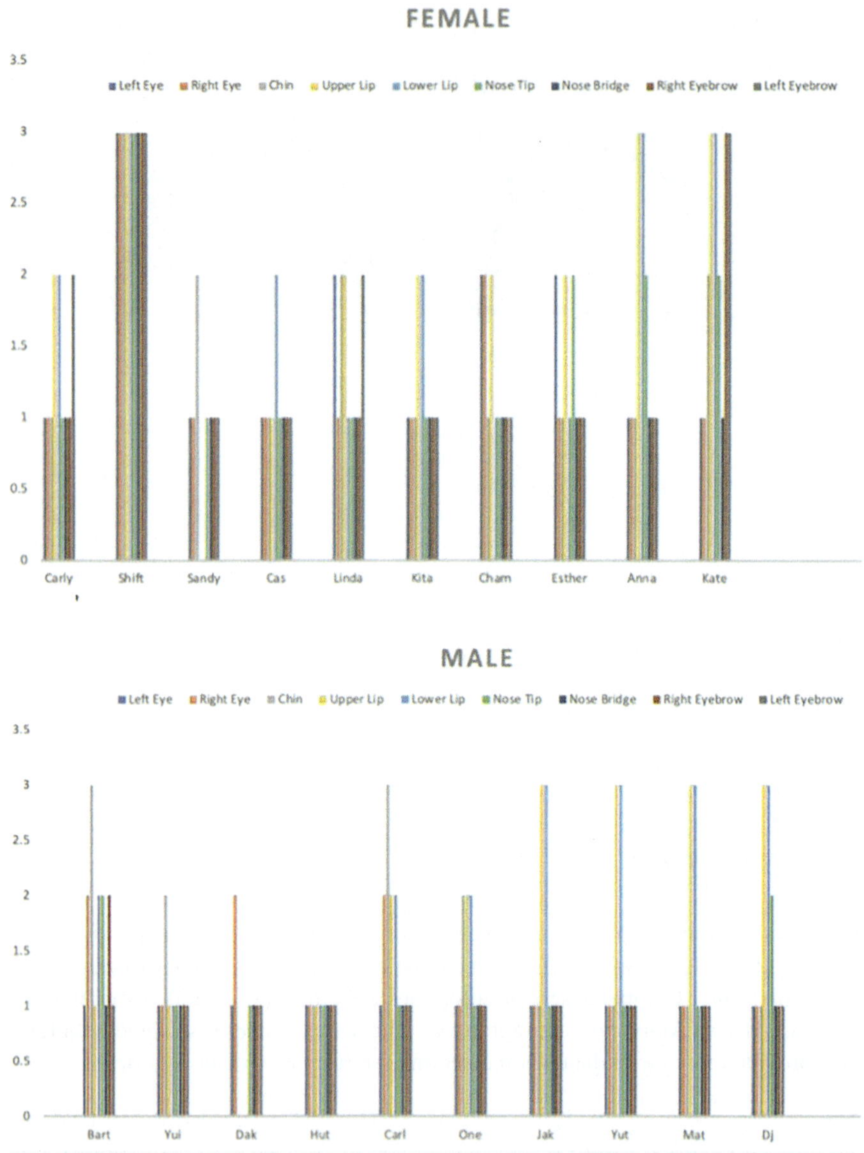

Fig. 5 RGB vs RYB results

two score (2) showed moderate changes, and a three (3) score showed significant changes from the facial landmarking placed on the RYB image. Both male and female images of RGB and RYB were evaluated during this experiment (Fig. 5).

The use of RYB images showed significant improvement on all images regardless of demographic, but there was a higher significance in images with darker colored

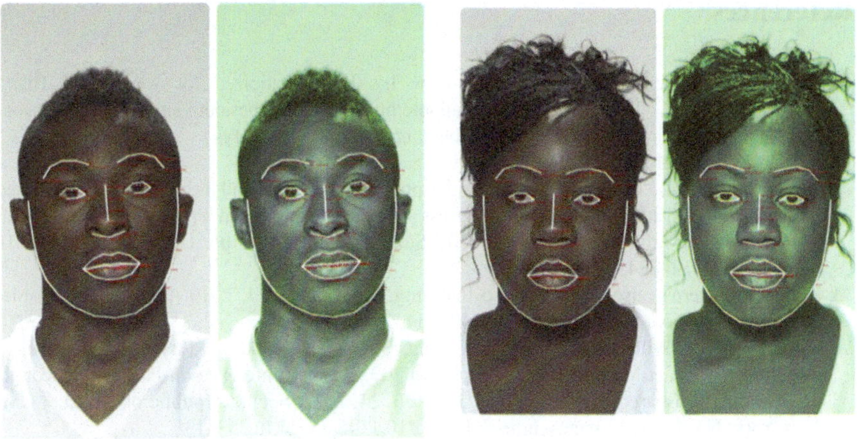

Fig. 6 Male RGB to RYB and female RGB to RYB [21]

pixels. In each feature of the face, the facial landmarking improved to a rating change of at least a score of 1. Significant changes not only appeared more in the African American demographic but for every demographic solely on adjustment of the landmarking in the upper and lower lip as well as the chin areas. Therefore, the use of RYB may significantly increase different feature areas of the men and women of color demographic, but it may lead to reduced bias in all demographics (Fig. 6).

5 Conclusion

The red color pixels of an RYB image may contain more facial data in African Americans. The higher the facial data, the higher the accuracy to detect African American faces without using bias. Computer screen colors detect more light colors when using RGB color pigments when associated with light. It is possible that skin tones of African Americans and women of color reflects differently with light changing the composition of colors associated with RYB far more than can be perceived in the RGB color palette. In conclusion, this area of RYB should be studied for future enhancements in decreasing bias in facial recognition systems as well as the colors associated with other image processing used in facial recognition algorithms and computer systems.

Acknowledgments This work was supported by Sophia Bohorquez Armenta and Quang Lee.

The research was sponsored by the Army Research Office and was accomplished under Grant Number W911NF-21-1-0264. The views and conclusions contained in this document are those of the authors and should not be interpreted as representing the official policies, either expressed or implied, of the Army Research Office or the US Government. The US Government is authorized to reproduce and distribute reprints for government purposes notwithstanding any copyright notation herein.

References

1. A. Johnson, "Racism and AI: Here's How It's Been Criticized for Amplifying Bias," Forbes, May 25, 2023. [Online]. Available:https://www.forbes.com/sites/ariannajohnson/2023/05/25/racism-and-ai-heres-how-its-been-criticized-for-amplifying-bias/?sh=1dfa1d11269dT
2. H. Xia and C. Li, "Face Recognition and Application of Film and Television Actors Based on Dlib," 2019 12th International Congress on Image and Signal Processing, BioMedical Engineering and Informatics (CISP-BMEI), Suzhou, China, 2019, pp. 1–6, https://doi.org/10.1109/CISPBMEI48845.2019.8965869
3. "Ethical concerns raised over AI use in hiring," BBC, Dec 24, 2019. [Online]. Available: https://www.bbc.com/news/technology-50865437
4. Andrew Hundt, William Agnew, Vicky Zeng, Severin Kacianka, and Matthew Gombolay. 2022. Robots Enact Malignant Stereotypes. In 2022 ACM Conference on Fairness, Accountability, and Transparency (FAccT '22), June 21- 24, 2022, Seoul, Republic of Korea. ACM, New York, NY, USA, 14 pages. https://doi.org/10.1145/3531146.3533138
5. Johnson, "Racism and AI: Here's How It's Been Criticized for Amplifying Bias," Forbes, May 25, 2023. [Online]. Available: https://www.forbes.com/sites/ ariannajohnson/2023/05/25/racism-and-ai-heres-how-its-been-criticized-for-amplifying-bias/?sh=324ea591269d
6. Innocence Project, "When Artificial Intelligence Gets It Wrong," [Online]. Available: https://innocenceproject.org/when-artificial-intelligence-gets-it-wrong/
7. Detroit Woman Sues City After Being Falsely Arrested While 8 Months Pregnant Due to Facial Recognition," NBC News, [Online]. Available: https://www.nbenews.com/news/us-news/detroit-woman-sues-city-falsely-arrested-8-months-pregnant-due-facial-rcna98447
8. I Did Nothing Wrong. I Was Arrested Anyway.'," [Online]. Available: https://www.aclu.org/news/privacy-technology/i-did-nothing-wrong-i-was-arrested-anyway
9. Innocence Project, "When Artificial Intelligence Gets It Wrong," [Online]. Available: https://innocenceproject.org/when-artificial-intelligence-gets-it-wrong/
10. M. Zhang, "Google photos tags two African-Americans as gorillas through Facial Recognition Software," Forbes, https://www.forbes.com/sites/mzhang/2015/07/ 01/google-photos-tags-two-african-americans-as-gorillas-through-facial-recognitionsoftware/?sh=7de63c5e713d (accessed Jan. 14, 2024).
11. S. E. Ozer, "A Gentle Introduction to Face Recognition in Deep Learning," [Online]. Available: https://sefiks.com/2020/05/01/a-gentle-introduction-to-face-recognition-in-deeplearning/
12. Backprop Labs, "Face Recognition Pipeline Clearly Explained," [Online]. Available: https://medium.com/backprop-labs/face-recognition-pipeline-clearly-explained-f57fc0082750
13. A. Adouani, W. M. Ben Henia and Z. Lachiri, "Comparison of Haar-like, HOG and LBP approaches for face detection in video sequences," 2019 16th International Multi-Conference on Systems, Signals & Devices (SSD), Istanbul, Turkey, 2019, pp. 266–271, https://doi.org/10.1109/SSD.2019.8893214.
14. Backprop Labs, "Face Recognition Pipeline Clearly Explained," [Online]. Available: https://medium.com/backprop-labs/face-recognition-pipeline-clearly-explained-f57fc0082750
15. Davis E. King, "Dlib-ml: A Machine Learning Toolkit." [Online]. Available: http://dlib.net https://towardsdatascience.com/facial-mapping-landmarks-with-dlib-python160abcf7d672#:~:text=What is Dlib%3F,iBUG300W dataset was used
16. C. Sagonas, E. Antonakos, G, Tzimiropoulos, S. Zafeiriou, M. Pantic. 300 faces In-the-wild challenge: Database and results. Image and Vision Computing (IMAVIS), Special Issue on Facial Landmark Localisation "In-The-Wild". 2016.
17. M. Zhang, "Google photos tags two African-Americans as gorillas through Facial Recognition Software," Forbes, https://www.forbes.com/sites/mzhang/ 2015/07/01/google-photos-tags-two-african-americans-as-gorillas-through-facial-recognitionsoftware/?sh=7de63c5e713d (accessed Jan. 14, 2024).

18. Colors on a computer screen. Available at: https://www.chem.purdue.edu/gchelp/cchem//RGB Colors/body_rgbcolors.html (Accessed: 04 August 2023)
19. Briggs, D. (2013) The Dimensions of Color, The dimensions of colour, traditional colour wheel. Available at: http://www.huevaluechroma.com/072.php
20. RGB and ryb color, http://www.compworks.faithweb.com/electronics/theory/light/rgbryb001.html (accessed Jan. 15, 2024).
21. DeBruine, Lisa; Jones, Benedict (2017). Face Research Lab London Set. figshare. Dataset. https://doi.org/10.6084/m9.figshare.5047666.v5

Part IV
Sustainable and Medical Forensics Innovations

Integrating Local Therapy in the Management of Oligometastatic Cancer and the Emerging Role of Artificial Intelligence

Rohit V. Mantena, Shivaek Venkateswaran, Deena Gold, Anthony Tudisco, Tej A. Patel, Abhinav Komanduri, Asher Shin, Tidie Song, Yue Helen Zhang, Jonas Willmann, Edward Christopher Dee, and Puneeth Iyengar

1 Part I: The Role of Local Therapy in Oligometastatic Disease

1.1 Introduction

Oligometastatic disease (OMD) may represent an intermediate state between localized cancer and widespread metastases, characterized by a limited number of metastatic lesions. This concept, first introduced by Hellman and Weichselbaum [1], has evolved with advancements in imaging and therapeutic modalities, leading to a paradigm shift in the management of metastatic cancer. The concept has also evolved with our understanding of metastatic biology. Importantly, the potential for curative-intent treatment in select patients with OMD has garnered significant interest, particularly in non-small cell lung cancer (NSCLC) [2, 3].

R. V. Mantena · A. Tudisco
Department of Radiation Oncology, Memorial Sloan Kettering Cancer Center, New York, NY, USA

School of Medicine, University of Pittsburgh, Pittsburgh, PA, USA

S. Venkateswaran · T. A. Patel
Department of Radiation Oncology, Memorial Sloan Kettering Cancer Center, New York, NY, USA

University of Pennsylvania, Philadelphia, PA, USA

D. Gold
Department of Radiation Oncology, Memorial Sloan Kettering Cancer Center, New York, NY, USA

New York University, NY, New York, USA

S. S. Iyengar et al. (eds.), *Artificial Intelligence Driven Forensics*, Signals and Communication Technology, https://doi.org/10.1007/978-3-031-98036-7_21

1.2 Defining Oligometastatic Disease

The definition of OMD varies across studies, but a consensus has emerged around the presence of one to five metastatic lesions, such that all sites are amenable to safe and definitive local therapy; of note, others have defined OMD as the presence of metastatic lesions amenable to curative/radical metastasis-directed therapy (MDT) regardless of volume or number of lesions [1, 4]. The ESTRO-ASTRO consensus recommends treatment for all metastatic sites that are safely treatable, and high-resolution imaging is crucial for accurate assessment [3]. We note, however, that the definition of oligometastatic disease is a moving target and is nuanced by tumor growth velocity, size, mutational status, and anatomical involvement, in the context of advances in systemic therapy control as well as patient functional status and quality of life.

1.3 Evidence Supporting Local Therapy

1.3.1 Prospective Trials

Several randomized trials have demonstrated the benefits of local consolidative therapy (LCT) in patients with OMD, some of which are reviewed below:

Iyengar et al. [5] conducted a phase II randomized trial comparing stereotactic ablative radiotherapy (SAbR) followed by maintenance chemotherapy versus maintenance chemotherapy alone in patients with limited metastatic NSCLC. The study showed a significant improvement in progression-free survival (PFS) with the addition of SAbR (9.7 months vs. 3.5 months; $P = 0.01$).

Gomez et al. [6] reported long-term results of a multi-institutional phase II randomized study where patients with oligometastatic NSCLC who did not progress

A. Komanduri
Department of Radiation Oncology, Memorial Sloan Kettering Cancer Center, New York, NY, USA

University of Southern California, Los Angeles, CA, USA

A. Shin · Y. H. Zhang · E. C. Dee · P. Iyengar
Department of Radiation Oncology, Memorial Sloan Kettering Cancer Center, New York, NY, USA
e-mail: iyengarp@mskcc.org

T. Song
Department of Radiation Oncology, University of Texas Southwestern, Dallas, TX, USA

J. Willmann
Department of Radiation Oncology, University Hospital Zurich, University of Zurich, Zurich, Switzerland

Department of Medical Physics, Memorial Sloan Kettering Cancer Center, New York, NY, USA after front-line systemic therapy were randomized to receive LCT or maintenance therapy/observation. The LCT arm demonstrated a median overall survival (OS) of 41.2 months compared to 17.0 months in the control arm ($P = 0.017$).

The **SABR-COMET trial** [7] was a landmark phase II randomized study that evaluated the impact of stereotactic ablative radiotherapy (SABR) in patients with controlled primary tumors and one to five metastatic lesions across various cancer types. Patients receiving SABR to all metastatic sites experienced a significant improvement in both progression-free and overall survival compared to those receiving standard palliative care alone, with a 5-year overall survival of 42.3% versus 17.7%. While the treatment was associated with higher rates of grade ≥ 2 toxicity and a small risk of treatment-related death (4.5%), the trial provided the first randomized evidence that aggressive local therapy could extend survival in carefully selected patients with oligometastatic disease.

1.3.2 Meta-Analyses

A meta-analysis utilizing restricted mean survival time (RMST) demonstrated the efficacy of metastasis-directed therapy (MDT) with stereotactic body radiotherapy (SBRT) for oligometastatic disease (OMD) using restricted mean survival time (RMST)—a more interpretable alternative to traditional survival metrics like hazard ratios [8].

By reconstructing data from Kaplan-Meier curves of six randomized controlled trials (426 patients), the analysis revealed meaningful RMST gains across multiple studies. For surrogate endpoints, RMST benefits ranged from 4.6 months (Iyengar et al.) to 11.1 months (SABR-COMET), while overall survival gains were 12.6 months (SABR-COMET), 15 months (Gomez et al.), and 7.9 months (SINDAS) [8]. These findings support the survival benefit of SBRT in OMD and highlight RMST as a robust, patient-centered method to communicate treatment outcomes.

1.3.3 Randomized Phase II/III Trial

The NRG-LU002 trial was a randomized phase II/III study evaluating the addition of local consolidative therapy (LCT)—comprising radiation and/or surgery—to maintenance systemic therapy in patients with limited metastatic non-small cell lung cancer (NSCLC) who had achieved at least stable disease after first-line systemic treatment. In the phase II portion, 215 patients were enrolled and randomized in a 1:2 ratio to receive either maintenance systemic therapy alone or LCT followed by maintenance therapy [9].

The primary endpoint, progression-free survival (PFS), showed no statistically significant improvement with the addition of LCT; the hazard ratio was 0.93 (95% CI 0.66–1.31), and the 2-year PFS rates were 36% in the maintenance-only arm versus 40% in the LCT arm. Overall survival and toxicity profiles were also comparable between the two groups, although there was a higher incidence of grade

3 or higher pneumonitis in the LCT arm (10% vs. 1%). These findings suggest that, within the context of modern immunotherapy-based systemic regimens, the addition of LCT may not confer a significant PFS benefit for patients with oligometastatic NSCLC [9].

Ultimately, the results of LU-002 suggest the importance of improved patient selection. Identifying patients who stand to benefit the most from metastasis-directed local therapy is critical.

1.3.4 Evidence from Other Disease Sites

Recent trials in oligometastatic disease demonstrate improved outcomes with MDT. In prostate cancer, STOMP and ORIOLE showed prolonged androgen deprivation therapy (ADT)-free survival and PFS with SBRT [10], while EXTEND reported a 75% 5-year OS benefit for MDT + systemic therapy [11]. The ARTO phase II trial demonstrated that adding SBRT to abiraterone acetate and prednisone significantly improved biochemical response rates (92% vs. 68.3%) and progression-free survival (HR 0.35, $P < 0.001$) in patients with oligometastatic castration-resistant prostate cancer [12]. Complete biochemical response was also more frequent with SBRT (56% vs. 23.2%), supporting its role as a beneficial addition to systemic therapy in this setting.

The RAPPORT phase I/II trial found that combining stereotactic ablative body radiotherapy (SABR) with short-course pembrolizumab in patients with oligometastatic clear cell renal cell carcinoma was well tolerated and achieved high local control (92% at 2 years) and a disease control rate of 83%. Despite the single-arm design, the regimen showed promising objective response (63%) and progression-free survival rates (60% at 1 year, 45% at 2 years), supporting further study [13].

1.4 Clinical Practice Guidelines

The ASTRO/ESTRO clinical practice guideline provides recommendations for the use of local therapy in extracranial oligometastatic NSCLC [14]. It emphasizes a patient-centered, multidisciplinary approach and suggests that definitive local therapy is relevant if all disease sites (defined as five or fewer) are technically feasible and clinically safe to treat.

1.5 Patient Selection and Disease States

Identifying patients who will benefit most from local therapy is critical. The concepts of oligoprogression and oligopersistence have been introduced: [2]

Synchronous oligometastatic disease is defined as the presence of a limited number (typically ≤5) of metastases detected at the time of initial cancer diagnosis.

The SINDAS trial (NCT02893332) was a randomized phase III study evaluating the addition of radiotherapy (RT) to first-line tyrosine kinase inhibitor (TKI) therapy in patients with EGFR-mutated synchronous oligometastatic non-small cell lung cancer (NSCLC) [15]. Patients received either TKI alone or TKI combined with RT to all metastatic sites and the primary tumor. The study found that the addition of RT significantly improved progression-free survival (20.2 months vs. 12.5 months) and overall survival (25.5 months vs. 17.4 months) compared to TKI alone. The trial was halted early due to these positive outcomes, suggesting that upfront RT enhances outcomes in this patient population.

In a study from 2024, Hongfu Sun and colleagues conducted a randomized trial assessing the impact of local consolidative therapy (LCT) in patients with EGFR-mutated oligometastatic NSCLC [16]. The trial compared outcomes between patients receiving standard systemic therapy alone and those receiving systemic therapy plus LCT to all known disease sites. The addition of LCT resulted in a significant improvement in both progression-free and overall survival, reinforcing the role of aggressive local therapy in managing oligometastatic NSCLC with EGFR mutations.

Oligoprogression refers to limited metastatic areas progressing in the background of otherwise controlled disease [17].

The CURB trial (NCT03808662) was a phase II randomized study that evaluated the role of local ablative therapy (LAT) in patients with oligoprogressive metastatic non-small cell lung cancer (NSCLC) or breast cancer who were otherwise controlled on systemic therapy [18]. Patients with ≤5 progressing metastatic lesions were randomized to either LAT (typically stereotactic radiotherapy) or standard-of-care systemic therapy continuation. In the NSCLC cohort, LAT significantly improved progression-free survival (PFS) compared to standard care, supporting the strategy of targeting resistant clones while maintaining effective systemic treatment. However, no significant PFS benefit was observed in the breast cancer cohort. These results support LAT as a promising approach in the management of oligoprogressive NSCLC [18].

Oligopersistence describes a state where formerly polymetastatic disease responds to treatment, decreasing to fewer than five active disease sites [19].

Trials specifically addressing oligopersistent NSCLC—where initially widespread disease responds to systemic therapy leaving ≤5 residual lesions—are emerging, with studies like NRG-LU002 and ongoing prospective registries exploring the benefit of local consolidative therapy in this evolving disease state.

1.6 Challenges and Future Directions

Despite promising results, challenges remain:

1.6.1 Heterogeneity in Definitions

Variability in defining OMD complicates study comparisons and treatment decisions.

One of the central challenges in advancing the clinical management of oligometastatic disease (OMD) lies in the persistent variability in its definition across studies, institutions, and disease sites. The ESTRO-ASTRO consensus led by Lievens et al. emphasized that OMD should not be treated as a monolithic entity, but rather as a spectrum of disease states that reflect both tumor biology and clinical context—including synchronous, metachronous, oligopersistent, and oligoprogressive presentations [20]. However, even with this taxonomy, operational definitions vary significantly between trials: some define OMD based strictly on lesion count (e.g., ≤ 3 vs. ≤ 5 metastases), while others include or exclude factors like prior systemic therapy, lesion location, or molecular subtype [21–23].

This heterogeneity limits cross-trial comparisons, impedes pooled analyses and meta-analyses, and complicates guideline development. Moreover, inconsistency in imaging modalities and surveillance intervals further obscures disease classification, potentially mislabeling polymetastatic disease as oligometastatic. To address these issues, the consensus advocates for standardized reporting of disease burden, systemic therapy status, and imaging techniques in OMD trials [20]. A unified, biologically informed framework is essential for both research rigor and real-world treatment decision-making.

1.6.2 Biomarker Development

The lack of validated biomarkers hinders the ability to predict which patients will benefit from local therapy.

A major barrier to refining the role of local therapy in oligometastatic disease (OMD) is the absence of validated biomarkers capable of reliably predicting which patients are most likely to benefit from metastasis-directed treatment. While lesion count and anatomical distribution remain the dominant selection criteria, they offer only crude surrogates for underlying tumor biology. Recent advances in circulating tumor DNA (ctDNA) analysis offer promise in bridging this gap, enabling real-time, minimally invasive assessment of tumor burden, clonal evolution, and residual disease [24]. For example, in patients with non-small cell lung cancer or colorectal cancer, undetectable ctDNA after systemic therapy has been associated with favorable outcomes and may indicate a state of minimal residual disease in which local ablation of residual lesions could confer prolonged disease control [25].

Conversely, persistently detectable or rising ctDNA levels may signify occult micrometastatic spread [26], identifying patients unlikely to derive long-term benefit from local interventions alone. However, despite these theoretical advantages, ctDNA has not yet been validated prospectively in the oligometastatic context, and no standardized thresholds or assays currently guide clinical decision-making.

Moreover, the temporal dynamics of ctDNA shedding in response to radiation, surgery, and immunotherapy are still being elucidated [27].

1.6.3 Integration with Systemic Therapy

Determining the optimal sequencing and integration of local and systemic therapies remains one of the most pressing and unresolved challenges in the management of oligometastatic disease (OMD) [14]. While randomized trials such as SABR-COMET [7], Gomez et al. [6], and SINDAS [15] have demonstrated survival benefits from adding local therapy to systemic treatment, the ideal timing—whether upfront, concurrent, intercalated, or delayed—has yet to be clearly defined. This question is further complicated by the heterogeneity of systemic therapy regimens, which vary by disease site, molecular subtype, and prior treatment exposure [28]. The introduction of immunotherapy and targeted agents, with their unique mechanisms of action and immune-modulatory effects, adds further complexity; for example, local radiation may synergize with immune checkpoint inhibitors through the abscopal effect [29], but may also increase toxicity if delivered too close to systemic dosing [30]. Conversely, premature interruption of systemic therapy to accommodate local ablative interventions could risk systemic disease flare or resistance.

Moreover, patient-specific factors—such as performance status, burden and tempo of disease, and treatment goals—must be carefully weighed in sequencing decisions. There is also limited real-world guidance on how to manage situations like oligoprogression on targeted therapy, where escalating systemic therapy versus applying localized treatment to progressing sites both represent plausible strategies. Compounding this, most trials to date have focused on fixed-treatment sequences rather than adaptive, response-driven approaches, leaving clinicians without a robust framework for tailoring timing to individual trajectories.

Ongoing and future trials are expected to address these issues, refining patient selection and treatment strategies.

2 Part II: The Role of Artificial Intelligence in Oncology Decision-Making

2.1 Integration of Artificial Intelligence

The integration of artificial intelligence (AI) into oncology holds the promise of transforming cancer care by enhancing diagnostic accuracy, predicting treatment outcomes, and personalizing therapy. AI's ability to analyze complex, multimodal datasets can aid in addressing the challenges associated with OMD management.

For instance, a study comparing AI models to medical oncologists in predicting 3-month mortality in patients with advanced cancer found that the AI model achieved an area under the receiver operating characteristic curve (AUC) of 81.2%, outperforming oncologists in overall accuracy [31].

In metastatic colorectal cancer, a meta-analysis of 21 AI predictive models demonstrated the efficacy of machine learning algorithms in discriminating responders from non-responders to therapy, highlighting the potential of AI in guiding treatment decisions [32].

Furthermore, the development of AI-based prognostic biomarkers, such as the ArteraAI MMAI biomarker, has shown promise in predicting overall survival and response to metastasis-directed therapy in patients with oligometastatic castration-sensitive prostate cancer [33].

2.2 Applications of AI in Cancer Care with a Focus on Oligometastatic Cancer

AI has been applied across various domains in oncology. Examples of AI's role in oligometastatic cancer are myriad.

2.2.1 Diagnostic Imaging

AI algorithms can improve the interpretation of imaging studies, aiding in the detection and characterization of tumors.

Artificial intelligence has significantly enhanced diagnostic imaging in oncology, particularly in the detection and characterization of metastatic lesions. Deep learning algorithms, especially convolutional neural networks (CNNs), have demonstrated superior performance in image segmentation tasks, leading to more accurate delineation of tumors and metastatic sites [34].

A study by Gibbons et al. compared deep learning-based auto-segmentation with traditional atlas-based methods for contouring organs at risk (OARs) in radiation therapy. The deep learning approach showed statistically significant improvements in accuracy across multiple anatomical regions, including the head and neck, thorax, and pelvis. Specifically, the mean editing time for deep learning contours was reduced by up to 61% compared to atlas-based methods, highlighting the efficiency gains achievable with AI integration [35].

These advancements in AI-driven imaging not only improve the precision of tumor and metastasis detection but also facilitate more personalized treatment planning, particularly crucial in managing oligometastatic disease where accurate lesion identification is paramount.

2.2.2 Predictive Modeling

Machine learning models can predict treatment responses and survival outcomes, assisting in therapy selection.

Predictive modeling using AI has emerged as a powerful tool in oncology, enabling the forecasting of treatment responses and survival outcomes based on complex datasets. Machine learning models can integrate diverse data types, including clinical parameters, imaging features, and genomic information, to generate individualized prognostic assessments [36, 37].

Banerjee et al. developed a deep learning model known as Probabilistic Prognostic Estimates of Survival in Metastatic Cancer Patients (PPES-Met), which analyzes free-text clinical notes to estimate short-term life expectancy. The model achieved an area under the receiver operating characteristic curve (AUC) of 0.89, demonstrating high accuracy in survival prediction. Importantly, the model incorporates explainability features, providing clinicians with insights into the factors influencing each prediction [38].

Such predictive models are instrumental in identifying patients with oligometastatic disease who are most likely to benefit from aggressive local therapies, thereby optimizing treatment strategies and improving outcomes.

Despite the growing promise of predictive modeling in oncology, several challenges hinder its full clinical integration—chief among them being the need for more granular, high-quality data. Many existing models rely on retrospective datasets that lack standardization, completeness, or sufficient representation of diverse populations [39]. This limits the generalizability and robustness of predictions, particularly in complex settings like oligometastatic disease. Accurate prediction of outcomes such as progression, response to therapy, or survival requires comprehensive inputs—including longitudinal clinical data, imaging, treatment details, comorbidities, and molecular or genomic profiles. Yet, many models are constrained by missing variables or inconsistently recorded real-world data [39].

Furthermore, there is often uncertainty about which variables truly drive outcomes, and model interpretability remains a critical barrier to clinical trust and adoption. Without integrated, multimodal datasets and clearer frameworks for feature importance, predictive models risk being either underpowered or overfitted, offering limited actionable insights. Addressing these challenges will require collaborative efforts across institutions to standardize data collection, share resources, and ensure transparency in AI development and validation.

2.2.3 Radiotherapy Planning

AI has revolutionized radiotherapy planning by automating and enhancing various aspects of the treatment workflow. Deep learning-based auto-contouring tools have significantly reduced the time required for delineating target volumes and OARs, leading to increased efficiency and consistency in treatment planning [40].

A study evaluating the Organs-RT deep learning auto-contouring tool demonstrated high agreement with expert-defined contours, particularly for structures like the bladder, heart, lungs, and femoral heads, with Dice similarity coefficients (DSC) $\geq$0.92. The tool achieved time-saving efficiencies of up to 84% in thoracic cases, underscoring its potential to streamline radiotherapy workflows [40].

Furthermore, the integration of AI in online adaptive radiation therapy (ART) systems has shown promise in improving target coverage and sparing OARs. For instance, the Ethos ART system demonstrated clinically acceptable accuracy in automatic contouring for lung cancer treatments, leading to enhanced dosimetric outcomes [41].

These advancements in AI-driven radiotherapy planning are particularly beneficial for patients with oligometastatic disease, where precise targeting of limited metastatic sites is critical for treatment success.

AI-driven advancements in radiotherapy (RT) planning hold transformative potential for improving global access to high-quality cancer care, particularly in low- and middle-income countries (LMICs) where shortages of trained personnel and infrastructure are common [42]. Automated contouring tools and treatment planning algorithms can significantly reduce the time, expertise, and resources needed to deliver complex RT, enabling task-shifting and remote collaboration in settings with limited specialist availability. AI can also support quality assurance and standardization, ensuring consistent treatment delivery across centers regardless of local experience levels. [43]

Moreover, cloud-based platforms and federated learning approaches can facilitate the deployment of AI models without requiring extensive local data infrastructure. By streamlining workflows and reducing dependence on highly specialized labor, AI-enhanced RT planning could democratize access to precision radiation therapy and narrow the global disparities in cancer treatment delivery.

3 Next Steps in Integrating AI and Local Therapy in Metastatic Disease

3.1 Modeling Metastatic Trajectories

A critical frontier in metastatic cancer research is the development of dynamic models that can anticipate metastatic trajectories over time. Traditional models often treat metastases as static snapshots, yet disease evolution is inherently temporal [44]. Leveraging longitudinal data—including sequential imaging, circulating tumor DNA (ctDNA) [45], and serial treatment responses—can enable machine learning models to forecast not only *where* metastases will develop but *when*. Such predictive temporal modeling could support proactive rather than reactive intervention strategies, helping clinicians time systemic or local therapies more effectively.

Recent research efforts have begun incorporating metastatic "velocity" and "acceleration" into prognostic frameworks, potentially identifying patients at higher risk of widespread progression who may benefit from early systemic intensification or closer surveillance [46, 47]. Developing AI models that account for evolving tumor biology, immune escape, and therapy-induced selection pressures will be essential for guiding durable, adaptive treatment strategies.

Advances in AI now allow for spatiotemporal forecasting of metastatic spread, which could revolutionize patient selection for ablative therapy [48]. By integrating longitudinal imaging (e.g., interval CT/MRI), ctDNA kinetics, and treatment response data, deep learning models can predict the likelihood and anatomic distribution of future metastases—distinguishing patients whose disease will remain truly oligometastatic from those with impending polymetastatic progression.

For example, recurrent neural networks (RNNs) or transformer-based architectures can analyze temporal patterns in lesion growth or ctDNA clearance to estimate the "window of opportunity" for curative-intent ablation. Such models may also identify metastatic dormancy signatures, helping clinicians target lesions at risk of imminent reactivation while avoiding overtreatment of indolent sites. Early work in prostate and hepatic cancer has demonstrated that AI-derived "time-to-progression" metrics outperform static clinical criteria in predicting ablation success, suggesting that dynamic risk stratification could optimize the timing and aggressiveness of local therapy [49, 50].

A novel frontier lies in adaptive AI systems that continuously refine ablation strategies based on real-world data streams. Imagine a scenario where AI assimilates post-ablation imaging changes, ctDNA fluctuations, and immune biomarkers (e.g., PD-L1 dynamics) to adjust surveillance intervals or recommend repeat ablation for emergent lesions. Reinforcement learning algorithms—trained on outcomes from thousands of oligometastatic patients—could propose personalized ablation thresholds (e.g., "treat up to 3 lesions if progression velocity remains below X mm/month") [51].

This paradigm shifts the focus from reactive lesion-by-lesion management to proactive metastasis interception, where AI anticipates geographic or biologic failure patterns and guides preemptive local control. Challenges remain, including the need for prospective validation in heterogeneous cohorts and the integration of AI tools into clinical workflows. However, as temporal modeling matures, AI may unlock the potential of ablation as a dynamic component of precision oncology, tailoring its use to the evolving narrative of a patient's metastatic disease.

3.2 Identifying Optimal Targets for Ablative Radiotherapy

Another major application of AI lies in refining the selection of metastatic targets for ablative radiotherapy. While local therapy has shown benefit in oligometastatic disease, not all metastases confer equal prognostic or biological weight [52]. Radiomic and radiogenomic profiling can help differentiate indolent from aggressive lesions based on imaging phenotypes, metabolic activity, and genomic signatures.

For instance, AI models can integrate FDG-PET uptake, lesion size, growth kinetics, and anatomical location to stratify lesions by their likelihood of seeding further metastases or contributing to symptomatic progression [53]. AI-driven decision-support tools may soon assist multidisciplinary teams in prioritizing which lesions to ablate, defer, or monitor—especially in patients with limited RT capacity or borderline performance status. In addition, incorporating AI into adaptive RT platforms could optimize fractionation schedules in real-time, tailoring treatment to the evolving landscape of metastatic burden.

AI can further refine target selection by quantifying the metastatic virulence of individual lesions—moving beyond simple counts or size criteria. For example, graph neural networks (GNNs) can model inter-lesion relationships (e.g., spatial clustering, vascular drainage patterns) to predict which tumors act as "hub" lesions driving systemic dissemination [54]. Meanwhile, radiogenomic models trained on paired imaging and biopsy data may identify imaging surrogates for resistance markers (e.g., hypoxia signatures on MRI predicting radioresistance), enabling clinicians to prioritize ablation of high-risk lesions while sparing others for surveillance or systemic therapy.

Such tools could prove invaluable in borderline oligometastatic cases (e.g., 5–10 lesions), where conventional selection is arbitrary [55]. AI-defined "high-impact" lesions—those with proliferative radiomics features or ctDNA-detected subclonal dominance—may correlate with poorer ablation outcomes unless aggressively targeted, underscoring the potential for AI to transform ablation from an anatomic to a biologically guided intervention [49].

In the near future, the integration of AI into adaptive ablative radiotherapy platforms could enable real-time treatment personalization [49]. Deep learning models analyzing daily CBCT or MR-linac imaging during a course of SBRT could detect subtle lesion responses (e.g., early necrosis on diffusion-weighted MRI) and dynamically adjust dose escalation or de-escalation—maximizing control for resistant tumors while minimizing toxicity to adjacent organs.

Reinforcement learning algorithms, trained on outcomes from prior ablations, might also optimize fractionation (e.g., switching from single-fraction to hypofractionated regimens for perivascular lesions) or even trigger adjuvant therapies (e.g., immunotherapy) based on predicted immune microenvironment changes post-ablation. Prototype systems, such as AI-driven "tumor-aware" RT planning engines, are already demonstrating feasibility in prostate oligometastases, where lesion-specific dose painting improved local control [56]. As these tools mature, AI could shift ablative therapy from a static, pre-planned intervention to a responsive, biology-aware modality that evolves with the patient's disease.

3.3 *Incorporating Social Determinants, Functional Status, and Cachexia: Toward Holistic Precision Oncology*

Precision oncology must move beyond genomics and imaging to integrate the full context of a patient's life and biology. Social determinants of health (SDoH)—including socioeconomic status, geographic access, housing stability, and social support—play a profound role in cancer outcomes and treatment adherence, yet are rarely captured in structured form [57, 58].

Similarly, functional status and syndromes like cancer cachexia remain underrepresented in most AI training datasets, despite their major prognostic value [39, 59–62]. Deep learning could integrate serial cachexia biomarkers (e.g., CRP, albumin trends) with CT-derived muscle mass analytics to predict tolerability of aggressive local therapy, prompting prehabilitation referrals when needed.

Moving forward, next-generation AI models should incorporate patient-reported outcomes, wearable-derived metrics (e.g., mobility, nutrition, fatigue), and structured clinical assessments (e.g., ECOG performance status, frailty indices) to deliver a more holistic view of risk and resilience [63, 64]. Such models would help clinicians and health systems tailor not just which therapies to give, but *whether*, *when*, and *how* to deliver them, in a way that aligns with patient capacity and priorities. Bridging the technical with the social and biological will be key to ensuring AI serves as a tool for equity and not amplification of existing disparities.

4 Conclusion

In conclusion, the convergence of advanced local therapies and artificial intelligence offers a transformative opportunity to reshape the management of metastatic cancer—not through a one-size-fits-all model, but by embracing the complexity and individuality of each patient's disease and lived experience. Realizing this vision will require deep, sustained multidisciplinary collaboration among oncologists, data scientists, computer scientists, ethicists, and—most importantly—patients and their advocates.

By anchoring innovation in the realities of diverse clinical contexts and prioritizing transparency and co-design, we can build predictive frameworks and therapeutic strategies that are not only smarter but also more patient-centric. The future of metastatic cancer care lies not only in technological precision, but in compassionate, patient-centered science that listens, adapts, and empowers.

References

1. Hellman, S. & Weichselbaum, R. R. Oligometastases. *J Clin Oncol* **13**, 8–10 (1995).
2. Ravella, R., Feliciano, E. J. G., Dee, E. C., Gomez, D. R. & Iyengar, P. Management of oligoprogressive and oligopersistent disease in advanced NSCLC. *Clin Adv Hematol Oncol* **23**, 40–50 (2025).

3. Guckenberger, M. *et al.* Characterisation and classification of oligometastatic disease: a European Society for Radiotherapy and Oncology and European Organisation for Research and Treatment of Cancer consensus recommendation. *Lancet Oncol* **21**, e18–e28 (2020).
4. Niibe, Y., Jingu, K. & Onishi, H. Oligometastases: history and future vision of breast cancer. *Transl Cancer Res TCR* **9**, 5028–5031 (2020).
5. Iyengar, P. *et al.* Consolidative Radiotherapy for Limited Metastatic Non-Small-Cell Lung Cancer: A Phase 2 Randomized Clinical Trial. *JAMA Oncol* **4**, e173501 (2018).
6. Gomez, D. R. *et al.* Local Consolidative Therapy Vs. Maintenance Therapy or Observation for Patients With Oligometastatic Non-Small-Cell Lung Cancer: Long-Term Results of a Multi-Institutional, Phase II, Randomized Study. *J Clin Oncol* **37**, 1558–1565 (2019).
7. Palma, D. A. *et al.* Stereotactic Ablative Radiotherapy for the Comprehensive Treatment of Oligometastatic Cancers: Long-Term Results of the SABR-COMET Phase II Randomized Trial. *J Clin Oncol* **38**, 2830–2838 (2020).
8. Averbuch, I., Moore, A., Ludmir, E. B., Markel, G. & Meirson, T. Assessing the effect of metastasis-directed therapy in oligometastatic disease using the restricted mean survival time. *Br J Cancer* **130**, 1929–1935 (2024).
9. Iyengar, P. *et al.* NRG-LU002: Randomized phase II/III trial of maintenance systemic therapy versus local consolidative therapy (LCT) plus maintenance systemic therapy for limited metastatic non-small cell lung cancer (NSCLC). *JCO* **42**, 8506–8506 (2024).
10. Deek, M. P. *et al.* Long-Term Outcomes and Genetic Predictors of Response to Metastasis-Directed Therapy Versus Observation in Oligometastatic Prostate Cancer: Analysis of STOMP and ORIOLE Trials. *J Clin Oncol* **40**, 3377–3382 (2022).
11. Tang, C. *et al.* Addition of Metastasis-Directed Therapy to Intermittent Hormone Therapy for Oligometastatic Prostate Cancer: The EXTEND Phase 2 Randomized Clinical Trial. *JAMA Oncol* **9**, 825–834 (2023).
12. Francolini, G. *et al.* Stereotactic Body Radiation Therapy and Abiraterone Acetate for Patients Affected by Oligometastatic Castrate-Resistant Prostate Cancer: A Randomized Phase II Trial (ARTO). *J Clin Oncol* **41**, 5561–5568 (2023).
13. Siva, S. et al. Stereotactic Radiotherapy and Short-course Pembrolizumab for Oligometastatic Renal Cell Carcinoma-The RAPPORT Trial. Eur Urol 81, 364–372 (2022).
14. Iyengar, P. *et al.* Treatment of Oligometastatic Non-Small Cell Lung Cancer: An ASTRO/ESTRO Clinical Practice Guideline. *Pract Radiat Oncol* **13**, 393–412 (2023).
15. Wang, X.-S. *et al.* Randomized Trial of First-Line Tyrosine Kinase Inhibitor With or Without Radiotherapy for Synchronous Oligometastatic EGFR-Mutated Non-Small Cell Lung Cancer. *J Natl Cancer Inst* **115**, 742–748 (2023).
16. Sun, H. *et al.* Thoracic Radiotherapy Improves the Survival in Patients With EGFR-Mutated Oligo-Organ Metastatic Non-Small Cell Lung Cancer Treated With Epidermal Growth Factor Receptor-Tyrosine Kinase Inhibitors: A Multicenter, Randomized, Controlled, Phase III Trial. *J Clin Oncol* **43**, 412–421 (2025).
17. Nguyen, K. T., Sakthivel, G., Milano, M. T., Qiu, H. & Singh, D. P. Oligoprogression in non-small cell lung cancer: a narrative review. *J Thorac Dis* **14**, 4998–5011 (2022).
18. Tsai, C. J. *et al.* Standard-of-care systemic therapy with or without stereotactic body radiotherapy in patients with oligoprogressive breast cancer or non-small-cell lung cancer (Consolidative Use of Radiotherapy to Block [CURB] oligoprogression): an open-label, randomised, controlled, phase 2 study. *Lancet* **403**, 171–182 (2024).
19. Nugent, K. & Good, J. The oligometastatic paradigm and the role of radiotherapy. *Clin Med (Lond)* **23**, 61–64 (2023).
20. Lievens, Y. *et al.* Defining oligometastatic disease from a radiation oncology perspective: An ESTRO-ASTRO consensus document. *Radiother Oncol* **148**, 157–166 (2020).
21. Tan, V. S. & Palma, D. A. Top Ten Lessons Learned from Trials in Oligometastatic Cancers. *Cancer Res Treat* **55**, 5–14 (2023).
22. Freedman, G. M., Jones, J. A. & Taunk, N. K. Stereotactic Radiation for Oligometastatic and Oligoprogressive Stage IV Breast Cancer: A Case-Based Review. *Curr Oncol* **30**, 2510–2523 (2023).

23. Chandy, E. T. J., Saxby, H. J., Pang, J. W. & Sharma, R. A. The multidisciplinary management of oligometastases from colorectal cancer: a narrative review. *Ann Palliat Med* **10**, 5988–6001 (2021).
24. Sánchez-Herrero, E. *et al.* Circulating Tumor DNA as a Cancer Biomarker: An Overview of Biological Features and Factors That may Impact on ctDNA Analysis. *Front Oncol* **12**, 943253 (2022).
25. Andrews, H. S. et al. ctDNA Clearance as an Early Indicator of Improved Clinical Outcomes in Advanced NSCLC Treated with TKI: Findings from an Aggregate Analysis of Eight Clinical Trials. Clin Cancer Res 31, 2162–2172 (2025).
26. Malla, M., Loree, J. M., Kasi, P. M. & Parikh, A. R. Using Circulating Tumor DNA in Colorectal Cancer: Current and Evolving Practices. *J Clin Oncol* **40**, 2846–2857 (2022).
27. Andersen, L. *et al.* Exploring the biology of ctDNA release in colorectal cancer. *Eur J Cancer* **207**, 114186 (2024).
28. Murray, J. M. & Coldman, A. J. The effect of heterogeneity on optimal regimens in cancer chemotherapy. *Math Biosci* **185**, 73–87 (2003).
29. Pevzner, A. M., Tsyganov, M. M., Ibragimova, M. K. & Litvyakov, N. V. Abscopal effect in the radio and immunotherapy. *Radiat Oncol J* **39**, 247–253 (2021).
30. Verma, V., Cushman, T. R., Tang, C. & Welsh, J. W. Toxicity of radiation and immunotherapy combinations. *Adv Radiat Oncol* **3**, 506–511 (2018).
31. Zachariah, F. J., Rossi, L. A., Roberts, L. M. & Bosserman, L. D. Prospective Comparison of Medical Oncologists and a Machine Learning Model to Predict 3-Month Mortality in Patients With Metastatic Solid Tumors. *JAMA Netw Open* **5**, e2214514 (2022).
32. Russo, V. *et al.* Artificial Intelligence Predictive Models of Response to Cytotoxic Chemotherapy Alone or Combined to Targeted Therapy for Metastatic Colorectal Cancer Patients: A Systematic Review and Meta-Analysis. *Cancers (Basel)* **14**, 4012 (2022).
33. Wang, J. H. *et al.* Validation of an artificial intelligence-based prognostic biomarker in patients with oligometastatic Castration-Sensitive prostate cancer. *Radiother Oncol* **202**, 110618 (2025).
34. Jiang, X., Hu, Z., Wang, S. & Zhang, Y. Deep Learning for Medical Image-Based Cancer Diagnosis. *Cancers (Basel)* **15**, 3608 (2023).
35. Gibbons, E. *et al.* Clinical evaluation of deep learning and atlas-based auto-segmentation for critical organs at risk in radiation therapy. *J Med Radiat Sci* **70 Suppl 2**, 15–25 (2023).
36. Lotter, W. *et al.* Artificial Intelligence in Oncology: Current Landscape, Challenges, and Future Directions. *Cancer Discov* **14**, 711–726 (2024).
37. Dercle, L. *et al.* Baseline Radiomic Signature to Estimate Overall Survival in Patients With NSCLC. *J Thorac Oncol* **18**, 587–598 (2023).
38. Banerjee, I. *et al.* Probabilistic Prognostic Estimates of Survival in Metastatic Cancer Patients (PPES-Met) Utilizing Free-Text Clinical Narratives. *Sci Rep* **8**, 10037 (2018).
39. Kolla, L. & Parikh, R. B. Uses and limitations of artificial intelligence for oncology. *Cancer* **130**, 2101–2107 (2024).
40. Maduro Bustos, L. A. *et al.* Feasibility evaluation of novel AI-based deep-learning contouring algorithm for radiotherapy. *J Appl Clin Med Phys* **24**, e14090 (2023).
41. Mao, W. *et al.* Evaluation of Auto-Contouring and Dose Distributions for Online Adaptive Radiation Therapy of Patients With Locally Advanced Lung Cancers. *Pract Radiat Oncol* **12**, e329–e338 (2022).
42. Christ, S. M. & Willmann, J. Measuring Global Inequity in Radiation Therapy: Resource Deficits in Low- and Middle-Income Countries Without Radiation Therapy Facilities. *Adv Radiat Oncol* **8**, 101175 (2023).
43. Baroudi, H. *et al.* Automated Contouring and Planning in Radiation Therapy: What Is 'Clinically Acceptable'? *Diagnostics (Basel)* **13**, 667 (2023).
44. Gui, P. & Bivona, T. G. Evolution of metastasis: new tools and insights. *Trends Cancer* **8**, 98–109 (2022).
45. Zengin, Z. B. *et al.* Illustration of temporal evolution in patients with metastatic renal cell carcinoma (mRCC) using both circulating tumor DNA (ctDNA) and tissue-based genomic data. *JCO* **39**, 347–347 (2021).

46. McTyre, E. R. *et al.* Multi-institutional validation of brain metastasis velocity, a recently defined predictor of outcomes following stereotactic radiosurgery. *Radiother Oncol* **142**, 168–174 (2020).
47. Willmann, J. *et al.* Distant Metastasis Velocity as a Novel Prognostic Score for Overall Survival After Disease Progression Following Stereotactic Body Radiation Therapy for Oligometastatic Disease. *Int J Radiat Oncol Biol Phys* **114**, 871–882 (2022).
48. Li, Z. *et al.* 4D-ACFNet: A 4D Attention Mechanism-Based Prognostic Framework for Colorectal Cancer Liver Metastasis Integrating Multimodal Spatiotemporal Features. Preprint at https://doi.org/10.48550/ARXIV.2503.09652 (2025).
49. Matsui, Y. *et al.* Applications of artificial intelligence in interventional oncology: An up-to-date review of the literature. *Jpn J Radiol* **43**, 164–176 (2025).
50. Brisbane, W. G. *et al.* Focal therapy of prostate cancer: Use of artificial intelligence to define tumour volume and predict treatment outcomes. *BJUI Compass* **6**, e456 (2025).
51. Eckardt, J.-N., Wendt, K., Bornhäuser, M. & Middeke, J. M. Reinforcement Learning for Precision Oncology. *Cancers (Basel)* **13**, 4624 (2021).
52. Jongbloed, M. *et al.* Prognostic and Predictive Biomarkers of Oligometastatic NSCLC: New Insights and Clinical Applications. *JTO Clin Res Rep* **5**, 100740 (2024).
53. Sadaghiani, M. S., Rowe, S. P. & Sheikhbahaei, S. Applications of artificial intelligence in oncologic 18F-FDG PET/CT imaging: a systematic review. *Ann Transl Med* **9**, 823 (2021).
54. Gogoshin, G. & Rodin, A. S. Graph Neural Networks in Cancer and Oncology Research: Emerging and Future Trends. *Cancers (Basel)* **15**, 5858 (2023).
55. Berzenji, D. *et al.* A comprehensive approach to defining the cutoff value of oligometastasis in head and neck squamous cell carcinoma. *Cancer* **131**, e35632 (2025).
56. Fredén, E., Tilly, D. & Ahnesjö, A. Adaptive dose painting for prostate cancer. *Front Oncol* **12**, 973067 (2022).
57. Zavala, V. A. *et al.* Cancer health disparities in racial/ethnic minorities in the United States. *Br J Cancer* **124**, 315–332 (2021).
58. Dee, E. C., Pierce, L. J., Winkfield, K. M. & Lam, M. B. In pursuit of equity in cancer care: Moving beyond the Affordable Care Act. *Cancer* **128**, 3278–3283 (2022).
59. Corti, C. *et al.* Artificial intelligence in cancer research and precision medicine: Applications, limitations and priorities to drive transformation in the delivery of equitable and unbiased care. *Cancer Treatment Reviews* **112**, 102498 (2023).
60. Corti, C. *et al.* Artificial intelligence for prediction of treatment outcomes in breast cancer: Systematic review of design, reporting standards, and bias. *Cancer Treatment Reviews* **108**, 102410 (2022).
61. Ravella, R., Dee, E. C., Corti, C., Celi, L. A. & Iyengar, P. Broadening the scope of artificial intelligence in oncology. *Lancet Reg Health Am* **25**, 100573 (2023).
62. Celi, L. A. *et al.* Sources of bias in artificial intelligence that perpetuate healthcare disparities—A global review. *PLOS Digital Health* **1**, e0000022 (2022).
63. Lapen, K. *et al.* Electronic patient-reported outcomes for CAR T-cell therapies. *Lancet Oncol* **26**, e6 (2025).
64. Aiyegbusi, O. L. *et al.* Key considerations to reduce or address respondent burden in patient-reported outcome (PRO) data collection. *Nat Commun* **13**, 6026 (2022).

Securing Medical Images in Cognitive IoT Networks with SCSOA-Based SIMON Encryption on Hyperledger Blockchain

Priyan Malarvizhi Kumar and Bharat S. Rawal

1 Introduction

The relationship between the fields of medical imaging and cognitive Internet of Things (IoT) networks has emerged as a crucial area of innovation in an age characterized by the expanding deployment of the IoT and the pressing need for safe data handling [1]. The urgent need for improved medical image security in healthcare systems is addressed by this convergence. By combining cognitive IoT with medical imaging, cutting-edge technologies like machine learning, artificial intelligence, and blockchain are used to build a strong ecosystem that can protect private patient data and pictures. In addition to guaranteeing the security and integrity of medical pictures, this integrated architecture also maximizes their accessibility for rapid and precise diagnosis and treatment. Real-time monitoring, secure transmission, and intelligent processing of medical images are crucial components of this paradigm shift that enhances patient care while bolstering data privacy and integrity.

Medical imaging is a crucial pillar in the diagnosis, care, and monitoring of a wide range of medical disorders in today's modern healthcare environment [2]. The development of digital imaging technology has completely changed the medical industry, allowing medical practitioners to treat patients more precisely and effectively. The sensitive medical image data that is a crucial component of healthcare operations must be protected, which means that this digital transformation has also brought about a fundamental requirement for strong security measures [3]. In this setting, the importance of patient privacy, data integrity, and adherence to healthcare standards cannot be overstated, making security a top priority for

P. M. Kumar
University of North Texas, Denton, TX, USA

B. S. Rawal (✉)
Grambling State University, Grambling, LA, USA
e-mail: rawalb@gram.edu

S. S. Iyengar et al. (eds.), *Artificial Intelligence Driven Forensics*, Signals and Communication Technology, https://doi.org/10.1007/978-3-031-98036-7_22

the healthcare industry. Encryption, a method that uses mathematical and logical methods to transform data, including photos, into unintelligible forms, is one of the primary tools in protecting medical image data [4]. Encryption protects data and information from unauthorized access, guaranteeing that private and intact sensitive medical photos are transmitted and stored. In this thorough investigation, it digs into the complex function of encryption in the field of medical imaging, illuminating its significance, methods, and ramifications for both patients and healthcare professionals [5].

Medical images, such as X-rays, MRIs, and CT scans, are crucial in the diagnosis and management of a wide range of illnesses. The accessibility and information sharing among healthcare practitioners are greatly facilitated by these photographs, which are normally stored and communicated in digital formats. However, because they are digital, they are also exposed to a wide range of security risks [6]. At this point, encryption acts as a strong defender of patient data. A crucial barrier that guarantees only authorized people or systems can access medical images is encryption. It works by encoding the data with a secret key, rendering it unintelligible to anyone without the key. This impenetrable layer of protection efficiently protects patient privacy and data integrity by preventing unauthorized access or modification with the images [7]. While encryption is proving to be a powerful tool in the fight against data breaches and unauthorized access, using it with medical imaging poses special difficulties [8, 9]. When dealing with the vast sizes of medical photographs, conventional encryption methods, such the commonly used Advanced Encryption Standard (AES), have disadvantages. High-resolution images like those from MRIs and CT scans include enormous amounts of data, necessitating creative solutions that can effectively manage these digital assets [10, 14]. While encryption is proving to be a powerful tool in the fight against data breaches and unauthorized access, using it with medical imaging poses special difficulties [11]. When dealing with the vast sizes of medical photographs, conventional encryption methods, such the commonly used Advanced Encryption Standard (AES), have disadvantages [12, 13].

The main contributions of this paper are:

- Creation of a specific security-based cognitive IoT network with a blockchain-driven Hyperledger architecture for safe management of medical data
- Implementation of a reliable and effective encryption technique to protect various medical image modalities, such as CT, MRI, and X-rays
- Improvement of the SIMON block cipher method with the use of the Sand Cat Swarm Optimization Algorithm (SCSOA) for optimum key generation
- Establishment of a ground-breaking technique for secure data exchange of medical information through a multi-channel Hyperledger blockchain in cognitive IoT networks, which achieved outstanding ratings on the Feature Similarity Index (FSIM) and Structural Similarity Index (SSIM), verifying the model's efficiency

The remaining sections of the study are organized in the form of shadows: The relevant works are summarized in Sect. 2, the suggested model is briefly explained in Sect. 3, the results and validation analysis are shown in Sect. 4, and the summary and conclusion are provided in Sect. 5.

2 Literature Survey

Author name	Methodology employed	Advantages	Limitations
Lai, Q et al. [15]	It created a two-dimensional Logistic-Gaussian hyperchaotic map (2D-LGHM) with a variety of hyperchaos	It is highly resistant to attacks and data loss in a variety of medical image formats by utilizing a potent 2D-LGHM hyperchaotic map, boosting security, and maintaining data integrity	It might ignore some security risks because it lacks a thorough investigation of potential flaws and attacks that are specific to medical picture encryption techniques
Wang, X., and Wang, Y [16]	Create a region-of-interest (ROI)-based medical picture encryption technique employing chaotic sequences from the Logistic-Tent chaotic system (LTS)	Enhanced security is achieved via chaotic sequences, region-of-interest-based encryption, and faster encryption for medical imaging	Risk of losing non-ROI data, dependency on hash value, and sensitivity to LTS parameters. Complexity of ROI extraction
Abdelfatah, R. I et al. [17]	Multi-chaotic maps and adaptable DNA coding. Chaos is produced by combining Henon, Gaussian, and Logistic maps	Combines adaptive DNA coding with a new multi-chaotic map architecture to provide increased security for medical data in WBANs, ensuring strong defenses against diverse threats while maximizing computing effectiveness and data transport capabilities	One drawback is the potential complexity and difficulties of practical application of adaptive DNA coding in genuine WBANs
Trujillo-Toledo, D. A et al. [18]	Use statistical tests, MQTT, the Lyapunov exponent, and four chaotic maps to encrypt medical images	High-data throughputs on the desktop PC and Raspberry Pi platforms. huge MIMO techniques	There may be compatibility problems and integration difficulties when integrating the suggested cryptosystem into current healthcare IoT systems
John, S., and Kumar, S. N [19]	LFSR-based medical picture encryption that is proposed, tested using DICOM images, and IoT	Enhances the security of medical data transported across a cloud network by using a linear feedback shift register (LFSR)	In light of emerging cyber threats, it might not offer the greatest level of protection necessary for sensitive medical data
Hazzazi, M. M [20]	Bit-level encryption employs DWT, bit-plane extraction, and cubic-logistic map	Data security and little information loss are guaranteed by strong "sandwich encryption"	Under specific attacks like noise and cropping attacks, the encryption system may show decreased resistance to content exposure
Singh, K. N., et al. [21]	Use GDWCN-PSO to optimize random sequences; combine with the Lorenz system to encrypt images; more secure than current techniques; and robust against attacks	Beyond state-of-the-art techniques, the EiMOL encryption algorithm provides strong security for medical pictures, ensuring data integrity in intelligent healthcare	For healthcare devices with limited resources, the computational complexity and resource needs of the proposed EiMOL encryption technique may be relatively high

3 Proposed Methodology

3.1 Speculative Cognitive IoT-Cloud Scenario for Healthcare

Within the framework of a networked smart city setting, smart healthcare systems provide various advantages to both patients and healthcare providers. To speed up access to electronic health records, these systems depend on intelligent sensor technology and use cloud-based cognitive Internet of Things (IoT) solutions. Patients may easily update their medical information with the help of integrated smart wearables and seamless communication. In order to make wise judgements and provide patients with individualized treatments based on their particular requirements, the cognitive system analyzes this data in real-time. Additionally, medical experts can remotely view and examine this uploaded data, allowing them to provide prompt aid when necessary. Patient-centric healthcare monitoring is of utmost importance for achieving the core goals of smart healthcare, which include cost reduction, improved efficiency, improved accessibility, accurate diagnostics, decreased hospital and medical staff visits, and an overall improvement in the quality of life. Below is a more elaborate example for the suggested smart healthcare monitoring framework in a smart city setting:

For access to this cutting-edge healthcare system, residents are urged to register with a smart healthcare service provider and make use of the smart city infrastructure. This registration makes it easier for the cognitive system, medical personnel, and support staff to communicate effectively, providing remote access to a patient's medical records. The technology also keeps track of the user's position, which comes in handy during emergencies. Between patients and specialized medical specialists, such as those who specialize in illnesses like epilepsy, the smart healthcare service provider is an essential connection. Patients' psychological and physiological data are continually collected in real-time using a range of smart sensors in the healthcare industry. While the cognitive system continuously monitors the patient's current status, the sensors are in charge of recording the patient's motions, gestures, and facial expressions. A handy and transportable smart electroencephalogram (EEG) sensor in the shape of a skull cap is made accessible for people with epilepsy. The person's brainwave activity is regularly recorded by an EEG gadget. The gathered data are immediately sent to the cloud for in-the-moment processing when the cognitive system detects a seizure. The information is then sent to medical professionals for in-depth study. On the basis of the nature and intensity of the seizure, medical professionals may completely assess the data and provide patients with the right recommendations. The cognitive system uses the smart healthcare provider to alert the appropriate stakeholders and medical staff in emergency scenarios. Smart ambulances with necessary medical equipment can effectively maneuver traffic to guarantee quick response. With the help of an automated traffic signal system, these ambulances are given priority passage across crowded roads. As a result, the smart city healthcare framework successfully provides residents and patients with remote, prompt, and real-time healthcare services from the convenience of their own places.

3.2 System Architecture

The data flow inside our suggested architecture for smart healthcare monitoring is shown graphically in Fig. 1. It demonstrates the procedure for obtaining various multimedia health-related data and EEG signals using IoT devices. These sensors are arranged in this system at various levels. We have created a Local Area Network (LAN) layer that makes use of short-range smart communication tools to make it possible for signals like EEG to be seamlessly sent from IoT devices to the hosting layer. Computers, cellphones, and other smart devices make up the hosting layer, which acts as a conduit for data transmission to the cloud through the Wide Area Network (WAN) layer. The WAN interface makes use of cutting-edge connection methods like 4G, 5G, or Wi-Fi to guarantee quick and effective data transport to the cloud.

The system performs user authentication to confirm the validity of all health-related data, including EEG, after it has reached the cloud, before the data is processed by the seizure detection system. The vast spectrum of IoT sensors, which includes a number of healthcare-related sensor devices such headgear, smartwatches, wearable sensors, and wristbands, is also shown in Fig. 1. The extensive range of health-related characteristics that these intelligent sensors may measure includes heart rate, blood pressure, respiration rate, ECG, body temperature, body movement, and EEG. These gadgets may be worn by patients or easily incorporated into a variety of smart city settings, such as smart homes, healthcare institutions, workplaces, or transportation. Furthermore, these sophisticated sensor devices are capable of short-range communication among themselves. Smart communication protocols created for short- to medium-range communication and device interconnectivity serve as the foundation for the LAN interface layer. These protocols, which include technologies like Bluetooth, Z-wave, Zigbee, and LoWPAN, enable effective connection and communication between devices.

Laptops, workstations, personal digital assistants, cellphones, and tablets are just a few examples of the wide variety of smart devices that make up the hosting layer. Data gathering and preliminary local processing are handled by these devices. They are equipped with the processing power needed to use specialized software or programs to find irregularities in general health. These deviations include differences in measurements like blood pressure, heart rate, or body temperature. The WAN interface makes it possible to send this data to the cloud through long-distance networks like 4G, 5G, or Wi-Fi. The data is examined and patient health records are checked once it reaches the cloud-based healthcare service providers in order to help in making prompt judgements. The suggested seizure detection and categorization system also processes specialized data, such as EEG signals. Following that, the findings are presented to medical experts for in-depth analysis. Medical facility

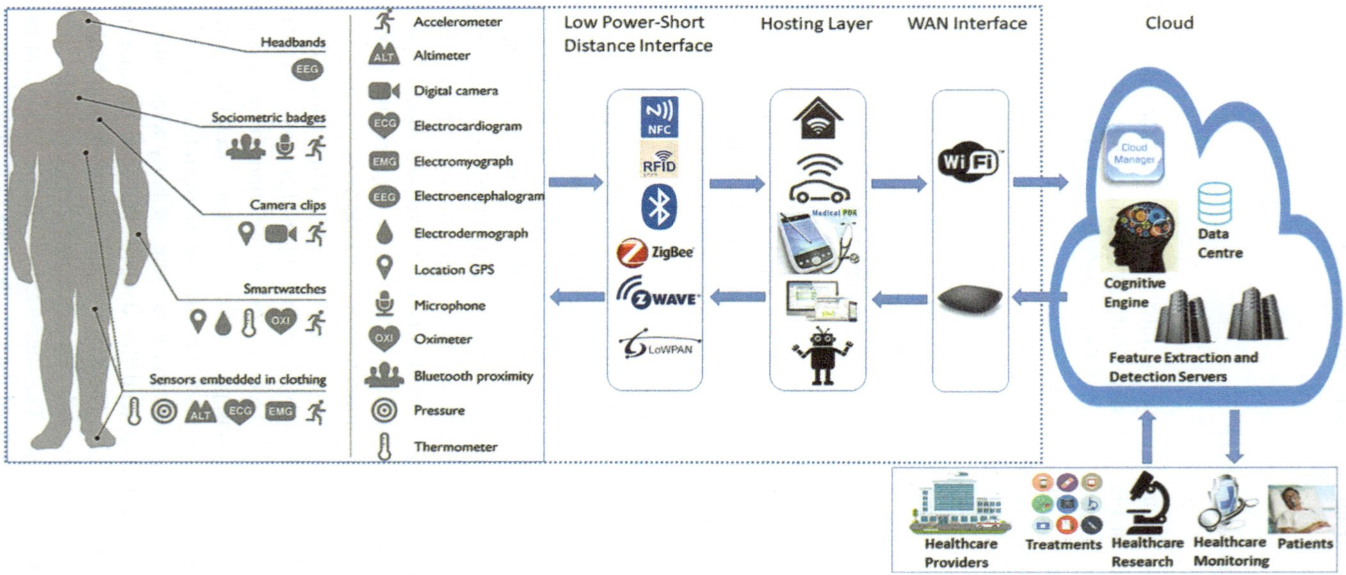

Fig. 1 Framework for cognitive IoT-cloud-based smart healthcare

visits may be time-consuming and expensive in non-urgent situations. The adoption of such a smart healthcare framework might thus result in savings in terms of money, time, and hospital resources.

Essential parts of the cloud architecture include the cloud manager, data center, feature extraction server, detection server, and classification server. A resident's registration with a smart healthcare provider is first checked by the cloud manager during user authentication. This duty includes making sure that everyone using the system is who they claim to be, including patients, hospital representatives, healthcare workers, and doctors. Controlling data transfer to and from other servers as well as resource allocation, storage, and communication is the responsibility of the cloud manager. Effective resource management is another aspect of it. It is sent to the cognitive engine, which analyzes multimodal data, including EEG, and physiological data to determine the patient's status. The patient's motions, gestures, and facial expressions are just a few of the many things that go into this assessment. Taking into account the patient's present condition, the cognitive system decides in real-time which tasks, treatments, and services are essential. Whether or not data should be sent to the deep learning module is also decided by this. The cognitive system contacts the appropriate parties and sends the EEG data to a deep learning module when it recognizes a seizure. To extract features, deep learning methods are used in the feature extraction server. Techniques for extracting important features are used in signal processing beyond the preprocessing step. The cloud manager receives the classification and identification of seizure data from the detection server, which then reports its findings. Lastly, the data center acts as the central location for all model parameters, features, and data. The richness of data from smart sensors and the classification of EEG data are used by healthcare practitioners to make decisions about the services to be provided to patients with epilepsy. Healthcare service data are made accessible to all smart city stakeholders through intelligent communication channels after analysis of patient health reports and improvements to resident care.

4 The Proposed Model

The operational workflow of the proposed system is shown in a model in Fig. 2, which includes various unique steps such as encryption, optimum key creation, secure data management utilising the Hyperledger blockchain, and diagnostics. The patient's medical records are encrypted in the early stage. Combining the SCSOA-based optimum key generation method with the SIMON block cipher approach, this encryption is carried out. The Hyperledger blockchain then enters the picture after encryption. A worldwide blockchain and several local blockchains customized for different medical institutions make up this blockchain architecture. In this system, the patient has the authority to give or remove access privileges to doctors or healthcare institutions. The real health records may then be accessed by authorized individuals by decrypting the ciphered data.

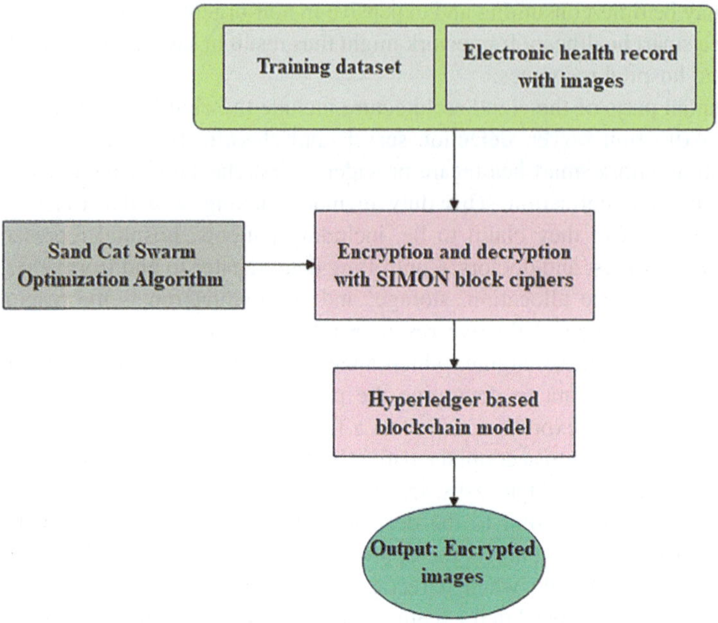

Fig. 2 Workflow of the proposed model

4.1 Data Encryption Process

Hash functions, block ciphers, validated decryption, and encryption modelling processes are examples of lightweight cryptographic approaches in this study. The use of block ciphers for secure healthcare record keeping is the specific subject of this work. The advancements in the Advanced Encryption Standard (AES) are responsible for the development of lightweight ciphers. Several ciphers are now available, with SIMON being a well-known lightweight block cipher. These include RECTANGLE, TWINE, KATAN, SPECK, and KLEIN. In particular when it comes to hardware implementation, SIMON, a lightweight block cipher, prioritizes efficiency influenced by hash functions. Each of the ten functions of this cipher family has unique properties, principally differing key sizes and block structures. Based on manipulation of picture pixels, the key changes per block. Block sizes range from 32 to 128 bits, usually about 16. It generates ciphertext blocks using fixed-size plaintext chunks for operation. With relation to block size and data security, the SIMON cipher mixes nonlinear and linear components. One may picture a tree structure for each variation throughout all rounds, producing several possible output variants, while taking into account diverse input variations and the potential for extending critical characteristics to future rounds. The block size of the SIMON cipher is not intended to be ideal, but the selected optimization approach seeks to get the best result in order to reduce the number of active S-boxes.

Encryption, rounds, bits, and decryption are essential elements in the context of safe data transfer using the cipher module. The critical operation of dividing an input plaintext block, normally sized at 2n, into two equally sized n-bit words is carried out by the SIMON cipher's round function. Three left shifts are performed to the left side of the block throughout this phase, along with bitwise AND logic operations. The right half block is XORed with the result of the XOR operation using the key from the current round for each round of the operations that follow. Each round's result is written back to the left block after the left half's initial value has been moved to the right block. As long as the selected configuration keeps the total number of rounds constant, this iterative round-based procedure continues. The SIMON cipher, created for use with 2n-bit blocks, may be expressed mathematically as follows:

$$\text{EncDI} = \text{Cipher}^{i}_{ql}, \ldots, \text{Cipher}^{n}_{q1}, i > 1 \tag{1}$$

Round functions and round keys are used to symbolize the cipher $C^{i}_{ql} \leq i \leq q$. Similar to an iterated block cipher, this function works in a similar way. The SIMON round function is described in terms of encryption as follows:

$$\text{RF}\left(w_l, w_r, k_{\text{round}}\right) = \left(\left(S^1\left(w_l\right) \& S^8\left(w_l\right) \bigoplus S^2\left(w_l\right) \bigoplus w_r \bigoplus k_{\text{round}}, w_l\right)\right) \tag{2}$$

According to Eq. (3), the inverse function is used to decode the data.

$$\text{RF}^{-1}\left(w_l, w_r, k\right) = \left(w_r\left(S^1\left(w_l\right)\right) \& S^8\left(w_l\right) \bigoplus S^2\left(w_l\right) \bigoplus w_l \bigoplus k_{\text{round}}\right) \tag{3}$$

The term in Eq. (3) is defined as follows: w_l denotes the left-most word in a block supplied, w_r denotes the right-most word, and k_{round} is the suitable round key.

Proposed Algorithm

Algorithm 1 is a suggested safe method of managing medical records.

4.1.1 Optimal Key Generation Process

In the SIMON cipher, the master key is used throughout the key expansion procedure to create a whole set of round keys. The selected configuration, SIMON64/128, tries to convert the original 128-bit master key into 32-bit round keys. This is accomplished by a methodical procedure that combines a dependable one-bit round operation with the "key words variable," which is made up of a group of previously saved round keys. In order to generate these round keys securely, the key expansion process entails a series of related actions.

Algorithm 1 Proposed SIMON block cipher addition for safe medical data management

Input: A patient P1's EHR is being read using images.
Output: Add blocks to the Patient P1 blockchain and perform Patient P1 diagnostic
1: EHR with images←Patient P1 and read EHR with images
2: Public Key and Private Key←generating keys for the SIMON block cipher
3: P1 is the encryption patient←Public key
4: Doctor←insurance company and shared private key for encryption
5: Encrypted EHR with images←EHR encryption using images Using public keys, SIMON
6: SIMON←SCSOA-based hash key for EHR encryption
7: the patient's user ID, password, and patient code are used to create a Hyperledger block for the Patient P1 blockchain.
8: Block←the encrypted EHR containing photos and the hash key value.
9: safe block to the Patient P1 blockchain
10: Stop

- Bitwise XOR is denoted by the symbol $a \oplus b$.
- Right circular shift, represented as F^{-j}, includes shifting by j bits to the right, whereas left circular shift requires shifting by j bits to the left, as shown by the symbol F^j.
- The round counter is written as $0 \leq i \leq T - 1$, and the constant sequence $j = 0$, 1, 2, 3, 4 is used to represent it.
- The number of cipher rounds, round keys (sub-keys), and constants are also essential elements.
- Right bitwise rotation, or ROR, is calculated by the formula $s^{-c}(a)$, where c stands for the number of rotations.

Equation (4) gives an expression for the key expansion procedure.

$$\mathrm{Key}_i\left(k, c, z_j\right) = F\left(k_{i+3}, k_{i+1}\right) \bigoplus S^{-1}\left(F\left(k_{i+3}, k_{i+1}\right)\right) \bigoplus k_i \bigoplus c \bigoplus \left(z_j\right)_i \tag{4}$$

An ideal key for data decryption may be selected from the large number of created keys. The use of SCSOA, which optimizes its value to either minimize or maximize it, results in the selection of this ideal key.

4.1.2 Optimal Key Selection Using Sand Cat Swarm Optimization Algorithm (SCSOA)

Initialize Population

Every sand cat is represented as a 1 × dim array in the context of the optimization problem described in [22], where "dim" stands for the dimension of the issue at hand. This array functions as a possible answer to the issue at hand, as seen in Fig. 3. Each "Pos" must adhere to lower and higher boundaries that are predetermined

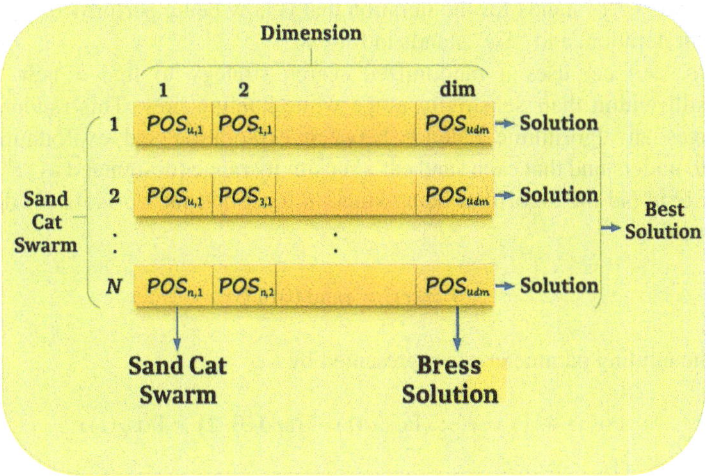

Fig. 3 Population initialization diagram

within a range of variable values (Pos_1, Pos_2, ..., Pos_{dim}). An initialization matrix is created to begin the procedure; it is sized to the size of the issue and produces an N-dim matrix. Every iteration also results in a related solution. The prior answer is replaced if the freshly produced one turns out to be better than the previous one. The answer from that particular iteration is not saved nor kept, however, if no better solution is found in the next iteration.

Search for Prey (Exploration Stage)

The SCSO algorithm uses "Posi" to represent each sand cat's location. This programme makes use of the amazing listening abilities of sand cats, especially their skill in identifying low-frequency noises. The capacity to detect frequencies below 2 kHz is shared by all sand cats. In order to define the sensitivity r_{G} mathematically, this paper use Formula (5). This sensitivity characteristic accurately reflects the sensitivity range of sand cats, which ranges from 2 kHz to 0 kHz. Additionally, this paper choose the parameter "R" in accordance with Formula (6), which is crucial for determining how well the algorithm balances its abilities for exploitation and exploration.

$$r_{\text{G}} = S_M - \left(\frac{S_M \times t}{T} \right) \tag{5}$$

$$R = 2 \times r_{\text{G}} \times \text{rand}\,(0, 1) - r_{\text{G}} \tag{6}$$

In this context, "t" stands for the iteration that is now being performed, "T" for the maximum iteration, and "S_M" stands in for two.

Every sand cat uses a randomized search strategy to find a new location that is still within their sensitivity range while hunting prey. This randomization encourages an algorithmic balance between exploration and exploitation. It is crucial to understand that each sand cat's sensitivity range (designated as "r") differs in order to avoid the algorithm being stuck in local optima. Formula (7) describes this variability.

$$r = r_G \times \text{rand}(0, 1) \tag{7}$$

where the guiding parameter r is represented by r_G.

$$\text{Pos}(t+1) = r \times (\text{Pos}_{bc}(t) - \text{rand}(0, 1) \times \text{Pos}_c(t)) \tag{8}$$

Each sand cat uses a variety of variables, such as the best candidate position (Pos_{bc}), the current position ($\text{Pos}_c(t)$), and its sensitivity range (r), to determine where the prey is. The formula given in (8) provides a clear definition of this procedure.

Attack Prey (Exploitation Stage)

The distance (Posrnd) between the sand cat and its prey is described by Formula (9), imitating the predatory chase made by the sand cat. Imagine the sensitivity range of the sand cat as a sphere to show this. The direction of the sand cat's movement is chosen using the Roulette Wheel selection approach. This technique chooses a chance angle "a," where "a" might be between -1 and 1. The values of "a" are next normalized to lie between 0 and 360 degrees. Therefore, as shown in Fig. 4, this method enables each sand cat to move around the search region in a wide variety of circumferential directions. Following the instructions outlined in Formula (10), the sand cat next launches its prey assault. The sand cat can more quickly reach its hunting location thanks to this tactic, which increases the effectiveness of its predatory pursuit.

$$\text{pos}_{\text{rnd}} = \left|\text{rand}(0, 1) \times \text{pos}_b(t) - \text{pos}_c(t)\right| \tag{9}$$

$$\text{pos}(t+1) = \text{pos}_b(t) - r \times \text{Pos}_{\text{rnd}} \times \cos(a) \tag{10}$$

SCSO Algorithm Implementation

It employs adaptive parameters in the SCSO approach, most especially r_G and "R," to precisely adjust the ratio of exploration to exploitation. According to Formula

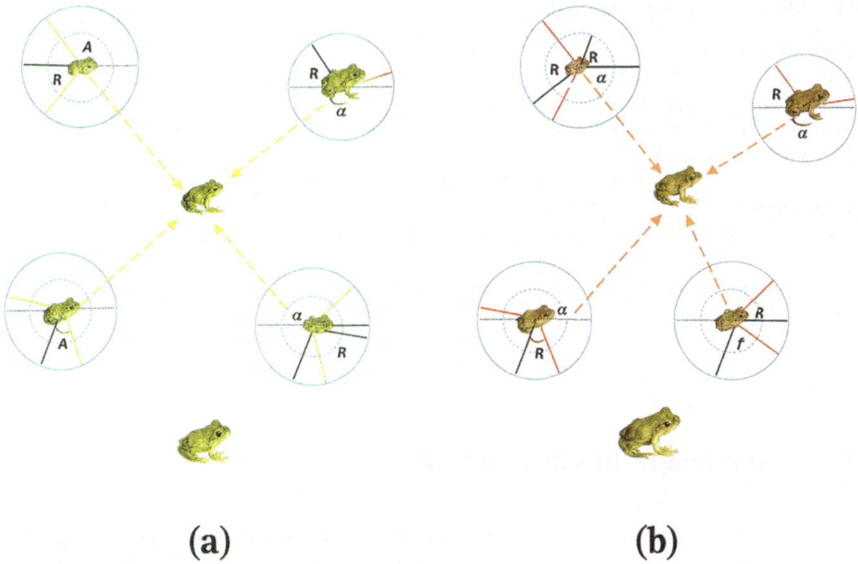

Fig. 4 SCSO algorithm's location updating mechanism. (**a**) The location of the group of sand cats in iteration. (**b**) The sand cat group's location in iteration $t + 1$

(5), over repetitions, r_G displays a systematic linear reduction, eventually declining from 2 to 0. On the other hand, a random value is chosen for the parameter "R" from the range $[-4, 4]$. The sand cat starts pursuing its victim when "R" drops to or reaches 1. Conversely, the sand cat continues its hunt for food when "R" exceeds 1, as explained in Formula (11). The algorithm's exploration and exploitation stages are seamlessly transitioned thanks to this dynamic control mechanism.

$$\text{Pos}\,(t+1)\begin{cases} r \times (\text{Pos}_{bc}(t) - \text{rand}\,(0,1) \times \text{Pos}_c(t)) & |R| > 1;\ \text{exploration} \\ \text{Pos}_b(t) - \text{Pos}_{\text{rnd}}(t) \times \cos(a) \times r & |R| \leq 1;\ \text{exploration} \end{cases} \tag{11}$$

Each sand cat updates its position throughout both the exploration and exploitation periods, as shown by Formula (11). The sand cat will attack its victim if the value of "R" is less than or equal to 1. However, if $R \leq 1$, the sand cat's goal is to look for fresh prey across the whole world. For further information, please see the pseudo-code under Algorithm 2.

Algorithm 2 SCSOA pseudo-code

```
Population initiation
Using the goal function as a foundation, calculate the fitness function.
Initialize the r, r_G, and R
While (t ≤ maximum iteration)
For every search agent
Using the selection from the roulette wheel, determine a random angle (0° ≤ α ≤ 360°)
If (abs(R) > 1)
The search agent position should be updated using Formula (8)
Else
Using Formula (10) update the location of the search agent
End
T = t + 1
End
```

4.2 *Hyperledger Blockchain (HB)*

In this research, a machine learning-based illness diagnosis model is implemented using a federated learning process inside a blockchain system. A shared ledger is built using blockchain technology, which has a number of benefits, such as suitability, availability, privacy, and decentralization. Blockchain's decentralization means that data is duplicated across several computers, removing the possibility of a single point of failure brought on by a central server. Decentralization improves system toughness. Even in the case of a few computer failures, availability ensures that data may be accessed when necessary. Integrity guarantees that data is safeguarded from unauthorized alterations and stays unchanged. All data saved inside the blockchain can be traced thanks to suitability. Technically speaking, a blockchain is made up of a chain of valid blocks, each having a header and data. The header is made up of a number of properties, such as a signature, an identity, and a reference to the block before it. The identifier is a value that represents each block of data and is globally unique. It is obtained from a mathematical formula. Because it establishes a logical chain of links between blocks, the reference to the preceding block is essential for preserving the blockchain's integrity.

One of the open source blockchains offered by the management framework Hyperledger is called Hyperledger Fabric. It consists of a number of parts that work together to create a decentralized environment, including endorser peers, ordering nodes, certificate authorities, clients, and committing peers. These parts connect with one another via channels, which allow for discreet and secret business dealings and demarcate various application areas. The Fabric certificate authority fulfils two functions. It does this in two ways. First, it makes sure that diverse parts, such users or smart contracts, follow the rules established for the system. Second, it verifies and gives permission to components for certain tasks like transaction execution or access to other components. Maintaining the chain connected to the channels built into the system is the responsibility of committing peers. According to the "individual chain per channel" strategy, they store several blockchains for the channels. As each

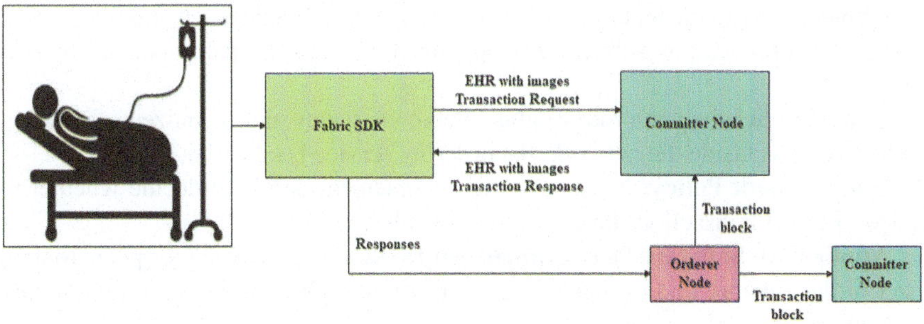

Fig. 5 Hyperledger blockchain

channel's data is kept separate from the others, this method improves scalability and privacy. In order to gain access to a chain via a committed peer linked to a distinct channel, components require access rights. The system can grow to meet increasing demand because of the "individual chain per channel" method, which enables the administration of many transactions and data across several committing nodes. Smart contracts are represented as chain code applications on the Hyperledger Fabric platform. The network-approved business logic is implemented using chain code. Each chain code maintains a private state that is off-limits to other chains, yet it is permitted to invoke another chain code with the right authorization. Chain code is divided into two categories: application chain code, which is in charge of maintaining application-specific states, and system chain code, which handles system-related operations. The private healthcare organization starts a registration function inside the smart contracts (chain code) to sign up individuals with the permission of the Fabric network manager. A private permissioned network is created by the healthcare authority only for registered stakeholders. Entry to the registration system is made possible over VPN connections to provide increased security. Patients must provide all necessary details, including their name, social security number, home address, and phone number, in order to finish the registration process. Primary care doctors, hospitals, labs, pharmacies, researchers, and insurers all go through the same registration procedure. The public health authority checks the records after registration and issues a chain code address. Transactions on the network may start as soon as all parties have successfully registered.

The complex procedure inside the HB infrastructure designed for the healthcare industry is shown in Fig. 5. The modular design approach used by the Hyperledger Fabric is distinguished by its focus on security, resilience, adaptability, and scalability. The many intricacies inherent in the economy may be accommodated by this framework, which also allows for the seamless integration of many aspects. Many important components in the structure of a block, such as the following, provide support for the essential concepts of Fabric technology:

Chain Codes: These operate as self-executing programmes, akin to smart contracts, and are presently written in the Go programming language.

Channels: With the main goal of allowing secret transactions, channels represent private communication subnets among specified network participants, such as hospitals.

Ordering Service: In order to guarantee consistency and organized scheduling of transactions inside the network, the ordering service is essential.

Endorsement Policy: Nodes use the endorsement policy to decide whether to allow or reject transactions based on a set of rules.

Application Software Development Kit (SDK): The SDK allows peers to connect to the network and establish communication, improving network functioning as a whole.

Endorsing Peers: In order to pre-approve operations based on the endorsement policy set inside the chain codes, endorsing peers are essential.

Committing Peers: Committing peers are essential to the block verification process. Then, after updating the blocks' status in the State database (DB), they update the ledger. They obtain blocks from the ordering service.

5 Results and Discussion

To demonstrate the effectiveness of the recommended optical colour medical image cryptosystem, several colour medical pictures with various characteristics are selected and studied. To be used as input for the cryptosystem, the employed colour medical pictures are first largely divided into their R, G, and B components. The simulation tests are carried out using a laptop with an i7-5200 Intel CPU and 8 GB of RAM. The MATLAB R2020b is the programme used in the simulation experiments.

5.1 Entropy Results

The ciphered colour medical image's level of unpredictable behaviour is described using the entropy measure. The optimal entropy value for the suggested model is almost equal to 8. The analysis results for the original, encrypted, and decrypted colour medical pictures are shown in Table 1 for the proposed model and the traditional optical cryptosystems. By attaining better values that are closer to the required ideal value than alternative models, the recommended model's superior entropy values demonstrate its resilience and dependability. Figure 6 displays the qualitative outcomes of the suggested approach utilising different medical photos.

From Table 1 and Fig. 7, Image 1 had an original image quality of 7.422, encrypted image quality of 7.488 without optimization, proposed model (encrypted

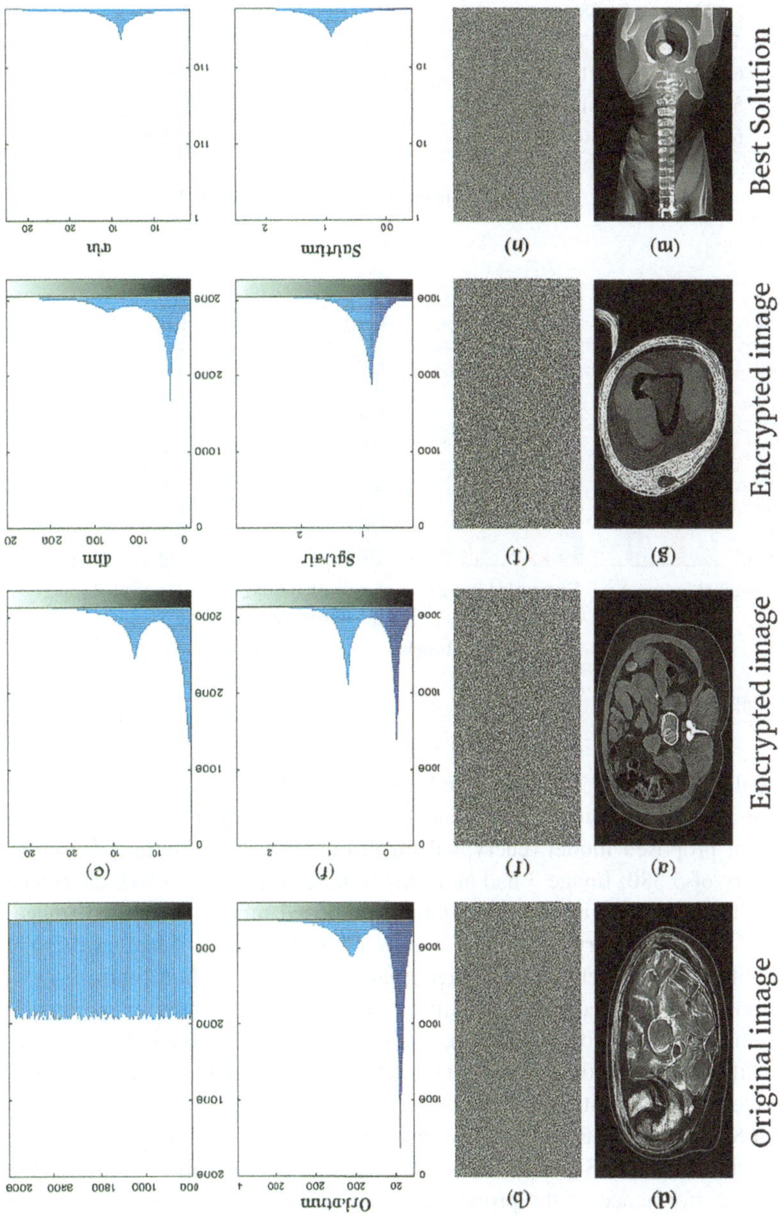

Fig. 6 Qualitative results of the suggested model. Rows top to bottom: brain MRI image, abdomen CT scan of the kidney, knee MRI, and X-ray of the spine. Columns left to right: original images, cipher counterparts, original image histograms, and cipher image histograms

Table 1 Entropy outcomes for the proposed model

Image	Original image	Encrypted image (without optimization)	Encrypted image (proposed model)	Decrypted image (all algorithms)
Image 1	7.422	7.488	7.821	7.422
Image 2	5.380	7.073	7.436	5.380
Image 3	6.864	7.391	7.694	6.864
Image 4	5.083	7.231	7.798	5.083
Image 5	6.055	7.536	7.807	6.055

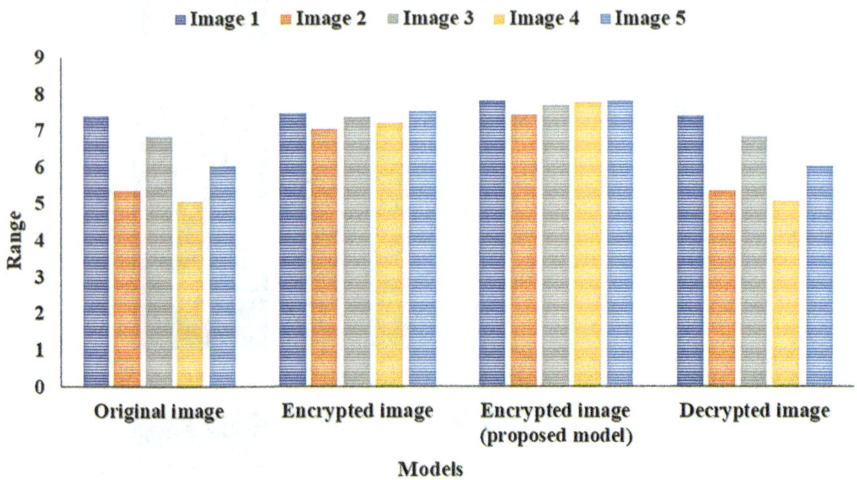

Fig. 7 Entropy analysis

image) quality of 7.821, and decrypted image quality of 7.422. Image 2 had an original image quality of 5.380, encrypted image quality of 7.073 without optimization, proposed model (encrypted image) quality of 7.436, and decrypted image quality of 5.380. Image 3 had an original image quality of 6.864, encrypted image quality of 7.391 without optimization, proposed model (encrypted image) quality of 7.694, and decrypted image quality of 6.864. Image 4 had an original image quality of 5.083, encrypted image quality of 7.231 without optimization, proposed model (encrypted image) quality of 7.798, and decrypted image quality of 5.083. Image 5 had an original image quality of 6.055, encrypted image quality of 7.536 without optimization, proposed model (encrypted image) quality of 7.807, and decrypted image quality of 6.055.

SSIM, FSIM, and PSNR analysis use important measures like SSIM (Structural Similarity), PSNR (Peak Signal-to-Noise Ratio), and FSIM (Feature Similarity) to evaluate the performance of the proposed model. We determine these parameters in our security research by contrasting the encrypted medical pictures with their original, unencrypted forms. More effective encryption is indicated by smaller

Table 2 Results of the suggested models' SSIM, FSIM, and PSNR tests on medical images in plain and encrypted formats

	Without optimization			Proposed model		
	SSIM	FSIM	PSNR (dB)	SSIM	FSIM	PSNR (dB)
Image 1	0.0103	0.5129	10.5681	0.0143	0.4920	10.6102
Image 2	0.0190	0.5210	13.7794	0.0225	0.5203	13.8057
Image 3	0.0120	0.5007	11.7526	0.0119	0.5003	11.7475
Image 4	0.0017	0.4459	8.4068	0.0024	0.4450	8.3806
Image 5	0.0016	0.4386	7.8848	0.0030	0.4371	7.9084

metric values. Table 2 shows the results of the suggested model, which are obviously superior to conventional models since they provide considerably better metric values.

The assessment findings, which are shown in Table 2 and the accompanying Figs. 8, 9, and 10, demonstrate the significant enhancements brought about by optimization for each of the five photos. Image 1 had an SSIM of 0.0103, FSIM of 0.5129, and PSNR of 10.5681 without optimizations. These parameters increased upon optimizations to an SSIM of 0.0143, FSIM of 0.4920, and PSNR of 10.6102. Similar to Image 1, Image 2 showed baseline scores of 0.0190 SSIM, 0.5210 FSIM, and 13.7794 PSNR. These values were then improved via optimizations to 0.0225 SSIM, 0.5203 FSIM, and 13.8057 PSNR. Before optimizations, Image 3 produced an SSIM of 0.0120, FSIM of 0.5007, and PSNR of 11.7526. After optimizations, these values slightly changed to 0.0119 SSIM, 0.5003 FSIM, and 11.7475 PSNR. Additionally, Image 4 was optimized from 0.017 SSIM, 0.4459 FSIM, and 8.4068 PSNR to 0.024 SSIM, 0.4450 FSIM, and 8.3806 PSNR. Last but not least, Image 5 had a non-optimized SSIM of 0.0016, FSIM of 0.4386, and PSNR of 7.8848, which was later enhanced to 0.0030 SSIM, 0.4371 FSIM, and 7.9084 PSNR. These results highlight the significant improvement made possible by the optimization procedure for every image.

5.2 Encryption/Decryption Time Analysis

Figure 11 shows the results of the investigation of encryption and decryption timings, which are reported in Table 3 and show a significant time decrease when optimizations are used. Without optimizations, the encryption and decryption timings for Images 1 and 2 were 4.6586 and 4.8272 s, respectively. For these procedures, Image 3 needed 4.495 s, Image 4 required 5.6271 s, and Image 5 required 5.7937 s. However, when the suggested model with optimizations was used, significant time savings were made. The encryption and decryption times for Images 1 and 2 were 3.2527 and 3.1596 s, respectively. Images 3 and 4 displayed times of 3.2047 and 3.3576 s, respectively, for these procedures. The

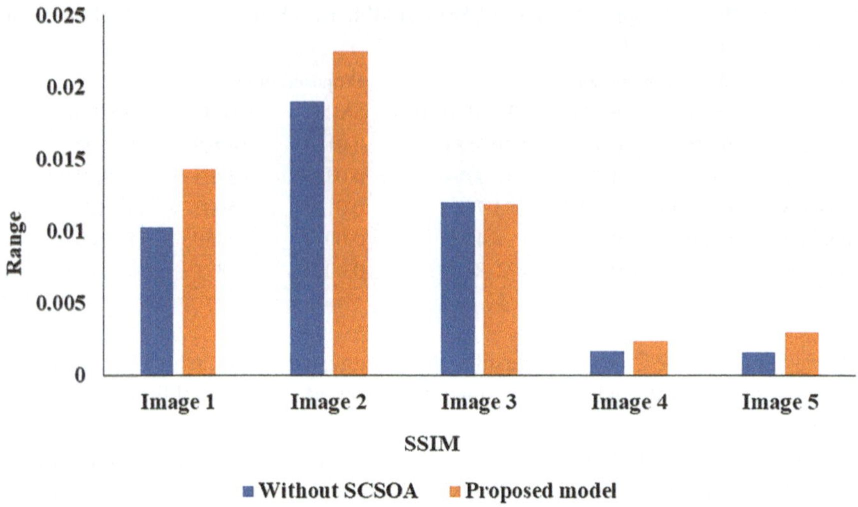

Fig. 8 SSIM analysis

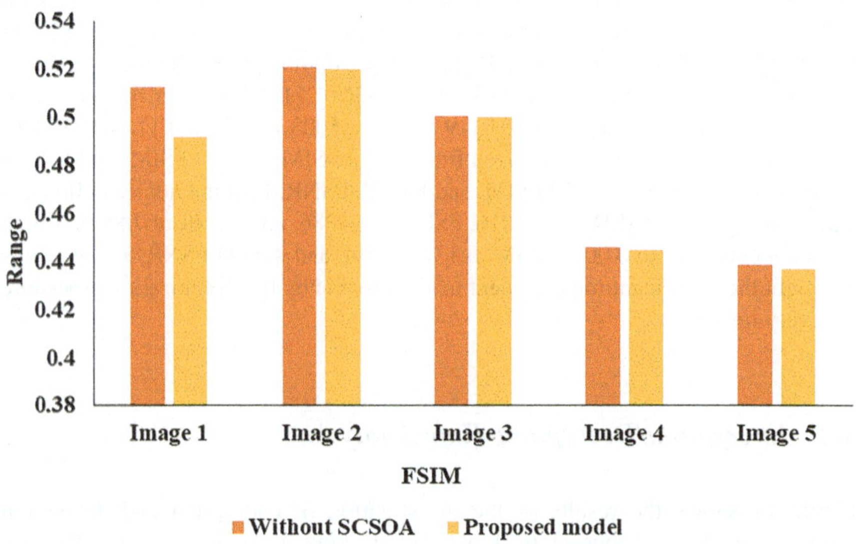

Fig. 9 FSIM analysis

encryption and decoding of Image 5 took 3.2129 s. These results demonstrate the efficiency benefits made possible by the optimization procedure, with encryption and decryption durations for all photos being drastically slashed in seconds.

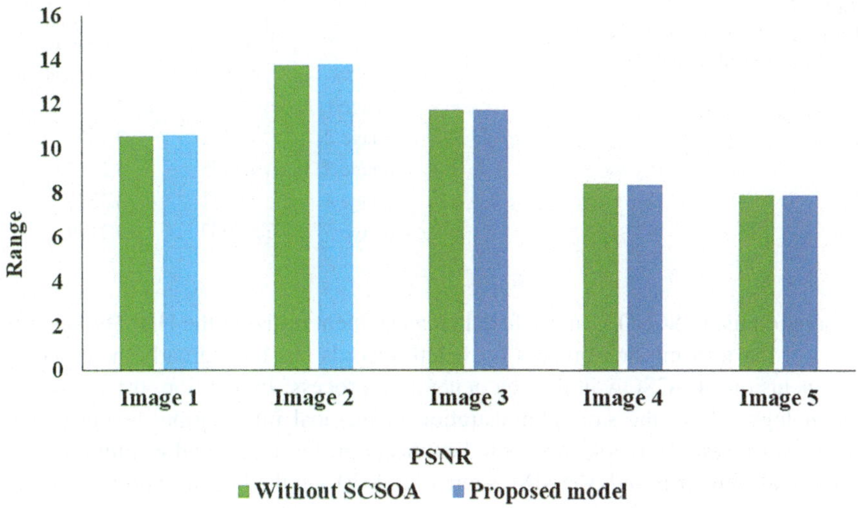

Fig. 10 PSNR analysis

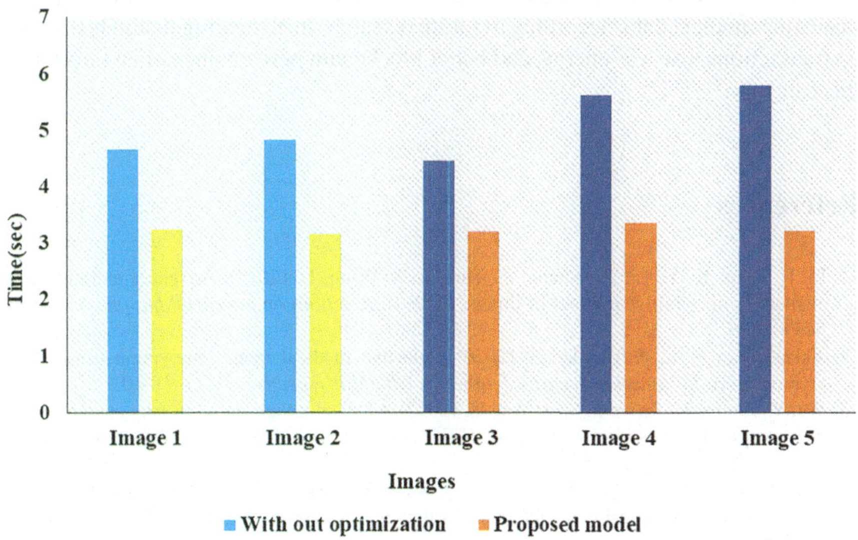

Fig. 11 Encryption/decryption analysis

6 Conclusion

To guarantee safe data transfer and streamline the diagnostic procedure, a thorough model has been developed in this study. The concept is composed of many operational stages, such as encryption using the SIMON block cipher, optimum key

Table 3 Encryption/decryption times for the proposed model

Images	Without optimization	Proposed model with optimization
Image 1	4.6586	3.2527
Image 2	4.8272	3.1596
Image 3	4.4495	3.2047
Image 4	5.6271	3.3576
Image 5	5.7937	3.2129

generation using SCSOA, and safe data management utilizing the HB. The security level of the transmission process for health records has been greatly increased by the inclusion of SCSOA in the key generation process. In addition, the use of HB technology allows the safe administration of medical information, giving people the power to give or revoke access to healthcare professionals and institutions. The Structural Similarity Index (SSIM) score of 0.0030 and the Feature Similarity Index (FSIM) score of 0.4371 show that the suggested model has accomplished outstanding achievements. These results represent a significant improvement demonstrating the effectiveness of the suggested paradigm. The HB should be optimized for managing medical data, according to future research. Investigating methods to speed up transactions, use less energy, and better blockchain performance often falls under this category.

References

1. Li, F., Lam, K. Y., Li, X., Sheng, Z., Hua, J., & Wang, L. (2019). Advances and emerging challenges in cognitive internet-of-things. *IEEE Transactions on Industrial Informatics*, *16*(8), 5489–5496.
2. Akkasaligar, P. T., & Biradar, S. (2020). Selective medical image encryption using DNA cryptography. *Information Security Journal: A Global Perspective*, *29*(2), 91–101.
3. Wu, Y., Zhang, L., Berretti, S., & Wan, S. (2022). Medical image encryption by content-aware dna computing for secure healthcare. *IEEE Transactions on Industrial Informatics*, *19*(2), 2089–2098.
4. Lin, H., Wang, C., Cui, L., Sun, Y., Zhang, X., & Yao, W. (2022). Hyperchaotic memristive ring neural network and application in medical image encryption. *Nonlinear Dynamics*, *110*(1), 841–855.
5. Jain, K., Aji, A., & Krishnan, P. (2021). Medical image encryption scheme using multiple chaotic maps. *Pattern Recognition Letters*, *152*, 356–364.
6. Ravichandran, D., Aashiq Banu S, Murthy, B. K., Balasubramanian, V., Fathima, S., & Amirtharajan, R. (2021). An efficient medical image encryption using hybrid DNA computing and chaos in transform domain. *Medical & biological engineering & computing*, *59*, 589–605.
7. Banik, A., Shamsi, Z., & Laiphrakpam, D. S. (2019). An encryption scheme for securing multiple medical images. *Journal of Information Security and Applications*, *49*, 102398.
8. Kaissis, G., Ziller, A., Passerat-Palmbach, J., Ryffel, T., Usynin, D., Trask, A., … & Braren, R. (2021). End-to-end privacy preserves deep learning on multi-institutional medical imaging. *Nature Machine Intelligence*, *3*(6), 473–484.
9. Belazi, A., Talha, M., Kharbech, S., & Xiang, W. (2019). Novel medical image encryption scheme based on chaos and DNA encoding. *IEEE access*, *7*, 36667–36681.

10. Deb, S., & Bhuyan, B. (2021). Chaos-based medical image encryption scheme using special nonlinear filtering function based LFSR. *Multimedia Tools and Applications*, *80*, 19803–19826.
11. Lakshmi, C., Thenmozhi, K., Rayappan, J. B. B., Rajagopalan, S., Amirtharajan, R., & Chidambaram, N. (2021). Neural-assisted image-dependent encryption scheme for medical image cloud storage. *Neural Computing and Applications*, *33*, 6671–6684.
12. Khashan, O. A., & AlShaikh, M. (2020). Edge-based lightweight selective encryption scheme for digital medical images. *Multimedia Tools and Applications*, *79*(35–36), 26369–26388.
13. Chen, Y., Tang, C., & Ye, R. (2020). Cryptanalysis and improvement of medical image encryption using high-speed scrambling and pixel adaptive diffusion. *Signal Processing*, *167*, 107286.
14. Pankaj, S., & Dua, M. (2021). A novel ToCC map and two-level scrambling-based medical image encryption technique. *Network Modeling Analysis in Health Informatics and Bioinformatics*, *10*, 1–19.
15. Lai, Q., Hu, G., Erkan, U., & Toktas, A. (2023). High-efficiency medical image encryption method based on 2D Logistic-Gaussian hyperchaotic map. *Applied Mathematics and Computation*, *442*, 127738.
16. Wang, X., & Wang, Y. (2023). Multiple medical image encryption algorithm based on scrambling of region of interest and diffusion of odd-even interleaved points. *Expert Systems with Applications*, *213*, 118924.
17. Abdelfatah, R. I., Saqr, H. M., & Nasr, M. E. (2023). An efficient medical image encryption scheme for (WBAN) based on adaptive DNA and modern multi chaotic map. *Multimedia Tools and Applications*, *82*(14), 22213–22227.
18. Trujillo-Toledo, D. A., López-Bonilla, O. R., García-Guerrero, E. E., Esqueda-Elizondo, J. J., Cárdenas-Valdez, J. R., Tamayo-Pérez, U. J., ... & Inzunza-González, E. (2023). Real-time medical image encryption for H-IoT applications using improved sequences from chaotic maps. *Integration*, *90*, 131–145.
19. John, S., & Kumar, S. N. (2023). IoT based medical image encryption using linear feedback shift register–Towards ensuring
20. Shaik Kashif Hussain & S. Saheb Basha. A Novel Chaotic Map-Based Noise-Resistant Image Encryption Scheme for Medical Images. In: Advances in Computational Intelligence and Informatics (ICACII 2023), Lecture Notes in Networks and Systems, vol. 993, Springer, pp. 147–157, 2024. https://doi.org/10.1007/978-981-97-4727-6_15
21. Singh, K. N., Singh, O. P., Singh, A. K., & Agrawal, A. K. (2023). EiMOL: a secure medical image encryption algorithm based on optimization and the Lorenz system. *ACM Transactions on Multimedia Computing, Communications and Applications*, *19*(2s), 1–19.
22. Wu, D., Rao, H., Wen, C., Jia, H., Liu, Q., & Abualigah, L. (2022). Modified sand cat swarm optimization algorithm for solving constrained engineering optimization problems. *Mathematics*, *10*(22), 4350.

Smart Agriculture: AI-Powered Green Sensor-Based Automated Pesticide Sprayer

K. R. Sowmya, Z. Ghouse Ahamed, and G. A. Srinidhi

1 Introduction

1.1 Problem Statement: Issues with Conventional Pesticide Spraying

The driver for this research undertaking, "Automated Pesticide Sprayers Working on Green Sensors Powered By AI," stems from the proliferating relevant grounds, which demand sustainability and resilience in agriculture. There are numerous complaints about pesticide chemical use from conventional, less effective pesticides, which contribute to excessive chemical use, higher operational costs, and adverse environmental impacts, such as soil health, water pollution, or unintended impacts on non-target organisms [1]. In addition to these issues, advances in agricultural mechanisms will include the incorporation of smart technologies, which add another layer of vulnerabilities associated with data security and system integrity. Traditional systems for spraying pesticides are vulnerable to unauthorized data access and/or modification. In this scenario, the harm may go beyond economic losses if a manager cannot verify the system's controls, and pesticides are sprayed inappropriately or abused. Due to the integration of within precision agriculture involving AI-based decision-making systems dependent on IoT devices, shielding sensitive digital agricultural data, along with a variety of potential cyber threats,

K. R. Sowmya (✉) · Z. Ghouse Ahamed
Department of ECE, SSIT, Sri Siddhartha Institute of Technology, SSAHE University, Tumkur, Karnataka, India

G. A. Srinidhi
Department of CSE (Cyber Security), SSIT, Sri Siddhartha Institute of Technology, SSAHE, Tumkur, Karnataka, India

S. S. Iyengar et al. (eds.), *Artificial Intelligence Driven Forensics*, Signals and Communication Technology, https://doi.org/10.1007/978-3-031-98036-7_23

combined with the loss of trust in AI-based systems from poorly conceived protocols to maintain confidentiality with data ownership.

Makes it all vital to engage with digital cyber/cybersecurity and environmental sustainability [2, 3]. Therefore, issues of digital and environmental sustainability are indisputable and significant in the context of modern agriculture. Improper use of pesticides, including too much pesticide and poor targeting, causes excessive overuse of chemicals by pest control professionals, pesticide resistance among targeted and non-targeted pests, and impacts on non-target organisms and their ecosystems. Additionally, environmental influences such as soil and water contamination and high operation costs, including labor and regulatory pressures, put pressure on pest control professionals. These challenges demonstrate the need for an AI-driven automated pesticide sprayer to optimize the practical application of pesticide spraying and reduce the potential environmental and economic effects of spraying pesticides.

The technologies needed for precision agriculture are AI, green sensors, and cybersecurity. Inefficiencies of traditional pesticide spraying are still a major issue. Over applying pesticides causes pollution to our environment and serious harm to human life. It wastes and significantly increases costs for unwarranted applications of chemicals, particularly when compared to smart spraying systems that only spray the pesticide to the identified targets. Manual and time-based sprayers are largely ineffective and static and do not have decision-making capabilities at the time of the application.

Precision spraying assures the lowest amount possible of chemicals, making it more environmentally responsible. It allows use in organic growing and helps promote sustainable agricultural practices that are environmentally responsible. The manner in which precision spraying allows you to achieve sustainable goals through reduced waste and conservation of resources. Deep learning provides up to expert level detection to marry pest detection to precision spraying. Real-time image recognition using the ESP32 camera and the machine learning model assures real-time reporting so chemicals are only applied where needed. IoT capabilities allow you to monitor and manage the farm remotely on a continuous data cycle [4].

An essential part of smart agriculture has become cybersecurity. The IoT sensor networks must be protected from cyber-attacks and other threats to maintain the proper data flow and reliability of any Artificial Intelligence (AI) systems, that is processing the information, making decision [5]. If the data flow can be disrupted, or data compromised, there will be issues. The system uses end-to-end encryption, multi-factor authentication, firewall protections, and intrusion detection systems to maintain confidentiality of communications from all IoT sensors, so they do not hold malicious actors (cyberthreats) hostage. Ultimately, these protections are from data breaches and help protect confidential agricultural information. It's important as it maintains system integrity, protects against unauthorized access, and improves the effectiveness of automated pesticide control from being compromised [6]. Unauthorized access can lead to overuse, underuse, or incorrect misuse of agricultural chemical substances. Since cybersecurity is continuous and there is an ongoing concern for data threats and breaches, the company will continue to develop their digital defense system in agriculture by conducting sophisticated security audits and performing continuous monitoring of potential threats. Certainly, the system saved on pesticide expenditures, proved useless on paying to prevent unnecessary spraying, and scampered known and unknown pest under control automatically with AI. Also, there was little to no reliance human labor which increased operational efficiency and became scalable. The usual results of data-driven precision ensured healthy crops, higher yields, and economical, sustainable, farming.

2 Objectives and Contribution

This study focuses on the development of an AI-driven automated pesticide sprayer to utilize deep learning algorithms for real-time pest identification and utilizes the ESP32 cameras and IoT-based technologies for increasing applicability.

- **For Environmental Sustainability** Pesticide use should be optimized to avoid any excess pesticide application that would ruin the environment. Stop soil degradation, water decontamination, and ecological distress.
- **Safeguard and Increase Security in Smart Farming**
  Integrate cybersecurity measures in smart pesticide sprayers to cater to cybersecurity threats. Ensure an uninterrupted communication connection between on-field devices and the AI decision systems. Avoid manipulative sabotage that would interfere with pesticide use and compromise farm information.
- **Far Less Operating Cost for Greater Efficiency**
  Pesticides would be saved on more precision-based management. Depend less on labor and let farm management work free. It would foster plant health, production, and yield.
- **Foster Engagement in Sustainable and Smart Agricultural Practices**. Support organic and environmental-friendly applications at a very low rate of pesticide use. Donative and focused data-driven information enable better management.

3 System Design and Methodology

3.1 Architecture of the AI-Driven Automated Pesticide Sprayer

The AI-driven automated pesticide sprayer is a deep learning, IoT, green sensor, and a cyber-secure model that optimizes pesticide application in modern agriculture with precision, efficiency, and security. To achieve this, there are different components of the system working together.

The ESP32 camera module receives real-time images of crop areas for processor input to a deep learning model that accurately detects pest-infected areas. The environmental parameters which include humidity, temperature, and soil moisture are continuously monitored by green sensors to synchronize optimal conditions for spraying.

The ESP32 microcontroller is intended to serve as the controller responsible for making AI decisions over the spraying very effectively. The cybersecurity layer secure the information and systems against cyber threats by utilizing secured communication channels, preventing unauthorized access, and protecting data of related operations on the farm. Such an intelligent and secured environment will enable smart pesticide application, thus creating a smart, sustainable, and resilient approach to farming.

3.2 Workflow of the AI-Driven Pesticide Sprayer

Image acquisition real-time crop images are obtained using the ESP32 camera. In pest detection and decision-making (AI Processing), the deep learning model is implemented to detect pest-infected areas. It marks out the affected regions and their required pesticide volume. Integration of Sensor Data (Smart Spraying conditions) Green sensors keep an eye on the surrounding conditions for optimizing spraying. Spraying in Auto Mode. The system only activates the sprayer in case of an infected area. IoT-Based Monitoring & Cybersecurity. IoT permits the monitoring and controlling of farms remotely through a mobile application or web dashboard. Cybersecurity protocols secure communications and avoid hacking attempts (Fig. 1).

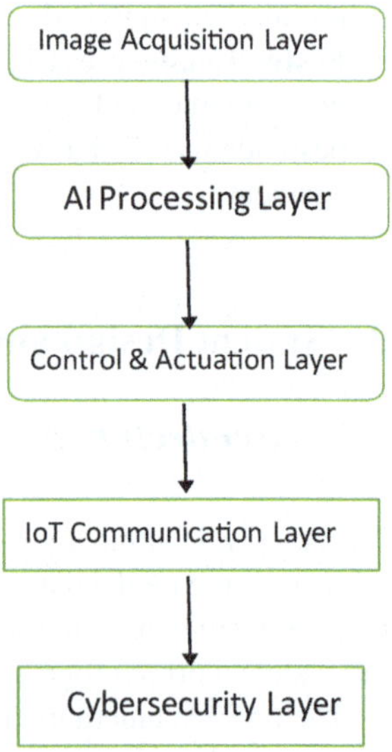

Fig. 1 System architecture diagram

3.3 Dataflow (Fig. 2)

3.4 Hardware Components

Architecture of the AI-Driven Automated Pesticide Sprayer: The AI-driven automated pesticide sprayer is a deep learning, IoT, green sensor, and a cyber-secure model that optimizes pesticide application in modern agriculture with precision, efficiency, and security. To achieve this, there are different components of the system working together.

1. **ESP32 Camera**
   The ESP32 camera is a compact, low-power microcontroller with built-in Wi-Fi and Bluetooth capabilities. In real time, it helps to acquire images that will be used to grade crop infestation. In these cases, high-resolution images will be taken of the crops to detect pest infestations. Images were fed to a deep learning model, trained to identify the affected areas and optimize pesticide application. It is low cost and highly efficient, making it ideal for deployment in remote areas and instances of farming.
2. **Integration of the IoT**
   The IoT is integrated with the system for real-time monitoring, data collection, and remote control. IoT connectivity allows farmers to access crop health information and pesticide application status through a dashboard or mobile application. The system comprises wireless communication protocols (Wi-Fi or LoRa) to send data from the ESP32 camera and sensors to a central processing unit. Automation is further enhanced by IoT, reducing human intervention and optimizing pesticide use based on AI-driven analysis.
3. **Green Sensors for Environmental Monitoring**
   This gadget uses green sensors to monitor some environmental constraints such as temperature, humidity, and soil moisture levels. These sensors will

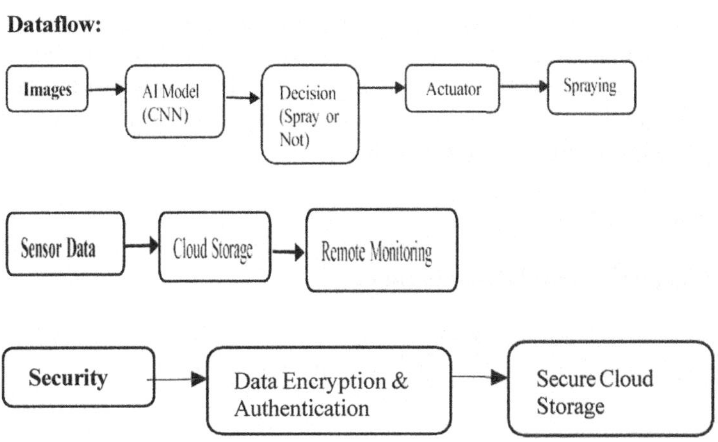

Fig. 2 Data flow

help ensure that pesticide spraying can only occur in optimum conditions, thus avoiding excessive usage of chemicals and halving the environmental impacts as well. The incorporation of green sensors aids precision agriculture through the real-time data provided to adjust the rate of spraying taking place to ensure that only required quantities of pesticides are used.

4. **Cybersecurity in IoT-Based Agricultural Systems**
   Establish ongoing security monitoring through anomaly detection systems and real-time notifications to incite peculiar behaviors that could threaten data integrity and pesticide application. Improve counter cyber threats that evoke the IoT-based agricultural devices and threaten data confidentiality and device's reliability.
5. **Protection from Cyber Attacks and Data Breaches**
   Apply advanced encryption algorithms and firewalls to protect sensitive farm data from cyber criminals. Preventing outside threats and maintaining uptime depend on ensuring area of reliable systems and procedures. Monitoring and updating the systems and adding an intrusion detection system help secure the farm from outside threats and interruptions on farm operations.

3.5 AI Model and Deep Learning Approach: Algorithm for Pest Detection, Dataset Used, Training Process

3.5.1 Algorithm for Pest Detection

In this model for pest detection, the deep-learning framework is based on a CNN that is adapted for tasks of image classification and object detection related to precision agriculture. The architecture has an input layer to train RGB images from an ESP32 camera resized to 224*224 pixels followed by convolutional layers extracting pest-related features including edges, color changes, and disease patterns. In order to prevent overfitting, pooling layers will gradually reduce spatial dimensions, and a fully connected layer will classify them into either healthy or infected images. The softmax activation layer outputs the final probability distribution of infected pest presence. The trained architecture is converted to TF Lite for real-time inference to run on the low-memory embedded processor ESP32 triggering a PWM motor to spray pesticides covering the infected area.

Mathematical Model:
For an input image **I(x, y, c)** with spatial dimensions (x, y) and color channels (c), the CNN applies a **convolution operation**:

$$F(i, j) = \sum_{m=0}^{M-1} \sum_{n=0}^{N-1} I(i + m, j + n) \cdot K(m, n)$$

where $\boldsymbol{K(m, n)}$ is the convolution kernel.

Table 1 Number of datasets

Category	Number of images
Healthy crops	10,000
Pest-infected crops	12,500
Fungal diseases	8000

3.5.2 Dataset Used for Training

The collection used for the introduction chiefly consisted of healthy and pest-infected leaf images from corn, soybean, and tomato taken from the Plant Village Dataset, custom datasets via computer vision with the ESP32 camera, and augmentations through rotation, brightness value change, and addition of Gaussian noise to complete robust statistics of the dataset, which could generalize well while training on the deep learning model (Table 1).

3.5.3 Training Process

This model has been trained on high-performance GPUs (NVIDIA RTX 3090) using TensorFlow, Keras, and some configuration that includes batch size 32, Adam optimizer, categorical cross-entropy loss function, adaptive learning rate of 0.001, and 50 epochs. The training pipeline first preprocesses the model by resizing, normalizing, and augmenting images and partitioning the dataset into 80% for training, 10% validation, and 10% testing; training the model is done by using CNN layers with the support of backpropagation and gradient descent optimized with early stopping to prevent overfitting; final evaluation metrics showed 92% accuracy, 94% precision, 91% recall, and 92.5% F1 score.

3.5.4 Deployment on ESP32

The model was quantized in FP32 to INT8 format for faster execution. In terms of cybersecurity, the model was thoroughly tested against adversarial noise attacks to ensure misclassification prevention, and encrypted training data using AES-256 before cloud storage provides integrity and privacy; PKI-based authentication enabling secured model deployment by allowing only trusted updates has thus been implemented; and "consequently, the pesticide spraying mechanism applies pesticides in a targeted manner based on these AI predictions."

3.6 Application

The AI-driven pesticide spraying mechanism integrates deep learning, IoT, and automation to precisely, efficiently, and minimally apply pesticides on the affected areas of the crop to reduce waste and environmental burden. Image-based pest inspection employs an ESP32 camera that can capture real-time crop pictures. Then a deep learning model inspects those images to highlight affected areas wherein pesticides have to be sprayed precisely, unlike traditional full-field spraying. Green sensors monitor humidity, temperature, and soil moisture in order to maintain the ideal spraying conditions, hence preventing the excess use of pesticides. As soon as an infestation is detected, the ESP32 microcontroller triggers an automated spraying mechanism that targets the infested areas, which reduces costs, chemical wastage, and the contamination of healthy crops. IoT-based remote monitoring allows farmers to watch and control operations via a mobile dashboard to ensure accuracy and efficiency for real-time data. Cybersecurity is also implemented to ensure that the data from sensors, AI models, and IoT communications are protected from malicious threats, guaranteeing that pesticide application remains seamless, secure, and that agricultural data integrity is ensured

3.7 Cybersecurity Aspects

Since different types of information are collected and processed in real time during image recognition or IoT-based communication for precise pesticide application, where very specific data integrity is important, during different types of processing, the data integrity concerned always plays a crucial role in information technology. Verification of image data authenticity with the use of checksums and cryptographic hash functions, such as SHA-256, by the Honouuido device for securing the images; and also works using adversarial hidden difficulties in image processing during the spraying operations to protect from erroneous spraying allows fugitive images. Furthermore, secure training and recognition of the image are carried on during the training of the deep learning model and training in a high specification computer before deploying it in the ESP32 microcontroller to ensure all their transfers unmodified and corrupted. Furthermore, control codes are implemented that can detect and correct an error; thus, during transmission, plant-detection signals along spraying commands are intact. Transmission security is an important aspect of the IoT-based Green Sensor-Based Automated Pesticide Sprayer: protection against unauthorized access, interception, or manipulation of data is critical. Employing end-to-end encryption such as TLS or AES to protect sensor data transmission between the ESP32 microcontroller and cloud or in-house farm management systems along with PKI or MFA provides an assurance that only authorized users or devices have access to the system. Additionally, the ESP32 supports MQTT, which is a commonly used IoT communication protocol; therefore, to minimize data breaches, it is recommended to use its secure version, MQTT over TLS, while OTA updates are protected by code signing mechanisms to avoid the risk of malicious firmware or software being injected.

4 Experimental Setup and Results (Table 2)

1. **Field testing: Testing environment**

   The green sensor-based automated pesticide sprayer tested in the field to ascertain the efficacy of such active spray in targeted application presents reliable operation by means of AI-based image recognition, IoT-oriented algorithms, secure data transmission, and robust cybersecurity measures. The hardware configuration includes an ESP32 camera module for real-time imaging of the plants, an ESP32 microcontroller capable of executing a trained deep-learning model to analyze image data and control the sprayer, and an IoT-integrated system that communicates remotely through Wi-Fi to transmit real-time data to cloud-based farm management software. The actuation system part contains relays and PWM-controlled motors to ensure accurate pesticide spraying based on AI's decision-making process. The field test was performed on the mid-size California vegetable farm, in which a mobile spraying vehicle scanned and treated those areas infected by pests. In consideration of preventing unauthorized access into the communication of devices via Wi-Fi network, cybersecurity measures were employed by enforcing AES-256 encrypted Wi-Fi communication, supplementing these with TLS-secured data-transfer protocols to ensure security and fluidity in the spraying action.
2. **Data Collection Process**

   Real-time image processing: An ESP32 camera continuously captured images of crops, and this was processed through an AI-driven deep learning model, locally. Check the data integrity: The images were hashed using SHA-256 to ensure their authenticity before they were sent. Decision by AI: A convolutional neural network was used to classify the health of the plants, which was further made with the conclusion of whether the crops are required to spray pesticides or not. Defensive measures against adversarial attacks: To bolster the model against data poisoning attacks, adversarial training techniques are employed. Secured communication of IoT: The data is sent over MQTT-on-TLS to the

Table 2 Experimental results

Metric	Result	Cybersecurity/AI relevance
Pesticide reduction	30% reduction in pesticide usage	AI ensures targeted spraying, reducing waste
Detection accuracy	92% precision in pest identification	CNN model trained on diverse datasets
Response time	<500 ms per image processing cycle	Real-time AI inference on ESP32
Data security	No recorded data breaches	AES-256 and TLS Encryption for secure transmission
System uptime	98% uptime in field testing	Secure OTA updates prevent malicious firmware Modifications

Table 3 Accuracy of metric

Metric	Value	Remarks
Detection accuracy	92%	Accurate detection of infected areas within a pest-infected leaf
False positives	5%	Moderate rate of healthy plants being detected and Classified as infested ones
False negatives	3%	A fair margin of misclassification of a genuine infected plant

Table 4 Parameters to measure up with traditional method

Metric	Value	Comparison with traditional methods
Pesticide reduction	30%	The pesticide consumption drastically reduced due to AI-based targeted spraying
Environmental impact	Lower runoff	Reduced soil and water sources contamination through pesticides
Cost savings	Up to 25%	Lower chemical cost due to targeted spraying Cybersecurity considerations

remote monitoring dashboard with an obvious end-to-end encryption guarantee. Access control: Modification of a certain set of spraying parameters was strictly by the authenticated personnel of the farm management.

The performance metrics include AI accuracy, pesticide usage reduction, and its efficiency compared with other pesticide sprayers.

The evaluation of the green sensor-based automated pesticide sprayer in detecting pests accurately and in reducing total pesticide applications while producing more efficient results than traditional spraying methods took into account the performance metrics provided, based on the test fields and experimental setup.

1. **Accuracy of pest detection** (Table 3)

   These measure the level of accuracy associated with pest detection.

   Deep learning model: An image detection technique based on convolutional neural network (CNN) has been applied and specifically trained with bug-infested parts in a dataset of plant disease images.

   Real-time image processing: Designed for near real-time operation, the ESP32 camera takes images and completes all processing within 0.5 s for each frame.

   Considerations for cybersecurity comprise cautious training to secure the AI model from data poisoning attacks, which could compromise pest detection by generating false positives or negatives. SHA-256 hashing is also introduced to certify that images sent from the field have not been interfered with, so that such images, adhering to the frame of reference of pest detection and the parameters of AI processing, are real and clean.
2. **The Reduction of Pesticide Utilizations** (Table 4)

   One of the key strategic objectives for the system is the targeted spraying of the infected spots with the intent of reducing excessive pesticide application. In three paragraphs: spray gives the uniform application of pest control agents from

Table 5 Efficiency comparison

Metric	Automated sprayer	Traditional spraying	Improvement
Time per field scanning	15 min/acre	30 min/acre	2× faster
Labor requirement	1 operator	2–3 workers	50% reduction
Spray precision	Targeted areas	Full field spraying	Minimizing waste
System uptime	98%	N/A	High reliability with OTA updates

Table 6 Impact on sustainability factors

Sustainability factor	Impact
Reduction in pesticide use	**30% decrease** in pesticide consumption, minimizing chemical runoff into soil and water sources
Environmental protection	Targeted spraying reduces contamination, supporting eco-friendly farming practices
Cost efficiency	Decreased pesticide waste leads to **25% cost savings** for farmers
Labor reduction	Automated pest detection and spraying reduce dependency on manual labor, improving farm productivity

one end to the other, thus excessive chemical use and environmental damage will take place. The system sprays pesticides only into the areas infested by insects and pests, permitting, therefore, higher efficiency against other methods due to noteworthy reduction of the applied amount.

3. **Efficiency Comparison** (Table 5)

   Efficiency includes uptime, speed of operation, and accuracy while working on real fields in agriculture.

   Cybersecurity considerations include secure OTA firmware updates to keep the system operational and protected from malware, while real-time data encryption ensures that external interference is prevented during the automated spraying process, maintaining system integrity and security.

5 Discussion and Future Works

5.1 Implication: Contribution toward Sustainable Agriculture

This process significantly enhances precision agriculture using AI-based pest detection and real-time IoT connectivity with sustainability (Table 6).

Cybersecurity measures focus on securing IoT transmission to ensure real-time data integrity and prevent cyberattacks from altering the pest application process, while AI model resilience against adversarial attacks safeguards accuracy in pest detection, preventing false spraying and ensuring reliable pest control operations.

5.2 *Future Enhancements: AI and Sensor Integrations*

To enhance the accuracy, efficiency, and cybersecurity of pest control technology, several advancements are in progress. AI model improvements include a self-learning AI system that leverages reinforcement learning to adapt to evolving pest patterns in real time without retraining, along with a federated learning approach that ensures decentralized, privacy-preserving model updates without relying on the cloud. Additional sensor integrations, such as multi-sensor data fusion, will combine thermal, humidity, and soil moisture data to better correlate environmental factors with pest infestations, while LIDAR and hyperspectral imaging will enhance plant health diagnostics by detecting early-stage infections beyond the visual spectrum. Cybersecurity-focused enhancements involve AI-driven threat detection utilizing fuzzy-logic-probabilistic anomaly detection to safeguard IoT devices, blockchain-based data integrity ensuring tamper-proof sensor logs, and quantum cryptography readiness to future-proof the encryption of sensitive agricultural information with quantum-resistant algorithms.

6 Conclusion

The green sensor-based automated pesticide sprayer has made sweeping progress in precision agriculture by integrating AI-based pest detection and secure IoT connectivity, achieving 92% accuracy in identifying pest-infested areas while minimizing false positives and negatives through deep learning and real-time image recognition; it reduces chemical consumption by 30%, cutting farmer expenses by 25% and drastically lowering environmental harm, while its automated system covers fields twice as fast as conventional methods and reduces labor requirements by 50% through IoT-based remote sensing and monitoring. In addition, robust cybersecurity measures—such as end-to-end encryption (TLS, AES-256), adversarial training to enhance AI resilience against data poisoning attacks, and secure OTA updates—ensure data integrity and protect against cyber threats, thereby enhancing the overall operational workflow in pest control.

References

1. J. Seol, J. Kim, and H. I. Son, "Field Evaluations of a Deep Learning-Based Intelligent Spraying Robot with Flow Control for Pear Orchards," *IEEE Access*, vol. 9, pp. 8420–8430, 2021. https://doi.org/10.1109/ACCESS.2021.3050310
2. H. M. Rouzbahani et al., "Communication Layer Security in Smart Farming: A Survey on Wireless Technologies," *IEEE Internet of Things Journal*, vol. 9, no. 15, pp. 13538–13557, 1 Aug. 2022. https://doi.org/10.1109/JIOT.2022.3156160

3. K. S. Babu, S. S. Kumar, and S. S. Kumar, "Agriculture Robotic Vehicle Based Pesticide Sprayer with Efficiency Optimization," *2015 International Conference on Innovations in Information, Embedded and Communication Systems (ICIIECS)*, Coimbatore, India, 2015, pp. 1–5. https://doi.org/10.1109/ICIIECS.2015.7358532
4. O. Friha, M. A. Ferrag, L. Shu, L. Maglaras, and X. Wang, "Internet of Things for the Future of Smart Agriculture: A Comprehensive Survey of Emerging Technologies," *IEEE/CAA Journal of Automatica Sinica*, vol. 8, no. 4, pp. 718–752, April 2021. https://doi.org/10.1109/JAS.2021.1003925
5. S. R. A. Balaji, S. P. Rao, and P. Ranganathan, "Cybersecurity Challenges and Solutions in IoT-Based Precision Farming Systems," *2023 IEEE International Conference on Advanced Networks and Telecommunications Systems (ANTS)*, 2023, pp. 1–6. https://doi.org/10.1109/ANTS57881.2023.10116154
6. M. A. Ferrag, L. Shu, O. Friha, and X. Yang, "Cyber Security Intrusion Detection for Agriculture 4.0: Machine Learning-Based Solutions, Datasets, and Future Directions," *IEEE/CAA Journal of Automatica Sinica*, vol. 9, no. 3, pp. 407–436, March 2022. https://doi.org/10.1109/JAS.2021.1004344

Virtual Reality as a Rehabilitation Tool for Stroke Patients: Tolerance, Safety, and Preliminary Outcomes

Sławomir Michalak, Paweł Śniatała, Ewa Kozielewska-Zwierska, Joanna Biskup-Danielak, Jakub Moskal, Konrad Śniatała, and Szymon Baliński

1 Introduction

The field of healthcare has been identified as a critical component of the global public health agenda, given its profound impact on the physical, social, and mental well-being of people around the world. Consequently, the advent of emerging technologies, such as Metaverse [1], finds application in the healthcare sector when feasible [2].

According to recent data presented in UN Population Division Data Portal [3], there are approximately 830 million people aged 65 and older, and this number is projected to rise to 1.7 billion by 2054. With this aging population comes an increased prevalence of comorbidities, with cerebrovascular (stroke) and neurodegenerative diseases (such as Parkinson's disease and dementia) being the most common. The incidence of stroke escalates with age, making post-stroke care a critical need for older adults. Stroke is a leading cause of disability among individuals over 50 years old. Up to 75% of stroke patients experience dysfunction, and 15–30% of stroke survivors endure severe disabilities. Additionally, around 60% of post-stroke survivors develop permanent disabilities, which can limit their mobility, vision, speech, and swallowing abilities. Some post-stroke patients may experience rapid recovery, while others face time-consuming rehabilitation processes. Currently, there are no reliable methods to predict individual recovery profiles. The most significant spontaneous recovery usually occurs within the first 3 months after a stroke, with minimal improvements seen after 6 months. However,

S. Michalak (✉) · E. Kozielewska-Zwierska · J. Biskup-Danielak · J. Moskal
Institute of Neurological Disorders, Poznan University of Medical Sciences, Poznan, Poland
e-mail: swami@ump.edu.pl

P. Śniatała (✉) · K. Śniatała · S. Baliński
Institue of Computing Science, Poznan University of Technology, Poznan, Poland
e-mail: pawel.sniatala@put.poznan.pl

S. S. Iyengar et al. (eds.), *Artificial Intelligence Driven Forensics*, Signals and Communication Technology, https://doi.org/10.1007/978-3-031-98036-7_24

all patients require neurorehabilitation and care since improvements can continue during the first year post-stroke and beyond. Despite some level of recovery, many patients are left with chronic deficits. Factors influencing post-stroke recovery include lesion location and size, the effectiveness of reperfusion and edema resolution during the acute phase, demographic factors (such as age, sex, handedness, and education), ethnicity, socioeconomic status, acute stroke severity (as evaluated using the National Institutes of Health Stroke Scale, NIHSS), and the severity of acute aphasia. Neural reorganization and related factors also play a role in language recovery following a left hemisphere stroke. The occurrence of aphasia in 30–40% of patients who have experienced a first ischemic stroke highlights the significant impact strokes have on language function, which is essential for communication and identity. Post-stroke language disorders can lead to a loss of independence, as individuals may struggle with daily tasks requiring communication. Patients may also face emotional challenges, including frustration, depression, or social isolation, due to their difficulty in expressing themselves and understanding others. For those who were professionally active before the stroke, language impairments may result in a loss of professional identity and the inability to continue in their careers. Moreover, language difficulties can strain social relationships by altering the dynamics of conversations and the nature of social support. Family members and caregivers often need to learn new ways to communicate with the stroke survivor, relying on non-verbal cues, gestures, or alternative communication aids. Caregivers may experience emotional stress while caring for disabled post-stroke patients, feeling helpless or grieving the loss of their loved one's previous capabilities. The family's lifestyle can change dramatically as they adjust their schedules, responsibilities, and even living arrangements to support the stroke survivor. The financial burden of ongoing care, rehabilitation services, and potential loss of income can also be significant for families of post-stroke patients. Several barriers limit post-stroke survivors' access to rehabilitation. On the healthcare system side, challenges include competing demands on health systems, non-transferable clinical practice guidelines, a lack of rehabilitation facilities and equipment, a shortage of stroke rehabilitation specialists, severely limited resources in rural areas, and inadequate acute stroke care. Individual factors that hinder access include limited transportation options, lost wages for both patients and caregivers, reliance on traditional medications instead of evidence-based therapies, low health literacy, and a lack of formal caregivers. In high-income countries (HICs), there are more varied options and skilled specialists available for stroke rehabilitation. In contrast, low- and middle-income countries (LMICs) face a scarcity of such resources, even though stroke incidence and fatalities are higher in these regions than in HICs [4]. Thus, there is a pressing need to enhance the range and accessibility of neurorehabilitation services for the elderly population, particularly stroke survivors. The integration of neurotechnology like human-computer interfaces or VR in rehabilitation can help achieve goals such as physical and cognitive restoration, enrichment of rehabilitation techniques, increased effectiveness, broader societal access, and the development of tele-neurorehabilitation.

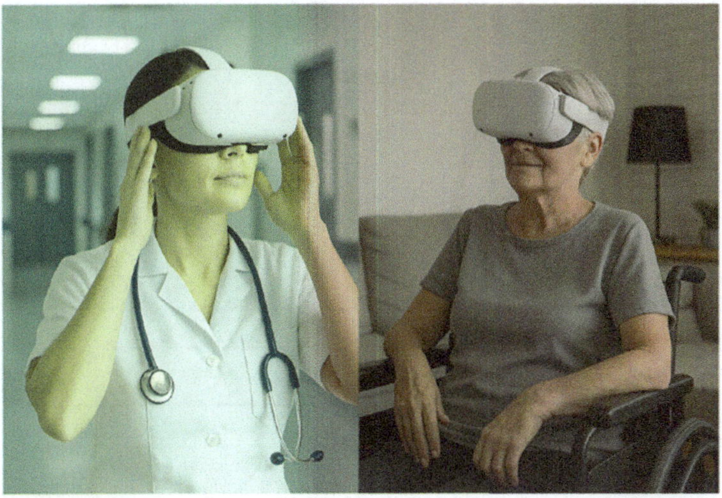

Fig. 1 VR medical interview with the patient

One of the typical examples of VR application in healthcare is hidden in the term telehealth (illustrated in Fig. 1). It is a medical practice that allows patients to have video consultations with expert physicians and receive diagnoses. It is especially valuable in cases where both parties may be geographically distant. In critical situations, it provides rapid intervention and access to specialist advice. The advent of telemedicine has effectively obliterated the concept of physical boundaries. Telehealth has become a prevalent practice during pandemics, replacing in-person consultations with virtual settings. This paradigm shift signifies a profound transformation in the manner in which healthcare services are delivered within immersive, virtual, and collaborative domains.

Applications of the metaverse in the healthcare field can be found on many levels. Let us start with teleconsultation (illustrated in Fig. 1), where doctors physically at different locations can consult together on a specific case.

Another useful application of VR is illustrated in Fig. 2. In the illustration shown, we see doctors analyzing an image of the brain. They can be physically at different locations and consult on a specific case together. Several studies have demonstrated the impact of immersion for better understanding of data, such as in brain tumors (e.g. BRAINLAB system) [6], shape perception [7], and MRI [8]. Today we can use many applications to visualize MRI and CT in VR using commercially available headsets such as Varjo, Questa, etc. [9]. Next, a more general example can be called virtual hospital as illustrated in Fig. 3.

Rapid development of virtual hospitals facilitates cloud technology, healthcare, AR/VR, and metaverse outreach in the healthcare industry. Here, doctors can remotely access their patient's conditions and eliminate the need for an in-person visit.

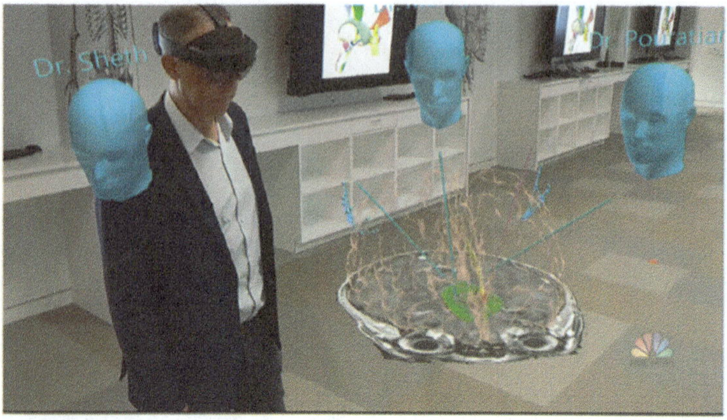

Fig. 2 VR tele-consultation (Source: [5])

Fig. 3 VR hospital

In the process of developing a treatment plan and continuously adjusting it to the patient's current condition, modeling with a digital twin (DT) is becoming more and more popular. The main components of a patient's DT consist of a physical entity, a virtual replica, and a connection between the two to enable bidirectional real-time impact on each other. The ever-evolving interactions between the physical entity and the digital twin could traverse from microscopic to macroscopic scale and last from birth to death for individuals.

The patient's digital twin should be:

- individualized,
- interconnected,

- interactive,
- informative,
- impactful.

These requirements allow us to create and keep updating the model to make it useful for the actual use. More information on the concept of DT in medicine can be found in article [10]. The authors define a digital twin for healthcare as a virtual representation of a person that allows dynamic simulation of the potential treatment strategy, monitoring and prediction of health trajectory, and early intervention and prevention, based on multiscale modeling of multimodal data such as clinical, genetic, molecular, environmental, social factors, etc. This is illustrated by these authors in a diagram presented in Fig. 4. Finally, we reach another important healthcare area where metaverse can support traditional approaches, which is digital therapeutics (DTx). DTx is defined as evidence-based therapeutic interventions driven by software to prevent, manage, or treat a medical disorder or disease. In other words, DTx are patient-facing software applications that help patients treat, prevent, or manage a disease and that have proven clinical benefit. DTx is expanding greatly, providing metaverse healthcare opportunities. This field offers great use of cognitive therapy, support groups, psychiatric assessments, and rehabilitation with the help of haptic sensors. Physical therapy is easy and responsive with AR and VR in the metaverse of healthcare (Fig. 5).

The key disease areas that are revolutionizing by DTx are listed below.

- Mental and Behavioral Health Disorders (Cognitive behavioral therapy, game-based therapies, mHealth digital treatments)
- Metabolic Diseases (Digital therapeutics around dietary and exercise, AI-enabled personalized offerings)
- Oncology (Symptom-monitoring tools during chemotherapy, evidence-based therapeutic interventions, medical adherence)
- Respiratory Diseases (Digital self-management programs/apps for COPD and asthma, digital therapies)
- Immunology/Immune Disorders (Self-assessment tools for disease monitoring, personalized recommendations apps to manage side effects)
- Gastrointestinal (Prescription digital therapeutics for abdominal pain, integrative approach to irritable bowel syndrome, Crohn's disease)
- Neurological Diseases (Videogame-based rehabilitation programs, cognitive rehabilitation, digital compensation therapies)
- Substance Abuse (Contingency management, cognitive behavioral therapy, and other behavioral interventions)
- Cardiovascular Diseases (Digital health apps to push healthier lifestyles, reduce stroke risks, and monitor health vitals)
- Other Chronic Conditions (Curative digital therapies for treatment adherence and lifestyle behavior, self-monitoring apps for health markers)

DTx has received approval from the Food and Drug Administration (FDA) for the treatment of attention-deficit/hyperactivity disorder (ADHD). EndeavorRx® is the

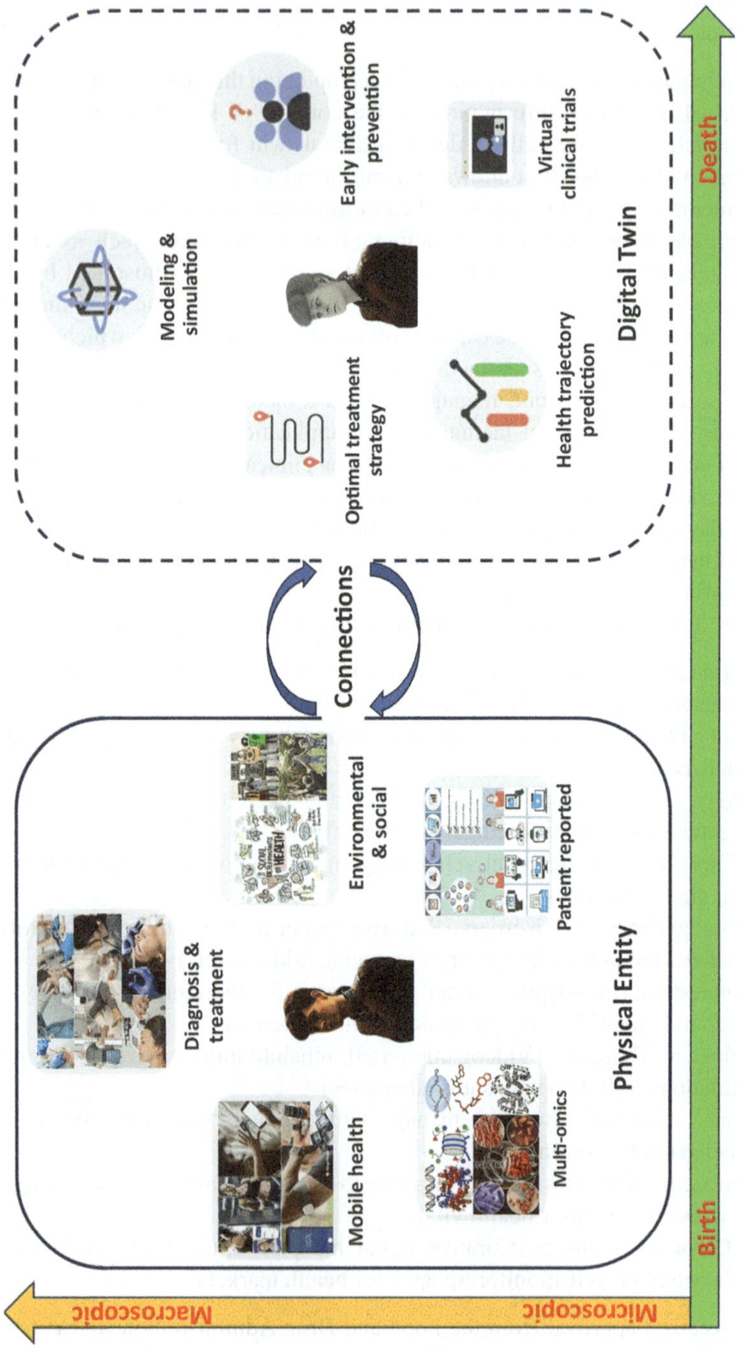

Fig. 4 Digital twin for health (DT4H) (Source: [10])

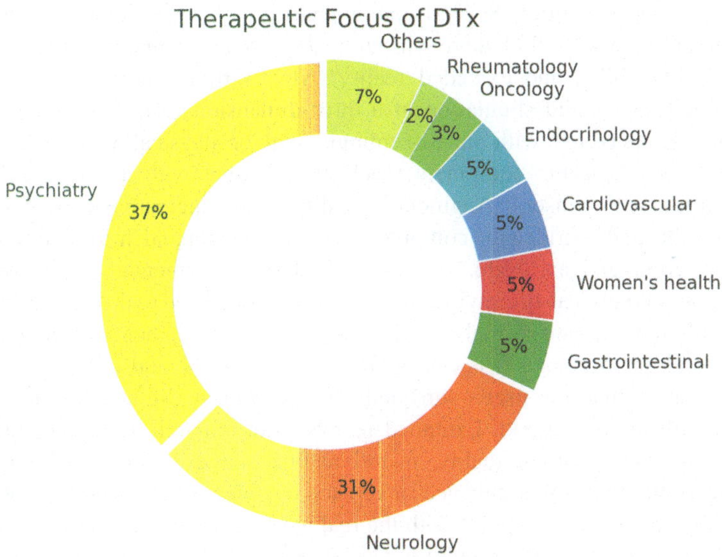

Fig. 5 Neurology and psychiatry as the main areas of DTx applications

first FDA-approved game-based digital therapeutic (DTx) that offers an alternative to traditional pharmacological treatments for ADHD [11].

In this paper, we focus mainly on neurological diseases. In particular, we present a metaverse application in the neurorehabilitation of post-stroke patients. Neurological disorders are one of the leading causes of disability and death worldwide. The burden of disability and mortality due to nervous system pathologies is increasing steadily, with recent data showing a 39% increase in deaths related to neurological diseases in the last three decades. In 2021, 43% of the global population, approximately 3.4 billion individuals, was reported to be affected by neurological conditions, a much higher figure than previously estimated [12]. These statistics underscore the urgent need for innovative solutions to address the growing global burden of neurological diseases.

The article is structured as follows. After an introduction, we explain the concepts related to metaverse. Next, the medical background of neurorehabilitation using VR will be briefly presented. In the following sections, we present our proposed solution which applies AR and VR in rehabilitation of post-stroke patients. The results then provide information on the effectiveness and tolerance of the proposed approach. Lastly, in conclusion, we summarize our results and propose future work.

2 Metaverse Components

There are many definitions of metaverse. One of them defines metaverse as a persistent virtual environment that allows access to and interoperability of multiple individual virtual realities. Several components/terms can be found in metaverse environment. We will find there virtual reality (VR), augmented reality (AR), mixed reality (MR), and extended reality (XR). Virtual reality is described as the computer-generated simulation of a three-dimensional image or environment that can be interacted with in a seemingly real or physical way by a person using special electronic equipment, such as a helmet with a screen inside or gloves fitted with sensors. Augmented reality is an interactive experience that enhances the real world with computer-generated perceptual information. Using software, apps, and hardware, such as AR glasses, augmented reality overlays digital content onto real-life environments and objects. Mixed reality is a technology that blends the real and virtual worlds, allowing digital elements to interact with the physical environment. It is essentially a fusion of VR and AR. MR headsets use cameras to map the user's surroundings and overlay digital content that can interact with the real world. Extended reality is an umbrella term encompassing augmented reality, virtual reality, mixed reality, and everything in between. It refers to technologies that enhance or replace our view of the world by blending digital information and graphics with the real world or by immersing us in entirely digital environments. In the context of virtual reality (VR), avatars refer to personal digital representations that individuals utilize for various purposes, such as social interaction and immersion in virtual environments. These avatars can be designed to mirror the appearance of the person controlling them. In this manner, they can signal to others in the virtual environment the appearance of their real-life controller. Alternatively, the avatar can appear as someone entirely different [13, 14]. The ability to be represented by an avatar that does not represent oneself affords a level of anonymity in VR interactions that may be appealing to some people in certain contexts, for example, for people with social anxiety in general interactions [15, 16] or to put people at ease in psychological therapy [17]. We can find additional devices to improve user interaction in metaverse such as head-mounted displays (HMD), non-hand-based input devices (eye-tracking, head tracking, voice input device), and motion input devices. VR can support patient training during gait rehabilitation using an anti-gravity treadmill, presented in Fig. 6. The metaverse environment can be enhanced using noninvasive interfaces based on eye movements, known as human-computer interfaces (HCI). Unlike brain-computer interfaces (BCI), HCIs do not directly sample signals from the brain, making them noninvasive. Systems such as Cyber-Eye track eye movements by monitoring the position and size of the pupil through a process called pupillometry. This technique collects information not only about where a person is looking and the amount of light but also about their momentary cognitive states and underlying brain processes. Pupillometry can track mental effort, memory load, conscious perception (which can also be evaluated with auditory signals), and memory processing. This capability allows for the effective monitoring of individual neurorehabilitation training. Examples of haptic devices for patients rehabiltation are presented in Fig. 7.

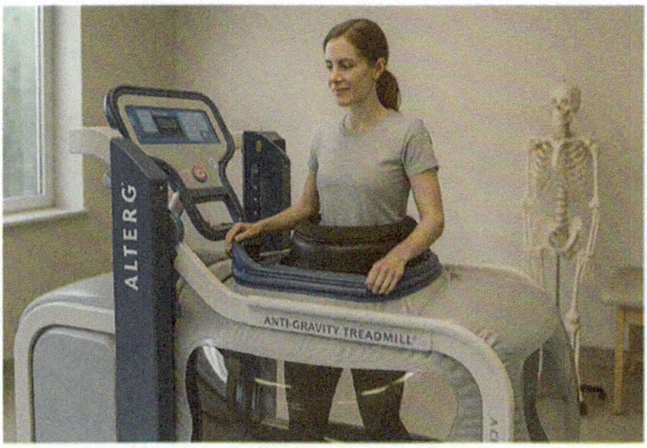

Fig. 6 Anti-gravity treadmill

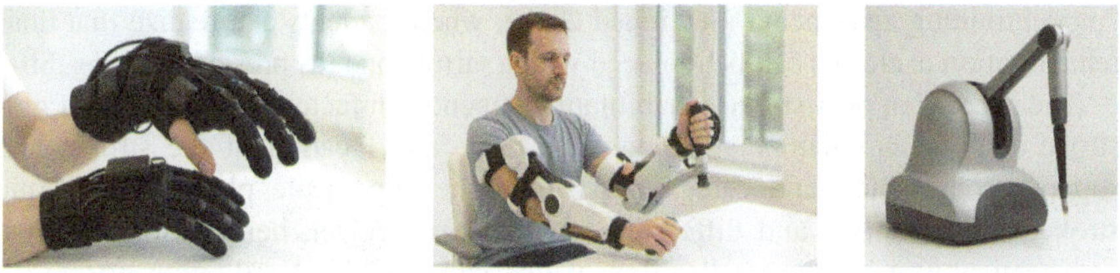

(a) (b) (c)

Fig. 7 Haptic devices for patients rehabilitation. (**a**) Haptic gloves. (**b**) Arm exoskeleton for active movement training with haptic support. (**c**) 3D haptic device used for stroke and fine motor rehab

3 Medical Background

The process of rehabilitation should be initiated as early as possible (usually the second day after the onset of the stroke) in the acute phase of cerebral ischemia. Later, it continues during the patient's stay in the stroke unit. Later, it continues during the patient's stay in the stroke unit. The disabilities caused by stroke vary; therefore, tailored post-stroke rehabilitation is needed that is appropriate for the needs of stroke patients and their daily activities at home. Most stroke survivors are discharged into their homes. Thus, the question of how to design rehabilitation training appropriate for use in the home should be raised. To initiate the adaptation process in some stroke centers (e.g., Institute of Neurological Disorders at the University Hospital of Poznan), "model/training apartments" are organized for initial patient training (shown in Fig. 8a). Using different objects such as balls, irons, spoons, cups, etc., as physical therapy tools in a "model/training apartment," supports the patient's daily activity at home. As technology becomes more prevalent and familiar, it can support rehabilitation in hospital and/or home settings. The real

(a) (b)

Fig. 8 "Model/Training apartment" for patient rehabilitation. (**a**) A real room. (**b**) Virtual room

"model/training apartment" mentioned above, where patients can realize that their rehabilitation exercises can be replaced with a virtual one (as presented in Fig. 8b).

The success of post-stroke rehabilitation depends on factors such as patient pre-stroke activity; circulatory sufficiency, severity, and location of the stroke; and support from family and caregivers. During the hyperacute phase of ischemic stroke, the evaluation and differentiation between neglect, hemisensory deficits, and hemianopsia can be challenging, which can hinder rehabilitation efforts. In addition, factors such as aphasia and disturbances of consciousness complicate the evaluation of neglect and hemisensory deficits using the NIHSS (National Institutes of Health Stroke Scale) to evaluate neurological status. In the pre-recanalization phase, clinical evaluations are often limited due to time constraints. However, in the post-recanalization phase, a comprehensive assessment of sensory modalities is crucial for planning effective neurorehabilitation. Active participation of the stroke patient in the rehabilitation process is crucial for optimal recovery. Thus, the need for tailored rehabilitation in post-stroke patients is clearly expressed. To meet this demand, the virtual rehabilitation system was prepared as a joint project between the Institute of Neurological Disorders of the Poznan University of Medical Sciences and the Poznan University of Technology. The pilot version of the system was tested at the Institute of Neurological Disorders.

The next sections describe the system and results achieved with a group of 102 patients.

4 Virtual Rehabilitation System

It is well-known that the rehabilitation process for neurological diseases is long, and after a period of hospitalization, the patient must perform various, often tedious, exercises at home. Some of these exercises can be performed in virtual reality. Performing these exercises in virtual reality helps to make them more attractive

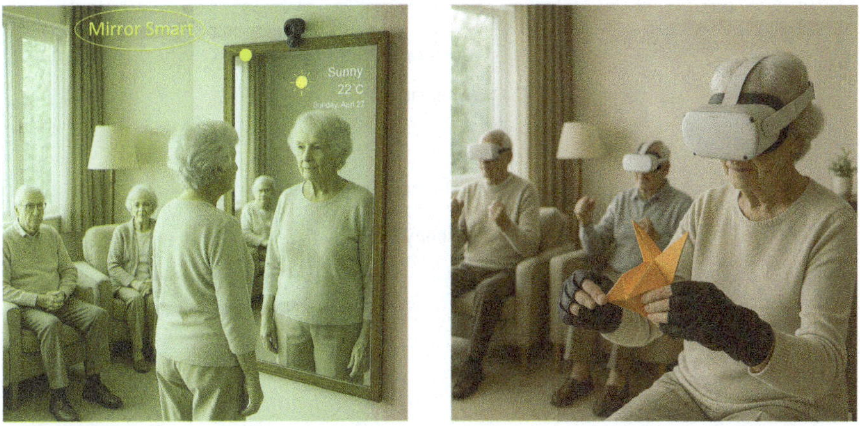

Fig. 9 Illustration of the proposed system implementation

and thus more effective (the patient is more likely to perform the exercises in a well-designed and attractive virtual reality). Of course, this assumes that the motor requirements appropriate to the set of exercises are met. The use of technology and virtual reality enables the dissemination of neurorehabilitation, allowing such goals as:

- physical and cognitive restoration;
- enrichment of rehabilitation techniques;
- increased effectiveness; wider access to society;
- teleneurorehabilitation.

The system we have developed uses virtual reality, but we also use augmented reality (AR) technology. AR technology has been implemented in the "smart mirror" device we made. We use it for a personalized interface with the system user. The patient approaches the mirror and is recognized, and then a dialogue begins which aims to offer the patient appropriate rehabilitation exercises. A set of these exercises is selected from a catalog developed in advance by doctors and physiotherapists. This set is dedicated/personalized for the patient. The illustration of the proposed system is shown in Fig. 9.

4.1 Smart Mirror: Personalized Interface

The smart mirror (presented in Fig. 10) allows the patient to be identified by facial recognition, and then there is a personalized dialogue with the patient. Upon recognition of the patient, as mentioned earlier, the system proposes to the patient a set of exercises dedicated to him. The proposal is not only personalized for the patient but also takes into account additional information such as the time of

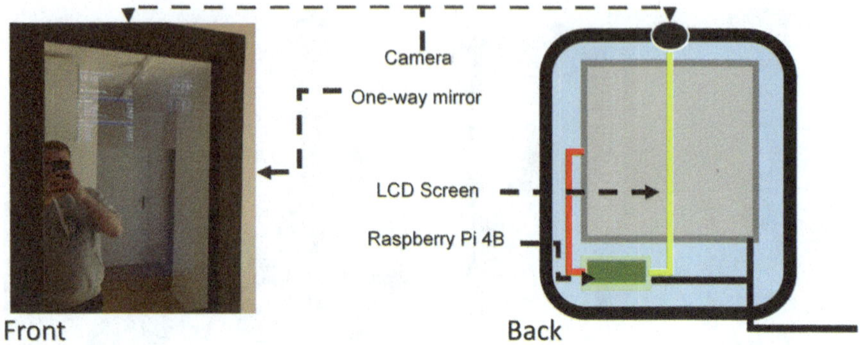

Fig. 10 Smart mirror implementation

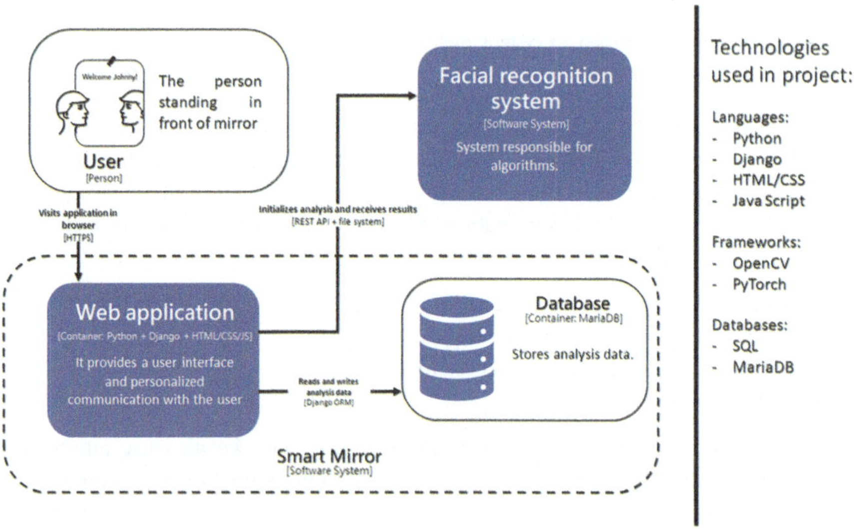

Fig. 11 Software environments used in the application

exercise, current weather conditions (weather forecast), the patient's mood, etc. The substantive scope of the system software is consulted with neurologists and neurorehabilitation specialists and implemented in the system accordingly. The system runs on a Raspberry Pi platform. However, the platform communicates with a more powerful computer that runs the main application. It is a web application implemented in the Django environment. Face recognition is implemented using the PyTorch and OpenCV libraries. The required data are stored in an SQL database (in our case, MariaDB). The software structure of the system is shown in Fig. 11. The facial recognition function in smart mirror is powered by algorithms based on artificial intelligence that can recognize faces and compare them with faces in the system database.

The virtual rehabilitation system will propose the most appropriate and/or expected virtual interaction with the recognized person.

4.2 Neurorehabilitation: Virtual Exercises

Alternative sensorimotor pathways that involve secondary motor areas can help facilitate motor control. Ischemic stroke can damage structures involved in motor control, such as the primary somatosensory areas and the thalamus, along with other subcortical structures. Consequently, the functional recovery of post-stroke patients can be delayed and challenging. In these cases, tailored neurorehabilitation should incorporate activities aimed at somatosensory stimulation to enhance stereognosis, proprioception, and the ability to discriminate and localize sensations. Combining sensory and motor techniques can improve recovery from limb dysfunction. One traditional and practical approach is mirror therapy, which has been developed for motor rehabilitation in post-stroke patients [18].

This therapy creates a visual illusion by allowing the patient to watch the reflection of the unaffected limb in a mirror. As the patient moves the unaffected limb, the reflection creates an image of the affected limb, stimulating interhemispheric communication and activating the brain regions responsible for controlling movement in the paralyzed limb (illustrated in Fig. 12). Kinaesthetic perceptional illusion through visual stimulation (KINVIS) is a neurorehabilitation protocol in which a person at rest experiences the sensation of a body part moving or the desire to move that part while watching a video depicting the same movement [19]. KINVIS has the potential to induce changes in the motor output system of the brain, as evidenced by the increased bioelectric activity in the affected hemisphere around the sensorimotor area following KINVIS. VR also presents opportunities to create

Fig. 12 Mirror therapy

immersive illusions that can help in the neurorehabilitation of stroke patients. The limb illusion generated by VR has been shown to enhance motor rehabilitation in post-stroke patients after they embody a virtual avatar. This approach stimulates interregional and interhemispheric communication, thereby improving movement control in a manner akin to mirror therapy. VR scenarios can be customized according to the range and type of neurological deficits experienced after a stroke. Additionally, the effectiveness of VR illusions can be amplified through the use of haptic feedback or tendon vibration via transcutaneous electrical nerve stimulation (TENS), such as in the wrist. By combining visual and proprioceptive stimuli, a conducive environment is created for the multisystemic activation of areas of the brain involved in movement control. At the Institute of Neurological Disorders at the Poznan University of Medical Sciences, we implement virtual reality illusions with haptics during the hyperacute phase of ischemic stroke. Our observations indicate a positive impact on motor recovery, including a reduction in tremors and the alleviation of neglect syndrome. However, we have also observed limitations in patients with Parkinson's disease, who exhibited increased dyskinesias during VR training. We attribute this phenomenon to impaired visual perception and the compromised multisensory integration characteristic of Parkinson's disease. Those who plan the rehabilitation process can create a library of virtual activities dedicated to a particular patient. These can be games/activities that improve the patient's motor skills or games that exercise mental activity. We have proposed to our patients simple games/exercises like a virtual fishing or origami. The illustration of our proposed system usage scenario is presented in Fig. 13. Origami (illustrated in Fig. 14) is a good example, as it requires hand-eye coordination, develops fine motor skills, and supports mental concentration, all of which stimulate the brain. This kind of training provides has following features:

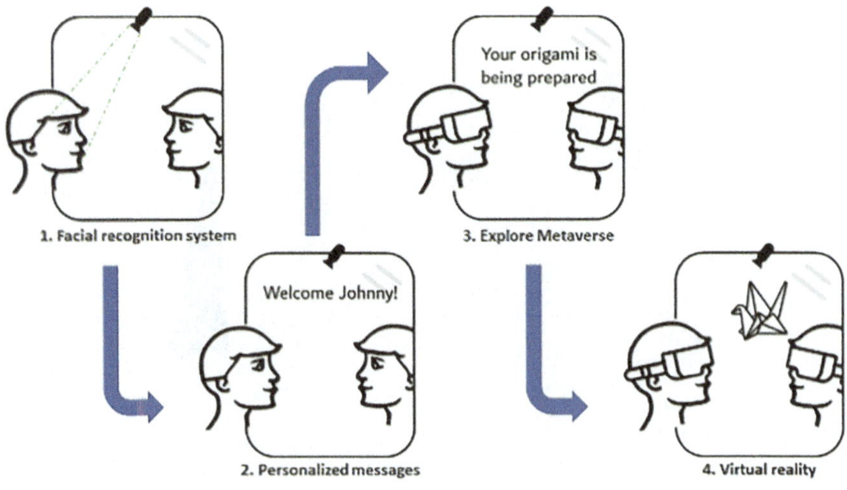

Fig. 13 Example of the system usage scenario

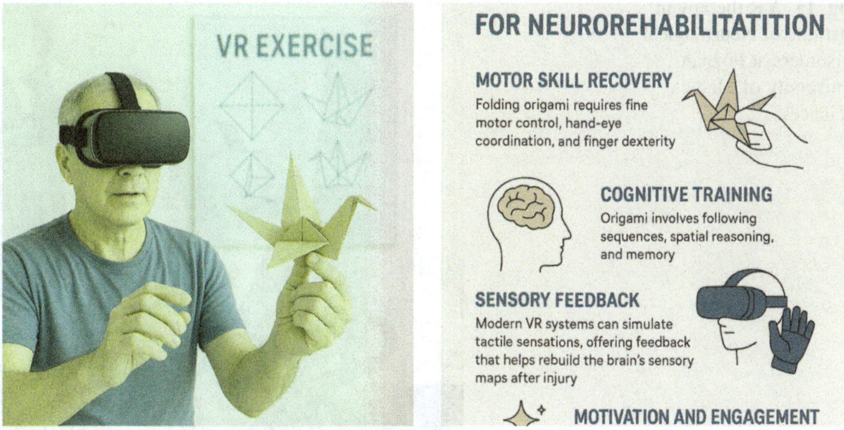

Fig. 14 Virtual origami to stimulate a brain

- Motor Skill Recovery: folding origami requires fine motor control, hand-eye coordination, and finger dexterity. VR-based origami exercises can simulate these movements safely, helping stroke patients, or people with neurological impairments, retrain these skills.
- Cognitive Training: origami involves following sequences, spatial reasoning, and memory—all critical cognitive skills. VR exercises can gradually increase complexity, supporting brain plasticity and cognitive rehabilitation.
- Sensory Feedback: Modern VR systems can simulate tactile sensations (with haptic gloves, for example), offering feedback that helps rebuild the brain's sensory maps after injury.
- Motivation and Engagement: origami is creative and satisfying. Using VR gamification makes repetitive rehabilitation exercises more enjoyable, increasing patient adherence to therapy programs.

The catalog of VR games/activities, which can be adopted to neurorehabilitation, is growing. Some information can be found in [20].

5 Results: Effectiveness of the VR Rehabilitation

As a result of the collaboration between Poznan University of Technology and the Institute of Neurological Disorders at Poznan University of Medical Sciences, Virtual Reality (VR) systems are now being used in daily practice in the stroke unit at Poznan University Hospital (Fig. 15). The disabilities caused by stroke vary, and personalized post-stroke rehabilitation is needed. The success of post-stroke rehabilitation depends on many factors among them: patient pre-stroke activity; circulatory sufficiency, severity, and location of the stroke; and support from family

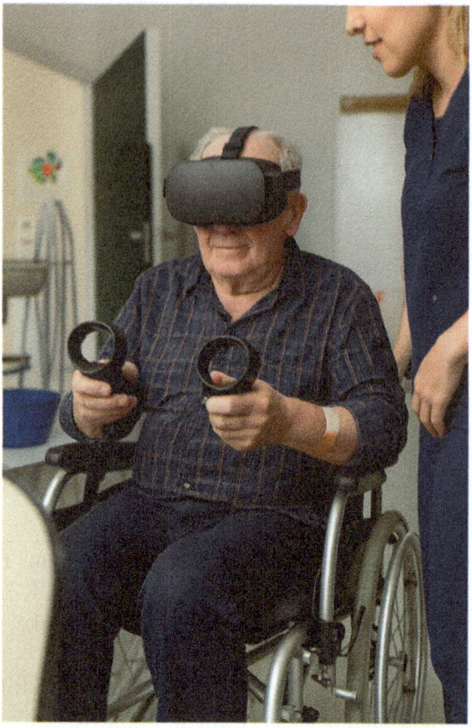

Fig. 15 VR therapy in Institute of Neurological Disorders at Poznan University of Medical Sciences

and caregivers. Active participation of the stroke patient in the rehabilitation process is crucial to optimal recovery. In our VR rehabilitation pilot program, we wanted to answer the following questions:

- What is the tolerance of this approach?
- What is its effectiveness?

Stroke patients are mainly older people who may not always be familiar with computer technology. Therefore, VR tolerance is currently being evaluated in a study involving stroke patients. A questionnaire has been created to assess the presence of symptoms such as vertigo, nausea, diplopia, headache, chest pain, arrhythmia, anxiety, and sweating both before and after VR training. The study also considers factors such as the use of computers, smartphones, and games by patients at home, as well as their educational level and profession. In addition, systolic blood pressure (SBP), diastolic blood pressure (DBP), heart rate (HR), and ECG are monitored before and after VR training. Patient evaluations using the questionnaire (Fig. 16) are performed at baseline and after 7 days of training. To date, 102 acute stroke patients have been evaluated. We have not found differences in systolic blood pressure (SBP, $P = 0.924$), diastolic (DBP, $P = 0.591$), and heart rate (HR, $P = 0.908$) before and after VR training during the hyperacute phase of stroke (Fig. 17).

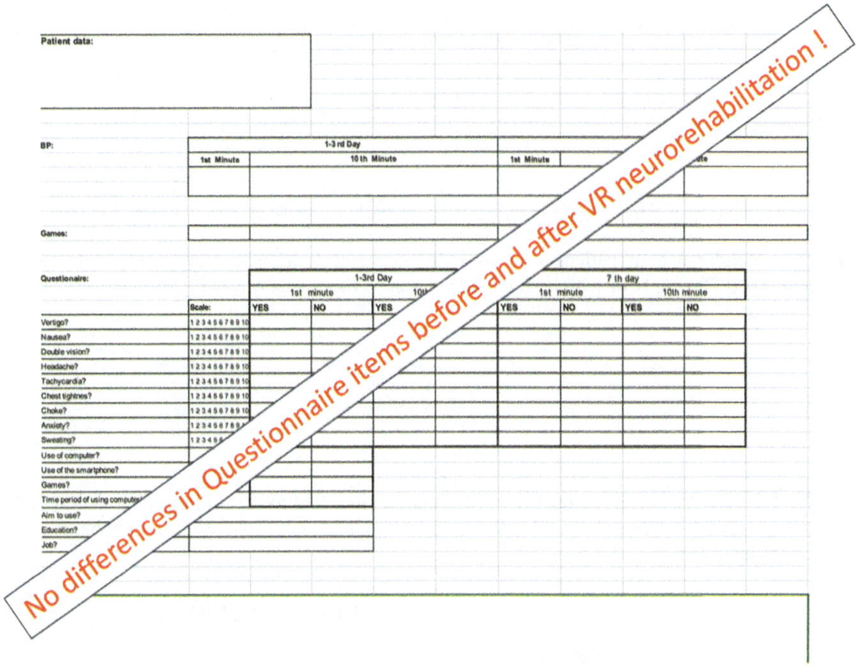

Fig. 16 The questionnaire for VR tolerance testing (first–third day and seventh day of the VR rehabilitation)

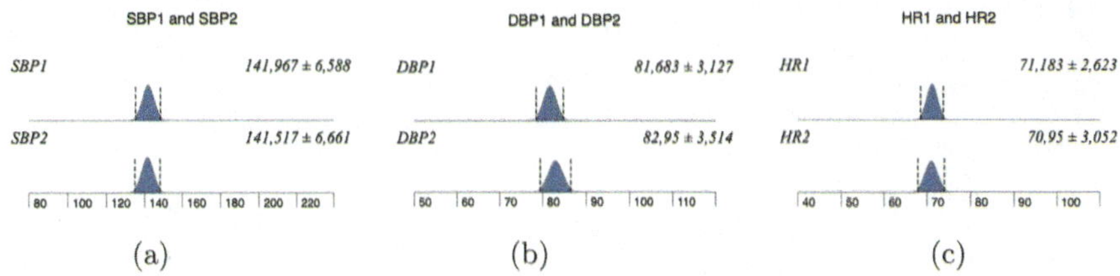

Fig. 17 Vital signs parameters before (1) and after (2) VR rehabilitation. (**a**) Systolic blood pressure. (**b**) Diastolic blood pressure. (**c**) Heart rate

There were also no differences in VR tolerance (SBP, $P = 0.817$; DBP, $P = 0.237$; HR, $P = 0.248$) in patients treated with mechanical thrombectomy (MT), intravenous thrombolysis (rTPA) only, or both recanalization methods (rTPA&MT) (Fig. 18). No significant changes in SBP ($P = 0.807$), DBP ($P = 0.895$), and HR ($P = 0.858$) (after VR training minus before VR training) were also observed in relation to the cerebral lesion visualized on computerized tomography (CT) (Fig. 19).

In conclusion, we noticed an excellent tolerance for VR training in stroke patients. Stable blood pressure and heart rate remain crucial for the clinical safety of procedures used during the hyperacute phase of ischemic stroke, and that was a

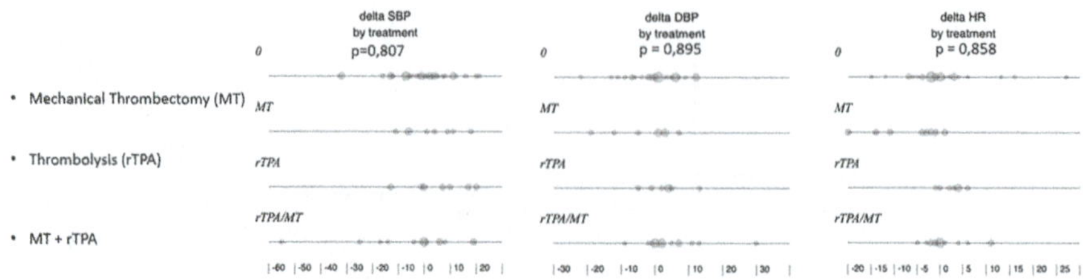

Fig. 18 VR tolerance in patients treated with mechanical thrombectomy (MT), intravenous thrombolysis (rTPA) only or with both recanalization methods (rTPA &MT)

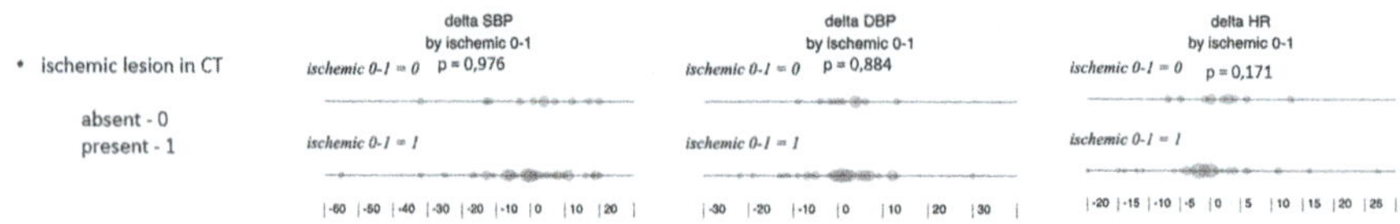

Fig. 19 SBP, DBP, and HR change (after VR training minus before VR training) in relation to the absence of ischemic lesion in head CT ("0"), its presence in left hemisphere ("1"), right hemisphere ("2"), or bilaterally ("3")

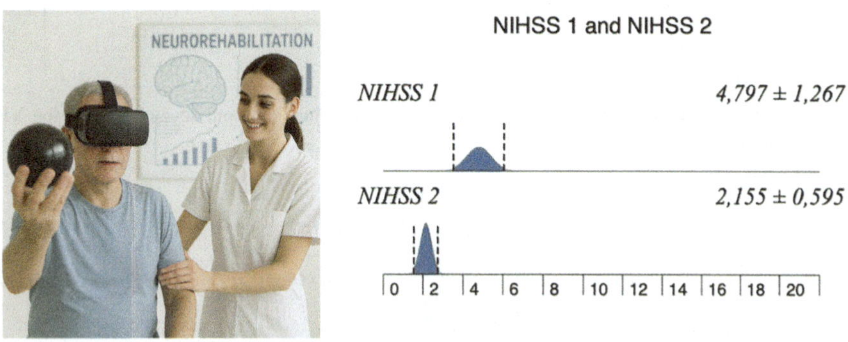

Fig. 20 NIHSS before and after VR trainings

case in our trial. Interestingly, some symptoms, such as intension tremor, that are present in real-world training disappeared in VR training. NIHSS after VR training was reduced in average from 4.797 to 2.155 as presented in Fig. 20.

6 Conclusion

Neurorehabilitation, supported with simple tools for everyday activities and advanced technology, opens possibilities to improve motor training and cognitive functions and enables the treatment of crippling symptoms such as neglect syndrome. VR-based therapy can provide a positive learning experience and be engaging and motivating. Enhancing typical exercises performed in a real environment, with virtual reality, individuals recovering from strokes can engage in practicing self-care tasks within an environment that is typically unfeasible to recreate within a hospital setting. Illusions generated by VR can activate "mirror neurons" and "re-programme" neuronal control of motor functions. Further expansion of the palette of exercises available in virtual reality will reach homes of patients (e.g., elderly persons) and will allow telerehabilitation. Telerehabilitation trainings can be developed and supported by the implementation of human-computer interfaces (HCI).

VR systems are currently used in everyday practice in the stroke unit of the University Hospital in Poznan. Stroke patients are mainly older people who are only occasionally familiar with computer technology. However, unexpectedly, their tolerance to VR is excellent. We noticed an excellent tolerance for VR training in stroke patients. Interestingly, some symptoms, such as intension tremor, that are present in real-world training disappeared in VR training. Access to rehabilitation centers is limited and many patients face challenges in staying motivated to follow rehabilitation programs. Our project directly addresses these challenges by using AI-driven telerehabilitation, which offers personalized, remote rehabilitation plans that adapt to each patient's progress, improving patient outcomes and increasing access to care for those affected by neurological disorders.

References

1. A. M. Al-Ghaili, H. Kasim, N. M. Al-Hada, Z. B. Hassan, M. Othman, J. H. Tharik, R. M. Kasmani, and I. Shayea, "A review of metaverse's definitions, architecture, applications, challenges, issues, solutions, and future trends," vol. 10, pp. 125 835–125 866.
2. R. Chengoden, N. Victor, T. Huynh-The, G. Yenduri, R. H. Jhaveri, M. Alazab, S. Bhattacharya, P. Hegde, P. K. R. Maddikunta, and T. R. Gadekallu, "Metaverse for Healthcare: A Survey on Potential Applications, Challenges and Future Directions," vol. 11, pp. 12 765–12 795. [Online]. Available: https://ieeexplore.ieee.org/document/10034994/
3. "UN Population Division Data Portal Interactive access to global demographic indicators." [Online]. Available: https://population.un.org/dataportal/
4. G. Kayola, M. M. Mataa, M. Asukile, L. Chishimba, M. Chomba, D. Mortel, A. Nutakki, S. Zimba, and D. Saylor, "Stroke rehabilitation in low- and middle-income countries: Challenges and opportunities," vol. 102, pp. S24—S32. [Online]. Available: https://europepmc.org/articles/PMC9846582
5. [Online]. Available: https://www.nbcnews.com/nightly-news/video/inside-the-metaverse-3d-detailed-anatomical-renderings-revolutionizing-.health-care-145680453700
6. S. Zhang, C. Demiralp, D. F. Keefe, M. DaSilva, D. H. Laidlaw, B. D. Greenberg, P. J. Basser, C. Pierpaoli, E. A. Chiocca, and T. S. Deisboeck, "An immersive virtual environment for dt-mri volume visualization applications: a case study," in *Proceedings Visualization, 2001. VIS'01.* Ieee, 2001, pp. 437–584.
7. C. Demiralp, C. D. Jackson, D. B. Karelitz, S. Zhang, and D. H. Laidlaw, "Cave and fishtank virtual-reality displays: A qualitative and quantitative comparison," *IEEE transactions on visualization and computer graphics*, vol. 12, no. 3, pp. 323–330, 2006.
8. J. Chen, H. Cai, A. P. Auchus, and D. H. Laidlaw, "Effects of stereo and screen size on the legibility of three-dimensional streamtube visualization," *IEEE transactions on Visualization and Computer Graphics*, vol. 18, no. 12, pp. 2130–2139, 2012.

9. H. Cecotti, Q. Richard, J. Gravellier, and M. Callaghan, "Magnetic resonance imaging visualization in fully immersive virtual reality," in *2020 6th International Conference of the Immersive Learning Research Network (iLRN)*, 2020, pp. 205–209.
10. E. Katsoulakis, Q. Wang, H. Wu, L. Shahriyari, R. Fletcher, J. Liu, L. Achenie, H. Liu, P. Jackson, Y. Xiao, T. Syeda-Mahmood, R. Tuli, and J. Deng, "Digital twins for health: A scoping review," vol. 7, no. 1, p. 77. [Online]. Available: https://www.nature.com/articles/s41746-024-01073-0
11. S. Oh, J. Choi, D. H. Han, and E. Kim, "Effects of game-based digital therapeutics on attention deficit hyperactivity disorder in children and adolescents as assessed by parents or teachers: A systematic review and meta-analysis," vol. 33, no. 2, pp. 481–493. [Online]. Available: https://link.springer.com/10.1007/s00787-023-02174-z
12. J. D. Steinmetz at al., "Global, regional, and national burden of disorders affecting the nervous system, 1990–2021: A systematic analysis for the Global Burden of Disease Study 2021," vol. 23, no. 4, pp. 344–381. [Online]. Available: https://doi.org/10.1016/S1474-4422(24)00038-3
13. R. Hooi and H. Cho, "Avatar-driven self-disclosure: The virtual me is the actual me," vol. 39, pp. 20–28. [Online]. Available: https://linkinghub.elsevier.com/retrieve/pii/S0747563214003549
14. C. Kyrlitsias and D. Michael-Grigoriou, "Social Interaction With Agents and Avatars in Immersive Virtual Environments: A Survey," vol. 2, p. 786665. [Online]. Available: https://www.frontiersin.org/articles/10.3389/frvir.2021.786665/full
15. M. T. Deighan, A. Ayobi, and A. A. O'Kane, "Social Virtual Reality as a Mental Health Tool: How People Use VRChat to Support Social Connectedness and Wellbeing," in *Proceedings of the 2023 CHI Conference on Human Factors in Computing Systems*. ACM, pp. 1–13. [Online]. Available: https://dl.acm.org/doi/10.1145/3544548.3581103
16. S. Zamanifard and A. Robb, "Social Virtual Reality Is My Therapist: Overcoming Social Anxiety Disorder Through Using Social Virtual Reality," in *Extended Abstracts of the 2023 CHI Conference on Human Factors in Computing Systems*. ACM, pp. 1–6. [Online]. Available: https://dl.acm.org/doi/10.1145/3544549.3585888
17. L. A. Baccon, E. Chiarovano, and H. G. MacDougall, "Virtual Reality for Teletherapy: Avatars May Combine the Benefits of Face-to-Face Communication with the Anonymity of Online Text-Based Communication," vol. 22, no. 2, pp. 158–165. [Online]. Available: https://www.liebertpub.com/doi/10.1089/cyber.2018.0247
18. H. Thieme, N. Morkisch, J. Mehrholz, M. Pohl, J. Behrens, B. Borgetto, and C. Dohle, "Mirror therapy for improving motor function after stroke," *Cochrane Database of Systematic Reviews*, vol. 2018, no. 7, Jul. 2018. [Online]. Available: http://doi.wiley.com/10.1002/14651858.CD008449.pub3
19. M. Okawada, F. Kaneko, K. Shindo, M. Yoneta, K. Sakai, K. Okuyama, K. Akaboshi, and M. Liu, "Kinesthetic illusion induced by visual stimulation influences sensorimotor event-related desynchronization in stroke patients with severe upper-limb paralysis: A pilot study," vol. 38, no. 6, pp. 455–465.
20. K. E. Laver, B. Lange, S. George, J. E. Deutsch, G. Saposnik, and M. Crotty, "Virtual reality for stroke rehabilitation," vol. 2018, no. 1. [Online]. Available: http://doi.wiley.com/10.1002/14651858.CD008349.pub4

Part V
Network, Supply Chain, and Cyber Forensics

Supply Chain Forensic with Distributed Ledger Technologies and Software Bill of Materials

Iwinosa Aideyan, Mert D. Pesé, and Richard Brooks

1 Introduction

This chapter introduces a forensic framework designed to secure software supply chains by integrating two emerging technologies: software bills of materials (SBoMs) and distributed ledger technologies (DLTs). In Sect. 1.1, we discuss software supply chain security challenges and describe how SBoMs and DLTs can be leveraged for forensic data collection, validation, and forensic investigation. Section 1.2 explains the contributions of our work, summarizing verification mechanisms, detection techniques, and audit trail capabilities that form the framework's core.

1.1 Overview

Compromised software packages affect multiple organizations simultaneously. We note that the sound software engineering practice of leveraging existing software packages introduces third-party dependencies into digital supply chains [1, 2]. High-profile incidents underscore the growing threat posed by software supply chain vulnerabilities. A detailed discussion of supply chain compromises including SolarWinds, Log4j, Okta, and Rules File Backdoor is in Sect. 2.1.

Digital forensics supply chain requires collecting, preserving, and analyzing evidence from the entire software ecosystem. This preserves provenance and chain of custody certification, which is essential for understanding an attack, establishing accountability, preventing future incidents and prosecuting wrong-doers. Existing

I. Aideyan (✉) · M. D. Pesé · R. Brooks
Clemson University, Clemson, SC, USA
e-mail: iaideya@clemson.edu; mpese@clemson.edu; rrb@clemson.edu

S. S. Iyengar et al. (eds.), *Artificial Intelligence Driven Forensics*, Signals and Communication Technology, https://doi.org/10.1007/978-3-031-98036-7_25

solutions include dynamic analysis frameworks like Buildwatch to detect malicious packages [2] and classify protective measures based on security mechanisms, information protection methods, and software development lifecycle stages [3]. Despite the growing urgency for secure software supply chains, integrating forensic evidence management into the process remains challenging [4]. Many organizations rely on outdated verification methods that are ill-equipped to address modern, complex threats. In response, recent technological advances, particularly DLTs and SBoMs, have emerged as promising tools to enhance forensic capabilities and strengthen supply chain defenses.

Blockchain technology provides decentralized, transparent, and immutable management of evidence artifacts to maintain security, integrity, and traceability throughout the evidence lifecycle [5]. Decentralized verification lets each node in a network independently attest to the authenticity of software components, removing single points of failure. Decentralization enhances supply chain security by ensuring data integrity and creating an immutable record of all transactions and changes. SBoMs, on the other hand, provide detailed inventory of a software package's components and dependencies [6, 7]. By offering a clear "tree of dependencies," SBoMs allow forensic analysts to trace the lineage of software artifacts and pinpoint where malicious injections or modifications occur. SBoMs certify software component security and enhance transparency in software supply chains [8].

SBoMs provide structured accounting of software dependencies, helping organizations identify vulnerabilities and manage license risks [7]. Their importance has been underscored by recent cybersecurity incidents and national directives such as Executive Order 14028 [9, 10].

DLTs and SBoMs provide unalterable audit trails. For instance, logging the signed hash of an SBoM on a blockchain allows investigators to verify software integrity and detect unauthorized changes. In our framework, we extend this idea by using SBoM analysis and decentralized verification to trace component changes and pinpoint tampering. Implementation details are in Section 3. By recording software update and verification on the blockchain, forensic investigators can trace the complete history of a compromised component and accurately pinpoint the origin of a breach.

1.2 Contribution

This chapter explores how DLTs and SBoMs can detect, trace, and mitigate software supply chain attacks. The study simulates a supply chain attack; specifically, an attacker inject a malicious dependency into a software update. It shows how the SBoMs and blockchain can detect tampered components, track unauthorized changes, and provide an immutable audit trail. This serves as a proof-of-concept highlighting the benefits of merging DLTs with SBoMs.

This chapter contributes:

- Component Verification: It demonstrates decentralized verification mechanisms where each node independently confirms software components' authenticity, which can mitigate the tampering risk.
- Early Detection: It explores how continuous SBoM analysis can facilitate the prompt identification of unauthorized injections or modifications within the supply chain.
- Audit Trail: It highlights the role of blockchain in establishing an immutable, distributed log of all supply chain transactions, providing verifiable proof of each action.
- Traceability: It showcases how tracking every step, from software procurement to integration, enables forensic investigators to reconstruct the chain of events in the aftermath of a breach.

  AI-Enabled Threat: Addresses emerging AI-assisted supply chain threats, such as Vibe Coding and the Rules File Backdoor, by demonstrating how the framework supports provenance, detection, and accountability even in the presence of AI-generated or manipulated code.
- Case Study: A case study uses Hyperledger Fabric and an SBoM tool to illustrate the application of these methods.

The remainder of this chapter is organized as follows: Sect. 2 provides the background and related work, discussing the supply chain attacks, the state of SBoMs, and challenges associated with forensic investigations in supply chains. Section 3 outlines the framework for supply chain forensics using DLTs and SBoMs. Section 4 presents our case study, showcasing the practical application of the framework using DLT and SBoM analysis. Section 5 concludes the chapter by summarizing our findings and reflecting on the implications for future work in digital forensic investigations.

This chapter introduces the cybersecurity challenges posed by modern software supply chain attacks and establishes the need for forensics tools to handle these vulnerabilities. It highlights the limitations of current security mechanisms and articulates the need for visibility, integrity, and verifiability across software artifacts. The core contribution of this work, a forensic framework combining SBoMs and DLTs, is outlined, along with a high-level roadmap of the chapter's structure.

2 Background

This section introduces the current state of software supply chain security and forensics. Section 2.1 discusses attacks and software supply chain risks. Section 2.2 reviews solutions and challenges. Section 2.3 discusses forensic investigation methods. In Sect. 2.4, we explore how SBoMs use the concept of the tree of dependencies. Finally, Sect. 2.5 presents the role of DLTs in supply chain security, with a focus on our GUIXCHAIN research. This section lays the groundwork for understanding our innovations.

2.1 Supply Chain Attacks and Risks

Over time, the focus of cyberattacks has shifted dramatically from individual endpoints to centralized servers and now to software supply chains. This progression reflects the increasing efficiency and impact of attacking components that influence broader ecosystems. While early attacks primarily targeted user devices or specific servers, modern attackers now exploit the interconnectivity of software supply chains, where a single compromised dependency can cascade into thousands of downstream products. This shift has amplified both the number of affected victims and the potential profit or disruption achievable through a single attack. For example, In 2020, attackers compromised SolarWinds' Orion software, a widely used network management solution, by injecting malicious code into its updates. This breach affected over 18,000 customers, including government agencies and Fortune 500 companies, and remained undetected for over a year, ultimately resulting in recovery costs exceeding $100 million [23]. These incidents highlight how a compromised component can propagate through an interconnected ecosystem, affecting hundreds or thousands of organizations.

The Kaseya VSA ransomware attack in 2021 [25] incident targeted managed service providers (MSPs) and their clients by exploiting a vulnerability in Kaseya's VSA software. A fake update embedded with ransomware was deployed, directly impacting 60 primary clients and indirectly affecting over 1,500 businesses globally. The attackers initially demanded a ransom of $70 million, later reduced to $50 million, demonstrating the cascading impact that supply chain attacks can have on downstream entities [24].

Supply chain attacks increasingly target the continuous integration/continuous deployment (CI/CD) pipelines that underpin modern software development. When attackers inject malware into the CI/CD process often by compromising its operating environment, they can distribute malicious software updates at scale. For example, the 2021 Codecov attack compromised a widely used Bash Uploader script by tampering with its version in the official Codecov GitHub repository [26]. AI-assisted coding tools have now been embedded into CI/CD workflows. Tools such as GitHub Copilot, Cursor Composer, Replit Agent, and Claude enable what is often termed *Vibe Coding* [27]. This approach encourages developers to rely heavily on AI-generated code suggestions, sometimes without critically validating the logic. Although this can accelerate development, it also introduces serious security risks. AI models trained on vast code bases may inadvertently propose insecure code patterns, thereby increasing exposure to vulnerabilities [29]. A notable example is the *Rules File Backdoor* attack, which leveraged hidden unicode characters and advanced evasion techniques to exploit AI code editors like Copilot and Cursor [28]. Attackers manipulated these tools to insert malicious logic into projects in ways that bypassed human review. These injected scripts silently propagated through supply chains, affecting numerous downstream consumers as illustrated in Fig. 1.

Similarly, the Log4j vulnerability, known as *Log4Shell*, exemplified the global risks tied to insecure default configurations in widely adopted libraries. This flaw

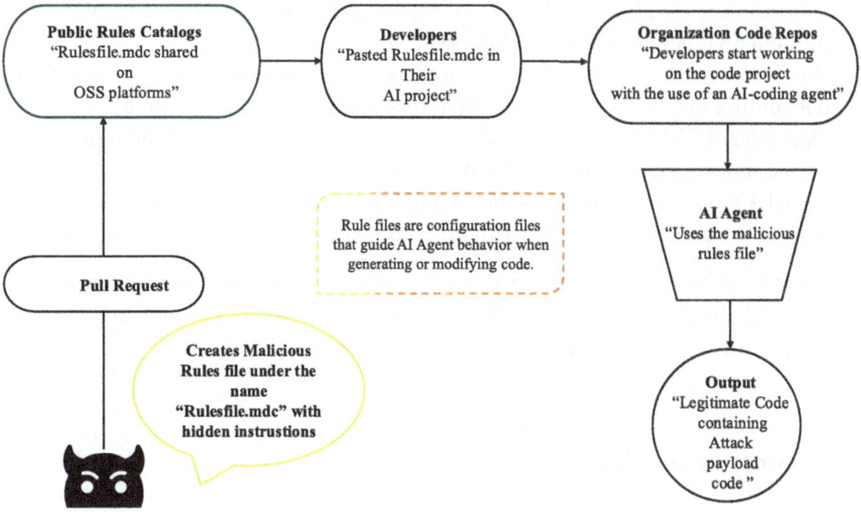

Fig. 1 How attackers weaponize AI-agents through compromised rule files

in Apache Log4j's JNDI lookup mechanism enabled remote code execution in thousands of services that used the library by default [30].

The Okta breach drew attention to third-party vendor risk. The Lapsus$ hacking group compromised a support vendor linked to Okta, a critical provider of identity and access management services including single sign-on (SSO), multi-factor authentication (MFA), and user lifecycle controls [31, 32]. Although the incident affected less than 3% of Okta's customers, it posed outsized risks because Okta solutions are central to authentication across many government and enterprise systems. Agencies like the Directorate of Defense Trade Controls (DDTC) and the Centers for Medicare and Medicaid Services (CMS) depend on Okta for secure identity management. As such, any breach can disrupt essential services and endanger sensitive national data [33].

More recently, the XZ Utils backdoor (CVE-2024-3094) demonstrated the risks of compromised compression libraries. Attackers manipulated the XZ compression toolchain to introduce a backdoor that jeopardized services like OpenSSH. We revisit this case in Sect. 2.4 to show how SBoMs can support forensic investigations.

These examples demonstrate the increasingly diverse and sophisticated nature of software supply chain attacks. While AI tools like Copilot can boost productivity, they can also be weaponized to introduce and propagate malicious code. Forensic investigations must adapt to detect and analyze these AI-facilitated threats, using advanced techniques to examine AI-generated code and its implications for software integrity. Common methods include:

- Injection of Malicious Dependencies: Attackers introduce compromised libraries during development or update processes [26].

- Vibe Coding Exploits: Adversaries exploit AI-generated code suggestions to introduce vulnerabilities or backdoors [29].
- Ransomware: Adversaries deploy this type of malware that encrypts a victim's data until a ransom is paid, disrupting operations and extorting funds.
- Dependency Confusion: Attackers exploit naming conflicts between internal and public repositories to insert malicious packages [28].
- Social Engineering: Threat actors manipulate open-source maintainers or vendors to weaken security or insert malicious code.

The implications of these attacks are profound. They can lead to widespread operational disruptions, financial losses, reputational damage, and a significant erosion of trust in software providers and their update mechanisms.

2.2 State of the Art

The software supply chain spans the full lifecycle from code development to deployment, involving internal processes, third-party components, and external dependencies [11, 23]. Vulnerabilities at any point can cascade downstream, affecting numerous products. Attackers increasingly exploit these gaps to insert malicious code or disrupt services across industries [12]. To counter such threats, organizations must adopt secure lifecycle management and monitor the security posture of both their products and suppliers [13].

Recent statistics stress the urgency of this issue. Ninety-one percent of organizations experienced at least one SSC incident in the last 2 years, with research indicating that 95% of organizations harbor at least one high or critical risk within their SSC. On average, organizations face around nine of such risks [14]. Consequently, the cost of SSC attacks is projected to escalate from $46 billion in 2023 to an estimated $138 billion by 2031. In response, securing the SSC has become a top priority, with forecasts suggesting that 80% of organizations will adopt specialized processes and tools by 2027. According to Synopsys's 2024 Open Source Security and Risk Analysis report, their findings from more than 1,000 commercial codebases across 17 industries in 2023 showed that 96% of the total codebases contained open source and 84% of codebases assessed for risk contained vulnerabilities [15].

One promising solution to enhancing SSC security is the use of software bills of materials (SBoMs), which have become central to transparency in tracking software components. An SBoM is an inventory list that details a software package's components, libraries, and dependencies. Modern development practices often rely heavily on open-source and third-party packages, making visibility into these components essential [17]. As emphasized by NIST in response to Executive Order 14028, SBoMs are expected to support vulnerability detection, license compliance, and informed risk-based decision-making across the software lifecycle. NIST envisions

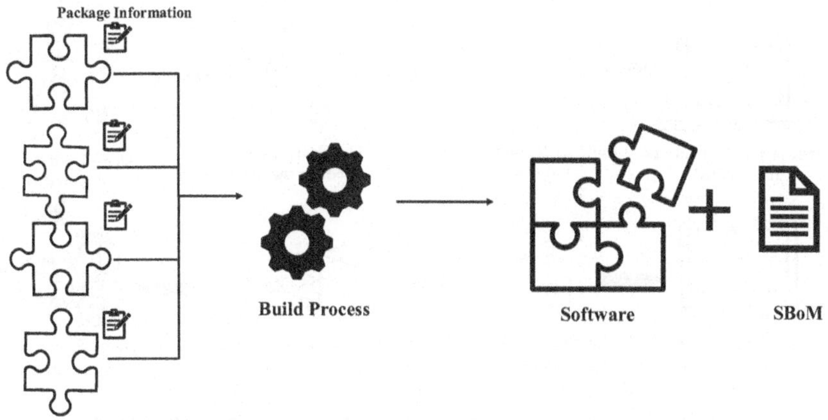

Fig. 2 Visualization of the software supply chain and SBoM generation process

a model in which SBoMs are consistently generated by developers, verified by suppliers, and consumed by procuring organizations to assess software integrity.

SBoMs offer a historical record of an application's creation, including the origins of third-party code, and provide ongoing visibility into the software composition. This enables DevOps teams to identify vulnerabilities and assess potential risks while equipping forensic teams with the information needed to understand what went wrong and how an incident occurred [18]. Figure 2 illustrates how software is constructed from multiple interdependent components. As the build process assembles these components into a final product, SBoM is simultaneously generated. The SBoM acts as a detailed inventory of all components used, supporting transparency, vulnerability tracking, and compliance.

Data formats, CycloneDX and SPDX, standardize the format and content of SBoMs, promoting interoperability and consistency across the industry [19]. SBoMs have limitations. There are no validation mechanisms to ensure SBoM integrity, both in terms of the data itself (i.e., ensuring it has not been tampered with) and the tooling used to generate it. Studies have shown that a significant number of stakeholders (approximately 49.3% agree and 26.2% disagree) see the lack of reliable SBoM validation as a critical gap. In comparison, 75.4% acknowledge the lack of practical SBoM consumption tools, without adequate tools to consume and analyze SBoMs, software procurers often question how to process and act on the information provided effectively [16].

This gap in SBoM integrity and consumption has led to the exploration of DLTs as an additional layer of verification. DLTs, particularly blockchain, aim to create a transparent, tamper-proof, and decentralized ledger of transactions. In a blockchain, transactions are grouped into blocks, each accompanied by a header containing a cryptographically signed hash referencing the previous block, thus forming a chronologically ordered, append-only chain [21].

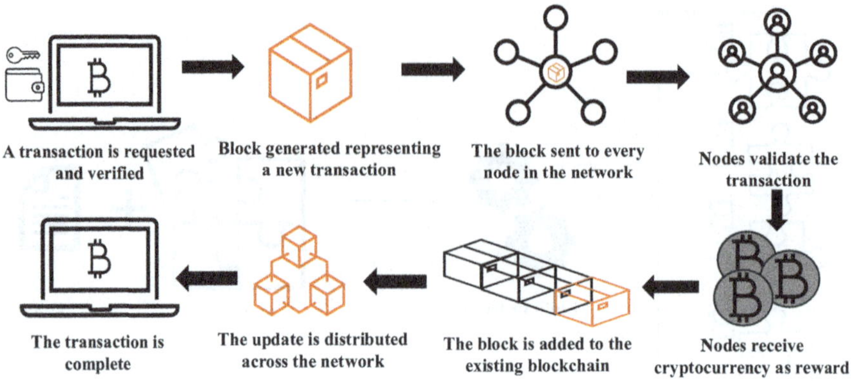

Fig. 3 General process of how transactions are added to a public blockchain, Bitcoin

Figure 3 illustrates how transactions are added to a public blockchain, such as Bitcoin, to introduce core concepts of blockchain architecture. In these systems, the order-execute mechanism ensures that each transaction is validated via consensus protocols, typically proof-of-work (PoW), where miners solve cryptographic puzzles, or proof-of-stake (PoS), where validators are selected based on staked assets [22]. These consensus mechanisms provide essential security properties such as immutability, data integrity, non-repudiation, and decentralized agreement, all of which contribute to tamper-evident audit trails critical for forensic applications. However, it is important to note that our framework uses Hyperledger Fabric, a permissioned blockchain platform with different consensus protocols and no reliance on mining or cryptocurrency. Despite its benefits, blockchain technology still faces challenges, particularly in performance and scalability, as propagation delays can affect overall system efficiency [20].

2.3 *Forensic Investigations in Supply Chains*

Forensic investigation in supply chains aims to unravel incidents of compromise and establish unassailable evidence of wrong doing [37]. While modern supply chains benefit from digital technologies that enhance traceability, they also become attractive targets for malicious actors. Digital forensics plays a critical role in reconstructing events, identifying the point of failure or breach, and preserving evidence that supports accountability and legal scrutiny [1].

The methodologies employed in forensic investigations are diverse and have evolved to address the dynamic challenges of modern supply chains. Dynamic code analysis tools integrated within continuous integration (CI) environments to monitor third-party dependencies, allowing real-time detection of abnormal behaviors resulting from malicious packages [38]. Forensic artifact extraction further enhances investigations by recording critical evidence during the software installation, tracking files created, network connections initiated, and processes launched to differentiate between benign and malicious software. System call monitoring [39], which logs detailed activities (e.g., calls to open or socket functions), provides additional layers of insight into software behavior during installation. In controlled sandbox environments, Cuckoo [40], suspicious files can be analyzed without endangering host systems. Moreover, observable artifact comparison, where forensic artifacts from compromised packages are compared with those of known-good versions, has proven effective in identifying unauthorized modifications. Despite these techniques, forensic investigations still face several challenges; one of the foremost challenges is **data collection**. The massive volume and heterogeneity of log and transaction data generated by modern digital devices can overwhelm standard forensic tools [42]. Forensic practitioners must filter through vast troves of data, where the noise can easily obscure critical evidence. This data volume explosion complicates the collection process and raises the risk of overlooking crucial details [46, 47].

Closely related to data collection are the issues surrounding **chain-of-custody** [43], as maintaining the integrity and authenticity of forensic evidence is essential yet complex. The potential for human error during the gathering, analysis, and interpretation of evidence poses a significant risk to the reliability of forensic outcomes [42]. Any misstep in handling evidence can compromise the entire investigation, rendering the results inadmissible or unreliable. Another critical challenge is **sample contamination**. Given that forensic evidence in supply chains is often available in limited quantities, even minor contamination during collection, storage, or processing can drastically affect the accuracy of the analysis [42]. Ensuring that evidence remains uncontaminated throughout the forensic process is a constant struggle, especially when dealing with small, sensitive data samples that may be easily compromised by extraneous material [41].

Handling and processing speed poses a challenge due to the massive data volumes generated, which require advanced storage and processing capabilities [44]. Traditional forensic tools, designed for smaller datasets, may encounter bottlenecks and delays, during which critical evidence can degrade or become irrelevant [48]. In real-time analysis scenarios, these delays could compromise an investigation's effectiveness. However, the use of distributed updates to counter network infections offers a superior approach, enhancing processing efficiency and ensuring timely, actionable evidence [45].

To improve forensic readiness, researchers have developed systems like digital forensic readiness (DFR), which prepare environments to collect reliable evidence and speed up incident response [49]. Live memory forensics also plays a key role by capturing time-sensitive data from volatile memory sources, such as autonomous systems [50]. Additionally, tools like Buildwatch analyze forensic artifacts from software builds to detect patterns in malicious open-source packages, offering new detection strategies [51].

2.4 SBoMs in Forensics: The Tree of Dependencies

SBoMs are an asset in digital forensics, enhancing transparency and traceability across software supply chains [52]. By providing a machine-readable inventory of all software components, SBoMs allow forensic analysts to map the intricate web of dependencies that form each application. This "tree of dependencies" is not merely a list; it represents a hierarchical structure where a single outdated or vulnerable subcomponent, buried deep within the dependency tree, can serve as a weak link, compromising the integrity of the entire system. When forensic investigators examine the dependency tree, they can pinpoint which libraries or modules may harbor known vulnerabilities [53]. This is crucial in open-source software, where dependencies are numerous and frequently updated. A comprehensive SBoM enables the construction of a detailed component dependency graph, which in turn allows for a more accurate assessment of risks associated with each node in the tree. For example, SBoM formats [54], such as CycloneDX or SPDX, allow for seamless integration with security tools and databases, thereby enabling real-time comparisons against known vulnerability repositories.

In digital forensics, the metaphor of a "tree of dependencies" highlights the layered complexity of modern software looking like a directed acyclic graph (DAG),[1] where dependencies often intersect due to shared libraries across projects. Just as a tree's health depends on the robustness of its roots, the security of an application is contingent upon the integrity of its deepest dependencies [55]. Forensic investigations leverage this concept by performing in-depth analyses that involve code scanning to identify embedded open-source artifacts and licenses, followed by manual interpretation of tool-generated data. In response to security incidents like the Apache Log4j vulnerabilities and SolarWinds hack, discussed in Sect. 2.1, SBoMs help improve the security of open-source software by mapping software dependencies; it tells forensic investigators where to go.

To illustrate the concept of the "tree of dependencies," Fig. 4 presents the dependency flow for NGINX software stack. At the apex, *NGINX* is a high-performance HTTP server and reverse proxy. It relies on several critical libraries: for instance, *OpenSSL* provides HTTPS support through TLS/SSL encryption; *libcurl* handles data transfers via URLs; and *zlib* offers compression, facilitating gzip support for web traffic. Delving deeper, *libcurl* itself depends on both *OpenSSL* and *zlib* to manage encrypted and compressed HTTP(S) traffic. *OpenSSL*, in turn, leverages libraries such as *libgcrypt* and *liblzma* for cryptographic operations and compression support. Meanwhile, *zlib* utilizes *libmagic* to detect file types prior to compression, and *libmagic* further depends on *libxml2* to parse file formats—a process that, in some cases, may involve compilation using *musl*, a lightweight *libc* implementation favored in static builds for Linux. This detailed breakdown clarifies

[1] Graph structure composed of nodes and directed edges, where the edges indicate a one-way relationship and no cycles are allowed.

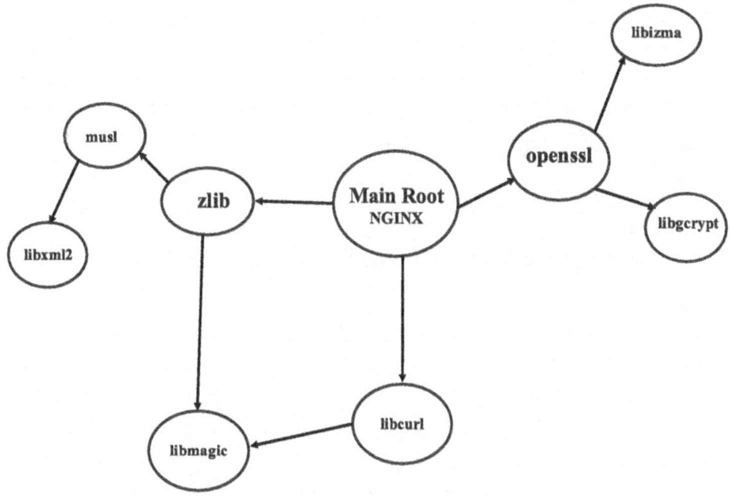

Fig. 4 DAG representation of Nginx dependencies

how components are interrelated and emphasizes that vulnerabilities in any one node can have cascading effects throughout the system.

A clear example that illustrates the forensic value of SBoMs is the recent vulnerability identified in *XZ Utils* (CVE-2024-3094) [56]. *XZ Utils*, a widely used data compression library, was discovered to contain a critical backdoor in versions 5.6.0 and 5.6.1. The vulnerability was introduced by injecting additional .m4 build files not part of the official repository. These files altered build instructions to extract a prebuilt object file, which in turn modified functions within the *liblzma* library and enabled an SSH authentication bypass [57]. In a forensic investigation, an SBoM that lists the exact version of *XZ Utils* used in a system would immediately flag these versions as suspect. Moreover, by examining the dependency tree, forensic analysts can trace how *XZ Utils* integrates with other components, such as its indirect usage via libraries linked to SSH services, thereby reconstructing the potential infection path. In this case, the SBoM not only serves as an initial indicator of vulnerability but also as a roadmap for further artifact analysis, including scrutiny of build logs, file hash comparisons, and even manual inspection of the additional .m4 files [58].

Beyond vulnerability detection, SBoMs significantly enhance incident response and vulnerability management. In a post-incident review, an SBoM provides critical corroborating evidence that helps determine the impact of a compromise on parallel systems or previous software versions. This detailed record allows forensic teams to establish a timeline, for instance, identifying when a vulnerable version of a component was introduced, correlating that event with suspicious log activity, and aligning it with subsequent security incidents. Such a timeline is invaluable in understanding the scope and sequence of an attack.

In conclusion, SBoMs provides a detailed "tree of dependencies" that improves incident response and vulnerability management and strengthens overall supply chain transparency. Their ability to map every component and interconnection within a software package makes them indispensable for tracing the origins of security breaches. As we transition to the next section, we will explore how integrating SBoMs with DLTs further augments forensic investigations, exemplified through our case study using GuixChain.

2.5 DLTs in Supply Chain Security (GUIXCHAIN)

Blockchain has emerged as a transformative technology that enhances supply chain security. Its decentralized and immutable ledger introduces features beneficial to forensic analysis, including comprehensive traceability and integrity. Every recorded transaction or software update becomes an unalterable component of the blockchain, providing an enduring and auditable history of software interactions. Additionally, the technology creates tamper-proof audit trails, ensuring that recorded evidence remains unchanged and trustworthy for forensic purposes [5].

Our research group is exploring how blockchain, SBoMs, reproducible builds, and version control systems can be integrated to enhance software supply chain security in the automotive sector through a system called GUIXCHAIN. GuixChain integrates Git, reproducible builds using Guix, and automated SBoM generation with blockchain technology to establish an immutable, decentralized record of software development and deployment actions. In this chapter, we focus on the forensic benefits of this integration, rather than on its reproducible build features. This framework strengthens software supply chain security and creates an auditable trail that is used for forensic investigations.

The main objectives of the GuixChain project are to:

- Integrate Git with Blockchain: Enhance code repository security by ensuring traceability and immutability throughout the software development lifecycle.
- Develop Smart Contracts for Automated Code Compilation: Automate the build process so that only verified actions can trigger compilation, reducing the risk of unauthorized code injection.
- Implement Reproducible and Deterministic Software Builds: Establish a controlled build environment that produces identical outputs, ensuring consistency and facilitating rapid detection of anomalies.

At its core, GuixChain uses Git to record every change in the source code during the development process into a blockchain ledger. Every code commit and push is permanently recorded, providing an unalterable history for forensic analysis. In the event of a security breach or malicious injection, investigators can trace the entire sequence of events back to its origin. This decentralized verification process, where each node in the blockchain independently confirms transactions, eliminates single points of failure, and ensures that any tampering is immediately evident [34].

GuixChain further reinforces its forensic capabilities by using Guix for reproducible builds. Guix's strict control over the build environment guarantees that identical source code consistently produces bit-for-bit identical binaries across different nodes. Any deviation in the build output suggests a potential compromise, such as unauthorized code injection. In our framework, smart contracts are employed to trigger builds and ensure that every compilation event adheres to these deterministic rules. As a result, even if an attacker compromises a build server, the discrepancy in outputs across nodes would serve as immediate evidence of tampering, supporting rapid forensic intervention [35].

Additionally, including SBoMs within GuixChain adds an additional layer of transparency and traceability. During the compilation phase, an SBoM is generated alongside the firmware, cataloging all the components, libraries, and dependencies incorporated into the final product. These SBoMs are then hashed, and their integrity is verified across multiple nodes. Only when the hashes match uniformly is the build advanced to the next stage. This process ensures the authenticity of the software artifacts and provides forensic investigators with a clear "tree of dependencies" to analyze in the event of a breach. The SBoM serves as a checkpoint, enabling investigators to pinpoint the exact stage at which any malicious modifications were introduced [36].

Figure 5 illustrates the overall architecture of GuixChain, showcasing how smart contracts, blockchain, Git, and SBoMs work in unison to secure the software development process from Git operations. The diagram emphasizes the continuous logging of each action, from code commits and build triggers, creating an immutable audit trail that is both comprehensive and transparent. This workflow mitigates the risk of attacks, as compromising the system now requires simultaneous corruption across multiple nodes rather than a single point of failure.

From a forensic standpoint, GuixChain provides several benefits. First, it creates a tamper-proof audit trail that captures every software change, making it possible to trace back to the origin of any malicious code. Second, the reproducible build environment ensures that any deviation from expected outputs can be quickly

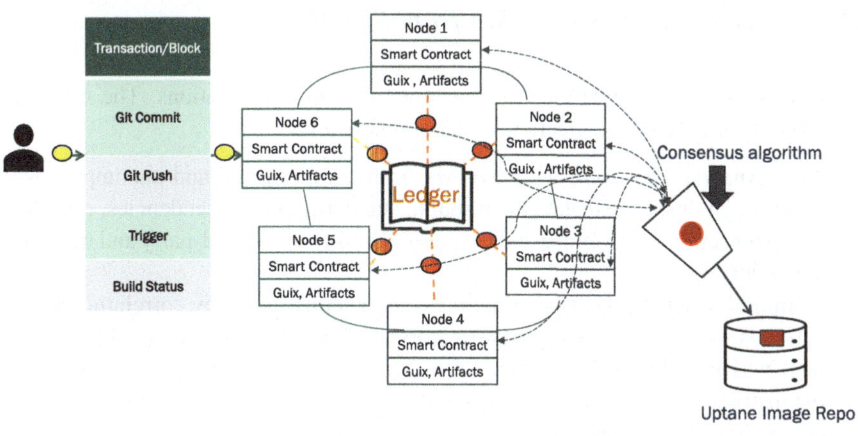

Fig. 5 GuixChain framework

identified, prompting further investigation. Third, the decentralized nature of the blockchain mitigates risks associated with centralized storage, such as data loss or targeted attacks like distributed denial of service and ransomware. Lastly is a history list of all the software components used in the compilation process and a map of the dependencies in the code. By distributing data across multiple nodes and securing it through consensus mechanisms, GuixChain ensures that forensic evidence remains intact, immutable, permanent, and verifiable.

This section provided an examination of the current landscape in software supply chain security. It began by contextualizing known attack vectors and incidents, establishing the severity and diversity of threats. The role of forensic investigations was explored, with emphasis on their limitations in dynamic software ecosystems. SBoMs were introduced as tools for exposing software composition and dependency trees, while DLTs were shown to address integrity and traceability gaps. The chapter concluded with the introduction of GuixChain.

3 Framework for Supply Chain Forensics Using DLTs and SBoMs

This section presents a framework for conducting forensic investigations in software supply chains using SBoMs and DLTs. It organizes the forensic process into three major functional components. Section 3.1 shows forensic data collection through SBoMs. Section 3.2 shows tamper-proof evidence storage using blockchain-based systems, and Sect. 3.3 real-time forensic audits powered by smart contracts. This chapter focuses on forensic data collection, tamper-proof evidence storage, and smart contracts for real-time auditability.

3.1 Forensic Data Collection from SBoMs

SBoMs serve as a foundational artifact for forensic investigations. The utility of SBoMs in forensic data collection is elaborated below:

- Taint Analysis: By comparing SBoMs from compromised and uncompromised systems, analysts can isolate the minimal standard set of dependencies, revealing likely points of compromise and granting visibility to all third-party and transitive dependencies.
- Temporal Tracking: Forensic timelines can be constructed by correlating SBoM versions with logs, allowing investigators to pinpoint when a vulnerable component was introduced. When anomalies are detected, SBoMs enable analysts to reconstruct the path of infection through hash comparisons, artifact tracking, and dependency flow visualization (as demonstrated in Fig. 4).

- Selective Disclosure and Verifiable Credentials: SBoMs can be shared via blockchain using cryptographic credentials that selectively expose only relevant portions of the component list, preserving sensitive information [59]. The XZ Utils backdoor illustrates this, as a forensic SBoM would have immediately flagged affected versions, and the dependency tree would reveal indirect connections to SSH authentication mechanisms.

3.2 *Using DLTs for Tamper-Proof Evidence Storage*

DLTs provide an immutable, decentralized ledger for maintaining forensic evidence integrity. In our framework, each action (e.g., code commit, component integration, SBoM generation) is cryptographically hashed and recorded on-chain. Smart contracts automate logging and verification processes, ensuring consistency and reducing the potential for human error. Tampering becomes practically infeasible since records must be altered across multiple nodes. Key forensic advantages include:

- **Tamper-Proof Storage**: According to *Ulybyshev et al.* (2018), permissioned blockchains like Hyperledger Fabric allow role- and attribute-based access control, ensuring only authorized users can write or query the forensic ledger [61].
- **Immutable Audit Trails**: *Lew et al.* (2024) demonstrate a blockchain-based assurance scheme with reproducible builds and SBoMs, achieving traceable builds with less than 1 second overhead [60].
- **Hash-Based Verification**: As *Ozkan et al.* (2024) proposed, storing cryptographic hashes of SBoMs enables decentralized verification. This mechanism can detect unauthorized component modifications without needing full SBoM access [6].

Furthermore, DLTs enhance the chain of custody for forensic artifacts [43]. By embedding hashes of SBoMs, build outputs, and version histories in the ledger; investigators can follow a hash chain that shows each transformation a component underwent, from inclusion in the source repository to deployment.

3.3 *Smart Contracts for Real-Time Forensic Audits*

Smart contracts act as autonomous forensic agents embedded within the DLT layer. They trigger actions based on pre-defined conditions. Smart contracts operationalize the forensic workflow by triggering automated audits when suspicious changes are detected, enforcing build policies to ensure only verified components are compiled, and enabling continuous integrity checks by comparing build outputs across multiple nodes. Smart contracts can verify that SBoMs match expected formats and known-good hash trees before allowing a build to proceed. This ensures consistency and blocks the propagation of tainted artifacts. Smart contracts can monitor incoming component updates and initiate forensic audits if anomalies (e.g., version mismatch, hash collision) are detected.

This section outlined a modular forensic framework integrating SBoMs and DLTs. We demonstrated how SBoMs, when generated at multiple stages of the software lifecycle, provide critical visibility for forensic investigations. DLTs were shown to offer tamper-proof storage and immutable audit trails. Smart contracts were presented as autonomous agents that enforce real-time integrity checks and build policy compliance. The chapter evaluated effectiveness and implementation considerations.

4 Case Study: Forensic Investigation of a Supply Chain Attack Using Hyperledger Fabric and SBoMs

In this case study, we demonstrate how a package (a C/C++ library under development) integrated with an application, combined with automated SBoM generation and HLF smart contracts, can create an immutable audit trail. Logging Git commits and SBoM differences on a blockchain ledger provides traceability and supports forensic analysis of the software supply chain discussed in Sect. 4.1. Section 4.2 shows the forensic capabilities of the study.

4.1 Case Study

The demonstration uses a custom C++ library called say (with a header say.h and implementation say.cpp) and a hello application that uses this library with Git and HLF. The say package is put into Conan's "editable" mode, meaning hello will use the local source code of say directly during development. Conan is a popular, open-source package manager for C and C++ developers, designed to simplify the management of dependencies and binaries in C++ projects [62]. This setup mirrors a real development scenario where an application depends on a library that is being actively developed. HLF is an open-source, permissioned blockchain framework specifically developed for enterprise applications. HLF has a special feature that allows an external application to interact with the fabric network. The system architecture is illustrated in Fig. 6.

Hyperledger Fabric is an open-source, modular blockchain framework designed for building enterprise applications with a focus on permissioned networks, modularity, and scalability, allowing for private and confidential transactions. Fabric presents an execute-order-validate mechanism segmenting the transaction flow into three steps: executing a transaction and verifying its correctness to endorse

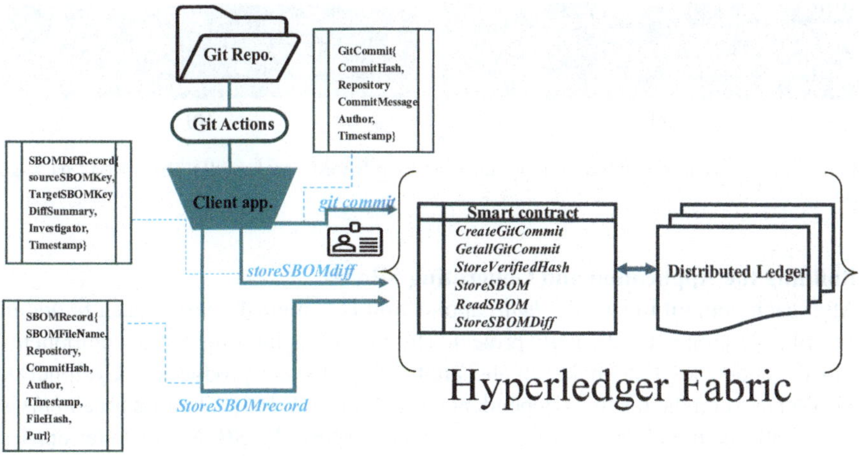

Fig. 6 System design of the case study

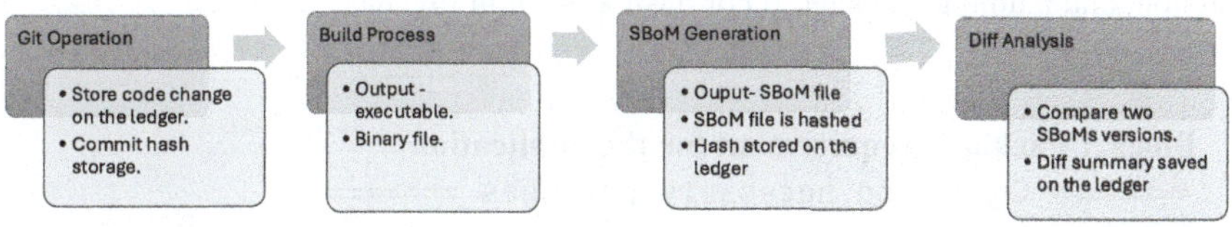

Fig. 7 Workflow of process

it, ordering transactions through a pluggable consensus protocol, and validating transactions against a specific endorsement policy before committing to the ledger.

Tracking Code Commits on Hyperledger Fabric Ledger
To ensure every code change in the *say* library is recorded, a Git post-commit hook is configured in the say repository. Whenever a developer commits changes to say.cpp (or any part of the *say* package), this hook automatically invokes a Hyperledger Fabric smart contract function called *CreateGitCommit*. The *CreateGitCommit* function creates a new transaction on the blockchain ledger that stores the commit metadata: details like the commit hash, author, timestamp, commit message, and identifier for the *say* package at that point. Using HLF for this means that each commit is recorded in an immutable ledger entry [63].

By having this on a blockchain, the history of changes to the say component is tamper-proof and auditable. No one can alter or delete past commit records, and all stakeholders (with permission) can verify the sequence of changes. This level of logging is crucial for forensics because it ties every code change to an accountable identity and a timeline of the software's development.

(a) (b)

Fig. 8 Comparison of graph hash representations. (**a**) SBoM1 package ID. (**b**) SBoM2 package ID

Building the Application and Generating SBoMs

After each commit to *say*, the *hello* application is compiled. During each build, an SBoM is generated for the *hello* project. The SBoM captures the list of components and dependencies that make up the application. Conan produces a CycloneDX SBoM based on the resolved dependency graph of *hello* [64]. The first time *hello* is built (with the initial version of *say*), we generate SBoM1. SBoM1 includes entries for the application itself and the *say* library and any other libraries present as seen in Block 1. Each component in the SBoM has identifying information such as name, version, and a unique package ID or hash as seen in Fig. 8a.

Block 1 Packages requirement for the application

```
======= Computing necessary packages =======
Requirements
openssl/1.1.1u#2cb3689d473c8582c989f95915d449c5:
519af2652d4d356cb8131159d7bac54ecb13b24
9#8a7718f79f0ca4a3907dd3db0b976a3 - Cache
say/1.0:308ccc1b7065dbcf54f86f902c83ab6fcd18d025 -
Editable
```

To secure this information, the smart contract's function, *storeSBoM*, is invoked after the build. The *storeSBOM* function can take the cryptographic hash of the SBoM and record it on the ledger. Storing the full SBOM ensures that the exact details of the build's components are preserved immutably; storing a hash ensures integrity (one can later prove a given SBOM file matches what was recorded). For instance, after the initial build, *storeSBOM* log a hash of SBoM1 along with its metadata. At this point, the ledger has a record of both the commit (from *CreateGitCommit*) and the resulting SBoM (from *storeSBOM*). This establishes a trusted baseline for the software: we know exactly what code and components were used.

Making a Code Change and Comparing SBoMs

To simulate an update in the supply chain, the say.cpp file is then edited. In our case study, this was a general, non-malicious edit. After modifying the code, the developer commits the change to Git, triggering the *CreateGitCommit* function to

Fig. 9 Highlighting the difference in the SBOM_diff.txt File

log this commit on the ledger (with a new commit hash, message, timestamp, and author info). Now the ledger has a chain of two commits, for say, the original and this new change. Next, the project is rebuilt with the updated say library code. After this rebuild, a new SBoM (SBoM2) is generated as seen in Fig. 8b and stored via another *storeSBOM* call, ensuring every build's full SBOM is archived on the ledger. SBOM2 should be largely similar to SBoM1, but it will reflect any changes in the say component's identity. Indeed, the SBOM's entry for *say* now carries a different package hash to denote the updated content. To verify what changed, SBoM1 and SBoM2 are compared using a diff tool. Running a diff on the two SBoM files highlights the differences in the dependency graph before and after the code edit, as illustrated in the Box below. The diff output reveals that the *say* component in SBoM2 has a different identifier than in SBoM1, as seen in Fig. 9.

Compare Two SBoMs

```
diff -side-by-side SBoM1.json SBoM2.json > SBoM_diff.txt
```

This confirms that a change in the say library occurred and was picked up in the product's composition. In a more complex scenario, an SBoM diff could also reveal if new dependencies were added or if versions of sub-components changed, but in our case, the primary difference is the updated say library artifact. After obtaining the diff, the final smart contract function, *storeSBOMDiff*, is executed. This function takes the diff result (essentially the changes between SBoM1 and SBoM2) and records it on the blockchain ledger. By storing the SBOM differences on the ledger, we achieve an immutable record of how the software's bill of materials evolved due to a particular code change. Just like commit metadata, this diff can be later retrieved to analyze what exactly changed in the software's contents at that point in time.

Smart Contract Functions and Their Roles

To summarize, the HLF'S smart contract in this case study provides three key functions to enable this forensic tracking:

1. *CreateGitCommit*—Called on each Git commit to the say package. It creates a blockchain asset/record containing the commit's metadata: commit hash, author, timestamp, commit message, and an associated package/version reference. This links a source code change to a unique ID on the ledger, establishing an immutable history of code changes.

2. *storeSBOM*—Called after each build of the application. It stores the cryptographic hash of the SBOM on the ledger. By doing so, it records the state of the software's components at a given time (post-build). Each SBOM entry on-chain can be tied back to a specific commit or build trigger, enabling one to retrieve the exact composition of a release or build.
3. *storeSBOMDiff*—Called whenever two SBoMs are compared to record the differences, typically the "before" and "after" of a change. The diff might include which component was added, removed, or changed. In our case, it logs that the say package's ID changed from the old value to the new value. Capturing diffs is useful for quickly pinpointing the impact of a particular change without needing to manually inspect full SBoMs.

Together, these functions create a comprehensive ledger-backed audit trail: *CreateGitCommit* ties to the code repository changes, *storeSBOM* captures the product of those changes in terms of software ingredients, and *storeSBOMDiff* highlights the delta between states. This design means that for every change, one can answer: What changed in the code (see commit via *CreateGitCommit*)? What was the software made of at that time (see SBoM via *storeSBOM*)? And what was the difference from before (see diff via *storeSBOMDiff*)?

4.2 Key Forensic Capabilities and Practical Considerations

The SBoM-DLT framework improves forensic readiness in the software supply chain. The case study demonstrates an approach to software supply chain traceability and auditability that remains effective, even as AI transforms software development. The following summarizes the key capabilities:

- **Traceability:** The framework records every software change, whether a code commit or a dependency update, on a blockchain. If investigators detect a vulnerability or incident in the application, they can trace back through the history to determine when the vulnerable component entered the system. By correlating this with on-chain commit records, they can identify the responsible commit and contributor, enabling precise forensic analysis.
- **Accountability:** Developer identities and timestamps are stored immutably. If a malicious actor or insider makes unauthorized changes, the system logs the event permanently, making it impossible to repudiate. This built-in accountability deters tampering and supports attribution.
- **Integrity of SBoM Data:** By recording SBoMs and diffs on an immutable ledger, we ensure the integrity of those SBoMs. Even if someone had access to the artifact repository or build system, they could not secretly swap out a dependency without the SBoM diff revealing the change. The blockchain is a secure witness to the software's ingredients over time.
- **Forensic Analysis:** Investigators can retrieve SBoM diffs to rapidly identify changes that occurred between builds. If only a single unexpected component

appears in a diff, it becomes a focal point for deeper inspection. In more complex cases, the diff reveals new or altered libraries that might suggest malicious activity, accelerating root cause analysis.
- **Tracking AI-Generated Changes:** Our framework records each code version, whether authored by a human or generated by an AI, along with its dependencies and metadata. This enables analysts to pinpoint when AI-generated components were introduced, detect deviations via reproducible builds, and preserve forensic records for post-incident attribution and rollback.
- **Compliance and Trust:** For highly regulated sectors, the framework supports compliance by creating a verifiable audit trail of code and dependency changes. This helps organizations demonstrate adherence to evolving standards and maintain trust across partners and regulators.

5 Conclusion

This chapter addressed the growing need to enhance forensic readiness in modern software supply chains. We proposed a forensic framework that integrates SBoMs with DLTs, creating a tamper-evident audit trail that supports the detection, analysis, and attribution of unauthorized changes in software ecosystems.

Our case study demonstrated how smart contracts (*CreateGitCommit*, *storeSBOM*, *storeSBOMDiff*) facilitate secure, distributed recording of every build artifact and its provenance. The integration of SBoMs with blockchain enabled investigators to trace code commits, component updates, and dependency diffs in a unified forensic timeline. These capabilities strengthen traceability, accountability, and transparency, making it significantly more difficult for malicious actors to conceal their activities.

Looking ahead, future work should explore extending this framework with smart contracts that enforce reproducible builds across distributed nodes. This would enable earlier detection of discrepancies during the build process. In parallel, integrating reverse engineering techniques such as binary inspection and behavioral pattern analysis into SBOM workflows can help flag anomalies and improve incident response precision. Further, combining these insights with AI-powered knowledge graphs could enable automated reasoning over changes in the supply chain, enhancing both detection and attribution capabilities [37].

As software development increasingly incorporates AI-generated code, forensic frameworks must evolve accordingly. Our system remains effective in this context by capturing all code changes, whether authored by humans or AI, and linking them to immutable metadata stored on-chain. This makes it possible to identify when AI-generated components were introduced, verify their integrity through reproducible builds, and preserve forensic records for post-incident analysis. By extending provenance tracking to include AI-assisted contributions, our framework addresses the emerging challenges of AI-enabled development while reinforcing forensic readiness across the software supply chain. In summary, our framework represents a step toward revolutionizing forensic investigations in software supply chains. By combining the strengths of SBoMs, DLTs, and smart contracts, we pave the way for systems that are more resilient against attacks and better equipped to uncover and trace the origins of security breaches.

Acknowledgments This work was supported by Clemson University's Virtual Prototyping of Autonomy Enabled Ground Systems (VIPR-GS), under Cooperative Agreement W56HZV-21-2-0001 with the US Army DEVCOM Ground Vehicle Systems Center (GVSC). Distribution A: Approved for public release; distribution unlimited. OPSEC#9720

Competing Interests The authors have no conflicts of interest to declare that are relevant to the content of this chapter.

References

1. Gudlur, V.V.R., Shanmugan, V.A., Perumal, S., Mohammed, R.M.S. (2019). IoT-Supply Chain Forensics and Vulnerabilities. In: *Proc. of the 1st International Conference on Artificial Intelligence and Data Sciences (AiDAS)*, pp. 106-109. Available: https://api.semanticscholar.org/CorpusID:210993454
2. Ohm, M., Sykosch, A., Meier, M. (2020). Towards Detection of Software Supply Chain Attacks by Forensic Artifacts. In: *Proc. of the 15th International Conference on Availability, Reliability and Security*. Available: https://api.semanticscholar.org/CorpusID:220872100
3. Barabanov, A.V., Markov, A.S., Tsirlov, V.L. (2020). On Systematics of the Information Security of Software Supply Chains. Available: https://api.semanticscholar.org/CorpusID:234673114
4. Attaran, M.: Digital technology enablers and their implications for supply chain management. Supply Chain Forum: An International Journal, **21**(3), 158–172 (2020)
5. Mustafa, M.M., Kishore, T.C., Krithika, N., Bharathi, L.M. (2024). Forensic Evidence Management Using Blockchain Technology. *International Journal of Innovative Science and Research Technology (IJISRT)*, IJISRT24APR354, pp. 268-273. 10.38124/ijisrt/IJISRT24APR354. Available: https://www.ijisrt.com/forensic-evidence-management-using-blockchain-technology
6. Ozkan, C., Zou, X., Singelée, D. (2024). Supply Chain Insecurity: The Lack of Integrity Protection in SBOM Solutions. *ArXiv preprint*, arXiv:2412.05138. Available: https://api.semanticscholar.org/CorpusID:274581772
7. Nocera, S., Romano, S., Di Penta, M., Francese, R., Scanniello, G. (2023). Software Bill of Materials Adoption: A Mining Study from GitHub. In: *Proc. of the 2023 IEEE International Conference on Software Maintenance and Evolution (ICSME)*, pp. 39-49. Available: https://api.semanticscholar.org/CorpusID:266197569
8. Haddad, I. (2024). Strengthening License Compliance and Software Security with SBOM Adoption: A Definitive SBOM Guide for Enterprises. Available: https://api.semanticscholar.org/CorpusID:274870392
9. Hendrick, S. (2022). Software Bill of Materials (SBOM) and Cybersecurity Readiness. Available: https://api.semanticscholar.org/CorpusID:274911699
10. The White House (2021). Executive Order 14028 on Improving the Nation's Cybersecurity. [Online; published 12-May-2021, accessed 24-Mar-2024]. Available: https://www.whitehouse.gov/briefingroom/presidential-actions/2021/05/12/executive-order-on-improving-thenations-cybersecurity/
11. Melara, M.S., Bowman, M. (2022). What is software supply chain security? *ArXiv preprint*, arXiv:2209.04006.

12. O'Donoghue, E., Reinhold, A.M., Izurieta, C. (2024). Assessing Security Risks of Software Supply Chains Using Software Bill of Materials. In: *2024 IEEE International Conference on Software Analysis, Evolution and Reengineering - Companion (SANER-C)*, pp. 134–140.
13. Wirth, A. (2020). CyberInsights: Talking About the Software Supply Chain.
14. OX Security. (2024). Supply Chain Security for Software: What You Need to Know. [Online; published 20-Nov-2024]. Available: https://www.ox.security/supply-chain-security-for-software-what-you-need-to-know/
15. Synopsys. (2024). *Open Source Security and Risk Analysis Report*. Available: https://github.com/jacobdjwilson/awesome-annual-security-reports/blob/main/Annual%20Security%20Reports/2024/Synopsys-Open-Source-Risk-Analysis-Report-2024.pdf
16. Xia, B., Bi, T., Xing, Z., Lu, Q., Zhu, L. (2023). An Empirical Study on Software Bill of Materials: Where We Stand and the Road Ahead. In: *Proc. of the 2023 IEEE/ACM 45th International Conference on Software Engineering (ICSE)*, pp. 2630–2642.
17. Herman, M. (2024). Software Bills of Materials in Production Environments.
18. Google Cloud, "Delivering software securely." [Online]. Available: https://cloud.google.com/resources/delivering-software-securely-whitepaper
19. B. Bensing, "History of the Software Bill of Materials (SBOM)." [Online]. Available: https://billbensing.com/software-supply-chain/history-software-bill-of-material-sbom/
20. Furneaux, N.: *Investigating Cryptocurrencies: Understanding, Extracting, and Analyzing Blockchain Evidence*. John Wiley & Sons, 2018.
21. Tschorsch, F., Scheuermann, B.: Bitcoin and beyond: A technical survey on decentralized digital currencies. IEEE Communications Surveys & Tutorials **18**(3), 2084–2123 (2016)
22. Aideyan, I., Bhat, N., Yu, L., Altarawnah, A., Pesé, M.D., Brooks, R.R.: Distributed Ledger Consensus Security: Correctness, Agreement, and Sustainability Threats. ACM Distributed Ledger Technologies (DLT) (2025), Under Review.
23. Martinez, J.J., Durán, J.M. (2021). Software supply chain attacks, a threat to global cybersecurity: SolarWinds' case study. *International Journal of Safety and Security Engineering*. [Online]. Available: https://api.semanticscholar.org/CorpusID:244056951
24. Fitzgerald, A. (2025). Software Supply Chain Security. Secureframe. [Online]. Available: https://secureframe.com/blog/software-supply-chain-security
25. CISA. (2025). Kaseya Ransomware Attack: Guidance for Affected MSPs and their Customers. Cybersecurity & Infrastructure Security Agency. [Online]. Available: https://www.cisa.gov/news-events/news/kaseya-ransomware-attack-guidance-affected-msps-and-their-customers
26. Codecov by Sentry. (2021). Post-Mortem / Root Cause Analysis (April 2021). [Online]. Available: https://about.codecov.io/apr-2021-post-mortem/
27. Frenkel, Britt, *Vibe Coding: Shipping Features or Shipping Vulnerabilities?*, Oligo Security, March 28, 2025, https://www.oligo.security/blog/vibe-coding-shipping-features-or-shipping-vulnerabilities?utm_source=chatgpt.com.
28. Paganini, Pierluigi, *Rules File Backdoor: AI Code Editors Exploited for Silent Supply Chain Attacks*, Security Affairs, March 19, 2025, https://securityaffairs.com/175593/hacking/rules-file-backdoor-ai-code-editors-silent-supply-chain-attacks.html?utm_source=chatgpt.com.
29. Miller, David J. and Xiang, Zhen, *Adversarial Learning and Secure AI*, Cambridge University Press, 2023.
30. P. Ferreira, F. Caldeira, P. Martins, and M. Abbasi, "Log4j vulnerability," in *International Conference on Information Technology & Systems*, Springer, 2023, pp. 375–385.
31. S. Snider, "Massive Okta breach: What CISOs should know," *InformationWeek*, Dec. 2023. Accessed: February 18, 2025. [Online]. Available: https://www.informationweek.com/cyber-resilience/massive-okta-breach-what-cisos-should-know
32. D. Bradbury, "Tracking unauthorized access to Okta's support system," *Okta Security Blog*, Oct. 2023. Accessed: February 18, 2025. [Online]. Available: https://sec.okta.com/articles/2023/10/tracking-unauthorized-access-oktas-support-system/
33. D. Bradbury, "Okta October 2023 security incident investigation closure," *Okta Security Blog*, Feb. 2024. Accessed: February 18, 2025. [Online]. Available: https://sec.okta.com/articles/harfiles/

34. K. R. K. Reddy, A. Gunasekaran, P. Kalpana, V. R. Sreedharan, and S. A. Kumar, "Developing a blockchain framework for the automotive supply chain: A systematic review," *Computers & Industrial Engineering*, vol. 157, p. 107334, 2021.
35. L. Courtès, "Building a secure software supply chain with GNU Guix," *arXiv preprint*, arXiv:2206.14606, 2022.
36. G. Guix, "GNU Guix," [Online]. Available: https://guix.gnu.org/. Accessed: August 27, 2024.
37. Iyengar, S. S., Nabavirazavi, S., Rathore, H., Hariprasad, Y., & Chaudhary, N. K. (2024). Advancing Forensic Science: AI and Knowledge Graphs Unlock New Insights.
38. M. Čarnogurský, *Attacks on Package Managers*, Master's thesis, Masaryk University, Faculty of Informatics, 2019.
39. strace, *strace - the Linux syscall tracer*, 2020. [Online]. Available: https://strace.io/. Accessed: March 21, 2025.
40. Stichting Cuckoo Foundation, *Cuckoo Sandbox - Automated Malware Analysis*, 2019. [Online]. Available: https://cuckoosandbox.org. Accessed: January 21, 2025.
41. Innocence Project. (2020). *Faulty forensic evidence*. [Online]. Available: https://www.innocenceproject.org/causes/faulty-forensic-evidence/
42. Sai, Y. R. K. M. (2022). Advancing forensic science: Addressing challenges and embracing emerging technologies. *Forensic Science Today*, 8(1), 001–005. https://doi.org/10.17352/fst.000023
43. Black, P., Gondal, I.: AFES: An advanced forensic evidence system. In: *Proceedings of the 2021 IEEE 25th International Enterprise Distributed Object Computing Conference (EDOC)*, pp. 67–74. IEEE (2021)
44. Fakhouri, H. N., AlSharaiah, M. A., Al Hwaitat, A. K., Alkalaileh, M., & Dweikat, F. F. (2024). Overview of Challenges Faced by Digital Forensic. In *2024 2nd International Conference on Cyber Resilience (ICCR)* (pp. 1–8). https://doi.org/10.1109/ICCR61006.2024.10532850
45. Griffin, C., Brooks, R.: A note on the spread of worms in scale-free networks. *IEEE Transactions on Systems, Man, and Cybernetics, Part B (Cybernetics)* **36**(1), 198–202 (2006)
46. Garfinkel, S. L., Parker-Wood, A., Huynh, D., & Migletz, J. (2010). An automated solution to the multiuser carved data ascription problem. *IEEE Transactions on Information Forensics and Security*, 5(4), 868-882.
47. Garfinkel, S. L., & Migletz, J. (2009). New XML-based files: implications for forensics. *IEEE Security & Privacy Magazine*, 7(2), 38-44.
48. Khan, A. A., Shaikh, A. A., Laghari, A. A., Dootio, M. A., Rind, M. M., & Awan, S. A. (2022). Digital forensics and cyber forensics investigation: security challenges, limitations, open issues and future direction. *International Journal of Electronic Security and Digital Forensics*, 14(2), 124-150.
49. Wagner, S. M., & Bode, C. (2006). An empirical investigation into supply chain vulnerability. *Journal of Purchasing and Supply Management*, 12, 301–312. https://doi.org/10.1016/J.PURSUP.2007.01.004
50. Dangi, S., Ghanshala, K., & Sharma, S. (2024). HOLMES: Artificial Intelligence Enabled Expert System for Efficient Acquisition and Analysis Using Live Memory Forensics. In *2024 International Conference on Innovations and Challenges in Emerging Technologies (ICICET)* (pp. 1–5). IEEE. https://doi.org/10.1109/ICICET59348.2024.10616275
51. Ohm, M., Sykosch, A., & Meier, M. (2020). Towards detection of software supply chain attacks by forensic artifacts. In *Proceedings of the 15th International Conference on Availability, Reliability and Security* (pp. 1–6).
52. Bi, T., Xia, B., Xing, Z., Lu, Q., Zhu, L. (2023). On the way to SBOMs: Investigating design issues and solutions in practice. *ACM Transactions on Software Engineering and Methodology*. https://doi.org/10.1145/3654442
53. Wu, W., Wang, P., Zhao, L., Jiang, W. (2023). An intelligent security detection and response scheme based on SBOM for securing IoT terminal devices. In: *Proceedings of the 2023 IEEE 11th International Conference on Information, Communication and Networks (ICICN)*, pp. 391–398. https://doi.org/10.1109/ICICN59530.2023.10393435

54. Sbom Scope. (2021). Standardized SBOM Formats. Available online: https://api.semanticscholar.org/CorpusID:236942115
55. O'Donoghue, E., Reinhold, A.M., Izurieta, C. (2024). Assessing Security Risks of Software Supply Chains Using Software Bill of Materials. In: *2024 IEEE International Conference on Software Analysis, Evolution and Reengineering - Companion (SANER-C)*, pp. 134–140. https://doi.org/10.1109/SANER-C62648.2024.00023
56. Red Hat, Inc. (2024). CVE-2024-3094 Detail. CVE Dictionary Entry, National Vulnerability Database (NVD), U.S. National Institute of Standards and Technology. Published: March 29, 2024; Last Modified: February 6, 2025. Available online: https://nvd.nist.gov/vuln/detail/CVE-2024-3094
57. DATADOG Security Labs. The XZ Utils backdoor (CVE-2024-3094): Everything you need to know, and more. April 3, 2024. [Online]. Available: https://securitylabs.datadoghq.com/articles/xz-backdoor-cve-2024-3094/
58. Red Hat Product Security. CVE-2024-3094. Public on 28 March 2024. Last Modified: 6 February 2025 at 08:25:00 UTC. [Online]. Available: https://access.redhat.com/security/cve/cve-2024-3094
59. Xia, B., Zhang, D., Liu, Y., Lu, Q., Xing, Z., Zhu, L.: Trust in Software Supply Chains: Blockchain-Enabled SBOM and the AIBOM Future. In: *Proceedings of the 2024 IEEE/ACM 4th International Workshop on Engineering and Cybersecurity of Critical Systems and 2024 IEEE/ACM Second International Workshop on Software Vulnerability (EnCyCriS/SVM)*, pp. 12–19 (2023). https://api.semanticscholar.org/CorpusID:259342108
60. Lew, K.; Sarker, A.; Wuthier, S.; Kim, J.; Kim, J.; Chang, S.-Y. Distributed Software Build Assurance for Software Supply Chain Integrity. *Applied Sciences* **2024**, *14*, 9262. https://doi.org/10.3390/app14209262
61. Ulybyshev, D. A.; Villarreal-Vasquez, M.; Bhargava, B. K.; Mani, G.; Seaberg, S. F. F.; Conoval, P.; Pike, R.; Kobes, J. (WIP) Blockhub: Blockchain-Based Software Development System for Untrusted Environments. In *Proceedings of the 2018 IEEE 11th International Conference on Cloud Computing (CLOUD)*, 2018, pp. 582–585. https://api.semanticscholar.org/CorpusID:3907047
62. Conan Team: Packages in editable mode in Conan. Conan Documentation. Available online: https://docs.conan.io/2/tutorial/developing_packages/editable_packages.html (accessed 4 Apr 2025)
63. E. Androulaki, A. Barger, V. Bortnikov, C. Cachin, K. Christidis, A. De Caro, D. Enyeart, C. Ferris, G. Laventman, Y. Manevich, S. Muralidharan, C. Murthy, B. Nguyen, M. Sethi, G. Singh, K. Smith, A. Sorniotti, C. Stathakopoulou, M. Vukolić, S. W. Cocco, and J. Yellick, "Hyperledger Fabric: A Distributed Operating System for Permissioned Blockchains," in *Proceedings of the Thirteenth EuroSys Conference (EuroSys '18)*, Porto, Portugal, 2018, Art. no. 30, pp. 1–15. Association for Computing Machinery. doi: 10.1145/3190508.3190538.
64. Conan, "SBOM Generation with Conan Tools," Conan Documentation, [Online]. Available: https://docs.conan.io/2/reference/tools/sbom.html. [Accessed: 3 Apr. 2025].

PCAP-Forensics: An Automated Network Traffic Analysis Framework

Sapna V. M., Darshan Krishna Hegde, Nikhil Kumar C, Pradeep Kumar, Prasad B Honnavalli, and Nagasundari S

1 Introduction

With the widespread adoption of digital technologies, ensuring the security and integrity of network communications has become a pressing challenge. Cyber threats such as denial-of-service (DoS) attacks, unauthorized access, data breaches, and malware infections have increased in complexity, making traditional network security solutions insufficient [3]. Organizations rely on intrusion detection systems (IDS) and firewalls to monitor and secure their networks. However, most of these systems depend on signature-based detection, which limits their ability to identify novel attacks [5]. As cybercriminals continue to exploit zero-day vulnerabilities and advanced evasion techniques, security solutions must evolve to become more adaptive and intelligence-driven [17].

Traditional signature-based IDS solutions such as Snort and Suricata are effective at detecting known threats but struggle against previously unseen attack patterns. Rule-based security mechanisms also generate a high volume of false positives, leading to alert fatigue and making manual analysis inefficient.

To address these shortcomings, machine learning-based network anomaly detection has gained prominence in cybersecurity research [21]. Unlike traditional IDS, anomaly detection models identify suspicious activities based on deviations from normal network behavior, making them effective against zero-day attacks and novel cyber threats [22]. Recent advancements in artificial intelligence applications for

These authors "Darshan Krishna Hegde, C. Nikhil Kumar, and Pradeep Kumar" contributed equally to this work.

Sapna V. M. (✉) · D. K. Hegde · Nikhil Kumar C · P. Kumar · P. B Honnavalli · Nagasundari S
Department of Computer Science and Engineering, PES University, Bangalore, Karnataka, India
e-mail: sapnavm@pes.edu; prasadhb@pes.edu; nagasundaris@pes.edu

S. S. Iyengar et al. (eds.), *Artificial Intelligence Driven Forensics*, Signals and Communication Technology, https://doi.org/10.1007/978-3-031-98036-7_26

cybersecurity further emphasize the importance of machine learning-driven network forensics and threat detection [3, 23].

This research presents a network forensic analysis framework that integrates real- time packet capture, interactive traffic visualization, and anomaly detection using the Isolation Forest algorithm. The system aims to enhance network security monitoring and forensic investigations by:

1. Capturing and storing real-time network packets for forensic analysis.
2. Visualizing network behavior using interactive graphs and charts.
3. Detecting anomalies using machine learning to identify suspicious activities.
4. Generating automated forensic reports summarizing network trends and security threats.

2 Related Work

The field of network traffic analysis and intrusion detection has been extensively studied in cybersecurity research [24]. Traditional approaches rely on signature-based and rule-based detection systems, while recent advancements focus on statistical anomaly detection and machine learning techniques [25].

2.1 Signature-Based Intrusion Detection

Signature-based IDS solutions, such as Snort and Suricata, compare incoming network traffic against a database of known attack signatures [4]. While these systems are effective at detecting well-documented threats, they are inherently incapable of identifying novel attacks [5]. The primary limitation of signature-based detection is its dependence on predefined rules, which must be constantly updated to keep pace with evolving cyber threats [6]. Additionally, these systems generate a high volume of false positives, requiring manual intervention from security analysts to verify alerts [7].

2.2 Deep Learning-Based Approach

To overcome the limitations of signature-based detection, researchers have explored statistical methods to identify anomalies in network traffic. Entropy-based and deviation-based models attempt to establish a baseline of normal network behavior and flag deviations as potential threats [8]. However, statistical methods often suffer from high false-positive rates, as normal fluctuations in traffic patterns can be misclassified as anomalies [9].

2.3 *Machine Learning for Intrusion Detection*

With advancements in artificial intelligence, machine learning-based network security solutions have gained popularity. Various supervised and unsupervised learning models have been applied to detect anomalies in network traffic [1].

- **Supervised learning approaches**, such as support vector machines (SVM) and random forest classifiers, require labeled datasets to distinguish between normal and malicious traffic [10]. However, obtaining labeled network traffic data is challenging and often impractical in real-world scenario's [15].
- **Unsupervised learning models**, such as autoencoders, K-means clustering, and Isolation Forest, are well-suited for anomaly detection since they do not require labeled datasets [11]. Among these, Isolation Forest has proven to be highly effective for identifying outliers in large-scale network traffic datasets.

2.4 *Isolation Forest for Anomaly Detection*

The Isolation Forest algorithm is an unsupervised learning technique designed for efficient anomaly detection in high-dimensional datasets [11]. Unlike traditional clustering methods, Isolation Forest isolates anomalous instances by recursively partitioning the feature space, making it computationally efficient for large-scale network monitoring. Several studies have demonstrated that Isolation Forest outperforms traditional anomaly detection methods, such as Gaussian mixture models (GMM) and One-Class SVM, in detecting network intrusions [12].

2.5 *Objective*

The primary objective of this project is to develop a comprehensive network forensic analysis framework that integrates real-time packet capture, traffic visualization, and machine learning-based anomaly detection to enhance network security monitoring and forensic investigations.

The key objectives of this research are:

1. **Real-Time Packet Capture and Storage:** Implementing a packet sniffing mechanism using Scapy to capture live network traffic and store it in PCAP format for further analysis.
2. **Network Traffic Analysis and Visualization:** Developing interactive visualization techniques, including protocol distribution charts, traffic volume histograms, top talkers analysis, and port monitoring, to provide insights into network behavior and potential security threats.

3. **Anomaly Detection Using Machine Learning:** Integrating the Isolation Forest algorithm to identify suspicious network activities, such as SYN flood attacks, port scanning attempts, and anomalous traffic spikes, by analyzing deviations from normal network behavior.
4. **Automated Forensic Reporting:** Generating structured reports summarizing key network events, detected anomalies, and potential security threats to assist in forensic investigations and incident response.

3 Methodology

The network forensic analysis framework consists of three main components: packet capture and storage, traffic analysis and visualization, and anomaly detection using machine learning [1, 26]. With AI-driven approaches playing a critical role in modern network security solutions, frameworks leveraging artificial intelligence for forensic investigations are becoming increasingly sophisticated [2, 28, 29]. Each component plays a crucial role in ensuring comprehensive network monitoring and forensic analysis [10, 12] .

Network traffic analysis typically follows a structured workflow beginning with packet capture from the network, saving the packets for later inspection, and performing both preliminary and deep packet analysis. Based on the findings, the system may optionally generate visualizations such as protocol distribution, traffic volume histograms, identification of top talkers, packet length distribution, and traffic by ports. The results are then compiled into automated reports and further processed using machine-learning-based anomaly detection to identify suspicious patterns and attacks. This sequential process enables comprehensive monitoring and forensic investigation of network data (Fig. 1).

3.1 Packet Capture and Storage

The foundation of network traffic analysis is packet capture, which involves monitoring, collecting, and storing network packets for forensic examination. Our system uses Scapy, a Python-based packet manipulation library, to sniff network packets in real time. Captured packets are stored in PCAP format, allowing investigators to analyze past network events [14].

For efficient data processing, PyShark is used to extract packet attributes from PCAP files. The extracted attributes include:

- Source and Destination IP addresses
- Protocol Type (TCP, UDP, ICMP, etc.)
- Packet Length and Payload Size
- Timestamp Information

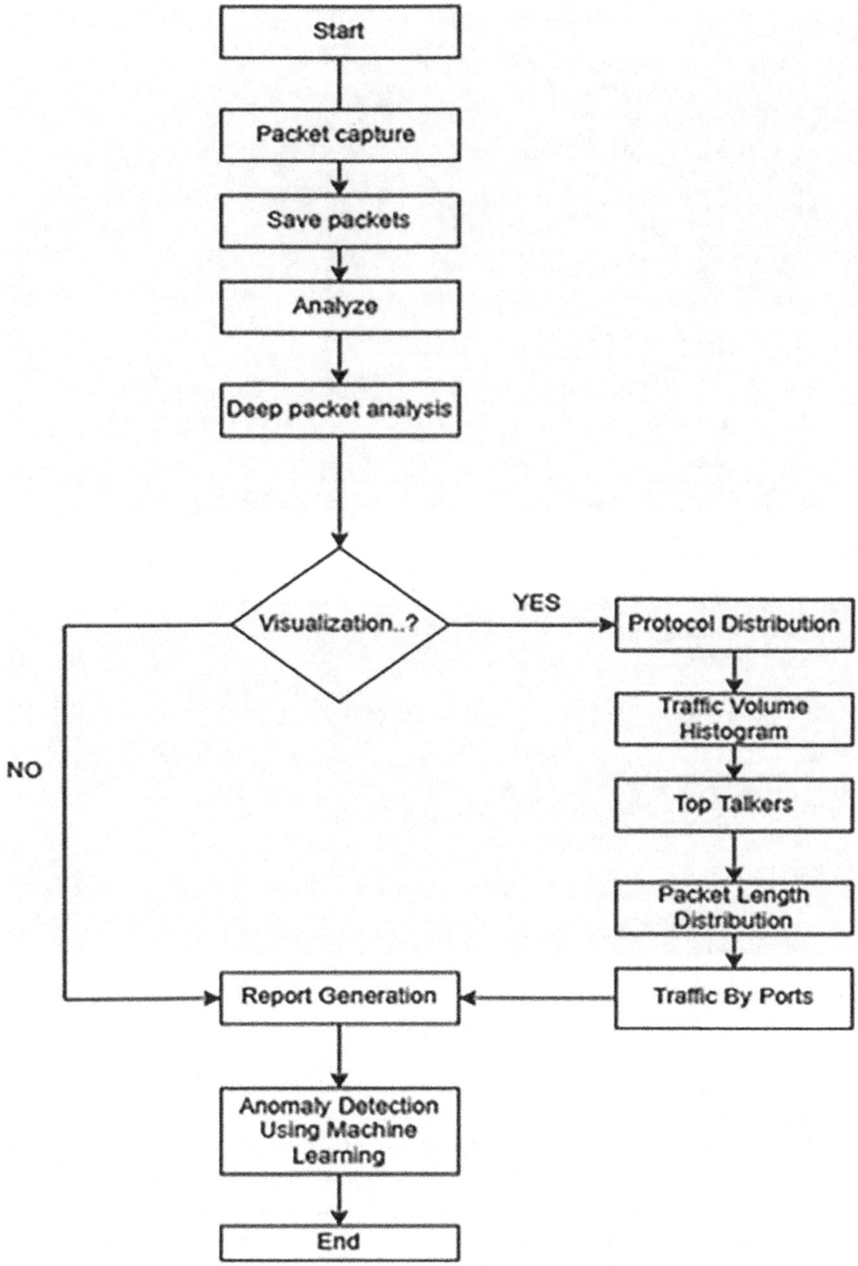

Fig. 1 Steps to analyze network traffic

Fig. 2 Detailed analysis of packets

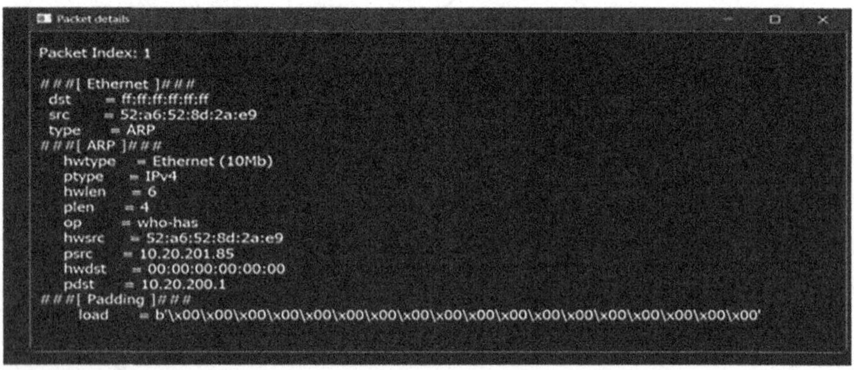

Fig. 3 Analysis of one packet

By storing packets systematically, our system enables in-depth forensic analysis of historical network data.

The interface first displays a detailed list of captured packets, showing key attributes such as source address, destination address, and protocol type. This high-level view enables quick identification of patterns, anomalies, or suspicious activity across multiple frames, providing an overview of the overall network (Figs. 2).

For deeper inspection, the tool allows the analyst to open an individual packet and view its full contents broken down field-by-field for each protocol layer, including Ethernet, ARP, IP, and others. This granular display reveals header values, payload size, and source/destination hardware addresses, offering a comprehensive understanding of the packet structure (Figs. 3).

3.2 *Traffic Analysis and Visualization*

To facilitate network monitoring and forensic investigations, our framework provides interactive visualizations of network traffic behavior. Each visualization method provides unique insights into network behavior, making it easier to detect suspicious activity before it escalates into a security incident.

3.2.1 Protocol Distribution Analysis

The protocol distribution analysis provides a pie chart representation of different network protocols present in the captured traffic, including Transmission Control Protocol (TCP), User Datagram Protocol (UDP), Internet Control Message Protocol (ICMP), and other network-layer protocols (Fig. 4).

3.2.2 Traffic Volume Monitoring

A histogram representation of network traffic volume is generated to track total data transmission over time. Traffic patterns are analyzed over a specific time window, allowing analysts to observe sudden surges in data transmission. By correlating traffic volume with protocol distribution and port activity, analysts can pinpoint the source of anomalous traffic and implement mitigation strategies such as rate limiting, IP blocking, or deep packet inspection (DPI) to prevent further damage (Fig. 5).

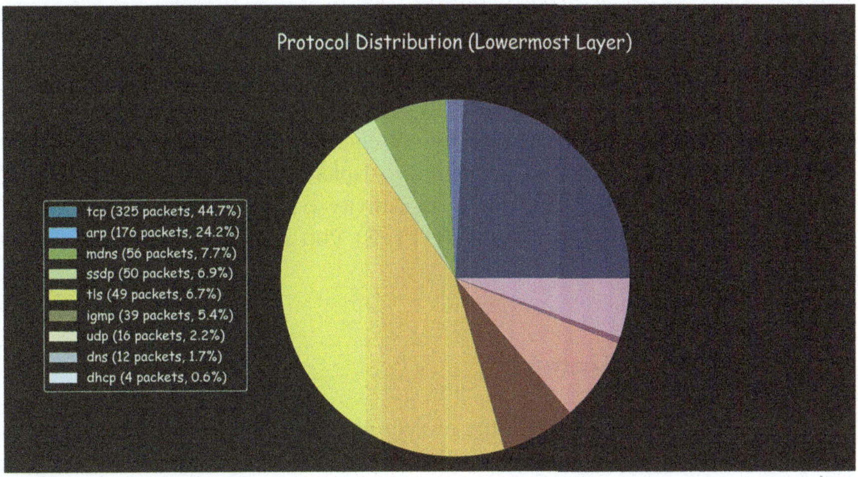

Fig. 4 Protocol distribution

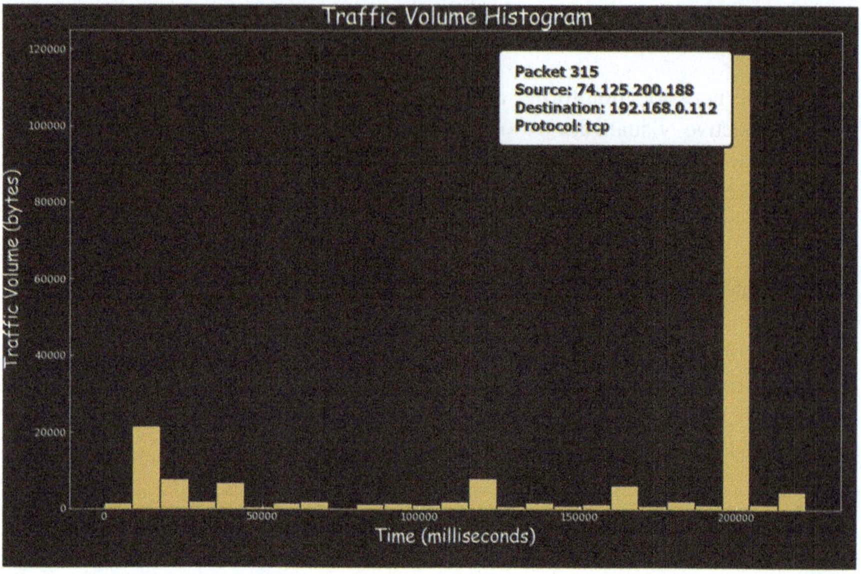

Fig. 5 Traffic volume histogram

3.2.3 Top Talkers

The top talkers analysis generates a bar chart listing the most active IP addresses in the network, ranked based on their traffic volume. This feature is useful for identifying compromised hosts, detecting unauthorized access, and analyzing network resource consumption (Fig. 6).

3.2.4 Port Usage Analysis

The port usage analysis generates a funnel chart depicting the most frequently accessed ports, allowing analysts to identify suspicious port activity and unauthorized scanning attempts. Normal network activity typically involves well-known ports, such as Port 80 (HTTP), Port 443 (HTTPS), Port 22 (SSH), and Port 53 (DNS) (Fig. 7).

3.2.5 Packet Length Analysis

The packet length analysis tab in the project visualizes the distribution of packet sizes captured from the network traffic. It provides insights into how data is transmitted by displaying a histogram of packet lengths. By hovering over the visualization, users can see detailed information about individual packets, including

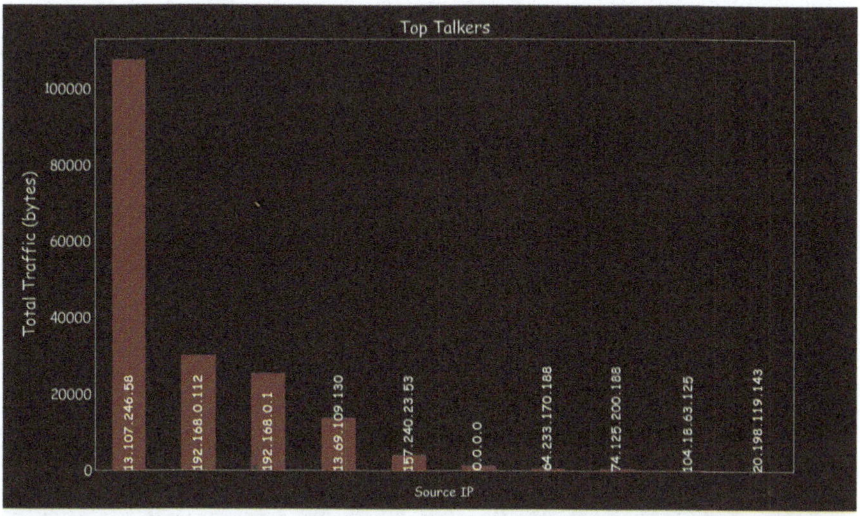

Fig. 6 Top talkers

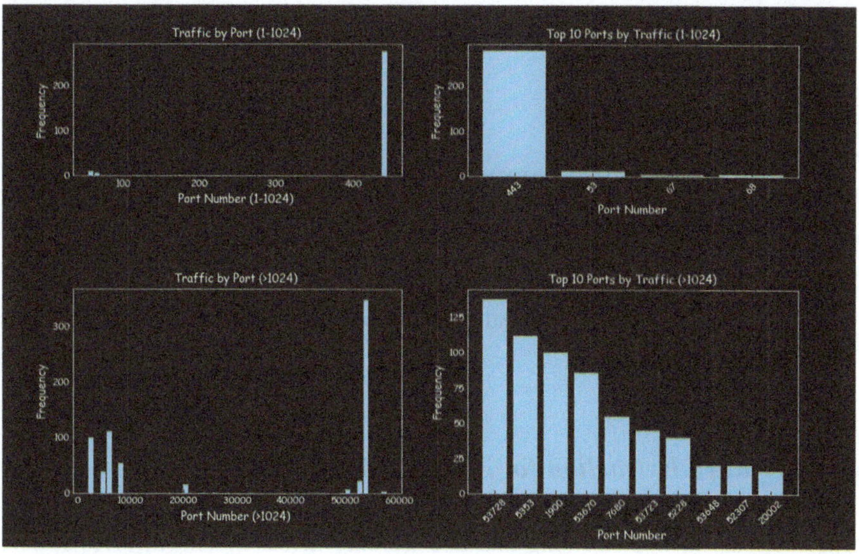

Fig. 7 Port usage analysis

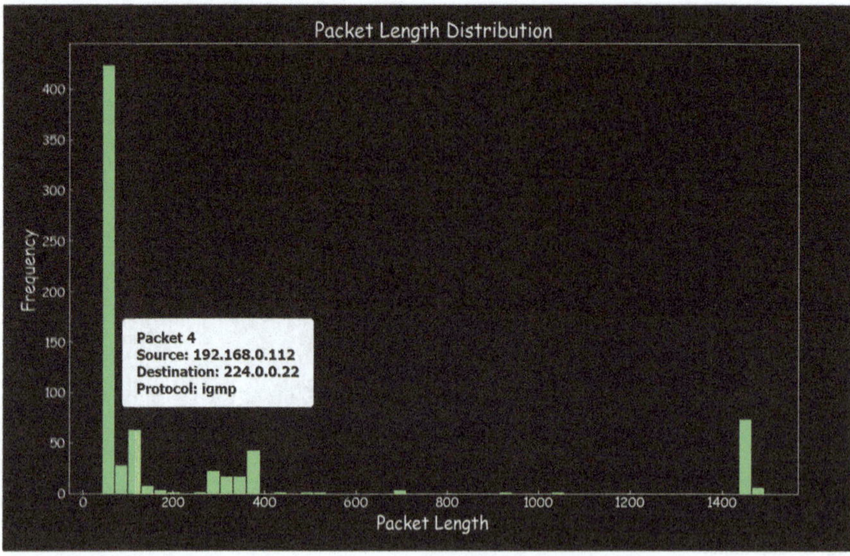

Fig. 8 Packet length distribution

their source, destination, and protocol type, aiding in forensic analysis and real-time network monitoring (Fig. 8).

4 Experimentation: Anomaly Detection Using Machine Learning

The methodology is detailed in sequential stages for each architecture, describing each key component's function and contribution to the anomaly detection process [1].

4.1 Feature Extraction for Anomaly Detection

The first step in implementing machine learning in our system involves feature extraction from network packets [11, 27]. Every network packet captured through Scapy and processed via PyShark contains multiple attributes that describe its behavior, including source and destination IP addresses, protocol type, timestamp, and payload size [13]. However, not all packet attributes are equally useful for anomaly detection [16]. To improve model performance, we extracted the following key network features that serve as input to the Isolation Forest algorithm:

- **Protocol Type (Numerically Encoded):** Different network protocols (TCP, UDP, ICMP) behave uniquely. Encoding them numerically allows the model to recognize protocol-based anomalies.
- **Relative Timestamp of Network Events:** The time difference between consecutive packets provides insight into packet burst patterns, which can help identify DoS attacks or scanning behaviors.

4.2 Training the Isolation Forest Model

Once the dataset is preprocessed, we train the Isolation Forest model to learn normal network behavior and detect anomalies. Isolation Forest works by randomly selecting features and partitioning the dataset recursively until each sample is isolated. Anomalous samples get isolated faster, leading to shorter path lengths, while normal samples remain within dense clusters with longer path lengths.

We utilize the Isolation Forest algorithm, a widely used unsupervised learning technique, to identify anomalous packets. The model is trained on normal traffic patterns and assigns an Anomaly Score to each packet:

- Normal packets are assigned a score of 1.
- Anomalous packets (potential threats) receive a score of -1.

A contamination factor of 5% is used, assuming a small percentage of network traffic is anomalous.

4.3 Anomaly Reporting and Analysis

The detected anomalies are saved into an `anomalies.csv` file for further forensic analysis. This dataset provides insights into:

- Unusual packet lengths or protocol usage.
- Spikes in network traffic that may indicate a DDoS attack.
- Uncommon communication between source and destination IPs, potentially signaling malware activity.

Our anomaly detection system enables real-time threat detection, assisting in network security monitoring and cyber forensic investigations.

5 Results Obtained During Experimentation

To evaluate the effectiveness of our network forensic analysis framework, we conducted extensive experiments on real-world network traffic and simulated cyberattacks [3]. The primary goal was to assess the system's accuracy, efficiency, and reliability in detecting network anomalies [17]. The evaluation focused on anomaly detection performance, visualization effectiveness, and forensic analysis capabilities [5].

The experiments were conducted in a controlled environment, where normal network traffic and malicious activities were generated to assess how well the system could distinguish between the two [14]. **Legitimate Network Traffic:**

- Regular browsing activities, file transfers, and video streaming.
- Normal communication between devices using TCP, UDP, and ICMP protocols.

5.1 Visualization Effectiveness

The visualization dashboard played a crucial role in helping analysts quickly interpret network behavior and detect suspicious activities.

5.1.1 Protocol Distribution Analysis

- Helped identify an unexpected spike in ICMP traffic, indicating a potential network scanning attempt.
- Showed that 80% of normal traffic was TCP-based, while anomalous traffic had higher UDP and ICMP usage.

5.1.2 Traffic Volume Monitoring

- Successfully detected traffic surges corresponding to SYN flood attacks.
- Allowed analysts to correlate sudden spikes with attack timestamps for forensic analysis.

5.1.3 Top Talkers (Source IP Analysis)

- Clearly highlighted the IP addresses generating the most traffic.
- Allowed easy identification of malicious IPs performing port scans.

5.1.4 Port Usage Analysis

- Showed unusual activity on non-standard ports, helping detect possible exploitation attempts.

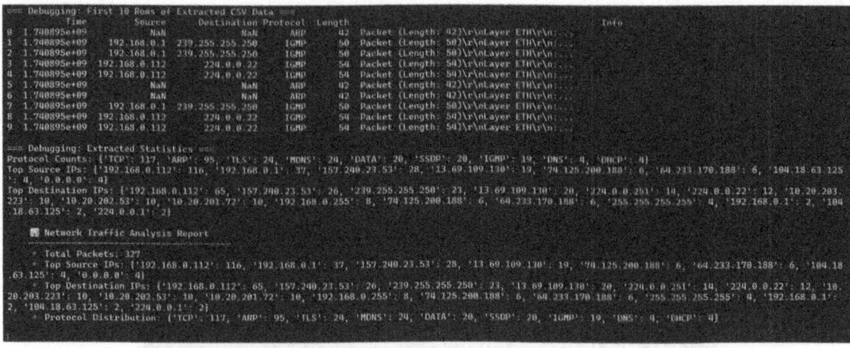

Fig. 9 Report generation

- Identified port scanning activity due to frequent access to multiple ports in a short time.

5.2 *Forensic Report Generation*

The system automatically generated structured forensic reports summarizing key findings, including:

- Timestamps of detected anomalies.
- List of top suspicious IP addresses.
- Breakdown of protocol anomalies and port activity.
- Packet metadata, such as source/destination IPs, payload size, and flags.

These reports were valuable for:

- Incident investigation and response.
- Tracking attack trends over time.
- Providing evidence for forensic analysis (Fig. 9).

5.3 *Real-Time Anomaly Detection and Classification*

The Real-Time Anomaly Detection and Classification module enables continuous monitoring and analysis of network traffic to identify abnormal behaviors and potential security threats as they occur [1]. This system leverages machine learning techniques, statistical analysis, and real-time packet inspection to classify anomalies into different categories, allowing for immediate response and forensic investigation (Fig. 10).

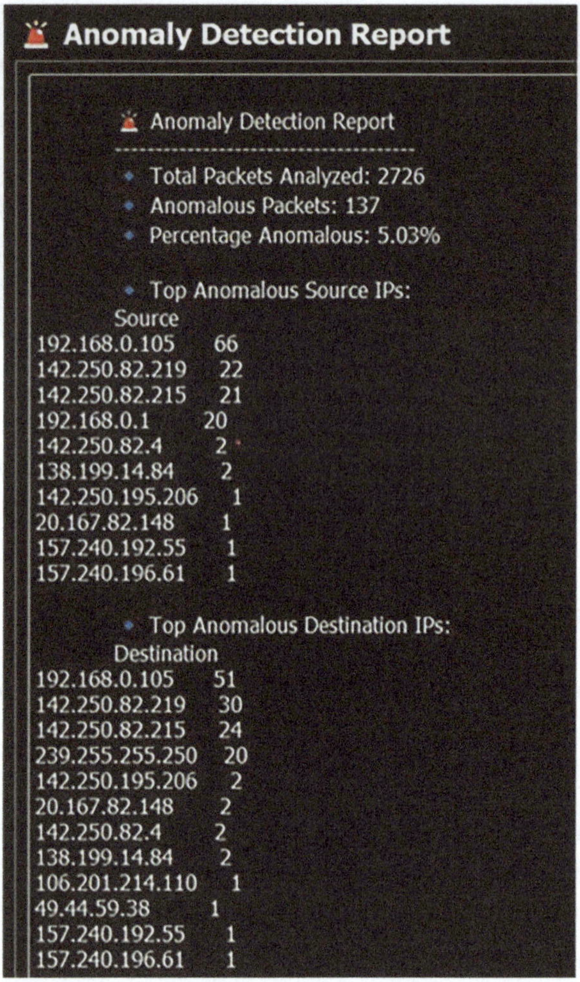

Fig. 10 Anomaly detection

5.4 *Comparison with Wireshark*

Network traffic analysis is a cornerstone of digital forensics, with Wireshark being the most widely used framework for packet inspection. However, as shown in Table 1, it lacks several features essential for forensic investigations–such as real-time monitoring with visualization, automated report generation, and machine-learning-based analysis.

Table 1 Comparision between wireshark and proposed framework

Feature	Proposed tool/framework	Wireshark
Real time monitoring	Provides live traffic visualization with an intuitive GUI	Requires extensive filtering and lacks built-in visualization
Packet capture	Uses scapy and pyshark for live capture	Uses winpcap/libpcap which offers deep packet capture
Packet analysis	Displays detailed packet information in a dedicated window for forensic review	Provides deep packet inspection, but in a raw text-based format
Protocol analysis	Supports protocol filtering with graphical insights	In-depth protocol breakdown, but lacks visual representation
Data visualization	Uses Matplotlib and seaborn to generate graphs, histograms and pie charts	No built-in visualization
Customization and extensibility	Can be extended with additional support	Requires scripting and plugin development
Report generation	Uses ML to generate the reports	Requires manual export and interpretation

6 Discussion

6.1 Framework Advantages

The network forensic analysis framework provides an integrated solution for network security monitoring by combining real-time packet capture, anomaly detection, and data visualization into a single framework [18]. The framework enhances forensic investigations by offering interactive visualizations such as protocol distribution analysis, traffic volume monitoring, and port activity tracking, allowing security analysts to quickly identify suspicious patterns [19]. The use of machine learning-based anomaly detection, specifically the Isolation Forest algorithm, automates the identification of threats like SYN flood attacks, port scans, and malformed packet injections, eliminating reliance on predefined signatures.

Furthermore, the automated forensic reporting feature provides detailed insights into network behavior and detected anomalies, improving the efficiency of cyber threat analysis and incident response [20].

6.2 Challenges and Limitations

Despite its advantages, the Network Forensic Analysis Tool (NFAT) faces several technical challenges, particularly in handling large-scale network traffic and optimizing processing efficiency. Memory usage and computational overhead can become significant concerns when analyzing high-volume network data, requiring further performance optimizations.

Additionally, the accuracy of anomaly detection depends on the quality of extracted network features, and tuning the Isolation Forest model to reduce false positives and false negatives required extensive refinement. Another limitation is the lack of a labeled dataset, which restricts the ability to train supervised learning models for more precise anomaly classification. The system also does not currently support real-time alerting, requiring manual intervention for security monitoring.

6.3 Future Work

Future work on this framework will focus on establishing a structure for continuous network traffic monitoring. Integrating advanced AI methodologies could further refine the anomaly detection process, reducing false positives and enhancing adaptability against emerging threats. This will enable proactive management of network performance and security, allowing users to detect and respond to threats in real-time.

The addition of continuous monitoring capabilities will further enhance the framework's utility, making it an even more valuable resource for digital forensic investigations.

7 Conclusion

With the rapid increase in cyber threats and sophisticated attack techniques, traditional signature-based intrusion detection systems (IDS) are no longer sufficient to effectively secure modern networks. The need for adaptive, intelligent, and real-time network monitoring solutions has led to the development of machine learning-based anomaly detection systems.

This research presented a network forensic analysis framework that integrates real-time packet capture, interactive traffic visualization, and machine learning-based anomaly detection to enhance network security monitoring and forensic investigations.

The proposed system successfully captured and analyzed network traffic using Scapy and PyShark, extracted meaningful features, and applied the Isolation Forest algorithm to detect anomalous network behaviors such as SYN flood attacks, port scanning, and malformed packet injections.

The visualization dashboard provided intuitive insights into protocol distribution, traffic volume trends, top talkers, and port activity, enabling security analysts to quickly identify potential security threats. The system's ability to generate detailed forensic reports further aids in post-incident investigations, making it a valuable framework for cyber-security professionals and forensic analysts.

Acknowledgments We would like to express our sincere gratitude to the US Army for their generous funding and unwavering support, which made the AI-Enabled National Conference on Forensic Investigations Network in Digital Sciences (FINDS) possible. Their commitment to advancing research and fostering collaboration is deeply appreciated.

We are profoundly grateful to Dr. S. S. Iyengar for his outstanding contributions to the field, which have been a constant source of inspiration to us. His pioneering work has shaped much of our research, and we are deeply thankful for the impact his research has had on our academic journey. His dedication and innovations continue to inspire us in our pursuit of knowledge.

References

1. Iyengar S. S., Nabavirazavi S., Hariprasad Y., Prasad H. B., Krishna Mohan C. *Artificial Intelligence in Practice: Theory and Applications for Cyber Security and Forensics*. Springer Nature, 2025.
2. Iyengar, S. S., & Geeta, S. *Data Auditing and Security in Cloud Computing: Issues, Challenges, and Future Directions*. Springer, 2023.
3. Shi, B., & Iyengar, S. S. *Mathematical Theories of Machine Learning: Theory and Applications*. Springer, 2019.
4. Roesch, M. "Snort: Lightweight Intrusion Detection for Networks." *Proceedings of USENIX LISA*, 1999, pp. 229–238.
5. Paxson, V. "Bro: A System for Detecting Network Intruders in Real-Time." *Computer Networks*, vol. 31, no. 23–24, 1999, pp. 2435–2463.
6. Sommer, R., & Paxson, V. "Outside the Closed World: On Using Machine Learning for Network Intrusion Detection." *IEEE Symposium on Security and Privacy*, 2010, pp. 305–316.
7. Garcia-Teodoro, P., Diaz-Verdejo, J., Macia-Fernandez, G., & Vazquez, E. "Anomaly-based Network Intrusion Detection: Techniques, Systems, and Challenges." *Computers & Security*, 28(1–2), 18–28, 2009.
8. Bhuyan, M. H., Bhattacharyya, D. K., & Kalita, J. K. "Network Anomaly Detection: Methods, Systems and Frameworks." *IEEE Communications Surveys & Tutorials*, 16(1), 303–336, 2014.
9. Lakhina, A., Crovella, M., & Diot, C. "Mining Anomalies Using Traffic Feature Distributions." *ACM SIGCOMM*, 2005, pp. 217–228.
10. Buczak, A. L., & Guven, E. "A Survey of Data Mining and Machine Learning Methods for Cyber Security Intrusion Detection." *IEEE Communications Surveys & Tutorials*, 18(2), 1153–1176, 2016.
11. Liu, F. T., Ting, K. M., & Zhou, Z. H. "Isolation Forest." *IEEE ICDM*, 2008, pp. 413–422.
12. Zhang, X., Li, Z., Wang, F., & Yuan, P. "Anomaly Detection in Network Traffic Using Isolation Forest." *Journal of Network and Computer Applications*, 154, 102538, 2020.
13. Sadreazami, H., Mohammadi, A., & Asif, M. S. "Deep Learning for Network Anomaly Detection: A Comparative Study." *IEEE Transactions on Information Forensics and Security*, 17, 2105–2117, 2022.
14. Ring, M., Wunderlich, S., Scheuring, D., Landes, D., & Hotho, A. "A Survey of Network-based Intrusion Detection Data Sets." *Computers & Security*, 86, 147–167, 2019.
15. Liu, F., & Chen, G. "Towards Network Anomaly Detection Using Graph Embedding." *IEEE Transactions on Network and Service Management*, 20(3), 1234–1245, 2023.
16. Kim, J., & Lee, S. "Network Forensics and Traffic Analysis With Machine Learning." *Cybersecurity*, 6, 22, 2023.
17. Wang, P., & Zhang, Q. "Cloud Network Anomaly Detection Using Machine and Deep Learning Techniques: Recent Research Advancements." *Journal of Cloud Computing*, 12, 56, 2023.
18. Chen, L., & Liu, Y. "A Comprehensive Investigation of Anomaly Detection Methods in Industrial Systems." *IET Cyber-Physical Systems: Theory & Applications*, 8(2), 123–134, 2023.

19. Jones, M., & Davis, S. "A Comprehensive Analysis of the Role of Artificial Intelligence and Machine Learning in Modern Digital Forensics and Incident Response." *Forensic Science International: Digital Investigation*, 45, 301–315, 2023.
20. Vert, G., Iyengar, S. S., & Phoha, V. *Introduction to Contextual Processing Theory and Application*. CRC Press, 2023.
21. Xavier, C., & Iyengar, S. S. *Introduction to Parallel Algorithms (Chinese)*. 263 pages, ISBN: 7-111-13390-0.
22. Xavier, C., & Iyengar, S. S. *Introduction to Parallel Algorithms*. John Wiley and Sons, 2000. 365 pages.
23. Soloway, E., & Iyengar, S. S. (eds.) *Empirical Studies of Programmers*. Ablex Pub. Co., Norwood, New Jersey.
24. Brooks, R. R., & Iyengar, S. S. *Multi Sensor Fusion: Fundamentals and Applications with Software*. Prentice Hall, New Jersey, 2000. 488 pages.
25. Patel, S. J., Chaudhary, N. K., Gohil, B. N., & Iyengar, S. S. *Information Security, Privacy and Digital Forensics - ICISPD 2022*. Springer Nature, ISBN: 978-98-19950-90-4.
26. Iyengar, S. S., & Boroojeni, K. G. *Oblivious Network Routing: Algorithms and Applications*. MIT Press, 2015, 176 pages. ISBN: 978-02-62029-15-5.
27. Iyengar, S. S., & Seetharaman, G. (in preparation). *The Design and Analysis of Algorithms for Processing Digital Satellite IR Images*.
28. Wang, C., Iyengar, S. S., & Sun, K. *AI Embedded Assurance for Cyber Systems*. Springer Nature, 2023. ISBN: 978-30-31426-36-0.
29. Singaram, J., Iyengar, S. S., & Madni, A. M. *Deep Learning Networks: Design, Development and Deployment*. Springer Nature, 2023. ISBN: 978-30-3139-24-3.

User Behavior Analysis Using Browsing History and to Support Forensics Investigation

Sapna V. M., Pradeep Kumar, Nikhil Kumar C, Pradeep Y. N., Harshith Reddy C, and Prasad B Honnavalli

1 Introduction

The rapid and continuous evolution of the digital landscape has profoundly transformed the way individuals interact with technology and consume information. With the proliferation of Internet-enabled devices and services, the Internet has become a vital part of daily life, facilitating activities such as social networking, communication, online shopping, education, entertainment, and information access. While this widespread dependence on digital platforms has unlocked unprecedented opportunities, it has also introduced numerous cyber threats that compromise users' online security, privacy, and trust [17].

The evolving cyber threat landscape is marked by increasing sophistication and complexity. Users are becoming more vulnerable to a wide range of malicious activities, including phishing attacks, malware infections, ransomware campaigns, and the exploitation of vulnerabilities in compromised websites [19]. Traditional security measures, which primarily depend on static, signature-based detection methods, struggle to keep pace with the dynamic and adaptive strategies employed by modern attackers. In this challenging environment, proactive and adaptive strategies are essential to safeguarding users (Iyengar et al., 2024).

Browsing history contains rich metadata that captures user interactions across diverse domains, revealing trends, patterns, and anomalies indicative of potential

These authors "Pradeep Kumar, Nikhil Kumar C, Pradeep Y. N, and Harshith Reddy C" contributed equally to this work.

Sapna V. M. (✉) · Pradeep Kumar · Nikhil Kumar C · Pradeep Y. N. · Harshith Reddy C · P. B Honnavalli
Department of Computer Science and Engineering, PES University, Bangalore, Karnataka, India
e-mail: sapnavm@pes.edu; prasadbh@pes.edu

S. S. Iyengar et al. (eds.), *Artificial Intelligence Driven Forensics*, Signals and Communication Technology, https://doi.org/10.1007/978-3-031-98036-7_27

risks or malicious activities. By leveraging this data, researchers and security professionals can develop systems capable of identifying and mitigating online threats before they result in significant harm [27].

1.1 Background

In recent years, cyberattacks have evolved significantly due to rapid technological advancements and the increasing interconnectedness of systems. Modern attackers employ sophisticated techniques such as automated bots, social engineering, and multi-layered strategies to bypass traditional security defenses [23]. For instance, phishing campaigns often use fraudulent emails or websites that mimic legitimate sources to deceive users into sharing sensitive information. Additionally, advanced persistent threats (APTs), fileless attacks, and zero-day exploits have become prevalent methods for malware propagation, often remaining undetected for extended periods [29].

The growing complexities of cyber threats have exposed the limitations of traditional security methods. Signature-based systems, which rely on predefined patterns to detect threats, are increasingly inadequate in addressing the dynamic and ever-evolving nature of cyber risks. To bridge this gap, there is a pressing need for advanced, data-driven approaches that can adapt in real-time to emerging threats. A key element of this transformation lies in understanding user behavior within the digital ecosystem [24, 25].

Users frequently interact with a diverse array of online platforms and domains, leaving behind digital footprints that reflect their habits, interests, and vulnerabilities [26]. Browsing history, in particular, serves as a valuable resource for identifying unusual behaviors and potential security risks. For instance, frequent visits to high-risk domains or anomalous URL access patterns could signal malicious intent or a susceptibility to cyberattacks [18]. Machine learning techniques have emerged as transformative tools for tackling these challenges [29]. Unlike traditional methods, ML algorithms can process vast datasets, identify patterns, and deliver highly accurate predictions.

1.2 Objective

The primary goal of this project is to develop and implement an automated system for analyzing user behavior, utilizing browsing history to classify users as either normal or abnormal. This classification serves as the foundation for identifying potential cybersecurity threats and enhancing online safety. By leveraging advanced machine learning techniques, particularly Random Forest and Extreme Gradient Boosting (XGBoost), the project aims to assess the risk associated with URLs visited by users and provide actionable recommendations to mitigate those risks.

The ultimate aim is to integrate robust machine learning models with intuitive data visualization tools to bridge the gap between technical threat detection and user-friendly reporting. This approach ensures that not only are malicious activities detected, but users are also equipped with the knowledge necessary to understand and improve their browsing habits.

2 Related Work

The increasing prevalence of malicious URLs and their impact on cybersecurity have significantly heightened interest in the fields of URL classification and user behavior analysis in recent years. Malicious URLs pose serious threats to individuals and organizations, serving as gateways for phishing attacks, malware dissemination, and data breaches. Consequently, substantial efforts have been dedicated to developing robust methods for accurately classifying URLs by leveraging advanced machine learning algorithms and detailed user behavior profiling.

2.1 URL Classification Using Machine Learning

Several studies have explored machine learning based approaches for URL classification, demonstrating their potential to outperform traditional rule-based and signature-based systems. Kundra et al. [1] proposed a comprehensive method for identifying and classifying benign and malicious URLs using machine learning classifiers. Their research highlighted the performance of various algorithms, emphasizing the importance of selecting optimal classifiers and features to enhance detection accuracy. Similarly, [2] conducted an in-depth analysis of machine learning models for malicious URL detection. Their findings underscored the critical role of feature selection techniques—such as mutual information analysis and recursive feature elimination—in improving model performance by isolating the most relevant features. He et al. [3] made another significant contribution in this field by analyzing the patterns of feature contribution in malicious URL identification. Their research highlighted the importance of particular characteristics in differentiating between safe and dangerous URLs, including domain age, URL length, and the presence of questionable keywords. Their method showed notable gains in classification accuracy by concentrating on targeted feature extraction, [10, 13] especially for real-world datasets. In a similar vein, [5] presented an intelligent URL identification system that combines several classifiers, including Random Forest, and uses ensemble learning approaches. This system's noteworthy detection rates demonstrated the effectiveness of ensemble techniques in tackling the many traits of malicious URLs [6, 8, 9].

2.2 AI in Cybersecurity and Forensic Analysis

Dr. S. S. Iyengar and his colleagues have made significant contributions to the application of artificial intelligence in cybersecurity and forensic investigations. In *Artificial Intelligence in Practice: Theory and Applications for Cyber Security and Forensics* [16] , the authors discuss AI-driven methodologies for digital forensics, including anomaly detection and behavioral analysis techniques that align closely with this study's focus on user behavior analysis through browsing history.

Similarly, *AI Embedded Assurance for Cyber Systems* (Wang et al., year) explores the integration of AI in securing digital ecosystems, emphasizing the importance of real-time monitoring and adaptive models. These principles are critical to the methodology adopted in our system, which leverages machine learning to classify browsing behavior as either normal or abnormal.

2.3 Contextual Processing for User Behavior Analysis

Contextual data processing has been a recurring theme in forensic studies. *Introduction to Contextual Processing—Theory and Application* (Vert et al., year) highlights the role of contextual cues in identifying behavioral anomalies. This aligns with our use of URL metadata (domain age, protocol type, and risk scores) to assess browsing behavior contextually, ensuring a more nuanced analysis of potential threats. Furthermore, in *Information Security, Privacy and Digital Forensics* [19], the authors discuss the ethical implications and methodologies in digital forensic investigations. This work supports our focus on privacy-conscious browsing behavior analysis, ensuring that forensic techniques align with ethical cybersecurity practices.

2.4 Machine Learning in Digital Forensics

In *Cyber Forensics* [20], the author explores methodologies for digital evidence collection and forensic analysis, emphasizing machine learning's role in cybersecurity. These principles align with our approach in analyzing user browsing history for forensic investigations. Additionally, *Video Origin Camera Identification using Ensemble CNNs of Positional Patches* (Iyengar et al., year) introduces an advanced machine learning technique for source identification, which parallels our work in tracking browsing behaviors for security assessment [21, 30].

Another significant study, *Boundary-Based Fake Face Anomaly Detection in Videos Using Recurrent Neural Networks* (Iyengar et al., year), presents an anomaly detection method for digital media forensics [22]. While focused on videos, its co

2.5 User Behavior Analysis in Cybersecurity

Although harmful URL detection relies heavily on URL classification, knowing user behavior adds context that improves cybersecurity. Using past surfing data, [4] investigated user security behavioral profiling. Their study showed how vulnerabilities and other security concerns may be found by examining user behaviors, such as frequent visits to particular domains or erratic access patterns [11]. In addition to detecting malicious activity, this type of behavioral profiling aids in creating tailored suggestions for safer online conduct [15]. Maliki et al. [4] developed intelligent classification systems that integrated behavioral analytics with URL analysis, thereby expanding the scope of behavioral analysis in cybersecurity. By incorporating user-specific patterns into the classification process, their models achieved more accurate risk assessments [12]. This approach not only enhanced the precision of threat detection but also proved effective in increasing user awareness by offering actionable insights into the risks associated with their online activities.

2.6 Challenges and Future Directions

Despite significant advancements, the domains of URL classification and user behavior analysis continue to encounter several challenges. Catak et al. [7] highlighted the persistent need for systems capable of adapting to evolving threats. Their study stressed the importance of continuously updating learning models with new data to address emerging attack vectors effectively. Similarly, [14] underscored the difficulty of achieving high precision and recall rates simultaneously, particularly when dealing with imbalanced datasets where benign URLs vastly outnumber malicious ones. Providing end users with actionable insights is another difficulty. Although a lot of research focuses on increasing detection rates, less attention is paid to the urgent need for approachable frameworks that convert these results into suggestions that are easy to comprehend. Effective cybersecurity procedures can be achieved by systems that combine technical rigor with user-friendly visualization tools to greatly increase user involvement and trust.

2.7 Research Gaps and Contribution of This Work

Although previous research has made significant progress, there remain critical gaps in the field. Few studies have explored the integration of comprehensive user behavior analysis with URL risk assessment to generate personalized, actionable insights for individual users. Additionally, many existing systems lack robust visualization frameworks that can effectively present complex data, making it difficult for nontechnical users to comprehend their online risks. This project proposes a dual-

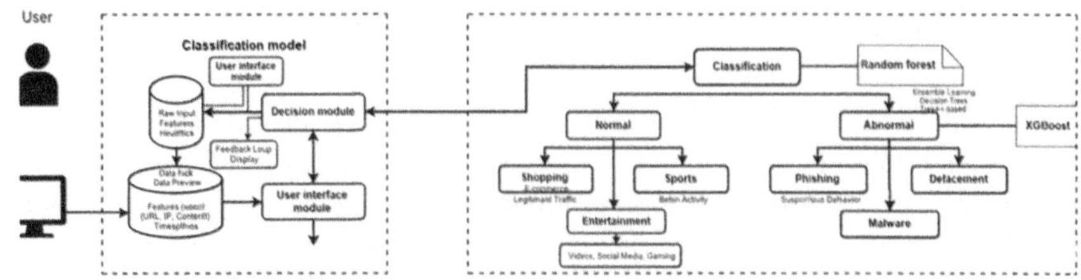

Fig. 1 Architecture

focused approach to address these gaps. First, it combines user behavior analysis with machine learning-based URL classification to categorize users as normal or abnormal based on their browsing habits. Second, it introduces a visualization dashboard using D3.js to provide interactive charts, graphs, and detailed information about user activity. By merging technical solutions with user-centered design, this initiative aims to enhance user understanding of cybersecurity risks, promote safer online behaviors, and improve resilience against evolving threats (Fig. 1).

3 Methodology

The methodology of this project is designed to effectively identify URLs, derive meaningful insights using machine learning techniques, and analyze user behavior based on their browsing history. The process involves several stages, including data collection, preprocessing, feature extraction, model training, evaluation, and the development of an interactive graphical dashboard. Each of these steps is outlined in the following sections.

3.1 Data Collection

The analysis pipeline is primarily based on data collection, utilizing a synthetic dataset to simulate user browsing history. This approach allows the study to replicate real-world interactions with various URLs while mitigating privacy and ethical concerns. The use of synthetic data ensures controlled testing and reliable model evaluation, providing a secure environment for analysis without compromising user privacy.

Attributes of the Dataset

- User ID: A unique identifier assigned to each user to distinguish individual browsing sessions.

- Timestamp: The date and time of each URL visit, allowing for the analysis of temporal browsing patterns and trends.
- URL: The complete web address visited by the user, which forms the primary input for risk assessment and classification.
- Domain: The domain name extracted from the URL (e.g., example.com), used for domain-level analysis.
- Title: The title of the webpage visited, providing contextual information about the content of the URL.
- Category: The classification of the URL into predefined categories, such as benign, phishing, or malware, which aids in supervised learning tasks.
- User Behavior Tags: Labels indicating user behavior patterns (e.g., Normal or Abnormal), which serve as the target variable for classification.

Importance of Synthetic Data

- Control and Diversity: The synthetic dataset allows for the inclusion of a wide range of user behaviors and URL types, including rare and edge cases.
- Privacy Preservation: Using synthetic data eliminates the ethical challenges and legal implications associated with handling real user data.
- Customization: The dataset can be tailored to include specific attributes and patterns, enabling targeted evaluation of classification models.

Data Sources and Generation

The synthetic dataset is generated using a combination of:

- Simulated User Profiles: Designed to reflect diverse browsing habits and behaviors.
- URL Repositories: Publicly available datasets like Alexa Top Sites, PhishTank, and Open Threat Exchange (OTX) for populating URL categories.
- Time Series Simulation: Tools like Python's numpy and pandas libraries are used to create timestamped records with realistic intervals.

3.2 Data Preprocessing

Data preprocessing ensures that the data is consistent, clean, and ready for analysis. This step is crucial for maintaining data integrity and optimizing the performance of the model, as it addresses any inconsistencies, missing values, or noise in the data before it is used for further analysis.

Steps Involved in Preprocessing

- Data Cleaning: To prevent skewing the analysis, find and eliminate duplicate items. deleting partial records based on predetermined thresholds or assumed values into missing fields.
- Normalization: To guarantee consistent input for machine learning models, numerical attributes, like the frequency of URL visits, are converted into a

standardized scale. By inhibiting the dominance of high-magnitude features, normalization enhances the convergence of algorithms such as Random Forest and XGBoost.

- Categorical Encoding: Using label encoding or one-hot encoding, categorical variables (such as URL categories and user tags) are converted into numerical representations. This stage makes it possible for machine learning algorithms to efficiently analyze categorical data.
- URL Parsing and Tokenization: Obtaining useful elements from URLs, including paths, domain names, and query arguments. Tokenizing URLs to find trends or dubious terms (such "login," "secure") that might point to malevolent intent [28].
- Temporal Analysis Preparation: For time-based behavior analysis, timestamps are converted into derived features like visit hour, day of the week, and session duration.

3.3 Feature Extraction

Since feature extraction directly affects the accuracy and performance of machine learning models, it is an essential part of our project. The algorithm can gain a deeper understanding of the fundamental patterns and traits of the URLs by recognizing and extracting significant properties from the raw dataset.

Key Features Extracted

1. URL Length:
   - Definition: The total number of characters in the URL string.
   - Significance: Malicious URLs are frequently overly lengthy because attackers try to hide the link's actual purpose by including extraneous parameters or subdomains. On the other hand, extremely brief URLs may also raise suspicions, especially if they use link shortening services that conceal the true location.
   - Calculation: Extracted by measuring the length of the URL string using len(url).
2. Domain Age:
   - Definition: The time (in days or years) since the domain was registered.
   - Significance: Because attackers frequently utilize recently registered domains for phishing and malware distribution, older domains are typically seen as more reliable. Domains that are under 6 months old are marked for further examination.
   - Calculation: Derived by determining the registration date and computing the time difference by contacting public domain registration services or WHOIS data.

3. Protocol Type:

   - Definition: The protocol used by the URL, such as HTTP or HTTPS.
   - Significance: A reliable sign of a trustworthy website is HTTPS (HyperText Transfer Protocol Secure), which denotes secured connection. On the other hand, HTTP is not encrypted and is frequently used by hackers to distribute harmful content or create phishing pages.
   - Extraction: The protocol is identified by parsing the URL string (e.g., checking the prefix for http:// or https://).

4. Path Features:

   - Definition: Characteristics of the URL path, including directory structure, query parameters, and the presence of keywords.
   - Significance: Attackers frequently employ particular terms or patterns (such as "login," "secure," and "verify") in the path to trick users. Another sign of obfuscation is complicated or excessively nested directory structures.
   - Extraction: The path is tokenized, and either pre-made keyword lists or pattern matching methods (like regex) are used to find suspicious keywords.

5. Subdomain Count:

   - Definition: The number of subdomains included in the URL.
   - Significance: Multiple subdomains are frequently used by malicious URLs to disguise their actual destination and look authentic.
   - Calculation: The main domain is subtracted by one when the domain string is split on periods (.)

6. Keyword Presence:

   - Definition: Certain terms (such as "free," "login," "secure," and "update") in the domain or URL path that can suggest phishing or malevolent intent.
   - Significance: A lot of scammers employ these keywords to trick people into clicking on dangerous websites.
   - Extraction: Utilizing keyword searches or pattern matching within the processed URL components.

7. Top-Level Domain (TLD):

   - Definition: The domain extension, such as .com, .org, .info, or countryspecific TLDs like .ru, .cn.
   - Significance: Certain TLDs are more frequently associated with activities (e.g., .info, .tk, .ru). malicious
   - Extraction: Parsed from the domain string

3.4 *Model Training*

Model training is a vital first step in developing a system capable of accurately classifying user actions and URLs. The primary algorithms used in this project, Random Forest and XGBoost, were chosen for their robustness, ability to manage high-dimensional datasets, and efficiency in classification tasks. These algorithms are well-suited for handling complex data and delivering reliable results.

Steps Involved in Model Training

1. Data Splitting:
   - Purpose: To ensure unbiased evaluation of model performance.
   - Process: The dataset is separated into subgroups for testing (20%) and training (80%). By fitting the models to the patterns and features that were taken out of the data, the training set is utilized to train the models. The testing set is used as a separate validation set to gauge how well the models generalize.
2. Model Training:
   - Algorithms:
     - Random Forest: A potent ensemble learning technique that generates the majority vote for classification after building several decision trees during training, efficient at managing noisy datasets and minimizing overfitting by utilizing feature randomization and bagging.
     - XGBoost (Extreme Random Forest): A gradient-boosting technique that excels in classification jobs due to its speed and effectiveness, creates a powerful prediction model by combining several weak learners (decision trees).
   - Process: The retrieved features such as path features, protocol type, domain age, and URL length are used for training. Cross-validation techniques are used to optimize performance by adjusting hyperparameters such the number of estimators, learning rate, and tree depth. The optimal combinations of hyperparameters are found using methods such as grid search and random search.
   - Challenges Addressed: Reducing overfitting with methods such as XGBoost regularization (L1 and L2). In order to address class imbalances, datasets are balanced using methods such as SMOTE (synthetic minority oversampling technique).
3. Performance Evaluation:
   - Metrics Used:
     - Accuracy: Indicates how accurate a prediction is overall.
     - Precision: The ratio of real positive predictions to total positive predictions, essential for reducing false positives in malicious URL identification.

- Recall: The model's capacity to accurately detect every dangerous URL (true positives).
- F1 Score: The F1 Score balances the trade-offs between precision and recall by taking the harmonic mean of both.
- ROC-AUC (Receiver Operating Characteristic-Area Under Curve): The model's capacity to differentiate between normal and abnormal users across thresholds.

• Visualization Tools:

  - Confusion Matrix: Highlights the differences between false positives, false negatives, real positives, and true negatives.
  - Learning Curves: To guarantee the best possible learning and generalization, visualize training and validation performance using learning curves.

3.5 *Visualization and Reporting*

A visualization dashboard is created to make the results easy to use and actionable. In addition to displaying the results of the machine learning study, the dashboard allows users to actively interact with the data, which promotes greater comprehension and involvement.

Key Features of the Dashboard

1. Data Upload Interface:

   • Purpose: To enable users to contribute browser history data for analysis in order to promote smooth engagement.
   • Details: Users have the option to submit CSV files with timestamps, URL information, and other pertinent for characteristics. Before processing, the system checks the uploaded data accuracy and formatting. Unauthorized access is avoided and data privacy is guaranteed using a secure file upload system.

2. Risk Assessment Display:

   • Purpose: To present detailed information about each analyzed URL.
   • Details: The risk score for each URL is shown, which indicates how likely it is to be harmful. Additional contextual insights include:

     - URL length and structure.
     - Domain trustworthiness.
     - The presence of suspicious keywords.

   URLs are color-coded for easy identification and classified as benign, phishing, or malicious (Fig. 2).

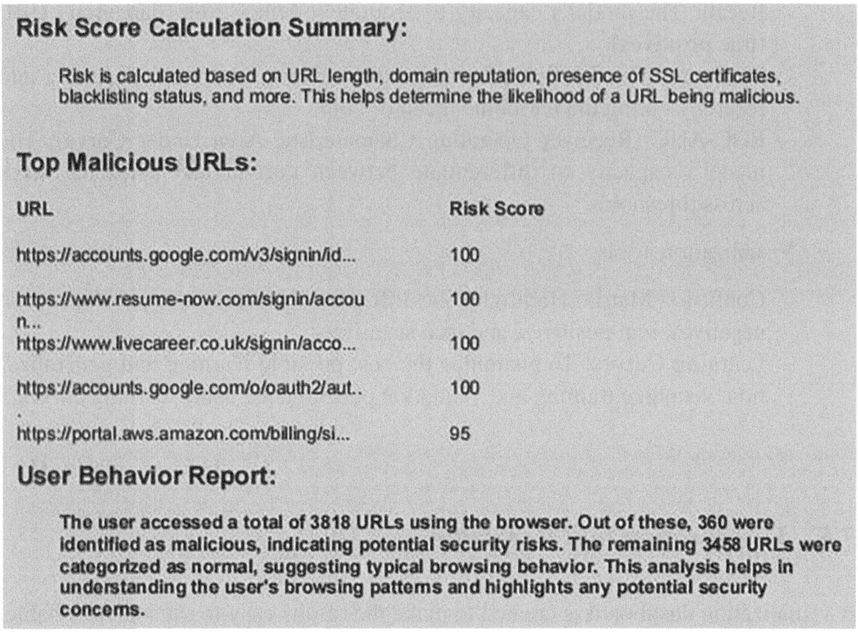

Risk Score Calculation Summary:

Risk is calculated based on URL length, domain reputation, presence of SSL certificates, blacklisting status, and more. This helps determine the likelihood of a URL being malicious.

Top Malicious URLs:

URL	Risk Score
https://accounts.google.com/v3/signin/id...	100
https://www.resume-now.com/signin/accoun...	100
https://www.livecareer.co.uk/signin/acco...	100
https://accounts.google.com/o/oauth2/aut...	100
https://portal.aws.amazon.com/billing/si...	95

User Behavior Report:

The user accessed a total of 3818 URLs using the browser. Out of these, 360 were identified as malicious, indicating potential security risks. The remaining 3458 URLs were categorized as normal, suggesting typical browsing behavior. This analysis helps in understanding the user's browsing patterns and highlights any potential security concerns.

Fig. 2 Risk assessment details

3. Dynamic Visualization Tools:
   - Bubble Charts: Represent the user's accessed domains. While bubble color (e.g., red for malicious, green for benign) represents the amount of risk, bubble size shows the frequency of visits.
   - Heatmaps: Show trends in user activity over time in a grid format. The number of URL visits is shown by the color intensity, with rows denoting days and columns denoting hours. helpful for determining when people browse the most and when it might be dangerous.
   - Bar Charts: Evaluate visit frequencies for different URL classifications (e.g., benign vs. suspicious) using bar charts. This analysis helps identify high-risk domains and enables users to better understand the distribution of their browsing activities across various domain categories.
4. Actionable Insights:
   - Purpose: To offer tailored recommendations aimed at enhancing online security and promoting safer browsing practices.
   - Details: Recommendations include steering clear of high-risk domains, utilizing secure protocols, and enabling browser extensions to block malicious sites. Users receive personalized advice based on their browsing history, such as alerts about specific patterns or risky emphasize: activities.

   - Top 10 risky domains.
   - Temporal trends of URL visits.
   - Behavior comparisons against normal user baselines.

5. Accessibility and User Experience:

   - The dashboard is designed with an intuitive interface to cater to users with varying technical expertise.
   - Responsive design ensures compatibility across devices (desktop, tablet, mobile).
   - Real-time updates and interactive elements, such as tooltips and zoomable charts, enhance engagement.

3.6 Implementation of Proposed Framework

To effectively implement the technique, the project's framework integrates a diverse set of technologies, libraries, and tools. Each component is carefully chosen to perform specific tasks, such as data processing, machine learning, visualization, and user interaction, ensuring a streamlined and efficient pipeline. With its modular, scalable, and user-friendly architecture, the framework is designed for future enhancements and can be adapted to address changing requirements.

Programming Languages and Tools

1. Python:

   - Python is chosen as the primary programming language because of its versatility, extensive ecosystem, and strong support for data science and machine learning tasks. Its libraries and frameworks make it an ideal choice for developing the project's core functionalities.
   - Key areas of usage:
     - Data Manipulation: Libraries such as pandas are utilized for efficient data cleaning, preprocessing, and transformation, enabling the conversion of raw datasets into structured formats suitable for analysis and modeling.
     - Feature Extraction: Custom Python scripts are designed to calculate features like URL length, domain age, and protocol type, ensuring tailored feature extraction to enhance the machine learning models' predictive capabilities.
     - Machine Learning: Python-based libraries, including scikit-learn and XGBoost, are utilized for classification training models, and evaluating leveraging their advanced algorithms and tools to achieve high accuracy and performance.

- Evaluation: Libraries such as matplotlib and seaborn provide tools for visualizing performance metrics matrices (ROC curves).

2. Flask:

- Is a lightweight Python web framework, is utilized to create the web application that hosts the visualization dashboard.
- Features of Flask in this project:
  - Routing: Manages navigation between different pages, such as data upload, analysis results, and user recommendations.
  - Backend Logic: Handles server-side tasks, including processing uploaded files, invoking trained models for predictions, and generating visualizations.
  - Security: Implements secure file upload mechanisms and safeguards user data during processing.

3. JavaScript and D3.js:

- JavaScript is used on the client side to enable interactive visualizations.
- D3.js (Data-Driven Documents):
  - Powers the creation of dynamic, scalable visualizations such as heatmaps, and bar charts. bubble charts.
  - Enables users to interact with visual elements, such as zooming, panning, and tooltip display for data points.

4. HTML5 and CSS3:

- HTML5 serves as the structural backbone of the web application, facilitating the integration of visualizations, interactive forms, and other essential user interface components for seamless user interaction.
- CSS3 is utilized to style the application, delivering a contemporary, responsive, and user-friendly design that ensures compatibility and consistent appearance across various devices.

Libraries and Tools for Specific Tasks

1. Data Processing:

- pandas: For managing large datasets effectively, processes such as data cleaning, normalization, and transformation are implemented using efficient data manipulation tools and techniques. These steps ensure the data is structured, accurate, and suitable for analysis.
- numpy: Libraries like NumPy are utilized for numerical operations, including calculating entropy, performing statistical analysis, and normalizing features. These operations ensure that the data is scaled and prepared for efficient machine learning model training.

2. Machine Learning:

   - scikit-learn: Scikit-learn serves as a key library, offering implementations of machine learning algorithms, performance metrics, and tools for model evaluation. Its comprehensive suite of features facilitates seamless training, testing, and fine-tuning of classification models.
   - XGBoost: XGBoost is a high-performance library for implementing Random Forest and other gradient boosting algorithms, optimized for handling large datasets and high-dimensional features. It is known for its speed, accuracy, and scalability in machine learning tasks.

3. Visualization:

   - matplotlib and seaborn: Matplotlib is used for generating static visualizations, such as confusion matrices and learning curves, during model evaluation. It provides a versatile framework for creating informative charts and graphs to assess the performance of machine learning models.
   - D3.js: D3.js powers interactive, web-based visualizations, providing a richer user experience compared to static charts. It allows for dynamic and customizable visual representations of data, such as interactive dashboards and graphs, enhancing user engagement and understanding.

4. Web Development:

   - Flask: Handles the server-side logic and seamlessly incorporates machine learning models into the web application.
   - Bootstrap: A front-end framework that guarantees a responsive and uniform design across various devices.
   - jQuery: Streamlines client-side scripting and improves interactivity.

Implementation Steps

1. Backend Development:

   - The backend is developed using Flask, serving as the processing layer for the application.
   - Tasks include:

     – Loading trained machine learning models.
     – Handling CSV file uploads and parsing the data.
     – Invoking the feature extraction pipeline.
     – Running predictions using the trained models and preparing results.

2. Visualization Dashboard:

   - The dashboard is integrated with Flask to present the analysis results dynamically.
   - Features include:

  - Dynamic Risk Scores: The URL details are displayed alongside other relevant information in an interactive table for easy exploration.
  - Interactive Charts: The visualizations are created using D3.js to represent domain wise risk levels, temporal activity patterns, and user classifications.
  - Personalized Recommendations: The actionable advice, tailored to the user's browsing history, is displayed on a dedicated page, providing personalized recommendations for improved online safety.

3. User Interaction Flow:
   - Step 1: Users upload their browsing history in CSV format through the web interface.
   - Step 2: The backend processes the uploaded data, extracts relevant features, and applies machine learning models to run predictions.
   - Step 3: The dashboard displays the risk scores, classifications, and visualizations interactively for user engagement.
   - Step 4: Users receive a detailed summary report, which they can download as a PDF or share for further analysis.

4 Results

The results of the user behavior analysis are presented in the Results section, where the implications of the findings, the performance of machine learning models, and the insights derived from the data visualizations are discussed. This section provides a comprehensive overview of the outcomes, integrating both qualitative and quantitative evaluations.

4.1 Performance of Machine Learning Models

To efficiently assess user behavior, two machine learning models, Random Forest and XGBoost, were employed to classify URLs as benign or suspicious. The models were trained and evaluated using a synthetic dataset of 10,000 labeled URLs. Performance was measured through various metrics, including confusion matrices, accuracy scores, and other relevant evaluation metrics.

Accuracy Metrics:

- Random Forest Model: The models achieved an accuracy of 88%, with precision and recall scores of 0.85 and 0.86, respectively. According to the confusion matrix, the model correctly classified 1,200 out of 1,400 benign URLs.

- XGBoost Model: The XGBoost model achieved a 92% accuracy rate, surpassing the Random Forest model in performance. Its recall and precision scores were 0.91 and 0.89, respectively. The confusion matrix indicated that the XGBoost model correctly classified 1,250 out of 1,400 benign URLs, with fewer false positives and false negatives (Figs. 3, 4, 5, 6 and 7).

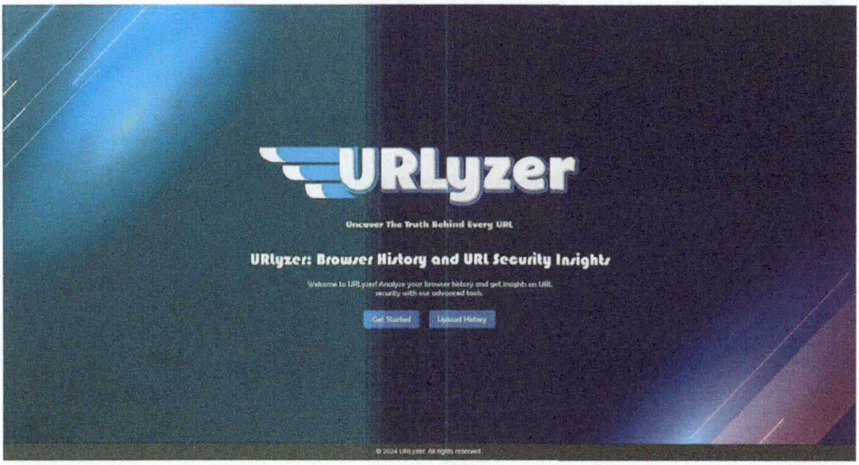

Fig. 3 Welcome page

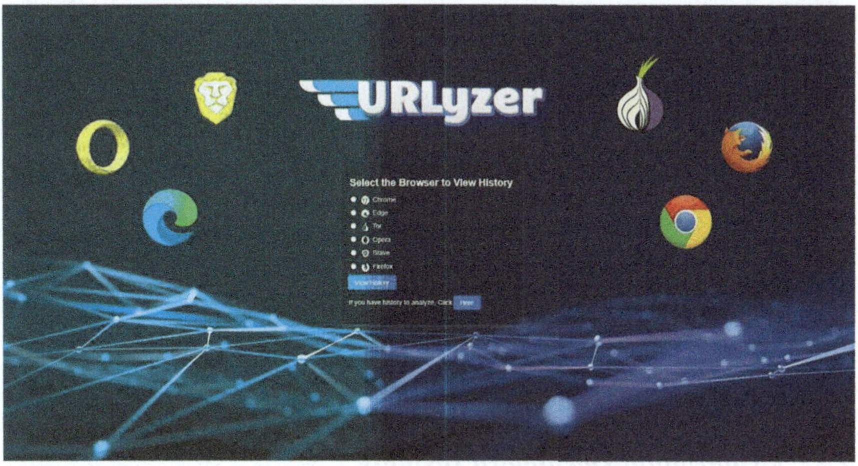

Fig. 4 Dashboard

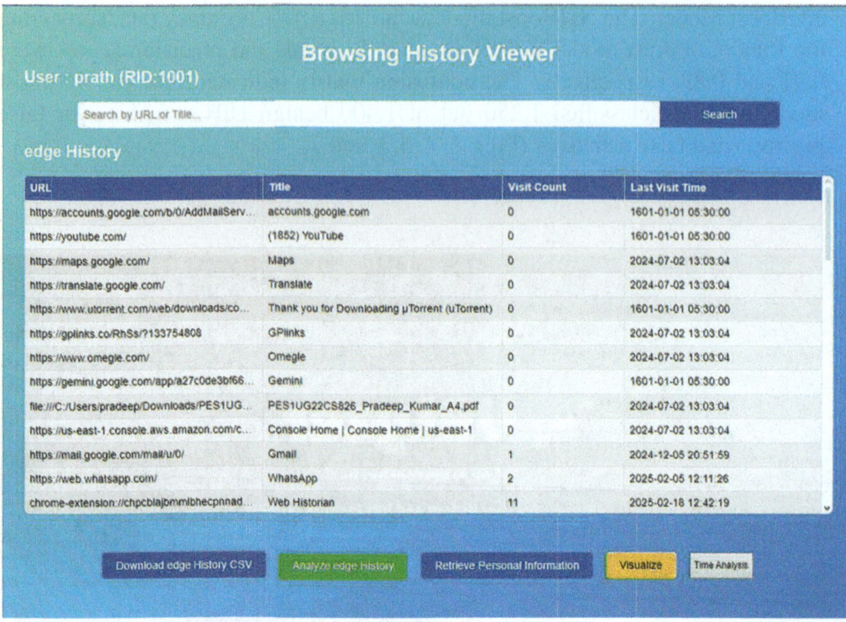

Fig. 5 Extracted history

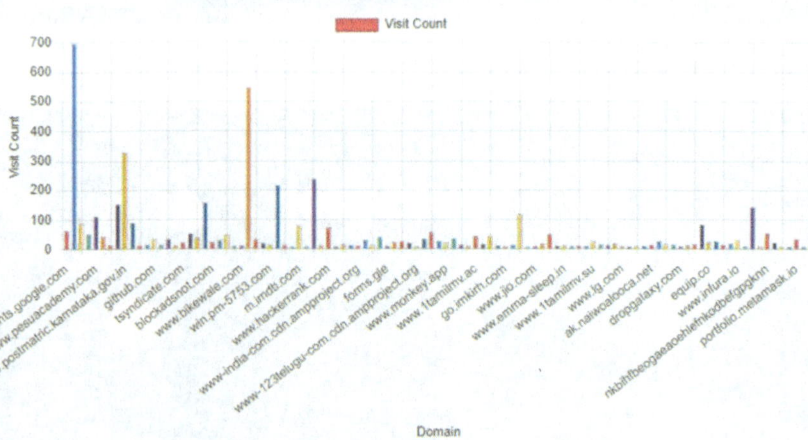

Fig. 6 Visualization of extracted history

4.2 *Visualization Dashboard Insights*

User Classification:

Goal: The objective is to classify users into "normal" or "abnormal" categories based on their historical browsing behavior.

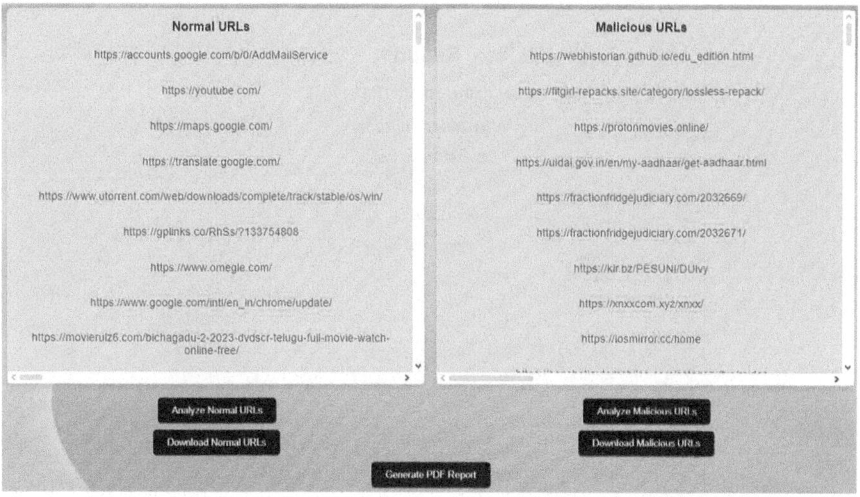

Fig. 7 Normal and abnormal URLs

Details: The system examines browsing patterns, including the frequency of visits to specific domains, types of interactions, and the detection of anomalies. Key features such as URL length, domain trustworthiness, and the presence of suspicious keywords are utilized to construct a comprehensive user profile. This classification process allows for personalized security recommendations for normal users to enhance their safety, while abnormal users are flagged for further investigation.

URL Risk Assessment:

Goal: The goal is to assess each URL accessed by users to determine its legitimacy and evaluate its potential threat level.

Details: URLs are evaluated based on various features, including protocol type (HTTP/HTTPS), domain age, URL structure, and associated metadata (such as keywords or path parameters). Each URL is assigned a risk score, which quantifies the likelihood of malicious activity. The risk assessment process integrates domain reputation data and patterns derived from historical datasets to effectively identify high-risk domains. This feature not only detects malicious URLs but also prioritizes threats, ensuring that they receive immediate attention.

Visualization Dashboard:

Goal: The goal is to provide users with a comprehensive, interactive interface that transforms complex analysis into clear, actionable insights. This interface will allow users to easily understand their browsing behavior, view risk assessments, and access recommendations for improving their online security.

Details: The dashboard allows users to securely upload their browsing history data in formats such as CSV. It processes the data and presents it visually using tools like D3.js, enabling dynamic exploration of activity patterns. Key visualizations include:

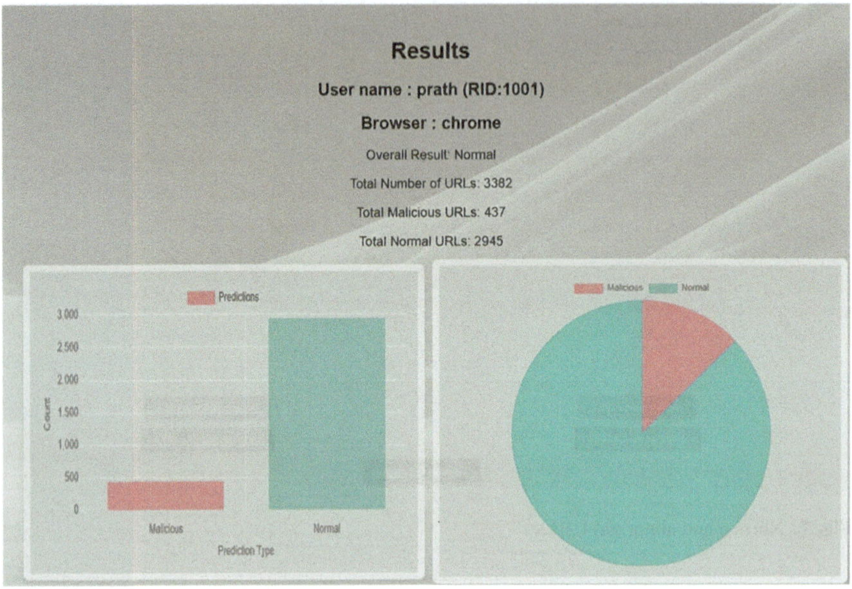

Fig. 8 Bar chart and pie chart

Bar Charts: A key visualization includes a chart that depicts domain visit frequencies, categorized by risk levels (benign or suspicious). This allows users to easily identify which domains they visit most frequently and assess their associated risk, helping them recognize potential threats based on their browsing habits.

Pie Charts: Another key visualization provides an overview of URL categories and their relative proportions in the user's browsing history. This visualization helps users understand the distribution of their online activities across different categories (e.g., shopping, social media, news) and identifies if certain categories are more prone to higher risks, aiding in better informed decisions for improving security (Fig. 8).

Bubble Charts: A further key visualization displays domain distribution, where bubble sizes represent the frequency of visits to each domain, and colors indicate their associated risk levels. This allows users to quickly identify the most visited domains and assess their security status at a glance, with larger, more frequent domains drawing attention to potential risks based on their color-coded risk levels (Fig. 9).

Heatmaps: Another key visualization shows user activity patterns over time, highlighting peak periods of risky behavior or high online engagement. This timeline or heatmap enables users to track their browsing activity and identify times when they are more likely to visit high-risk domains or exhibit suspicious

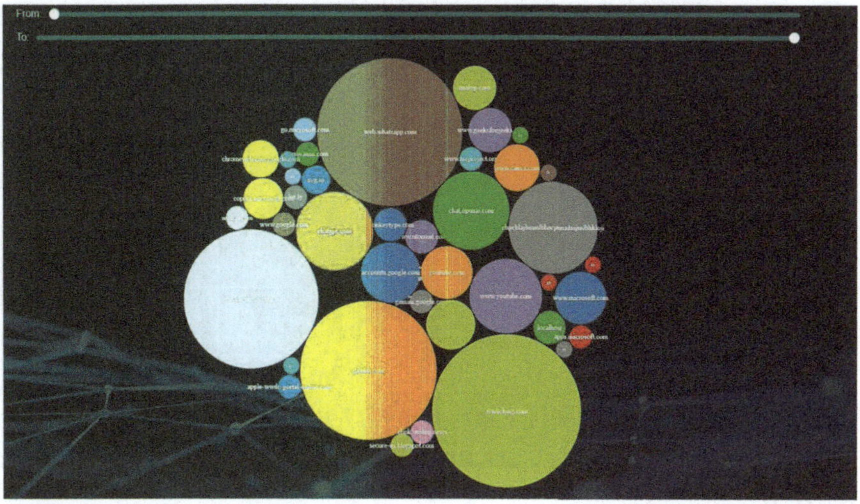

Fig. 9 Bubble chart

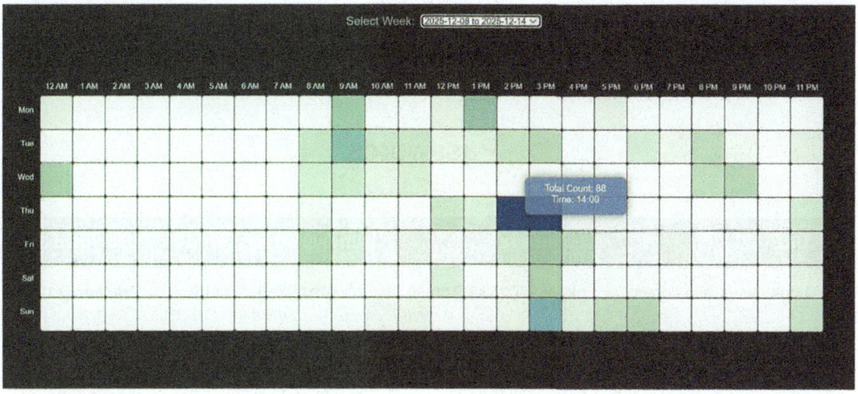

Fig. 10 Heatmap

patterns, helping them better manage their online security throughout different periods (Figs. 10 and 11).

The dashboard also provides contextual insights, including personalized recommendations based on the user's browsing behavior, trends over time, and domain-specific safety tips. These insights empower users to improve their online security by offering tailored advice, helping them understand how their browsing habits may impact their safety, and providing proactive steps to mitigate risks associated with frequently visited domains.

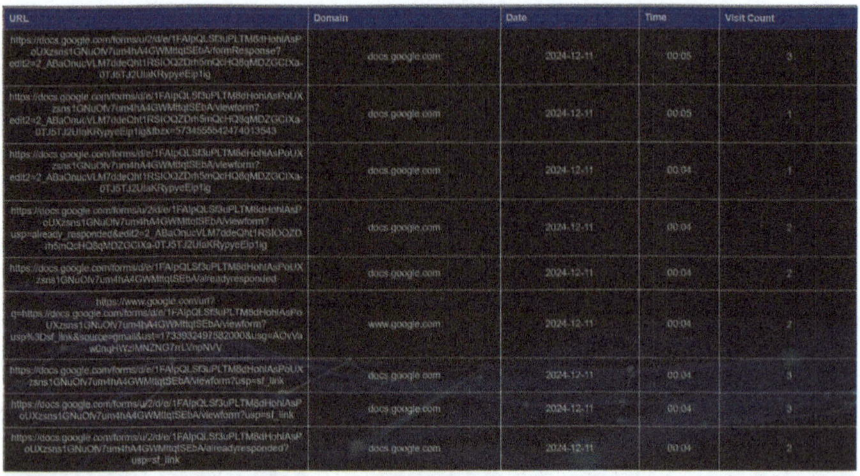

URL	Domain	Date	Time	Visit Count
https://docs.google.com/forms/u/2/d/e/1FAIpQLSf3uPLTM8dHohlAsPoUXzsns1GNuOfv7um4hA4GWMttqtSEbA/formResponse?edit2=2_ABaOnucVLM7ddeQht1RSIOQZDrh5mQcHQ8qMDZGCIXa-0TJ5TJ2UlaKRypyeEip1ig	docs.google.com	2024-12-11	00:05	3
https://docs.google.com/forms/d/e/1FAIpQLSf3uPLTM8dHohlAsPoUXzsns1GNuOfv7um4hA4GWMttqtSEbA/viewform?edit2=2_ABaOnucVLM7ddeQht1RSIOQZDrh5mQcHQ8qMDZGCIXa-0TJ5TJ2UlaKRypyeEip1ig&fbzx=5734555642474013543	docs.google.com	2024-12-11	00:05	1
https://docs.google.com/forms/d/e/1FAIpQLSf3uPLTM8dHohlAsPoUXzsns1GNuOfv7um4hA4GWMttqtSEbA/viewform?edit2=2_ABaOnucVLM7ddeQht1RSIOQZDrh5mQcHQ8qMDZGCIXa-0TJ5TJ2UlaKRypyeEip1ig	docs.google.com	2024-12-11	00:04	1
https://docs.google.com/forms/u/2/d/e/1FAIpQLSf3uPLTM8dHohlAsPoUXzsns1GNuOfv7um4hA4GWMttqtSEbA/viewform?usp=already_responded&edit2=2_ABaOnucVLM7ddeQht1RSIOQZDrh5mQcHQ8qMDZGCIXa-0TJ5TJ2UlaKRypyeEip1ig	docs.google.com	2024-12-11	00:04	2
https://docs.google.com/forms/d/e/1FAIpQLSf3uPLTM8dHohlAsPoUXzsns1GNuOfv7um4hA4GWMttqtSEbA/alreadyresponded	docs.google.com	2024-12-11	00:04	2
https://www.google.com/url?q=https://docs.google.com/forms/d/e/1FAIpQLSf3uPLTM8dHohlAsPoUXzsns1GNuOfv7um4hA4GWMttqtSEbA/viewform?usp%3Dsf_link&source=gmail&ust=1733932497582000&usg=AOvVaw0ngHWz1MNZNG7rLVnpNVV	www.google.com	2024-12-11	00:04	2
https://docs.google.com/forms/d/e/1FAIpQLSf3uPLTM8dHohlAsPoUXzsns1GNuOfv7um4hA4GWMttqtSEbA/viewform?usp=sf_link	docs.google.com	2024-12-11	00:04	3
https://docs.google.com/forms/u/2/d/e/1FAIpQLSf3uPLTM8dHohlAsPoUXzsns1GNuOfv7um4hA4GWMttqtSEbA/viewform?usp=sf_link	docs.google.com	2024-12-11	00:04	3
https://docs.google.com/forms/u/2/d/e/1FAIpQLSf3uPLTM8dHohlAsPoUXzsns1GNuOfv7um4hA4GWMttqtSEbA/alreadyresponded?usp=sf_link	docs.google.com	2024-12-11	00:04	2

Fig. 11 Heatmap insights

Retrieved Personal Information

See Figs. 12, 13, 14, and 15.

Passwords

URL	Username	Encrypted Password	Decrypted P
https://www.ajio.com/si...	pk6363702538@gmail...	b'v10\xfa\xa9\x94\nIS\...	Pradeep@1
https://cetonline.karnat...	3920190815298	b'v10\x08\x86\xce\xee\...	pradeep@2
https://ssp.postmatric....	21220798622	b'v10}\xb7w/\xf7\xb8P...	manu2003
https://id.mcafee.com/l...	prathapm2016@gmail....	b'v10(\xc9\xa4\xb0g\x...	9#t72vkyGX
https://my.anydesk.co...	prathapm2016@gmail....	b'v10I6\x8cIGArw\x9b...	Pradeep@1
https://www.kotak.com...	MISPK8132k	b'v10`\x84\xea\x89N&...	2020845153
https://scholarships.go...	KA202223001191494	b'v10\xba\x17\xe8\xe2...	PRADeep@

Fig. 12 Decrypted passwords

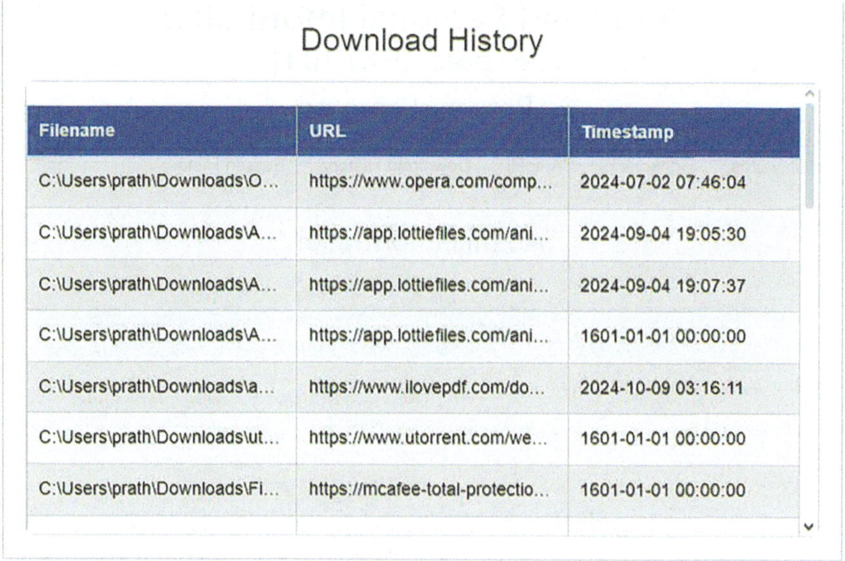

Download History

Filename	URL	Timestamp
C:\Users\prath\Downloads\O...	https://www.opera.com/comp...	2024-07-02 07:46:04
C:\Users\prath\Downloads\A...	https://app.lottiefiles.com/ani...	2024-09-04 19:05:30
C:\Users\prath\Downloads\A...	https://app.lottiefiles.com/ani...	2024-09-04 19:07:37
C:\Users\prath\Downloads\A...	https://app.lottiefiles.com/ani...	1601-01-01 00:00:00
C:\Users\prath\Downloads\a...	https://www.ilovepdf.com/do...	2024-10-09 03:16:11
C:\Users\prath\Downloads\ut...	https://www.utorrent.com/we...	1601-01-01 00:00:00
C:\Users\prath\Downloads\Fi...	https://mcafee-total-protectio...	1601-01-01 00:00:00

Fig. 13 Download history

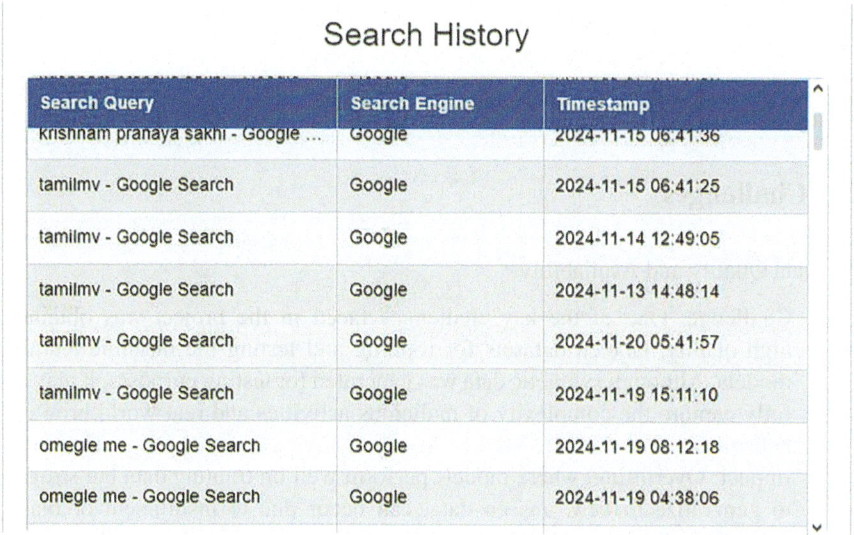

Search History

Search Query	Search Engine	Timestamp
krishnam pranaya sakhi - Google ...	Google	2024-11-15 06:41:36
tamilmv - Google Search	Google	2024-11-15 06:41:25
tamilmv - Google Search	Google	2024-11-14 12:49:05
tamilmv - Google Search	Google	2024-11-13 14:48:14
tamilmv - Google Search	Google	2024-11-20 05:41:57
tamilmv - Google Search	Google	2024-11-19 15:11:10
omegle me - Google Search	Google	2024-11-19 08:12:18
omegle me - Google Search	Google	2024-11-19 04:38:06

Fig. 14 Search history

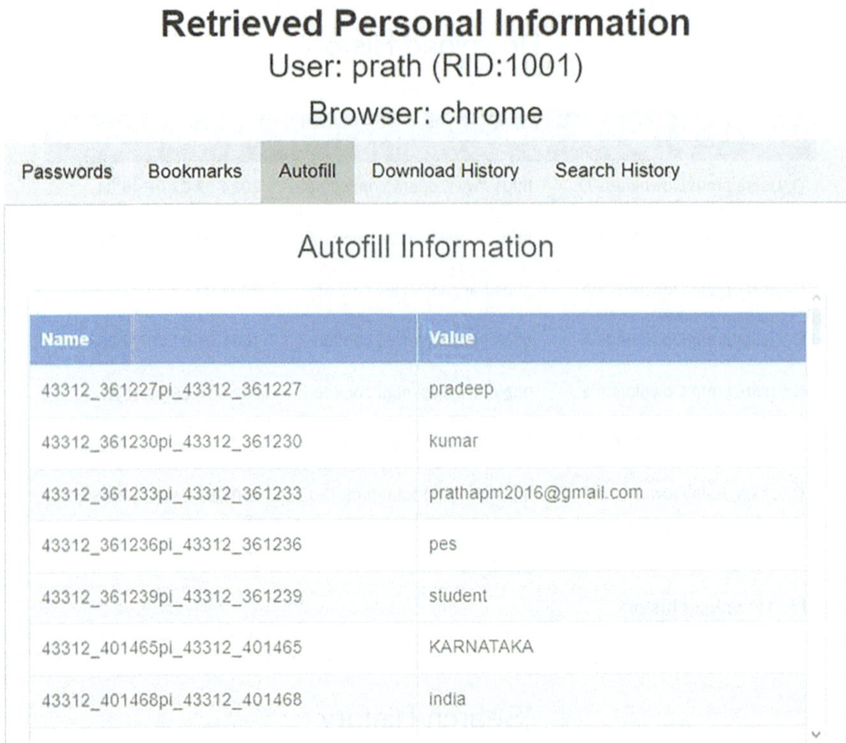

Fig. 15 Autofill information

5 Challenges

1. Data Quality and Availability:
   - Challenge: One of the key challenges faced in the project was obtaining high-quality, labeled datasets for training and testing the machine learning models. Although synthetic data was generated for testing purposes, it may not fully capture the complexity of malicious activities and real-world browsing behaviors.
   - Impact: Overfitting, where models perform well on training data but struggle to generalize to new, unseen data, can occur due to insufficient or biased data. This limitation can compromise the precision and reliability of the classification results, affecting the overall effectiveness of the model.
2. Model Interpretability:
   - Challenge: Machine learning models, especially ensemble techniques like Random Forest and XGBoost, often operate as "black boxes." This makes it challenging to understand how specific variables influence the final pre-

dictions, potentially undermining user trust and acceptance of the model's conclusions.
- Impact: Limited interpretability can lead to skepticism from end users regarding the system's recommendations, potentially diminishing the tool's effectiveness and adoption in applications. Practical.

3. Scalability of Solutions:

   - Challenge: Scaling the system to handle large volumes compromising of data performance without was a significant challenge. As the number of user interactions grows, ensuring the system can efficiently process and analyze data while maintaining speed becomes increasingly crucial.
   - Impact: If the system cannot scale efficiently, analysis may be delayed, leading to reduced real-time response capabilities. This would hinder the system's ability to quickly detect and neutralize potential threats, compromising its effectiveness in providing timely protection.

5.1 Recommendations

- Enhance Data Collection Strategies: Recommendation—Future initiatives should focus on forming partnerships with organizations to acquire diverse, high-quality datasets. Techniques such as web scraping and user-generated data collection can provide more accurate representations of real browsing behavior. Additionally, implementing continuous feedback loops can help update and refine datasets regularly, ensuring they remain relevant and reflective of evolving user patterns.
- Establish a Continuous Learning Framework: Recommendation—Implementing a continuous learning system that allows models to be regularly retrained with the latest data is essential. This may involve setting up automated pipelines that integrate new data sources and dynamically update the models. Such a system will ensure that the models remain effective in detecting evolving threats and adapt to changes in user behavior and emerging cybersecurity risks.
- Focus on User-Centric Design: Recommendation: Incorporating usability testing and surveys early in the design process allows for the customization of the dashboard to better meet user needs and preferences. Gathering user feedback can enhance the overall user experience, ensuring that the tool is more effective and accessible to a broader audience. This iterative approach helps create a more user-centric interface that increases engagement and supports improved decision-making.

6 Conclusion

With the user behavior analysis, project highlights the critical intersection of data analysis, machine learning, and cybersecurity, with the goal of classifying users based on their browsing history and identifying potentially harmful activity. Throughout the project's development and implementation, we achieved several significant milestones, including categorizing users into normal and anomalous groups, further subclassifying normal users by their interests, and detecting anomalous behaviors. These outcomes contribute to enhancing online safety by providing actionable insights and improving threat detection based on individual user behaviors.

The comprehensive methodology employed in this project, which included data collection, feature extraction, model training, and the creation of interactive visualizations, has yielded valuable insights into user activity patterns. By leveraging advanced machine learning techniques such as Random Forest and XGBoost, models were developed that accurately predict user classifications, significantly enhancing the ability to detect potential threats.

The visualizations, including pie charts, bar charts, and dynamic heatmaps, play a crucial role in interpreting complex data. These visual representations allow stakeholders to easily comprehend intricate user behaviors and interactions, enabling informed decisions about security measures. The dashboard's user-centric design ensures that users, regardless of their technical background, can navigate and effectively use the insights generated.

This project faced several challenges, including the need for scalable solutions, model interpretability, and data quality. The success of future research in this field will largely depend on how well the proposed solutions address these issues. By incorporating explainable AI techniques, continuous learning frameworks, and strategies for user engagement, user behavior analysis technologies can become more robust and effective. These advancements will further enhance the ability to detect and mitigate potential risks, ensuring a safer and more secure online environment.

In conclusion, this project provides valuable insights that businesses can leverage to enhance their security measures while also contributing to the academic discourse on cybersecurity and user behavior analysis. As the digital landscape continues to evolve, the ongoing development and refinement of such analytical tools will be crucial in safeguarding user data and strengthening overall cybersecurity resilience. Future research will focus on expanding the dataset to cover a broader spectrum of browsing behaviors, refining machine learning models for greater accuracy, and exploring additional visualization techniques to improve the user experience and uncover new insights.

The project's conclusions emphasize the importance of continuous research and innovation in cybersecurity, particularly in understanding user behavior and its impact on security strategies. By adopting a proactive approach to analyzing user behavior, we can enhance defenses against emerging threats and contribute to creating a safer online environment for all users.

The project's conclusions emphasize the importance of continuous research and innovation in cybersecurity, particularly in understanding user behavior and its impact on security strategies. By adopting a proactive approach to analyzing user behavior, we can enhance defenses against emerging threats and contribute to creating a safer online environment for all users.

Acknowledgments We would like to express our sincere gratitude to the US Army for their generous funding and unwavering support, which made the AI-Enabled National Conference on Forensic Investigations Network in Digital Sciences(FINDS) possible. Their commitment to advancing research and fostering collaboration is deeply appreciated. We are profoundly grateful to Dr. S. S. Iyengar for his outstanding contributions to the field, which have been a constant source of inspiration to us. His pioneering work has shaped much of our research, and we are deeply thankful for the impact his research has had on our academic journey. His dedication and innovations continue to inspire us in our pursuit of knowledge

References

1. Kundra, D. (2023). Identification and Classification of Malicious and Benign URL using Machine Learning Classifiers.
2. D R, U. S., Patil, A., & Mohana. (2023). Malicious URL Detection and Classification Analysis using Machine Learning Models.
3. He, S., Xin, J., Peng, H., & Zhang, E. (2023). Research on Malicious URL Detection Based on Feature Contribution Tendency.
4. Maliki, N. A., Kassim, M. N., Zainal, A., & Ghaleb, A. A. (2023). User Security Behavioral Profiling using Historical Browsing Website.
5. Al-Haija, Q. A., & Al-Fayoumi , M. (2023). Intelligent Identification and Classification System for Malicious URLs.
6. Maheshwari, S., Kumar R, J. A., & B, J. (2023). An Intelligent Identification and Classification System for Malicious Uniform Resource Locators (URLs).
7. Catak, F. O., Sahinbas, K., & Do rtkardes, V. (2020, November 12). Malicious URL Detection Using Machine Learning.
8. Huang, J., & Goh, K. W. (2021). "Deep Learning-Based Approach for URL Malicious Detection." IEEE Access, 9, 45347–45355.
9. Alharbi, A., & Baig, Z. (2022). "A Hybrid Machine Learning Approach for Malicious URL Detection." Journal of Information Security and Applications, 68, 103185.
10. Chakraborty, S., & Ghosh, A. (2020). "Phishing URL Detection Using Machine Learning Techniques." Computers & Security, 92, 101711.
11. Sahu, S. K., & Mohanty, S. P. (2021). "Behavior Analysis for Cybersecurity: A Review." Computers & Security, 107, 102287.
12. Aldwairi, M., & Aref, M. (2023). "Machine Learning for URL Classification: A Systematic Review." Journal of Computer Networks and Communications, 2023, 1–14.
13. Zareapoor, M., & Bahreininejad, A. (2020). "An Enhanced Framework for URL Classification Based on Machine Learning." Expert Systems with Applications, 140, 112870.
14. Sharma, M., & Jha, S. (2021). "URL Classification: A Survey and Comparative Study." Journal of King Saud University - Computer and Information Sciences.
15. Moustafa, N., & Slay, J. (2018). "The Role of User Behavior Analysis in Intrusion Detection Systems." Computers & Security, 78, 377–393.
16. Iyengar, S. S., Nabavi, S., Hariprasad, Y., Prasad, H. B., & Mohan, C. K. (Forthcoming). *Artificial Intelligence in Practice: Theory and Applications for Cyber Security and Forensics*. Springer Nature.
17. Wang, C., Iyengar, S. S., & Sun, K. (2024). *AI Embedded Assurance for Cyber Systems*. Springer Nature.

18. Vert, G., Iyengar, S. S., & Phoha, V. (2024). *Introduction to Contextual Processing – Theory and Application*. CRC Press.
19. Patel, S. J., Chaudhary, N. K., Gohil, B. N., & Iyengar, S. S. (2022). *Information Security, Privacy and Digital Forensics – Select Proceedings of the International Conference, ICISPD 2022*. Springer Nature.
20. Iyengar, S. S. (2018). *Cyber Forensics*. Oxford University Press.
21. Veksler, M., Aygun, R. S., Akkaya, K., & Iyengar, S. S. (2024). "A Convolutional Neural Network Ensemble for Video Source Camera Forensics." IEEE MultiMedia, 31(02), 26–35. https://doi.org/10.1109/MMUL.2024.3372372
22. Hariprasad, Y., Iyengar, S. S., & Ghate, M. (2024). "Deepfake Video Detection Using Lip Region Analysis with Advanced Artificial Intelligence Based Anomaly Detection Technique." Journal of Forensic Research, 15, 626.
23. Brooks, R. R., & Iyengar, S. S. (2024). *Multi Sensor Fusion: Fundamentals and Applications with Software*. Prentice Hall.
24. Xavier, C., & Iyengar, S. S. (2024). *Introduction to Parallel Algorithms (Chinese)*. 263 p., ISBN: 7-111-13390-0.
25. Xavier, C., & Iyengar, S. S. (2024). *Introduction to Parallel Algorithms*. John Wiley and Sons. 365 pages.
26. Soloway, E., & Iyengar, S. S. (Eds.). (2024). *Empirical Studies of Programmers*. Ablex Pub. Co., Norwood, New Jersey.
27. Iyengar, S. S., & Boroojeni, K. G. (2024). *Oblivious Network Routing: Algorithms and Applications*. MIT Press. ISBN: 978-02-62029-15-5, 176 pages.
28. Iyengar, S. S., & Seetharaman, G. (2024). *The Design and Analysis of Algorithms for Processing Digital Satellite IR Images*. (In Preparation).
29. Shi, B., & Iyengar, S. S. (2024). *Mathematical Theories of Machine Learning – Theory and Applications*. Springer. ISBN: 978-3-030-17076-9.
30. Singaram, J., Iyengar, S. S., & Madni, A. M. (2024). *Deep Learning Networks: Design, Development and Deployment*. Springer Nature. ISBN: 978-3-031-39243-6.

Part VI
Forensics in IoT, Wearables, and Environmental Surveillance

Forensic Analysis of Wearable Technology: Insights from Smartwatch Data

Sapna V. M., Navya Peram, Namita Patil, Vishwajeet Raut, Ankit Borkar, and Prasad B Honnavalli

1 Introduction

In today's world, wearable technology is no longer just a novelty—it has become a staple of daily life. Smartwatches, in particular, have emerged as essential gadgets that not only serve as fashion accessories but also as powerful tools for tracking health metrics, monitoring physical activities and staying connected with various services. These devices have transformed how individuals maintain their fitness, keep an eye on their health and engage with the digital world. With the recent advances in AI, they are now capable of supporting more sophisticated functionalities, including enhanced monitoring, threat detection, and digital forensic analysis, as highlighted in study [1].

Smartwatches are equipped with an impressive array of sensors, acting as extensions of the human body, continuously gathering and storing data. This information can offer valuable insights into daily routines, physical health and even whereabouts. Several studies [2–4] have demonstrated that smartwatches with their ability to collect and store health, location, and activity data, can serve as valuable sources of digital evidence. While the widespread adoption of smartwatches brings numerous benefits, it also poses significant challenges for digital forensic experts. Furthermore, emerging technologies in society, such as the Google Glass [5] and Meta Smart Glasses, are being studied as potential sources of digital evidence, thereby expanding the scope of digital forensics beyond traditional devices.

In contrast to prior studies [6–9] that have concentrated on global brands such as Sony, Samsung, Fitbit, and Garmin, this work focuses primarily on locally

These authors contributed equally to this work.

Sapna V. M. (✉) · N. Peram · N. Patil · V. Raut · A. Borkar · P. B Honnavalli
Department of Computer Science and Engineering, PES University, Bangalore, Karnataka, India

S. S. Iyengar et al. (eds.), *Artificial Intelligence Driven Forensics*, Signals and Communication Technology, https://doi.org/10.1007/978-3-031-98036-7_28

manufactured Indian brands, including NoiseFit and Boat, along with a global brand Amazfit. This paper delves into the world of smartwatch forensics, shedding light on the key challenges and evaluating existing methodologies that enhance forensic analysis of these devices. The study focuses on smartwatches like BoAt Xplorer RTL, Amazfit Band 5 and Noise Colorfit Pulse, using free and open-source tools such as mitmproxy and DB Browser for SQLite. By doing so, the extent to which data can be captured and analysed from these smartwatches and their associated mobile applications is assessed. Ultimately, the goal is to advance forensic investigation techniques in the context of wearable technology.

2 Literature Survey

Several papers were reviewed to gain insights into the forensic analysis and extraction of different watches. Some of the papers are described below.

A study [10] investigated the forensic implications of data generated and stored by the Fitbit Versa 2 smartwatch on a rooted Samsung Galaxy S6 (Android 7) smartphone. Using tools like MSAB XRY, Magnet AXIOM and DB Browser for SQLite, the researchers were able to extract sensitive data like heart rate, GPS location and even credit card details which were all stored in an unencrypted form. Additionally personal and health-related data was found stored in plaintext. Although this data could be used in criminal investigations, it could pose potential risks for phishing attacks. However, the study noted that social media message notifications were not stored. Some limitations include the use of an outdated Android 7 version smartphone and the reliance on proprietary, non-open-source forensic tools.

Another similar paper [11] focuses on the forensic analysis of the Fitbit Versa smartwatch by comparing the artefacts stored by the smartwatch in a rooted android (Google Pixel 2XL) and an iOS (iPhone 7 Plus) smartphone, running Android 10 and iOS 13, respectively. Cellebrite UFED, MSAB XRY and Genymotion were the tools used in identification and extraction of the data. In the iOS device, both the tools successfully extracted similar information regarding the smartwatch, user profile, GPS, heart rate, sleep data and even messages. Even deleted messages and logs were recovered by the tools. However, the tools weren't able to extract any significant data from the android device as the data was stored in a protected data folder. Genymotion was used to emulate the android smartphone and gain access into the protected folder and extract all the data as the iOS device except for the deleted logs and messages.

The paper [12] focuses on the forensic analysis of the fitness trackers Xiaomi Mi Band 2, Fitbit Charge 2 and Huawei Band 2 Pro. The paper focuses on the kind of data that can be collected from these devices such as the steps taken, heart rate, sleep pattern and the GPS locations. This information can provide crucial evidence in legal investigations. The researchers have developed an open-source tool to standardize the forensic analysis process. This tool is capable of extracting data from the devices using Bluetooth communication, analysing smartphone applications linked to the fitness devices and examining Internet traffic generated by the devices. This tool allows the investigators to reconstruct the user's daily activities and also detect unusual patterns or behaviours that could be relevant in a criminal case. Although some encrypted data was not extracted highlighting the limitation of this tool.

The paper [13] investigates the Zepp Life Android application used with the Xiaomi Mi Band 6. The study examines the data stored on two rooted android smartphones, namely, the Xiaomi Mi A2 (android 10) and the BQ Aquaris X Pro (android 8). The data found in databases and XML files were analysed. They are used to identify the types of data that could be used as digital evidence. These include the GPS coordinates, biometric data such as heart rate and sleep patterns and activity logs. It also focuses on the development of two open-source tools which can be used to assist forensic practitioners in analysing Zepp Life data. They include a python script named ZL_std and a module named ZL_autopsy for the Autopsy digital forensic software.

The paper [14] analyses the data stored by six fitness applications on a rooted android smartphone, Samsung A40 and a Garmin Vivosmart 4 smartband. The applications consist of Adidas Running, MapMyWalk, Nike Run Club, Pumatrac, Runkeeper and Strava. Data is extracted from SQLite databases, XML files and log files. This data is used to identify forensic artefacts such as GPS coordinates, timestamps, user account details and workout metrics. The extracted data is then processed using python scripts and incorporated into the Android Logs events and Protobuf Parser (ALEAPP) framework for efficient forensic analysis. However, the paper doesn't analyse the real-time data transmission between applications and cloud services.

The paper [15] investigates multiple fitness applications focusing on their forensic artefacts and mechanisms of data storage. It follows the standard digital forensic analysis process based on the National Institute of Standards and Technology (NIST) framework. This includes the identification, collection, organization and presentation of data. It examined the applications MapMyFitness, Fitbit, NoiseFit, Nike, StepSetGo and MyFitnessPal which were installed on the android smartphone, Lenovo ZUK Z2. These were analysed to identify digital evidence such as user profiles, step count, heart rate, GPS coordinates and timestamps. Some of the gaps in the paper include the limited scope of testing as it was tested only on android devices. It also did not address the encryption or obfuscation mechanisms used by some of the apps which could impact the reliability of forensic investigations.

These studies have shown that sensitive data such as personal account details, heart rate, sleep details, GPS logs and even deleted messages can be extracted from the smartwatches. iOS extractions revealed more information than android extractions which needed rooting to extract the information. However, gaps remain as many smartwatch brands are yet to be explored. These findings enhance forensic methodologies and expose security risks like unencrypted GPS logs and personal data storage which are privacy concerns.

Furthermore, integrating various technological domains can greatly enhance the precision and depth of smartwatch forensic analysis. Leveraging techniques from machine learning, artificial intelligence enables more comprehensive investigations and improves the accuracy of forensic findings. Recent research in both IoT security and digital forensics offer valuable methodologies that could be applied to smartwatch forensics. In a research [16], machine learning algorithms were used to detect attacks and anomalies in IoT, showing techniques to catch unusual patterns and possible security compromise. They could be applicable in detecting anomalies from smartwatch information. Similarly, another study [17] did a survey of data storage and retrieval methods on encrypted cloud data that also offers important considerations in managing and protecting sensitive forensic evidence, something that is particularly relevant when dealing with encrypted wearable device data. Lastly, the paper [18] discusses the convergence of AI and knowledge graphs in forensic science, providing unique solutions for correlating sophisticated datasets and augmenting evidence analysis. The paper [19] proposes the RSA-CRT scheme and efficient index-building algorithm, which can enhance smartwatch forensics by providing secure data sharing and retrieval from cloud storage. Another study [20] suggests the Domain and Range-Specific Multi-keyword Search (DRSMS) scheme, which enables the fast and secure retrieval of encrypted logs from cloud storage, so that the confidential investigation data remains protected. Forensic analysts can efficiently locate relevant digital evidence (e.g. timestamps, locations or activities) while preventing unauthorized access or data breaches by minimizing search time and index storage space. By integrating these techniques, forensic investigators can efficiently manage encrypted smartwatch data, streamline access control and improve forensic data retrieval speed while maintaining security. These techniques are promising for automating and optimizing data interpretation in smartwatch forensic analysis. Future research could focus on cloud data retrieval, forensic techniques for non-rooted devices and cross-platform comparisons to strengthen the reliability of smartwatch forensics.

3 Materials and Technology

To extract the data from the smartwatches, various hardware devices and software applications were used.

3.1 Hardware Devices

Three smartwatches, namely, Amazfit Band 5, BoAt Xplorer RTL and Noise Colorfit Pulse, were considered. A rooted phone, Xiaomi Poco F1 was used for extracting the information from the smartwatches.

The first smartwatch is the Noise Colorfit Pulse manufactured by Noise. The Noise company is known for its affordable products and has been consistently ranked first amongst the most popular smartwatch brands in India for several consecutive years. The Noise Colorfit Pulse smartwatch uses the NoiseFit app for synchronizing and tracking its data. It uses a proprietary OS developed by Noise. It offers several features such as heart rate monitoring, SpO2 monitoring, sports modes, activity data and sleep tracking. The second smartwatch that we explored was the BoAt Xplorer RTL from BoAt (Imagine Marketing Limited). The required information was collected from its mobile application, BoAt ProGear. A proprietary operating system is being used by the smartwatch. It consists of several health and activity features. The third smartwatch Amazfit Band 5 is a brand of the Zepp Health Corporation, formerly known as Huami. It uses the Zepp application, previously known as the Amazfit App, from which the forensic data was collected and extracted. It consists of multiple features such as a heart rate monitor, SpO2 sensor, sleep tracker and stress tracker. It runs on a proprietary operating system.

3.2 Software

Data was extracted from the mobile applications of the smartwatches: Zepp for the MI Band, boAt ProGear for the boAt Xplorer and NoiseFit for the Noise Colorfit Pulse. Open-source tools like mitmproxy were used to capture the data being sent from the smartwatch to the mobile app and DB Browser for SQLite to view the database's information. Since the smartwatches use proprietary operating systems, extracting information directly from the devices was not feasible. Therefore, the data had to be obtained through their corresponding mobile applications instead.

mitmproxy is an open-source, interactive HTTP/HTTPS proxy that allows the user to interact with, modify and replay network traffic. It's most commonly used for debugging, penetration testing and analysing network requests. It is able to capture the HTTP/HTTPS requests and responses making it an easy and resourceable tool for real-time interception and modification of traffic. One of its most important features is its ability to act as a man-in-the-middle (MITM) proxy to inspect encrypted HTTPS traffic. We used this feature to intercept the data between the smartwatch and its application. It can save and export the traffic logs which is extremely beneficial, as the data can be analysed later when required. Additionally, it offers both a web-based interface and a terminal-based UI increasing the ease of use.

DB Browser for SQLite is a free and open-source database tool designed for viewing and analysing SQLite database files. It provides a graphical user interface (GUI) that allows users to view the database tables. This tool is widely used in digital forensics and data analysis to examine the data stored in SQLite databases, such as those found in mobile apps and forensic extractions. It is able to view raw data, decode the stored information and reconstruct database structures, making it a valuable asset in data recovery and forensic investigations.

4 Methodology

This study follows the standard digital forensic analysis process based on the National Institute of Standards and Technology(NIST) framework. It consists of the steps—identification, collection, analysis and presentation similar to the framework used by Sinha et al. [sinha et al]. The first stage consists of identifying the incident or the device. The collection phase consists of acquiring the necessary evidence and identifying the required data while discarding any redundant or unrequired data. The analysis phase involves analysing the data and then making the corresponding conclusions. The final presentation phase involves presenting the obtained conclusions to the required personnel. Our primary goal is to systematically acquire and examine forensic evidence from smartwatches and their paired mobile applications. To achieve this, two complementary data extraction approaches were implemented on a Xiaomi Pocophone F1 running Android 10 (rooted). The workflow followed is given in Fig. 1.

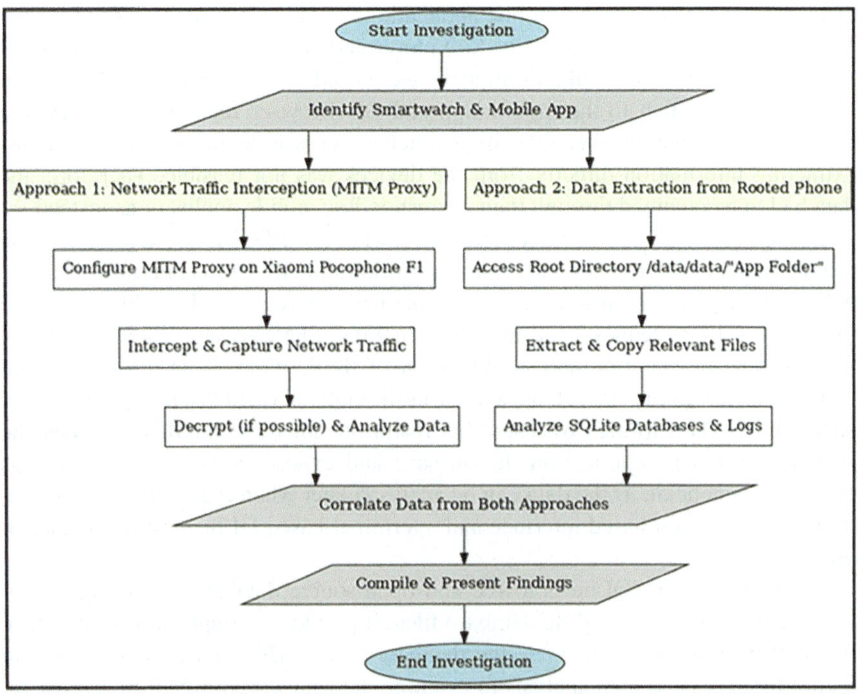

Fig. 1 Workflow for forensic analysis of smartwatch data

4.1 Identification

This phase involves identifying the necessary devices and setting up the environment for the collection and analysis of forensic data. A rooted phone, Xiaomi Poco F1, was selected for the investigation, and the respective companion applications for each smartwatch were installed from Google Play Store. Separate Gmail accounts were used for creating accounts. To facilitate the extraction and analysis of data, both the tools, mitmproxy and DB Browser for SQLite were installed and configured. Each smartwatch was paired with its corresponding application:

1. boAt Xplorer RTL with the boAt ProGear app
2. Amazfit Band 5 with the Zepp app
3. Noise ColorFit Pulse with the NoiseFit app.

4.2 Collection

After creating the required accounts, each smartwatch was then synced with its respective applications, for the transfer of the stored smartwatch data to the application. The necessary data was collected using two different approaches.

4.2.1 Data Extraction from Rooted Phones

The first approach involves accessing the mobile device at a deeper level by leveraging its rooted status, which provides full access to the file system. The details of this method are as follows:

1. Device Context: Using a Xiaomi Pocophone F1 running Android 10 (rooted) enabled the bypassing of standard security restrictions and access to the device's root directory where the smartwatch app stores its data.
2. Root Access and File System Exploration: A file explorer tool was employed to access the root file system. Specifically, the path /root/data/data/"Folder Name of the app" was navigated to, where all the data for each individual application is stored. By copying the files from this location, retrieval of local logs, cached files, configuration settings, and application databases that are not transmitted over the network was possible.
3. Utilization of Forensic Tools: In addition to the file explorer, tools such as DB Browser for SQLite were used to parse and analyse the locally extracted databases. This method effectively complemented the network interception approach by revealing locally cached information and system-level logs.
4. Advantages: Extracting data directly from the phone provides a more comprehensive view of forensic evidence, especially when some information is stored

exclusively on the device. This approach verifies and supplements the findings obtained through network analysis, resulting in a robust investigation.

4.2.2 Network Traffic Interception Using mitmproxy

This approach focuses on the communication between the smartwatch application and its associated cloud servers. By using mitmproxy, an open-source intercepting proxy, real-time network traffic generated during the synchronization process was captured and analysed. Key details of this approach include:

1. Device Context: The experiments were conducted on a Xiaomi Pocophone F1 running Android 10. The rooted status of the device ensured that all necessary configurations could be implemented without typical operating system restrictions.
2. Data Collection: mitmproxy was configured to intercept encrypted communications between the mobile app and the cloud. This setup enabled the capture of transmitted data including personal information, health measurements, call logs, and message histories.
3. Data Decryption and Analysis: Once intercepted, the traffic was decrypted (where possible) to reveal the underlying data. This process facilitated the analysis of transmission patterns, identification of potential vulnerabilities in encryption protocols, and extraction of forensic artefacts, which are critical when traditional evidence is limited.
4. Advantages: This method is particularly useful when direct access to device storage is challenging or when much of the data is maintained in the cloud. It provides insights into real-time data exchange and highlights security loopholes in data transmission.

4.3 Analysis

This phase involves the analysis of the data obtained by using open-source tools. The collected data can be further divided into separate parameters like health parameters which consist of heart rate and SpO2, the activity data consisting of the daily steps, calories and distance count. This also included analysing the monthly activity data in some watches. Furthermore, details of user profiles were present which were also analysed.

4.4 *Presentation*

The final phase involves compiling and presenting the results of the analysis in a clear and concise format. The conclusions drawn from the study can be used to support further investigations or legal proceedings.

5 Results

5.1 *BoAt Xplorer RTL*

The data for BoAt Xplorer RTL was extracted by going to the database directories of the BoAt progear app in a rooted phone. The databases were then opened in DB Browser for SQLite, and all the information was obtained from there.

The data stored in the database can be viewed in Fig. 2. It includes important data like user information which is stored in the UserDetails table, Blood oxygen

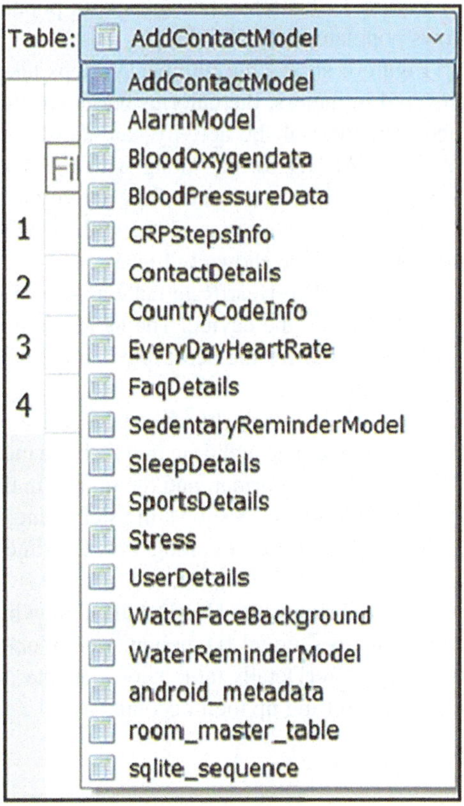

Fig. 2 All the information stored in the database

Fig. 3 No. of rows in filled tables

Table: sqlite_sequence

	name	seq
	Filter	Filter
1	CountryCodeInfo	224
2	CRPStepsInfo	138
3	EveryDayHeartRate	588
4	BloodOxygendata	2
5	SleepDetails	5
6	UserDetails	4
7	AddContactModel	4

levels which is stored in the BloodOxygendata table and blood pressure information which is stored in the BloodPressuredata table. It also includes some less sensitive data like steps information, sleep details and stress. Some common data like country code info, faq details and watch face background can also be seen.

Figure 3 illustrates the tables with populated rows and indicates the number of rows populated in each table.

Figure 4 shows the number of steps taken by the user along with the distance covered in meters, the calories burnt and the duration it took in seconds. The date and start time of the activity can also be seen. The date of the activity is in the YYYY/MM/DD format along with the device name and the mac address of the smartwatch. The start time of the activity is stored in the system time format.

The various countries along with their country codes are stored in the CountryCodeInfo table shown in Fig. 5.

The EveryDayHeartRate table as shown in Fig. 6 shows the heart rate measurements taken by the device. The table contains the mac address of the device along with the values for the heart rate and the date when the heart rate was measured in system time.

Figure 7, shows the blood oxygen value of the user measured by the smartwatch along with the date and time in which the measurement was taken. The date is in the YYYY/MM/DD format, and the time is in the system time format.

The table in Fig. 8 shows the sleep date, total time slept, amount of restful and light sleep the user had along with the time that they went to bed, slept and woke up.

Figure 9 shows the UserDetails table which shows the name, age, gender, DoB, phone number, email ID, height, weight and the daily step and calorie goals of the user. The UserDetails table shows all the information that the user entered while they were setting up their account.

Table: CRPStepsInfo

	Id	activityDate	deviceName	macAddress	starttime	steps	distance	calories	duration
	Filter	Filter	Filter	Filter	Filter	Filter	Filter	Filter	Filter
1	1	2024/08/06	3	AC:98:B1:17:22:CD	1722923793	2863	2207	111829	1491
2	2	2024/07/30	3	AC:98:B1:17:22:CD	1722326985	5827	4485	226498	3031
3	3	2024/07/30	3	AC:98:B1:17:22:CD	1722335274	5905	4544	229420	3067
4	4	2024/07/30	3	AC:98:B1:17:22:CD	1722338070	5971	4594	231892	3095
5	5	2024/07/30	3	AC:98:B1:17:22:CD	1722346140	6252	4811	242906	3246
6	6	2024/07/30	3	AC:98:B1:17:22:CD	1722349799	10182	7798	390108	5354
7	7	2024/07/30	3	AC:98:B1:17:22:CD	1722352559	10996	8417	420597	5795
8	8	2024/07/30	3	AC:98:B1:17:22:CD	1722356811	11377	8711	435453	5970
9	9	2024/07/30	3	AC:98:B1:17:22:CD	1722357911	11502	8806	440135	6042
10	10	1970/01/01	3	AC:98:B1:17:22:CD	0	0	0	0	0
11	11	2024/07/31	3	AC:98:B1:17:22:CD	1722389060	261	198	9775	134
12	12	2024/07/31	3	AC:98:B1:17:22:CD	1722392779	621	471	23260	323
13	13	2024/07/31	3	AC:98:B1:17:22:CD	1722395389	742	563	27792	398
14	14	2024/07/31	3	AC:98:B1:17:22:CD	1722400177	811	616	30376	430
15	15	2024/07/31	3	AC:98:B1:17:22:CD	1722402967	1067	810	39965	551
16	16	2024/07/31	3	AC:98:B1:17:22:CD	1722407095	1576	1197	59030	816
17	17	2024/07/31	3	AC:98:B1:17:22:CD	1722410996	4405	3355	166193	2327
18	18	2024/07/31	3	AC:98:B1:17:22:CD	1722414577	5522	4208	208590	2948
19	19	2024/07/31	3	AC:98:B1:17:22:CD	1722416763	5902	4508	224578	3165
20	20	2024/07/31	3	AC:98:B1:17:22:CD	1722421332	6266	4785	238212	3377
21	21	2024/07/31	3	AC:98:B1:17:22:CD	1722424848	6489	4955	246564	3506
22	22	2024/07/31	3	AC:98:B1:17:22:CD	1722426367	6661	5085	253007	3587
23	23	2024/07/31	3	AC:98:B1:17:22:CD	1722432567	6973	5333	266159	3742
24	24	2024/07/31	3	AC:98:B1:17:22:CD	1722435955	7123	5447	271778	3816
25	25	2024/07/31	3	AC:98:B1:17:22:CD	1722439731	7377	5644	281916	3931

Fig. 4 CPRStepsInfo table

5.2 *Amazfit Band 5*

The data for the Amazfit Band 5 was extracted in a similar manner to the BoAt smartwatch. First, the database directories of the Zepp mobile application were located on a rooted phone. Then, the database information was viewed using an sql viewer github repository [13].

Figure 10 shows the heart rate of the user measured at the particular time. The time is stored in the system time format.

Figure 11 shows the alarm data with the date in the system time format.

Figure 12 displays the user information that was entered while setting up the device. It includes the user ID, name, gender, height and weight of the user.

Figure 13 shows the date of the activity, total distance covered in meters, the pace of the user and the average heart rate.

Figure 14 shows that the smartwatch was set up in a region outside China and also displays a URL that might be used to share data. The row `IS_IN_MAINLAND`

CountryCodeInfo Filter in any column

d	Id1	country	code	fileUrl	min	max	supported	status
1	1	India	+91	https://prod-wearable-s3.s3.ap	10	10	1	active
2	2	Bangladesh	+880	https://prod-wearable-s3.s3.ap	7	15	0	active
3	3	United Arab Emirates	+971	https://prod-wearable-s3.s3.ap	7	12	0	active
4	4	Nepal	+977	https://prod-wearable-s3.s3.ap	7	12	0	active
5	5	Saudi Arabia	+966	https://prod-wearable-s3.s3.ap	2	15	0	active
6	6	United State of America	+1	https://prod-wearable-s3.s3.ap	2	15	0	active
7	7	United Kingdom	+44	https://prod-wearable-s3.s3.ap	2	15	0	active
8	8	Singapore	+65	https://prod-wearable-s3.s3.ap	2	15	0	active
9	9	Canada	+1	https://prod-wearable-s3.s3.ap	2	15	0	active
10	10	Australia	+61	https://prod-wearable-s3.s3.ap	2	15	0	active
11	11	Qatar	+974	https://prod-wearable-s3.s3.ap	2	15	0	active
12	12	Germany	+965	https://prod-wearable-s3.s3.ap	2	15	0	active
13	13	Kuwait	+965	https://prod-wearable-s3.s3.ap	2	15	0	active
14	14	Bhutan	+975	https://prod-wearable-s3.s3.ap	2	15	0	active
15	15	Netherlands	+31	https://prod-wearable-s3.s3.ap	2	15	0	active
16	16	Sri Lanka	+94	https://prod-wearable-s3.s3.ap	2	15	0	active
17	17	France	[illegible]	https://prod-wearable-s3.s3.ap	2	15	0	active

Fig. 5 CountryCodeInfo table

suggests that the Zepp app is currently running in global mode rather than Chinese mode. Even the MAC address of the device can be seen.

Figure 15 shows the smartwatch's information like the operating system version and level, the app version, the sdk version and even the resolution of the device.

Figure 16 presents an internal system table that keeps track of the last assigned auto-increment values for each table with an auto-increment primary key. This table indicates that no recent records have been modified in the `CONFIG` table. The entry `FW_DWON_SUCCESS_INFO` could relate to recent successful firmware downloads. The `ALARM` row suggests that ten alarms were recently created. `TRACKRECORD` shows that only one record has been logged so far.

In Fig. 17, the config table displays the default smartwatch settings, including notifications, step goals and call alerts. It shows that alarm and daily report notifications are currently turned off and that the step goal is set to 8000. It also shows that the notification light colour is set to blue.

5.3 *Noise Colourfit Pulse*

The data for the Noise Colourfit Pulse smartwatch was extracted by using the mitmproxy tool. The tool acted as a middle man between the smartwatch and the mobile application and received all the information that was being sent through the connection. mitmproxy was utilized to capture the data as the database information on the smartphone couldn't be found.

Table: EveryDayHeartRate

	Id	syncDate	deviceName	macAddress	startMeasureTime	heartRateIntegers
1	1	1722882600000	6	AC:98:B1:17:22:CD	0	[{"a":72,"b":1722784839000}]
2	2	1722882600000	6	AC:98:B1:17:22:CD	0	[{"a":84,"b":1722325516000}]
3	3	1722882600000	6	AC:98:B1:17:22:CD	0	[{"a":86,"b":1722326424000}]
4	4	1722882600000	6	AC:98:B1:17:22:CD	0	[{"a":101,"b":1722327325000}
5	5	1722882600000	6	AC:98:B1:17:22:CD	0	[{"a":104,"b":1722328225000}
6	6	1722882600000	6	AC:98:B1:17:22:CD	0	[{"a":101,"b":1722329124000}
7	7	1722882600000	6	AC:98:B1:17:22:CD	0	[{"a":99,"b":1722330025000}]
8	8	1722882600000	6	AC:98:B1:17:22:CD	0	[{"a":90,"b":1722330919000}]
9	9	1722882600000	6	AC:98:B1:17:22:CD	0	[{"a":87,"b":1722331818000}]
10	10	1722882600000	6	AC:98:B1:17:22:CD	0	[{"a":85,"b":1722332717000}]
11	11	1722882600000	6	AC:98:B1:17:22:CD	0	[{"a":82,"b":1722333618000}]
12	12	1722882600000	6	AC:98:B1:17:22:CD	0	[{"a":78,"b":1722334516000}]
13	13	1722882600000	6	AC:98:B1:17:22:CD	0	[{"a":72,"b":1722335417000}]
14	14	1722882600000	6	AC:98:B1:17:22:CD	0	[{"a":66,"b":1722336322000}]
15	15	1722882600000	6	AC:98:B1:17:22:CD	0	[{"a":68,"b":1722337225000}]
16	16	1722882600000	6	AC:98:B1:17:22:CD	0	[{"a":88,"b":1722338124000}]
17	17	1722882600000	6	AC:98:B1:17:22:CD	0	[{"a":75,"b":1722339023000}]
18	18	1722882600000	6	AC:98:B1:17:22:CD	0	[{"a":82,"b":1722339917000}]
19	19	1722882600000	6	AC:98:B1:17:22:CD	0	[{"a":83,"b":1722340818000}]
20	20	1722882600000	6	AC:98:B1:17:22:CD	0	[{"a":90,"b":1722341719000}]
21	21	1722882600000	6	AC:98:B1:17:22:CD	0	[{"a":81,"b":1722342617000}]
22	22	1722882600000	6	AC:98:B1:17:22:CD	0	[{"a":100,"b":1722343516000}
23	23	1722882600000	6	AC:98:B1:17:22:CD	0	[{"a":79,"b":1722344424000}]
24	24	1722882600000	6	AC:98:B1:17:22:CD	0	[{"a":72,"b":1722345321000}]
25	25	1722882600000	6	AC:98:B1:17:22:CD	0	[{"a":67,"b":1722346224000}]

Fig. 6 EveryDayHeartRate table

Table: BloodOxygendata

	Id	activityDate	b_oxygen_time	b_oxygen_value
1	1	2024/08/06	1722784923	99
2	2	2024/08/06	1722927773	99

Fig. 7 BloodOxygendata table

Figure 18 shows the device information that we accessed through the mitmproxy tool. It includes data like the smartphone device manufacturer, which was used for rooting and the smartwatch model, name, device id and its access token along with the country and language. It also displays the purpose of the request which in this case was the heart rate data and the format in which the data was obtained which

	Id	SleepDate	deviceName	macAddress	totalTime	restfulTime	lightTime	soberTime	details	actualWakeupTime	actualBedTime	actualSleepTime
	Filter	Filter	Filter	Filter	Filter	Filter	Filter	Filter	Filter	Filter	Filter	Filter
1	1	1722277800000	6	AC:98:B1:17:22:CD	36060	17100	18960	0	[{"a":2520,"b":0,"c":0,"d":1},{"a":...	1722733860	1722364620	1722369000
2	2	1722709800000	6	AC:98:B1:17:22:CD	38040	25440	12600	0	[{"a":2580,"b":0,"c":0,"d":1},{"a":...	1722897720	1722795480	1722806400
3	3	1722796200000	6	AC:98:B1:17:22:CD	7440	2940	4500	0	[{"a":3120,"b":0,"c":0,"d":1},{"a":...	1722905220	1722897780	1722903840
4	4	1722882600000	6	AC:98:B1:17:22:CD	37500	23700	13800	0	[{"a":2760,"b":0,"c":0,"d":1},{"a":...	1723060740	1722962880	1722970500
5	5	1722969000000	6	AC:98:B1:17:22:CD	16680	12300	4380	0	[{"a":2520,"b":0,"c":0,"d":1},{"a":...	1723077540	1723060860	1723064820

Fig. 8 SleepDetails table

	Id	gender	age	phoneNumber	emailId	firstName	lastName	height	height_unit	weight	weightUnit	goal_calorie	goal_steps	name	dob	profilePic
	Filter	Filter	Filter	Filter	Filter	Filter	Filter	Filter	Filter	Filter	Filter	Filter	Filter	Filter	Filter	Filter
1	4	Female	21	NULL	[redacted]@gmail.com	namita patil		5.7	Feet	57	kg	650	10000	namita patil	23/01/2003	NULL

Fig. 9 UserDetails table

HEART_RATE (3 rows)

```
SELECT * FROM 'HEART_RATE' LIMIT 0,30
```

TYPE	DEVICE_TYPE	DEVICE_SOURCE	TIME	HR	STATE	TIME_ZONE	DEVICE_ID
2	0	73	1724059724	109	1	19800000	D059B4FFFEEBB221
2	0	73	1724061744	96	1	19800000	D059B4FFFEEBB221
2	0	73	1724061890	88	1	19800000	D059B4FFFEEBB221

Fig. 10 Heart rate information

ALARM (10 rows)

```
SELECT * FROM 'ALARM' LIMIT 0,30
```

_id	CALENDAR	ENABLED	IS_UPDATE	M_DAYS	M_SMART_WAKEUP_DU...	VISIBLE	IS_SMART_WAKEUP	INDEX	IS_S
1	1724031015000	0	0	31	0	1	1	0	0
2	1724031015000	0	0	127	0	1	1	1	0
3	1724031015000	0	0	0	0	1	1	2	0
4	1724059275385	0	0	0	0	0	1	3	0
5	1724059275388	0	0	0	0	0	1	4	0
6	1724059275389	0	0	0	0	0	1	5	0
7	1724059275392	0	0	0	0	0	1	6	0
8	1724059275396	0	0	0	0	0	1	7	0
9	1724059275398	0	0	0	0	0	1	8	0
10	1724059275403	0	0	0	0	0	1	9	0

Fig. 11 Alarm data

USER_INFOS (1 rows)

```
SELECT * FROM 'USER_INFOS' LIMIT 0,30
```

USER_ID	NAME	BIRTHDAY	AVATAR_URL	AVATAR_PATH	GENDER	HEIGHT	SYNCED	WEIGHT	TARGET_WEIGHT	CITY	LONGI
3305351217	navyaperam16	2002-7			0	170	1	51	-1		

Fig. 12 User information

TRACKRECORD Export

```
SELECT * FROM 'TRACKRECORD' LIMIT 0,30
```

Execute

_id	SOURCE	TYPE	DATE	TRACKID	DISTANCE	COSTTIME	CAL	PACE	SFREQ	AVGHR	FFPERCENT	V	ENDTIME	MAXRTP	MINRTP
1	173	8	2024-08-19	1724059420	214	162	18	0.7559999823570251	88	150	-1	12	1724059587	3.8496580123901367	0.4953720867633

Fig. 13 Steps records

HM_PROPERTY

```
SELECT * FROM 'HM_PROPERTY' LIMIT 0,30
```

KEY	VALUE	STATUS
IS_IN_MAINLAND	false	1
huami.mifit.sport.shareRouter	https://fe-cdn.huami.com/sport-history/index.html#/	1
huami.mifit.user.settings.synctime	{"D0:59:B4:EB:B2:21":{"single_stress":1724062001000}}	0

Fig. 14 Property information

context Export

```
SELECT * FROM 'context' LIMIT 0,30
```

Execute

	appVersionCode	appVersionName	sdkVersionName	osLevel	osVersion	channel	locale	network	resolution	eventVersion	packageName	brand	model
...	151386	8.11.5-play	2.0.2	29	10	play	en_IN	wifi	1080*2027	1	com.huami.watch.hmwat...	Xiaomi	POCO F1

Fig. 15 Smartwatch information

sqlite_sequence

```
SELECT * FROM 'sqlite_sequence' LIMIT 0,30
```

name	seq
CONFIG	0
FW_DWON_SUCCESS_INFO	7
ALARM	10
TRACKRECORD	1

Fig. 16 Sequence table

CONFIG Export

```
SELECT * FROM 'CONFIG' LIMIT 0,30
```

Execute

_id	ALARM_NOTIFY_EN...	DAY_REPORT_NOTI	DISCONNECTED_RE...	ENABLE_CONNECTE...	GOAL_STEPS_COUNT	LIGHT_COLOR	IN_COMING_CALL
0	0	OFF		0	8000	BLUE	259

Fig. 17 User default settings

is JSON. This shows that the heart rate data is stored remotely on the Noise server and is obtained by using the access token. This token is used to verify the user's

Fig. 18 Device information

identity and authorize access to the smartwatch's data. If exposed it can be misused to access personal health data.

Figure 19 displays the profile data of the user, along with the authentication information. It also displays the access and refresh token used for authentication.

Figure 20 displays the user's profile data such as their personal details and their goals.

Figure 21 displays the heartbeat data collected over all the days of a week. It also consists of the cumulative heartbeat data containing the average, minimum and maximum rate over the week. This data is displayed as a JSON format and is obtained through mitmproxy. Similarly monthly data and daily data along with hourly breakups can also be found.

Figure 22 shows the daily data of multiple attributes such as steps, calories and distance. It displays the average calories burnt in a single day along with the average distance and steps covered. It stores the values in an hourly format. This could also help the forensic investigators understand the user's activities during a particular period of time. Similarly a weekly format can also be seen.

Figure 23 displays the number of steps that were taken that week along with the total distance covered in meters and the number of calories burnt.

Figure 24 presents the blood oxygen data of that particular day. Similarly, weekly and monthly blood oxygen data can also be obtained.

The type of activities performed by the user can be viewed in Fig. 25. Multiple parameters are described for each activity, these include the activity type and id,

```
{
    "data": {
        "is_pwd": false,
        "message": "OTP verified",
        "token": {
            "access_token": "eyJhbGciOiJIUzI1NiIsInR5cCI6IkpXVCJ9.e
            "refresh_token": "eyJhbGciOiJIUzI1NiIsInR5cCI6IkpXVCJ9.
        },
        "user": {
            "auth_token": "632fdbd5-fc6b-4696-aebc-cc8ccacac949",
            "country": "IN",
            "country_code": "",
            "email": "navyaperam16@gmail.com",
            "end_game": -1,
            "first_name": "Navy",
            "google_uid": "116592841337118985091",
            "id": 10501118,
            "image_url": "https://lh3.googleusercontent.com/a/AEdFT
            "interests": [
                {
                    "id": 1,
                    "name": "Swimming"
                },
                {
                    "id": 2,
                    "name": "Walking"
                },
                {
                    "id": 3,
                    "name": "Cycling"
                },
                {
                    "id": 5,
                    "name": "Better Sleep"
                },
                {
                    "id": 6,
                    "name": "Productivity"
                }
            ],
            "is_verified": 1,
            "language": "en",
            "last_name": "",
            "location": null,
```


Fig. 19 User authentication profile data

```
            "notifications_enabled": 1,
            "offset": 330,
            "password_digest": null,
            "timezone": "Asia/Kolkata",
            "user_goals": {
                "calories_goals": 240,
                "distance_goals": 6000,
                "id": 11210960,
                "sleep_goals": 8,
                "step_goals": 6000,
                "unit_system": "Metric",
                "user_id": 10501118
            },
            "user_info": {
                "dob": "[redacted],
                "gender": "Female",
                "height": 161,
                "step_length": 70,
                "weight": 50
            },
            "vendor_id": null,
            "vendor_ids": null
        }
    },
    "success": true
}
```


Fig. 20 User profile data

Fig. 21 Weekly heart rate data


```
"data": {
    "cumulative": {
        "avg": 90,
        "max": 123,
        "min": 69
    },
    "heart_rates": [
        {
            "avg": 92,
            "date": "2024-08-02",
            "max": 120,
            "min": 69,
            "resting_hr": 0
        },
        {
            "date": "2024-08-03"
        },
        {
            "date": "2024-08-04"
        },
        {
            "avg": 87,
            "date": "2024-08-05",
            "max": 97,
            "min": 78,
            "resting_hr": 0
        },
        {
            "date": "2024-08-06"
        },
        {
            "date": "2024-08-07"
        },
        {
            "avg": 93,
            "date": "2024-08-08",
            "max": 123,
            "min": 72,
            "resting_hr": 0
        }
    ],
    "history_type": "weekly"
},
"message": "",
"success": true,
"time": "1723109279459"
```

```
{
    "data": {
        "avg": {
            "calories": 132,
            "distance": 2557,
            "steps": 3874
        },
        "history_type": "daily",
        "step_activities": [
            {
                "active_time": 0,
                "calories": 225,
                "date": "2024-08-02",
                "distance": 4345,
                "hourly_breakup": [
                    {
                        "active_time": 0,
                        "calories": 0,
                        "date": "02/08/2024",
                        "distance": 195,
                        "hour_of_the_day": 8,
                        "steps": 296
                    },
                    {
                        "active_time": 0,
                        "calories": 0,
                        "date": "02/08/2024",
                        "distance": 765,
                        "hour_of_the_day": 9,
                        "steps": 1160
                    },
                    {
                        "active_time": 0,
                        "calories": 0,
                        "date": "02/08/2024",
                        "distance": 197,
                        "hour_of_the_day": 10,
                        "steps": 299
                    },
                    {
                        "active_time": 0,
                        "calories": 0,
                        "date": "02/08/2024",
                        "distance": 658,
                        "hour_of_the_day": 11,
                        "steps": 997
```


Fig. 22 Daily distance data

```
{
    "data": {
        "avg": {
            "calories": 132,
            "distance": 2557,
            "steps": 3874
        },
        "history_type": "weekly",
        "step_activities": [
            {
                "active_time": 0,
                "calories": 225,
                "date": "2024-08-02",
                "distance": 4345,
                "steps": 6583
            },
            {
                "active_time": 0,
                "calories": 0,
                "date": "2024-08-03",
                "distance": 9,
                "steps": 14
            },
            {
                "date": "2024-08-04"
            },
            {
                "active_time": 0,
                "calories": 119,
                "date": "2024-08-05",
                "distance": 2309,
                "steps": 3499
            },
            {
```


Fig. 23 Weekly distance data

calories burnt, distance, steps and the heart rate average during the activity. The activity type can be chosen from various activities in the list such as outdoor running, walking, outdoor cycling, trial run, trekking, treadmill, indoor cycling and yoga. It also includes the duration of the activity in system time. The time and date at which the activity was performed are also visible.

The sleep data of the user can be extracted through this method as well, as seen by the format in Fig. 26.

Figure 27 displays the challenge "Calorie Callin' Challenge," one of the multiple challenges undertaken by the smartwatch users. It contains the start and end dates

Fig. 24 Daily blood oxygen data


```
{
    "data": {
        "blood_oxygen": [
            {
                "date": "2024-07-30"
            },
            {
                "date": "2024-07-31"
            },
            {
                "date": "2024-08-01"
            },
            {
                "date": "2024-08-02"
            },
            {
                "date": "2024-08-03"
            },
            {
                "date": "2024-08-04"
            },
            {
                "count": 99,
                "date": "2024-08-05",
                "hourly_break_up": [
                    {
                        "date": "05/08/2024",
                        "time": "15:02",
                        "value": 99
                    }
                ],
                "max_count": 99,
                "min_count": 99
            }
        ],
        "cumulative": {
            "count": "99.0",
            "max": 99,
            "min": 99
        },
        "history_type": "daily"
    },
    "message": "",
    "success": true
}
```

```
{
    "data": {
        "activities": [
            {
                "activity_type": "walking",
                "calories": 105,
                "date": "2024-07-30",
                "distance": 202,
                "duration": 209633,
                "heart_rate_average": 104,
                "id": 248314911,
                "steps": 296,
                "time": "22:43:41"
            },
            {
                "activity_type": "treadmill",
                "calories": 131,
                "date": "2023-10-01",
                "distance": 0,
                "duration": 792,
                "heart_rate_average": 131,
                "id": 222745695,
                "steps": 1005,
                "time": "19:31:01"
            }
        ]
    },
    "message": "",
    "success": true,
    "time": "1722850028062"
}
```


Fig. 25 Activity data

and times of the particular challenge. It also contains information regarding the participants' profiles and their profile names.

Figures 28 and 29 display the messages posted by participants on the dashboard during past challenges and the corresponding comments by other participants. The commenters' id's and usernames along with the date and time at which their comments were posted can also be viewed.

Fig. 26 Sleep data


```
{
    "data": {
        "avg_awake": null,
        "avg_bedtime": "0:0",
        "avg_deep": null,
        "avg_duration": 0,
        "avg_light": null,
        "avg_readiness_score": 0,
        "avg_sleep_score": 0,
        "avg_sober": null,
        "bo_max": null,
        "bo_min": null,
        "breakup_duration": [
            {
                "date": "2024-07-30"
            },
            {
                "date": "2024-07-31"
            },
            {
                "date": "2024-08-01"
            }
```



```
content-security-policy: default-src 'self';base-uri 'self';block-all-mixed-content;font-src 'self' https: data:;frame-ancestors 'self';img-src 'self' data:;object-src 'none';script-src 'self';script-src-attr 'none';style-src 'self' https: 'unsafe-inline';upgrade-insecure-requests
x-dns-prefetch-control: off
expect-ct: max-age=0
x-frame-options: SAMEORIGIN
strict-transport-security: max-age=15552000; includeSubDomains
x-download-options: noopen
x-content-type-options: nosniff
x-permitted-cross-domain-policies: none
referrer-policy: no-referrer
x-xss-protection: 0
access-control-allow-headers: x-access-token, Origin, Content-Type, Accept
etag: W/"369-qo2t214Qvv6LMp96aUfm5oU7u0c"
JSON                                                    Edit  Replace  View: auto
{
    "data": {
        "challenge_list": [
            {
                "avatar": [
                    "https://images.gonoise.com/mobileApp/production/users/4535802/thumbnail-image-1686070447070.png",
                    "https://lh3.googleusercontent.com/a/ACg8ocIhIAJDmS8be7YesdPgVJK7CnuCtK-mirn9q7aMdMaFQrSSGw=s200",
                    "https://lh3.googleusercontent.com/a/ACg8ocL-claHQX94HEmyLlK8LpE4Zo292QFKco2IIU_JsKj=s96-c"
                ],
                "challenge_name": "Calorie Callin' Challenge",
                "challenge_type": "calories",
                "end_date": "2024-08-21 18:29:54",
                "id": 149,
                "image_url": "https://noise-images.s3.ap-south-1.amazonaws.com/challenges/prod/Calorie%20Callin%E2%80%99%20Challenge/images/list-1722598
                "participant_str": "Akhil Shaji and 13340 others",
                "start_date": "2024-08-11 18:30:00",
                "status": "upcoming"
            }
        ],
        "currentTime": "2024-08-08 09:36:46",
        "dashboard_image": "https://challenges-cdn.gonoise.com/custom_challenges/group_1321314710.png",
        "received_request": []
    },
    "success": true
```


Fig. 27 User challenge information

```
{
    "admin_post_cta": null,
    "caption": "Due to the demise of my naniji I won't be active for a week or so, I'll try to post some morning posts but there will be no activity log for some days."
    "comment": {
        "comment": "Ok",
        "comment_id": 1622135,
        "image_url": "https://images.gonoise.com/mobileApp/production/users/4722860/thumbnail-image-1685707169323.png",
        "is_my_comment": false,
        "name": "Bithika Pramanik",
        "post_id": 3555534,
        "updated_at": "2024-08-08T09:18:40.000Z",
        "user_id": 4722860,
        "user_type": "user"
```


Fig. 28 Challenge participant's messages


```
"admin_post_cta": null,
"caption": "A new day 🌞 \nGood morning 😃",
"comment": {
    "comment": "Aap 01",
    "comment_id": 1622111,
    "image_url": "https://lh3.googleusercontent.com/a/ACg8ocIFO2RncqhkAsXOfgaXCttNTu27UPrfbPAtyy1BOyuTmOC0Hg=s96-c
    "is_my_comment": false,
    "name": "Shailesh Kumar",
    "post_id": 3514358,
    "updated_at": "2024-08-08T08:41:37.000Z",
    "user_id": 29037499,
    "user_type": "user"
```


Fig. 29 Challenge participant's messages

The profile picture and details of the users who participate in challenges present in the app are visible through this method. Figure 30 shows the profile information of a user. It displays the person's profile image along the date they last modified their personal details. It can be seen that the user's profile picture is stored on Amazon S3, a cloud storage service, and is delivered via Amazon CloudFront. This is a CDN that caches and distributes the image efficiently. The image is encrypted using AES-256 when at rest. However, there are a large number of security gaps, such as token exposure, where an attacker upon gaining access to the token data can retrieve the private data, including the profile picture as seen. Misconfigured S3 permissions could also make the images publicly accessible, leading to unauthorized access or modification. Additionally, if the TLS encryption is not enforced, intercepted requests could expose the images, making them vulnerable to identity spoofing. CloudFront caching means that even if an image is deleted, it may still be accessible for a period of time.

Figure 31 shows data from various time frames like the total steps taken, distance covered and calories burnt. A message is also stored regarding the distance covered and the number of calories burnt by comparing the data of the current time frame to the previous time frame.

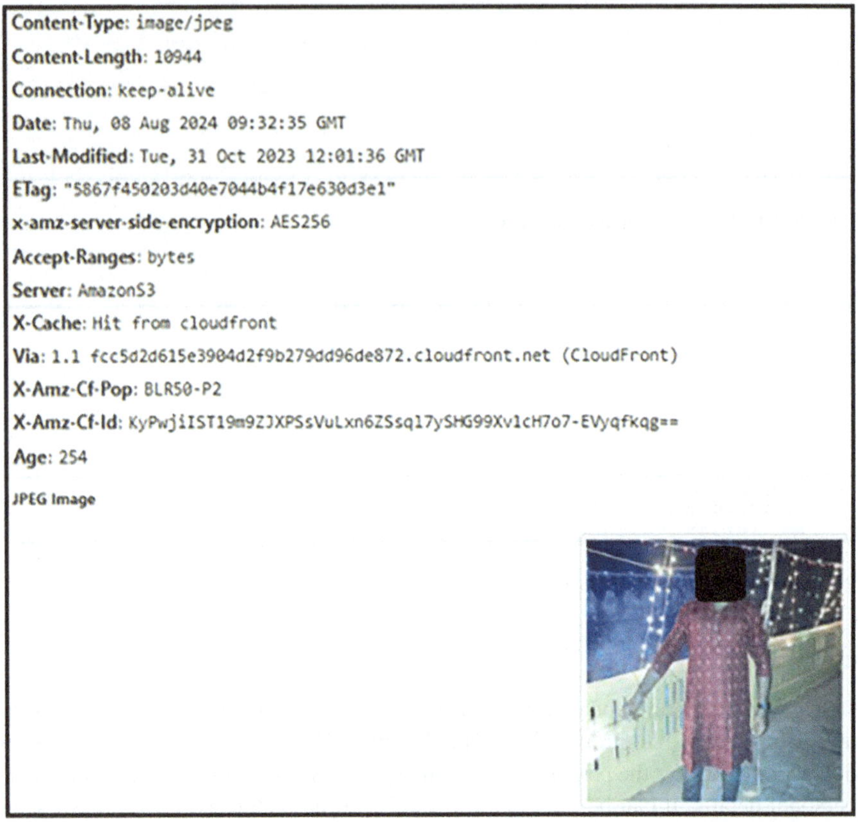

Fig. 30 Visible challenge participant's profile data

Figure 32 displays the multiple watchfaces present in the watch face gallery along with their links. Upon clicking on the links, the watchfaces designs can be viewed online.

Figure 33 shows one of the many watchfaces designs present in the application, as seen in Fig. 32. The user can choose from the multiple interfaces, which then loads onto the watch after the syncing process. The above image shows that the WebP image is hosted on Amazon S3 and is accessed through CloudFront.

Table 1 is a compilation of all the data that was obtained by studying the three smartwatches.

```
{
    "data": {
        "daily": {
            "calories_msg": "You burnt more calories today than yesterday.",
            "distance_msg": "Your total distance today is more than yesterday.",
            "msg": "Your total step count today is more than yesterday.",
            "today": {
                "calories": 184,
                "distance": 3565,
                "label": "Today",
                "steps": 5402
            },
            "yesterday": {
                "calories": 0,
                "distance": 0,
                "label": "2024-08-07",
                "steps": 0
            }
        },
        "max": null,
        "monthly": {
            "calories_msg": "You are burning less calories a day this month than last month.",
            "current": {
                "calories": 528,
                "distance": 10228,
                "label": "This Month",
                "steps": 15498
            },
            "distance_msg": "You covered less distance a day this month than last month.",
            "last": {
                "calories": 1133,
                "distance": 21939,
                "label": "Last Month",
                "steps":33239
            },
            "msg": "You are taking fewer steps a day this month than last month"
        },
        "weekly": {
            "calories_msg": "You are burning more calories a day this week than last week. ",
            "current": {
                "calories": 303,
                "distance": 5874,
                "label": "This Week",
                "steps": 8901
            },
            "distance_msg": "You covered more distance a day this week than last week.",
            "last": {
                "calories": 225,
                "distance": 4354,
                "label": "Last Week",
                "steps": 6597
            },
            "msg": "You are taking more steps a day this week than last week. "
        }
    },
    "message": "",
    "success": true,
    "time": "1723109331133"
}
```


Fig. 31 Comparative user activity data

```
data": {
    "customwatchface_list": {
        "artistic": null,
        "background": [
            {
                "category_id": 1,
                "image_url": "https://watchfaces-cdn.gonoise.com/watch_faces/production/custom_watchface/background+images/group_1321314748.webp",
                "title": "upload"
            },
            {
                "category_id": 1,
                "image_url": "https://watchfaces-cdn.gonoise.com/watch_faces/prod/custom_watchface/admin_upload/square/240%20X%20240/Father%27s%20Day-171
                "title": "Father's Day"
            },
            {
                "category_id": 1,
                "image_url": "https://watchfaces-cdn.gonoise.com/watch_faces/production/custom_watchface/background+images/Square/independence+day+Square
                "title": "Independence day"
            },
            {
                "category_id": 1,
                "image_url": "https://watchfaces-cdn.gonoise.com/watch_faces/undefined/custom_watchface/admin_upload/square/240%20X%20240/Earthday-1-1713
                "title": "Earthday-1"
            },
            {
                "category_id": 1,
                "image_url": "https://watchfaces-cdn.gonoise.com/watch_faces/production/custom_watchface/admin_upload/square/240%20X%20240/Shiv-170989099
                "title": "Shiv"
            },
            {
                "category_id": 1,
                "image_url": "https://watchfaces-cdn.gonoise.com/watch_faces/undefined/custom_watchface/admin_upload/square/240%20X%20240/Mothers%20Day-2
                "title": "Mothers Day-2"
```


Fig. 32 Watchface design information

6 Conclusion

In this study, the NIST digital forensic analysis was used to analyse three smartwatches: Noise Colorfit Pulse, BoAt Xplorer RTL and Amazfit Band 5. The primary objective was to examine the type of user data generated and stored by the application after user registration and login. Additionally, the app's security was analysed, focusing on the extent of information accessible to the investigators, including the user profiles and challenge details. We successfully extracted data like user information, steps taken and heart rate and verified it against the data shown on the smartwatches and in their apps. For all three smartwatches, we managed to retrieve user information along with stats such as heart rate. The data obtained provides valuable insights into the user's daily activities, which can be crucial in important investigations. For example, investigators can analyse whether the user engaged in any energy-intensive activities based on health parameters like heart rate and SpO2 rates. Additionally, sleep schedules can be reviewed based on recorded data. Also, though services like CloudFront enable seamless access to smartwatch data stored in cloud, this also involves several potential security risks. It was noted that the user's access token could be exploited to view user stats and data stored in the database, posing a security risk. Similarly, profiles of other users, along with their pictures, could be accessed, highlighting potential security threats. To mitigate these risks, it is essential to enforce strict IAM policies, signed URLs, and access logging to prevent unauthorized access and data leaks. Enhanced protection measures are crucial to ensure that sensitive user's information is protected from unauthorized access and breaches, especially in Cloud [22]. Paper [23–25] mentions about how many of the IoT devices are weak in security, lacking basic measures like

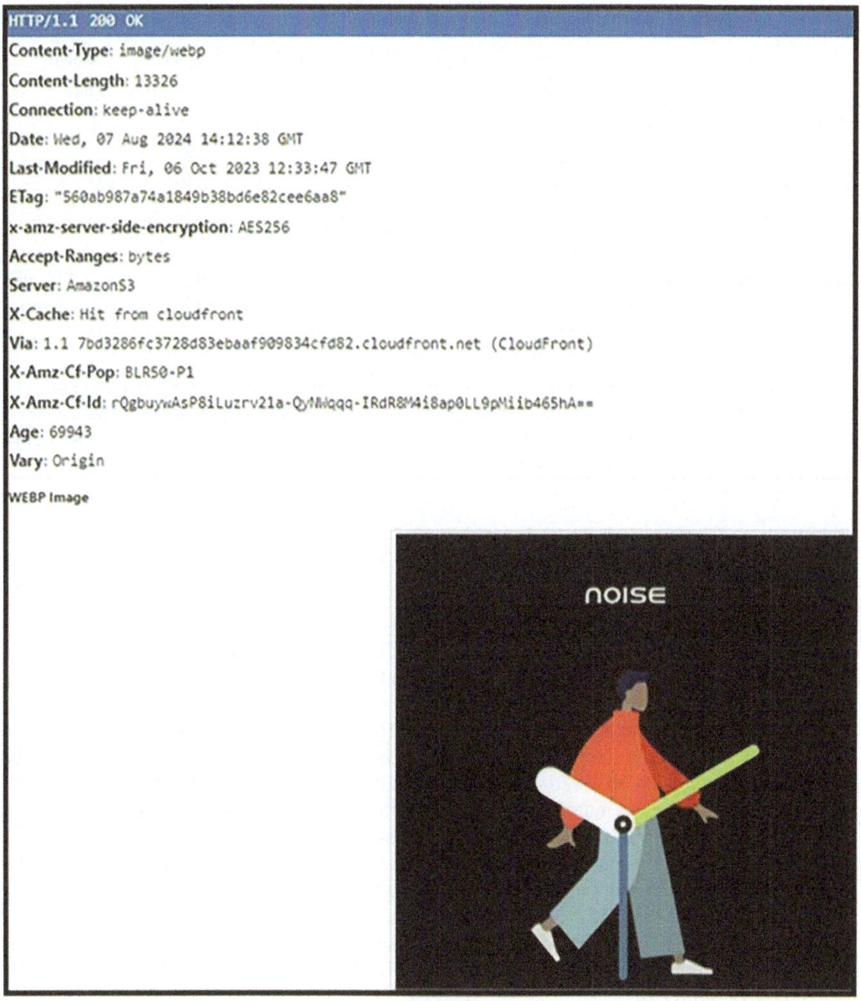

Fig. 33 Single watch face design information

encryption which could make it easy for hackers to collect personally identifiable information(PII). They also rightfully point out the vast majority of devices that exist in the market, and all of them could have different security needs and data collection methods. This would make it harder for forensic experts as they would have to identify the correct tools and methods that are needed to extract the necessary information. Future research could expand the scope by including more types of smartwatches. Further investigation is needed to determine if the data extraction remains consistent across different smartphone models, including iOS devices. Additionally, exploring data extraction through non-open-source tools or subscription-based tools could reveal additional data from the devices.

Table 1 Comparison of smartwatch features and results

	User Details	Device Details	Steps Info	Sleep Data	Heart Rate	SpO2	Other Profiles	Alarm Data
Noise Colorfit Pulse	✓	✓	✓	✓	✓	✓	✓	✗
BoAt Xplorer RTL	✓	✗	✓	✓	✓	✓	✗	✓
Amazfit Band 5	✓	✓	✓	✗	✓	✗	✗	✓

Acknowledgments We would like to express our sincere gratitude to the US Army for their generous funding and unwavering support, which made the AI-Enabled National Conference on Forensic Investigations Network in Digital Sciences(FINDS) possible. Their commitment to advancing research and fostering collaboration is deeply appreciated.

We are profoundly grateful to Dr. S. S. Iyengar for his outstanding contributions to the field, which have been a constant source of inspiration to us. His pioneering work has shaped much of our research, and we are deeply thankful for the impact his research has had on our academic journey. His dedication and innovations continue to inspire us in our pursuit of knowledge.

References

1. S.S. Iyengar, Sina Nabavi, Yashas Hariprasad, Prasad HB, Krishna Mohan C. "Artificial Intelligence in Practice: Theory and Applications for Cyber Security and Forensics", Springer Nature (Forthcoming)
2. J. P. van Zandwijk and A. Boztas, "The iPhone Health App from a forensic perspective: can steps and distances registered during walking and running be used as digital evidence?," Digital Investigation, vol. 28, pp. S126–S133, Apr. 2019. https://doi.org/10.1016/j.diin.2019.01.021.
3. A. W. CNN, "Police use murdered woman's Fitbit to charge her husband," CNN. https://edition.cnn.com/2017/04/25/us/fitbit-womans-death-investigation-trnd/index.html
4. C. Hassenfeldt, S. Baig, I. Baggili, and X. Zhang, "Map My Murder," Proceedings of the 14th International Conference on Availability, Reliability and Security, Aug. 2019. https://doi.org/10.1145/3339252.3340515.
5. R. J and G. Z, "Extraction and Forensic Analysis of Artifacts on Wearables," International Journal of Forensic Science & Pathology, pp. 312–318, Jan. 2017. https://doi.org/10.19070/2332-287x-1700070.
6. S. Hutchinson et al., "Investigating Wearable Fitness Applications: Data Privacy and Digital Forensics Analysis on Android," Applied Sciences, vol. 12, no. 19, p. 9747, Jan. 2022. https://doi.org/10.3390/app12199747.
7. A. Almogbil, A. Alghofaili, C. Deane, T. Leschke, A. Almogbil, and A. Alghofaili, "Digital Forensic Analysis of Fitbit Wearable Technology: An Investigator's Guide," 2020 7th IEEE International Conference on Cyber Security and Cloud Computing (CSCloud)/2020 6th IEEE International Conference on Edge Computing and Scalable Cloud (EdgeCom), Aug. 2020. https://doi.org/10.1109/cscloud-edgecom49738.2020.00017.
8. A. MacDermott, S. Lea, F. Iqbal, I. Idowu, and B. Shah, "Forensic Analysis of Wearable Devices: Fitbit, Garmin and HETP Watches," 2019 10th IFIP International Conference on New Technologies, Mobility and Security (NTMS), Jun. 2019. https://doi.org/10.1109/ntms.2019.8763834.
9. S. Parikh, D. Chavda, S. Chakraborty, Dr. P. H. Rughani, and Dr. M. S. Dahiya, "Analysis of Android Smart Watch Artifacts," International Journal of Scientific and Engineering Research, vol. 6, no. 8, pp. 920–930, Aug. 2015. https://doi.org/10.14299/ijser.2015.08.011.

10. Y. H. Yoon and U. Karabiyik, "Forensic Analysis of Fitbit Versa 2 Data on Android," Electronics, vol. 9, no. 9, p. 1431, Sep. 2020. https://doi.org/10.3390/electronics9091431.
11. J. Williams, A. MacDermott, K. Stamp, and F. Iqbal, "Forensic Analysis of Fitbit Versa: Android vs iOS," IEEE Xplore, May 01, 2021. https://ieeexplore.ieee.org/stamp/stamp.jsp?tp=&arnumber=9474285
12. F. Hantke and A. Dewald, "How can data from fitness trackers be obtained and analyzed with a forensic approach?," 2020 IEEE European Symposium on Security and Privacy Workshops (EuroS&PW), Sep. 2020. https://doi.org/10.1109/eurospw51379.2020.00073.
13. P. Domingues, J. Francisco, and M. Frade, "Post-mortem digital forensics analysis of the Zepp Life android application," Forensic Science International: Digital Investigation, vol. 45, p. 301555, Jun. 2023. https://doi.org/10.1016/j.fsidi.2023.301555.
14. F. Nunes, P. Domingues, and M. Frade, "The Digital Footprints on the Run: A Forensic Examination of Android Running Workout Applications," Future Internet, vol. 16, no. 9, pp. 304–304, Aug. 2024. https://doi.org/10.3390/fi16090304.
15. R. Sinha, Vikas Sihag, G. Choudhary, Manu Vardhan, and P. Singh, "Forensic Analysis of Fitness Applications on Android," Communications in computer and information science, pp. 222–235, Jan. 2022. https://doi.org/10.1007/978-981-16-9576-6_16.
16. S. J. Patel, N. K. Chaudhary, B. N. Gohil, and S. S. Iyengar, "Machine Learning Algorithms for Attack and Anomaly Detection in IoT," in Information Security, Privacy and Digital Forensics-Select Proceedings of the International Conference, ICISPD 2022, Springer Nature, 2022. https://doi.org/10.1007/978-981-99-5090-4.
17. S. Raghavendra, C. S. Reddy, C. M. Geeta, R. Buyya, K. R. Venugopal, S. S. Iyengar, and L. M. Patnaik, "Survey on data storage and retrieval techniques over encrypted cloud data," International Journal of Computer Science and Information Security, vol. 14, no. 9, p. 718, 2016.
18. S. S. Iyengar, Seyedsina Nabavirazavi, H. Rathore, Y. Hariprasad, and N. K. Chaudhary, "Advancing Forensic Science: AI and Knowledge Graphs Unlock New Insights," Journal of Forensic Research, vol. 15, no. 3, pp. 1–7, Jun. 2024. https://doi.org/10.37421/2157-7145.2024.15.615.
19. S. Raghavendra, K. Meghana, P. A. Doddabasappa, C. M. Geeta, R. Buyya, K. R. Venugopal, S. S. Iyengar, and L. M. Patnaik, "Index generation and secure multi-user access control over an encrypted cloud data," Procedia Computer Science, vol. 89, pp. 293–300, 2016.
20. S. Raghavendra, C. M. Geeta, R. Buyya, K. R. Venugopal, S. S. Iyengar, and L. M. Patnaik, "DRSMS: Domain and range specific multi-keyword search over encrypted cloud data," International Journal of Computer Science and Information Security, vol. 14, no. 5, p. 69, 2016.
21. Github repository: https://github.com/inloop/sqlite-viewer Website: https://inloop.github.io/sqlite-viewer/.
22. K. G. Boroojeni, M. H. Amini, and S. S. Iyengar, "Cloud Network Data Security," Springer eBooks, pp. 71–82, Oct. 2016. https://doi.org/10.1007/978-3-319-45050-65.
23. D. R. Garcia Avila, J. F. Miller, and S. S. Iyengar, 'Current Challenges in IoT Security and Forensics: Strategies for a Secure Connected Future', Key Issues in Network Protocols and Security [Working Title]. IntechOpen, Nov. 15, 2024. https://doi.org/10.5772/intechopen.1007766.
24. L. Caviglione, S. Wendzel, and W. Mazurczyk, "The Future of Digital Forensics: Challenges and the Road Ahead," IEEE Security & Privacy, vol. 15, no. 6, pp. 12–17, Nov. 2017. https://doi.org/10.1109/msp.2017.4251117.
25. A. MacDermott, T. Baker, and Q. Shi, "Iot Forensics: Challenges for the Ioa Era," 2018 9th IFIP International Conference on New Technologies, Mobility and Security (NTMS), Feb. 2018. https://doi.org/10.1109/ntms.2018.8328748.

Broadband and Forensics: Accelerating Digital Investigations in a Connected World

Veneeth Iyengar

1 Introduction

The digital revolution has profoundly reshaped the landscape of crime and criminal investigation [5–7]. The proliferation of broadband Internet—characterized by high-speed, always-on connectivity—has created new opportunities for both malicious actors and law enforcement agencies. On one hand, broadband facilitates a wide range of cybercrimes such as ransomware attacks, identity theft, phishing, and the dissemination of deepfake content [8, 9]. On the other, it enables unprecedented advancements in digital forensics, remote investigations, and real-time evidence processing [10].

Broadband, with its ability to support the transfer of large volumes of data and seamless communication across geographic boundaries, has become an essential tool in the modern forensic toolkit. It allows investigators to access and analyze digital evidence remotely, collaborate across jurisdictions, and employ AI-powered forensic tools that rely on cloud-based infrastructures. In rural and underserved areas, where traditional forensic facilities may be limited, broadband connectivity can bridge significant gaps by providing remote access to expert testimony, digital labs, and case management platforms.

Furthermore, broadband contributes to streamlining the judicial process. Digital case files, online legal portals, and virtual courtrooms powered by reliable Internet connectivity can enhance the speed and accessibility of justice delivery. It also opens avenues for training and capacity building in digital forensics through virtual platforms and e-learning tools.

This paper aims to explore the dual nature of broadband as both a challenge and a catalyst in the forensic sciences. It delves into the evolving crime landscape shaped

V. Iyengar (✉)
Executive Director, ConnectLA, New Orleans, LA, USA

S. S. Iyengar et al. (eds.), *Artificial Intelligence Driven Forensics*, Signals and Communication Technology, https://doi.org/10.1007/978-3-031-98036-7_29

by digital connectivity, outlines broadband-enabled forensic technologies, and highlights the transformative potential of broadband in ensuring equitable access to justice, especially in remote regions. By addressing the ethical, technical, and infrastructural considerations, this paper provides a comprehensive understanding of broadband's growing influence on forensic practices in a connected world.

2 What Is Broadband?

Broadband is a term used to describe high-speed Internet access that is continuously available and significantly faster than traditional dial-up connections. Unlike the dial-up systems of the past, broadband does not interrupt telephone usage and allows for the simultaneous transmission of large quantities of data. The core technologies used to deliver broadband include digital subscriber line (DSL), cable modems, fiber-optic connections, satellite services, and mobile networks such as 4G and the more recent 5G.

Each of these broadband types comes with its own strengths and limitations. Fiber-optic connections offer the highest speeds and reliability, ideal for handling data-intensive forensic applications such as live imaging, cloud-based analytics, and video surveillance streams. Satellite broadband, though slower and more prone to latency, can provide vital connectivity in remote areas lacking ground infrastructure. Mobile broadband, particularly 5G, introduces ultra-low latency and increased bandwidth, supporting real-time forensic analysis even on the move.

Broadband's significance in forensic investigations lies in its capacity to enable instantaneous communication, remote access to forensic tools, and the secure transfer of large digital evidence files. With growing adoption globally, broadband is increasingly becoming a foundational layer for modern digital forensics infrastructure [1].

Louisiana is considered a thought leader in both policies developed and tactics used to eliminate the digital divide. Leveraging funds from the Infrastructure Investment and Jobs Act, nearly $1.35 billion will be spent to ensure that approximately 140,000 households, small business, and community anchor institutions will receive access to high-speed Internet by 2028. While the state plans to leverage broadband investments as an asset, platform, and enabler to drive its competitive growth, they are beginning to focus efforts on better partnering with public safety and the Attorney General of Louisiana's office to address the unintended consequences of bad actors who will take advantage of people in their use of Internet.

3 The Digital Crime Landscape

The availability and ubiquity of broadband have dramatically transformed the crime landscape. High-speed connectivity has enabled the rise of sophisticated cybercrimes that can be orchestrated from virtually anywhere in the world. Threat actors exploit broadband networks to distribute malware, launch ransomware attacks, conduct phishing schemes, perpetrate identity theft, and spread misinformation using AI-generated content such as deepfakes.

Ransomware has become one of the most prominent broadband-enabled threats. Cybercriminals deploy malicious software that encrypts victims' data and demand cryptocurrency payments for restoration. These attacks are often executed through high-speed networks that allow the rapid dissemination of malware across multiple systems and geographies. Similarly, phishing attacks, which rely on deceiving users into revealing sensitive information, leverage broadband to propagate spoofed websites and mass email campaigns at scale.

Another disturbing trend is the emergence of deepfake technology—AI-generated synthetic media that convincingly alters audio or video content. Deepfakes pose a serious challenge to the authenticity of digital evidence, enabling false impersonation and undermining public trust in video surveillance and courtroom testimonies.

The forensic response to these digital crimes relies heavily on broadband to collect, preserve, and analyze digital traces. Logs of IP addresses, timestamps of activity, metadata from files and devices, and communication records such as emails and chat histories all form essential elements of digital evidence. As broadband enables perpetrators to mask their identities and operate across borders, forensic professionals must leverage the same connectivity to detect, attribute, and investigate these offenses [2].

4 Broadband-Enabled Forensic Tools

High-speed connectivity has enabled a new class of forensic tools that operate remotely and in real-time. Some of the key broadband-driven innovations include:

- **Cloud-Based Digital Evidence Management Systems (DEMS):** Securely storing, sharing, and analyzing digital evidence across jurisdictions.
- **Remote Mobile Forensics:** Broadband allows investigators to access and extract mobile data from distant locations using encrypted connections.
- **AI-Powered Network Traffic Analysis:** Detects anomalies and intrusions by analyzing live network data.
- **Live Disk Imaging:** Real-time duplication and analysis of storage devices via secure broadband channels [3].

5 Broadband Access and Rural Justice

Broadband access also plays a democratizing role in digital forensics. In rural and remote areas where physical access to forensic labs is limited, broadband enables:

- **Remote Expert Testimony:** Experts can provide insights via video conferencing.
- **Cloud-Based Laboratories:** Virtual labs for digital evidence analysis.
- **Fast Case File Access:** Immediate access to shared documents, improving judicial efficiency.

6 Challenges and Ethical Considerations

While broadband connectivity has enhanced forensic capabilities, it also introduces a spectrum of technical, ethical, and legal challenges that must be addressed to ensure responsible and equitable use.

Privacy and Security The handling of digital evidence over broadband networks raises serious concerns regarding confidentiality and data integrity. If not adequately encrypted or protected by secure transmission protocols, sensitive information such as personal communications, financial records, or digital surveillance footage may be intercepted, altered, or leaked. Cybersecurity protocols must be integrated into every stage of the forensic workflow to protect against unauthorized access and tampering.

Legal Admissibility For digital evidence to be admissible in court, it must maintain a clearly documented and unbroken chain of custody. However, remote handling and transmission of evidence over broadband networks complicate this requirement. Forensic investigators must adopt rigorous documentation and digital signing protocols to demonstrate the integrity and authenticity of evidence collected, transmitted, or analyzed remotely.

Infrastructure Disparity Access to high-speed Internet remains uneven, particularly in developing regions or marginalized communities. This disparity restricts the implementation of broadband-enabled forensic systems in areas that may benefit most from such capabilities. The uneven development of digital infrastructure risks creating a justice divide, where rural or underserved regions are unable to leverage the same forensic tools available to urban centers.

Digital Divide and Training In addition to infrastructure limitations, the lack of digital literacy and training among law enforcement and forensic personnel further hampers the effective utilization of broadband technologies. Specialized training in digital forensic protocols, cybercrime investigation, and secure communication is essential to equip professionals with the skills needed to operate in broadband-enhanced environments [4].

Addressing these challenges requires a coordinated effort involving policymakers, technology developers, forensic experts, and educators to ensure that the benefits of broadband in forensics are realized responsibly and inclusively.

7 Conclusion

Broadband technology stands as a transformative force in the field of forensic science, enabling faster, more decentralized, and widely accessible investigative processes. From real-time mobile forensics to cloud-based evidence management, broadband has reshaped how law enforcement agencies gather, analyze, and share digital evidence across jurisdictions. It facilitates collaboration between remote experts, enhances judicial efficiency through virtual case processing, and ensures that even rural or underserved communities have a pathway to justice through digital means.

However, the potential of broadband cannot be fully realized without addressing its associated challenges. Investments must go beyond mere infrastructure deployment to encompass robust cybersecurity safeguards, training programs for forensic professionals, and legal frameworks that adapt to the dynamic nature of digital evidence.

As we move into an era of AI-enabled and data-driven justice systems, broadband will remain at the heart of forensic innovation. It is imperative for stakeholders—governments, academic institutions, forensic agencies, and private sector partners—to work together in shaping a secure, inclusive, and ethically sound broadband-powered forensic ecosystem. With the right policies and partnerships in place, broadband can not only accelerate investigations but also uphold the core principles of fairness, transparency, and justice in the digital age.

References

1. Federal Communications Commission, “Broadband Speed Guide,” 2024. https://www.fcc.gov
2. Europol, “Internet Organised Crime Threat Assessment (IOCTA),” 2023. https://www.europol.europa.eu
3. Eoghan Casey, *Digital Evidence and Computer Crime: Forensic Science, Computers and the Internet*, Academic Press, 2019.
4. Nir Kshetri, “Cybersecurity and Digital Forensics in the Global South,” *Journal of Global Information Technology Management*, 2020.
5. Yashas Hariprasad, Latesh Kumar K. J., Suraj L., and S. S. Iyengar, “Boundary Based Fake Face Anomaly Detection in Videos using Recurrent Neural Networks,” *Springer Nature*, 2024.
6. Yashas Hariprasad, S. S. Iyengar, “AI Powered Correlation Technique to Detect Virtual Machine Attacks in Private Cloud Environment,” *Springer Nature*, 2025.
7. S. S. Iyengar, Yashas Hariprasad, Sina Nabavi, Prasad H. B., *Artificial Intelligence in Practice: Theory and Applications for Cyber Security and Forensics*, Springer Nature, forthcoming November 2024.
8. S. S. Iyengar and Richard Brooks, *Distributed Sensor Networks: Sensor Networking and Applications*, CRC Press, 2009.
9. Singaram, Jayakumar, S. S. Iyengar, and Azad M. Madni, *Deep Learning Networks: Design, Development and Deployment*, Springer Nature 2023.
10. Shi, Bin, and S.S. Iyengar, *Mathematical theories of machine learning-Theory and applications*, Springer Nature, 2020.

Innovations and Insights: A Visual Poster Exploration

1 ProDiscover Forensics, Software Company, Hyderabad, India

Nrupul Rao

Abstract
ProDiscover empowers forensic investigators and cybersecurity teams to analyze digital evidence and combat cyber threats with precision, speed, and ease. Whether you are recovering critical data from multiple devices, analyzing complex evidence, or managing electronic discovery, ProDiscover equips you with a comprehensive suite of tools to uncover the truth, streamline workflows, and secure justice. From small-scale incidents to large-scale investigations, ProDiscover adapts to your needs, making it your reliable ally to navigate the complexities of modern cybercrime and respond to cyber threats.

2 Question Hive: Bridging Gaps in Education with AI

G. R. Rohith Krishna • G. A. Srinidhi

Abstract
Traditional exam paper creation is inefficient, manual, and inconsistent. Question Hive uses artificial intelligence (AI) and machine learning (ML) to automate and streamline question paper generation. It offers vetted question banks, easy customization, and automatic PDF creation for educators. Beta tests showed high adoption rates, addressing flexibility and usability gaps for 70% of users. Future upgrades will add AI-driven question suggestions, strengthening its impact in EdTech.

S. S. Iyengar et al. (eds.), *Artificial Intelligence Driven Forensics*, Signals and Communication Technology, https://doi.org/10.1007/978-3-031-98036-7

3 WarNet Intelligence AI-Enabled Cyber Defense Military Network

K. Y. Sinchana • G. A. Srinidhi

Abstract

Modern military networks have evolved into complex global cyberspace environments, increasing exposure to cyber threats. This paper explores how artificial intelligence (AI) enhances cybersecurity by enabling real-time threat detection, autonomous defense, and predictive analytics. It highlights vulnerabilities in wireless tactical systems and proposes AI-driven solutions for resilient military operations. The research emphasizes that future military success will rely on AI-empowered cyber defense architectures. Continuous innovation and ethical AI deployment are crucial to maintaining battlefield superiority.

4 AI-Based Malware Detection from RAM Using Volatility and Machine Learning

S. Chaithanya Kumar • G. A. Srinidhi

Abstract

Fileless malware operates in system memory, evading traditional detection. This paper proposes detecting such malware using Volatility 3 for memory forensics and Scikit-learn for classification. Features like processes, network connections, and DLLs are extracted and analyzed to identify infections. The model achieves 80–90% accuracy and offers a user-friendly Streamlit interface for real-time analysis. This approach enhances cybersecurity by combining memory analysis with machine learning techniques.

5 Wear-It: Revolutionizing Fashion, E-Commerce with AI, AR, and VR

U. Priyanka • G. A. Srinidhi

Abstract

The rapid rise of fashion e-commerce has led to decision fatigue, high return rates, and poor personalization. Wear-It leverages augmented reality (AR), virtual reality (VR), and artificial intelligence (AI) to create a seamless, interactive digital shopping experience. It offers real-time virtual try-ons, personalized recommendations, and a unified brand interface. The platform boosts consumer confidence and significantly reduces return rates. Wear-It has the potential to revolutionize India's online fashion retail, especially for Gen Z and Millennials.

Index

S. S. Iyengar et al. (eds.), *Artificial Intelligence Driven Forensics*, Signals and Communication Technology, https://doi.org/10.1007/978-3-031-98036-7

The manufacturer's authorised representative in the EU is Springer Nature Customer Service Centre GmbH, Europaplatz 3, 69115 Heidelberg, Germany. If you have any concerns regarding our products, please contact ProductSafety@springernature.com

Printed and bound by CPI Group (UK) Ltd, Croydon, CR0 4YY
16/09/2026
02217780-0014